All-Russian Research-to-Practice Conference for Young Researchers, Postgraduate Students and Undergraduates 2017

Ecology and Safety in the Technosphere: Current Problems and Solutions

IOP Conference Series: Earth and Environmental Science
Volume 115

Yurga, Russia
6 – 7 March 2017

ISBN: 978-1-5108-5834-3
ISSN: 1755-1307

Printed from e-media with permission by:

Curran Associates, Inc.
57 Morehouse Lane
Red Hook, NY 12571

Some format issues inherent in the e-media version may also appear in this print version.

Printed by Curran Associates, Inc. (2018)

For permission requests, please contact the Institute of Physics
at the address below.

Institute of Physics
Dirac House, Temple Back
Bristol BS1 6BE UK

Phone: 44 1 17 929 7481
Fax: 44 1 17 920 0979

techtracking@iop.org

Additional copies of this publication are available from:

Curran Associates, Inc.
57 Morehouse Lane
Red Hook, NY 12571 USA
Phone: 845-758-0400
Fax: 845-758-2633
Email: curran@proceedings.com
Web: www.proceedings.com

Table of contents

Volume 115

All-Russian research-to-practice conference "Ecology and safety in the technosphere"

6–7 March 2017, Yurga, Russian Federation

Accepted papers received: 24 January 2018
Published online: 7 February 2018

Preface

Papers

PREFACE

All-Russian research-to-practice conference for young researchers, postgraduate students and undergraduates "Ecology and safety in the technosphere: current problems and solutions" will make a contribution to the rethinking of the relationship between man and nature in today's world, to finding the best ways to overcome the negative effects of the environmental crisis and to solve security issues.

The conference aims at discussing research development and introducing innovations in the field of environmental and technosphere safety; searching for collaborative research areas and strengthening cooperation between young researchers and professionals from educational and scientific institutions in Russia; creating a platform for creative communication and exchange of experiences between young researchers and professionals in the issues of environmental and technosphere safety. It will also focus on intellectual and creative development of young researchers and specialists; demonstrating the best achievements of young researchers and professionals in scientific and technological activities; acquisition scientific debate and public speaking skills by young researchers; increasing the prestige of education, research and innovative activity in the field of environmental and technosphere safety; systematization of scientific data on the topical issues of ecological and technosphere safety, developing evidence-based approaches to their solution.

The results of the conference for young researchers, postgraduate students and undergraduates can be applied in making improvements in the curricula of students in the field of technosphere security; developing measures for the organization of schools for young researchers; presenting the research results for further discussion with colleagues and leading scientists in the priority areas; issuing new monographs, textbooks, laboratory exercises, lecture notes.

The Organizing Committee of the Conference thank all the participants of the Conference for their fruitful work and personal contribution to the development of these Conference Proceedings.

Dmitry A. Chinakhov, PhD,
Conference Convenor,
Vice-Chancellor, Yurga Institute of Technology
of National Research Tomsk Polytechnic University

Peer review statement

All papers published in this volume of *IOP Conference Series: Earth and Environmental Science* have been peer reviewed through processes administered by the proceedings Editors. Reviews were conducted by expert referees to the professional and scientific standards expected of a proceedings journal published by IOP Publishing.

IOP Conf. Series: Earth and Environmental Science **115** (2018) 012001 doi:10.1088/1755-1315/115/1/012001

Effect of laser radiation on the cultivation rate of the microalga Chlorella sorokiniana as a source of biofuel

N Politaeva, Y Smyatskaya, V Slugin, A Toumi, M Bouabdelli

St. Petersburg state Polytechnical University, Saint Petersburg, Politeshnicheskaya, 29, Russian Federation.

E-mail: conata07@list.ru

Abstract. This article studies the influence of laser radiation on the growth of micro-algal biomass of Chlorella sorokiniana. The composition of nutrient medium and the effect the laser beam (2 and 5 cm diameter, 1, 5, 10, 15 and 20 minutes exposure time) for accelerated cultivation of microalgal biomass were studied. The source of laser radiation (LR) was a helium-neon laser with a nominal output power of 1.6 mW and a wavelength of 0.63 μm. The greatest increase in biomass was observed when LR was applied to a suspension of microalga Chlorella sorokiniana with a beam of 5 cm diameter for a time of 10, 15 and 20 minutes. The results of the microscopic study of the microalga cells show a significant increase in the number of cells after an exposure to LR with a beam diameter of 5 cm in diameter. These cells were characterized by a large vacuole, a thickened lipid shell and a large accumulation of metabolites prone to agglutination. This study proposed to obtain valuable components (lipids, carotenoids, and pectin) from the obtained biomass by extraction method and to use the residual biomass formed wastes, after the extraction of valuable components, as a co-substrate for anaerobic digestion to produce biogas. The composition of biogas consists mainly of methane and carbon dioxide. Methane is recommended to be used for economic needs in supplying the whole process with heat and electricity. The carbon dioxide formed during fermentation and after combustion of methane for energy production, is planned to be used as a carbon source in the cultivation of Chlorella sorokiniana for photoautotrophic biomass production.

1. Introduction

Microalgal biomass is seen as a promising renewable source of raw material for biofuel production, as well as, a source of valued nutrients in livestock feeds, dietary supplements and in cosmetology. The considered biomass source has a high reproduction rate and the ability to accumulate a significant amount of high-energy lipids due to the photosynthetic activity of the microalgae. As a source of biofuel, microalgae significantly exceeds the photosynthetic productivity of land oilseeds, such as canola, soybean, sunflower and palm oils (in the same land area under the influence of sunlight). Under optimal growth conditions, microalgal biomass is able to produce yields of 20 to 70% of lipids by dry weight besides other by-products with high added value such as proteins, pigments, carbohydrates and biopolymers [1]. Microalgae Chlorella sorokiniana among other species of Chlorella have the highest yield of nutrients. In order to maximize biomass growth in the shortest time, it is necessary to select the optimal growing conditions, such as temperature, illumination, agitation, sparging, and additional external physical actions [2-4].

Laser radiation (LR) is generated by an optical quantum generator. This technical device emits light in a very narrow spectral range in the form of a directed high-coherent monochromatic polarized beam, that is, in the form of a highly ordered electromagnetic one-color radiation in space and time. A report on laser radiation appeared for the first time in 1954 - these were the works of scientists of the Physical Institute of the USSR Academy of Sciences (N.G. Basova, A.M. Prokhorov) and the staff of

IOP Conf. Series: Earth and Environmental Science **115** (2018) 012001 doi:10.1088/1755-1315/115/1/012001

the Columbia University in the City of New York (Ch. Townes, A. Shavlov)- discovery for which they were awarded the Nobel Prize [5]. Laser has a very broad spectrum of application. Its sharp focusing combined with its high power of radiation makes it useful in technology processes for cutting, welding and piercing holes in solid materials. Laser radiation (LR) is also employed to conduct monitoring studies on air pollution [6].

The effects of lasers on the human body is widely studied and applied in the medical field. In fact, it is used for surgical interventions (laser scalpel), which made it possible to perform operations with minimum bleeding and to open new possibilities in eye microsurgery. It is also used in procedures to enhance the immune system, in cosmetology and in other therapeutic actions on the human body [7]. In agriculture, laser treatment is used to accelerate the growth and development of plants, which leads to an increase in productivity [8-9]. It is also known from the literature [10] that a low-intensity LR stimulates the metabolic activity of a cell. These processes are based on photophysical and photochemical reactions that arise in the body when exposed to laser radiation. The photo-physical reactions are primarily due to the heating of the object (within 0.1 - 0.3°C) and the spread of heat in biological tissues. The temperature difference is more pronounced in the biological membranes, which leads to the outflow of Na + and K + ions, the opening of protein channels and the increase of the molecules and ions transport. The photochemical reactions are due to the excitation of electrons in the atoms of a light-absorbing substance. At the molecular level, this is expressed in the form of photo-ionization of the substance, photo-reduction, photo-oxidation, photo-dissociation of molecules, or in their rearrangement – photo-isomerization [11]. In this case, water absorbs visible light and the red parts of the spectrum. The membranes change the structural organization of the water layer and the functions of the membranes' thermolabile channels [12]. At the optimum exposure doses of bio-objects to low-energy laser radiation, appropriate energy swap is realized. In response to this, the systems and organs undergo activation processes of self-regulation and mobilize their own sanogenesis reserves [13-14].

Laser radiation is, for any living organism, an unusual irritant that cannot be found in natural conditions. Its biological effect depends on the wavelength and intensity of the radiation. In this regard, the entire wavelength range is divided into several regions: from 380 to 780 nm-the visible region; from 780 to 1400 nm - the near infrared, over 1400 nm - the far infrared region. There are several types of effects of laser radiation on a living organism [14 ± 15]:

- Thermal (thermal) - the release of a significant amount of heat in a small volume and in a short period of time;
- Energy - high electric field strength, which causes the polarization of molecules and other effects;
- Photochemical - fading of many pigments;
- Mechanical - appearance of ultrasound oscillations in the radiated organism;
- The formation of an electromagnetic microwave field within the cell.

The article [15] examined the use of laser radiation as a bio-protector for plants growing in chemically contaminated land. The results of laboratory experiments on the effect of laser radiation on two plant species using arsenic solutions of different concentrations are presented. The laser emitter was a neon laser LGN-207A No. 1063 with a power of 1.8 mW and a wavelength of 0.63 μm. The experiments showed that the plants exposed to radiation had a longer life span than the neighbouring plants that had not been laser treated. These samples were more efficiently restored and less exposed to tissue necrosis. Thus, the positive effect of the helium-neon laser was confirmed [15]. According to the literature [16], laser radiation has a positive effect on the growth of protozoa (infusoria) when exposed for 45 seconds at a 1.3 μm wavelength.

Objective: the study of the effect of laser radiation (LR) on the biomass growth of microalga *Chlorella Sorokiniana.*

2. Results and Discussion

The object of the study was a suspension of microalgae *Chlorella sorokiniana* in a culture media. The composition of the culture media is given in Table 1.

Table 1. Composition of the culture media.

Substance	Concentration, mg / l
$ZnSO_4 \cdot 7H_2O$	100
$CuSO_4 \cdot 5H_2O$	10
$CoSO_4 \cdot 7H_2O$	100
$MnCl_2 \cdot 4H_2O$	500
$H_3BO_3 \cdot WF$	50
$Na_2MoO4 \cdot 2H_2O$	100
$FeCl_3 \cdot 6H_2O$	4.000

The radiation source was a single-mode red laser LGN 208V with a nominal output power of 1.6 mW and a wavelength of 0.63 μm. To measure the optical density of the Chlorella sorokiniana suspension, a UNICO 1201 spectrophotometer was used.

11 samples were prepared in 125 ml conical flasks. The volume of each suspension was 100 ml, with an initial optical density of 0.174 (at a wavelength of 750 nm). The biomass cultivation was carried out with a constant supply of air at a rate of 1.5 litters/min and with a constant exposure to daylight and short-term exposure to Laser radiation (LR).

The installation schema is presented in Figure 1. The lens is disposed almost close to the laser exit port. The sample is located at a 5.4 m distance from the lens, the laser ray diameter on the sample is 2 cm for samples No. 1-6, the radiation power density of the objective is 2.5 W/m^2 and the illumination is equivalent to 340 lux.

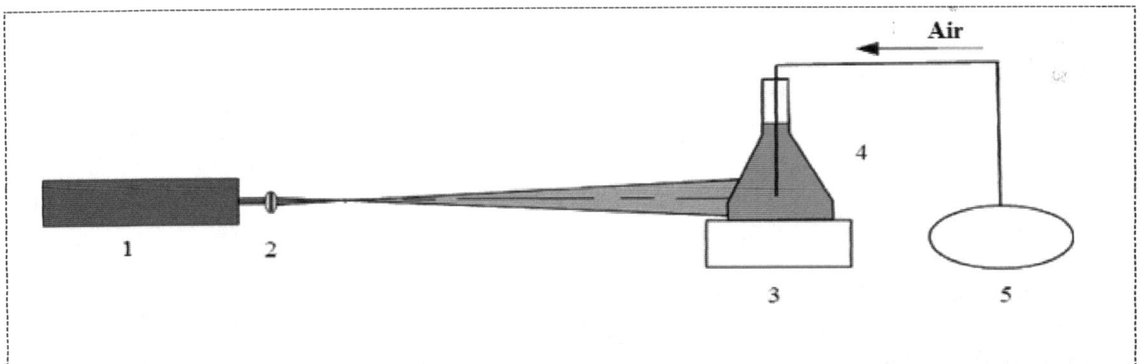

Figure 1. Schema of the installation put in place to investigate the effect of LR on the suspension of *Chlorella sorokiniana*: 1- laser LGN 208B, 2-convergent lens (F = 110 mm), 3- adjustable support, 4- test sample, 5- aerator.

A telescope was then used to increase the laser ray diameter on the sample to 5 cm (samples No. 7-11). The radiation power density on the sample was 0.3 W/m^2 and the illumination was equivalent to 40 Lux. The distance between the laser and the telescope (angular increase of 30×) was 2.1 m and the distance between the telescope and the sample was 0.1 m (Figure 2).

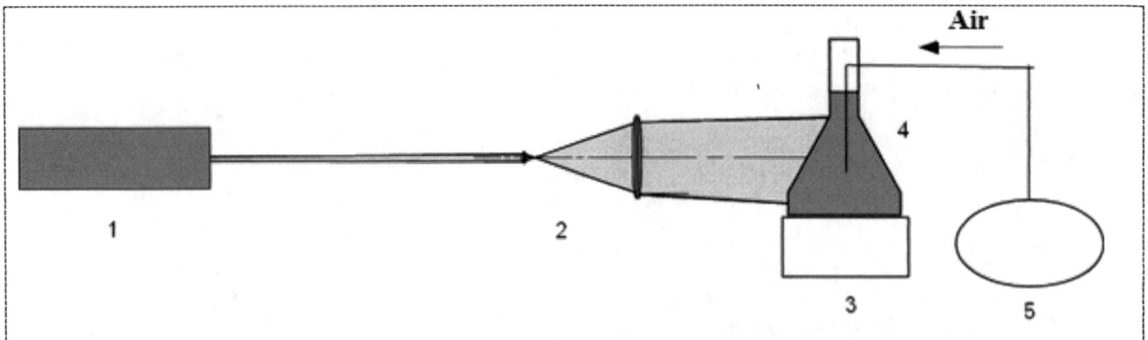

Figure 2. Schema of the installation put in place to investigate the effect of LR on the suspension of *Chlorella sorokinian*. 1 - laser LGN 208 V, 2 - ray expander (telescope), 3- stand, 4- sample.

The laser ray exposure time (2 and 5 cm) is shown in Table 2.

Table 2. Time of exposure of the LR beam on the suspension of *Chlorella Sorokiniana*.

D, cm	**2**						**5**				
No of the sample	1	2	3	4	5	6	7	8	9	10	11
t, min	1	5	10	15	20	0 (control)	1	5	10	15	20

t: time of exposure (min): D: diameter of the LR beam (cm)

3. Results and Discussion

After the exposure to LR (D = 2 cm), a daily monitoring of the optical density was carried out on the *Chlorella sorokiniana*'s suspensions. The results are shown in Figure 3. On the first day after exposure to LR, a decrease in the optical density of the suspension of *Chlorella sorokiniana* (except for the control sample) was observed, which is explained by the stress state of the biomass cells. However, in the following days, an increase in the number of cells was observed in proportion to the exposure time. After 7 days, the maximum biomass increment was achieved. Based on the obtained data, it can be concluded that the effect of LR (D = 2 cm) to accelerate the growth of biomass is not significantly advantageous.

After the exposure to LR (D = 5 cm), a daily monitoring of the optical density was carried out on *Chlorella Sorokiniana* suspensions. The results are shown in Figure 4. On the first day after exposure to LR, an increase in the optical density of the *Chlorella sorokiniana* suspensions was observed in comparison with the control sample. In the following days, there was a significant increase in the number of cells treated with LR. After 7 days, a maximum biomass (15 times) was achieved with the LR treatment (D = 5 cm) for 15 minutes, which indicates the benefit of LR treatment (D = 5 cm).

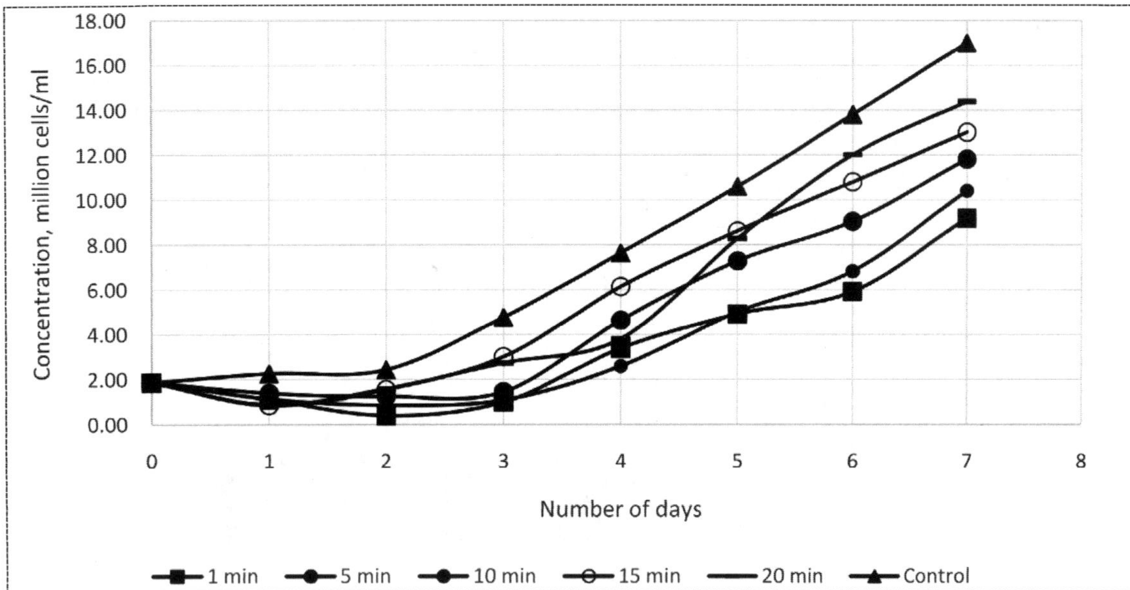

Figure 3. Relation between the cell concentration and the number days (with a 2 cm diameter laser ray).

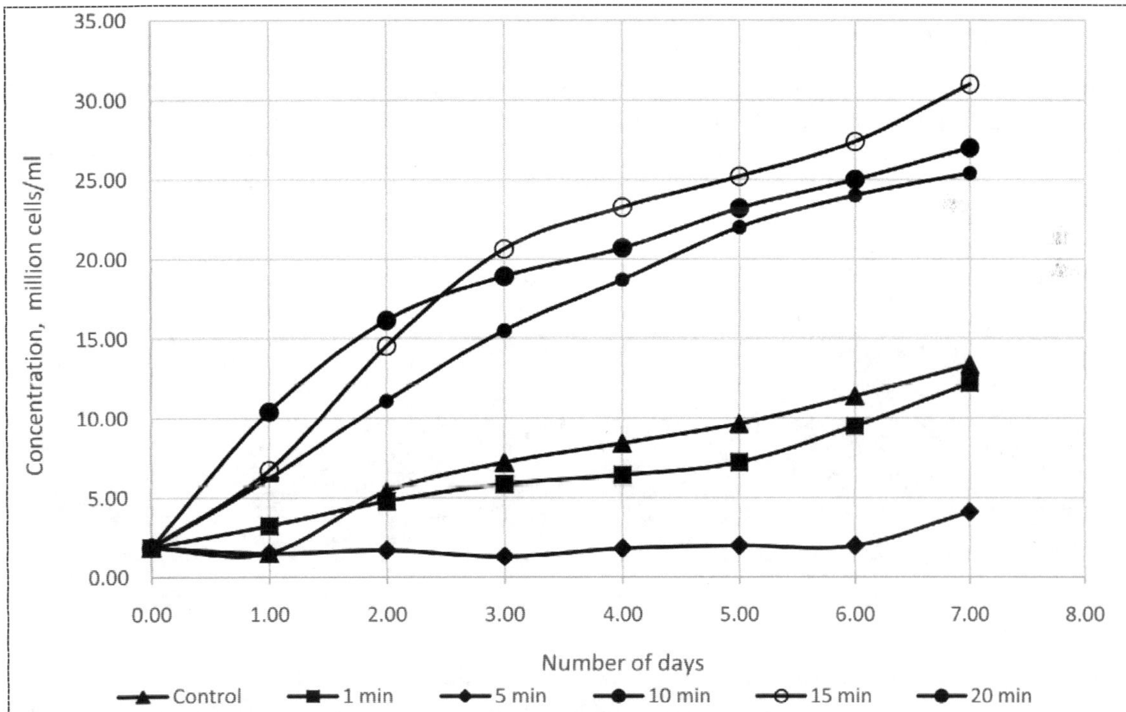

Figure 4. Relation between the cell concentration and the number days (with a 5 cm diameter laser ray).

Figure 5 shows the biomass's growth on the seventh day after exposures of the samples to a 2 cm and 5 cm diameter laser ray. The greatest increase in biomass was observed when exposed to the 5 cm diameter expanded ray and an optimum exposure time of 15 min. The D=5 cm ray was more efficient than the D = 2 cm focused laser ray. This is most probably due to a more uniform distribution of the LR and a power drop of 2.5 W/m² to 0.3 W/m² due to scattering. At a power higher than 0.3 W / m2, the microalgae cells are destroyed.

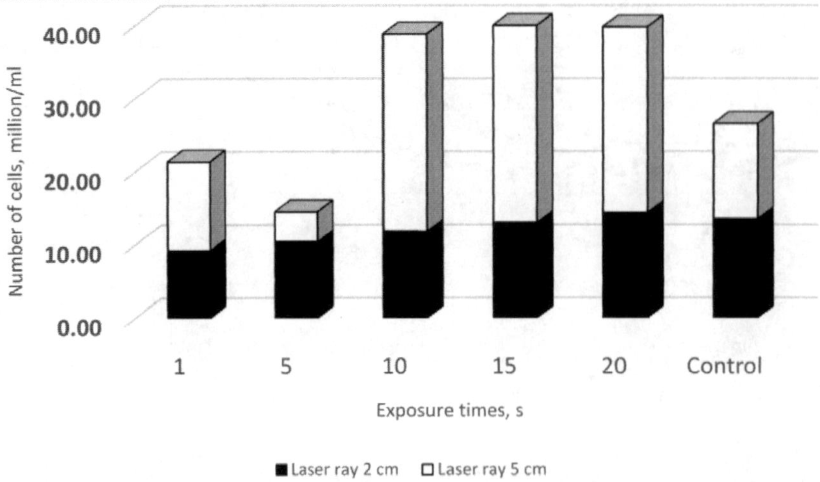

Figure 5. Increase in the number of cells after exposure to a laser of different diameters.

Microscopic studies were conducted to study the effects of laser radiation on the growth of biomass (as shown in Figure 6). To observe the shape of chlorella cells, photomicrographs were taken using a digital camera IS-500 and the FOTO microanalysis program in ten fields of vision. The analysis of the photomicrographs using the Levenguk computer program allows to contrast and change the brightness of photos and additionally enlarge the images.

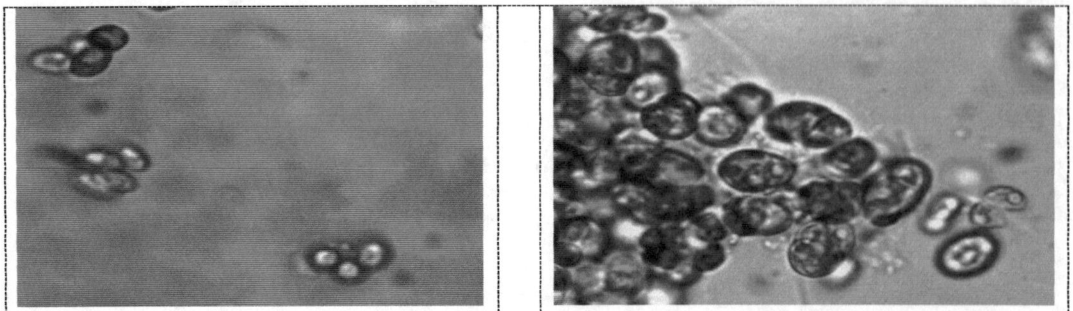

A B

Figure 6. Microstructure of microalga Chlorella sorokiniana before (A) and after (B) laser radiation

After the exposure to LR, a significant increase in the number of cells is observed. Cells are characterized by a large vacuole, a thickened lipid shell, a large accumulation of metabolites and are more prone to agglutination.

Later on, it is proposed to obtain valuable components (lipids, carotenoids, and pectin) from the obtained biomass by extraction method and to use the residual biomass formed wastes, after the extraction of valuable components, as a co-substrate for anaerobic digestion to produce biogas (Figure 7). The composition of biogas consists mainly of methane (CH_3) and carbon dioxide (CO_2) which should be separated. Methane is recommended to be used for economic needs in supplying the whole process with heat and electricity. The carbon dioxide formed during fermentation and after combustion of methane for energy production, is planned to be used as a carbon source in the cultivation of Chlorella sorokiniana for photoautotrophic biomass production.

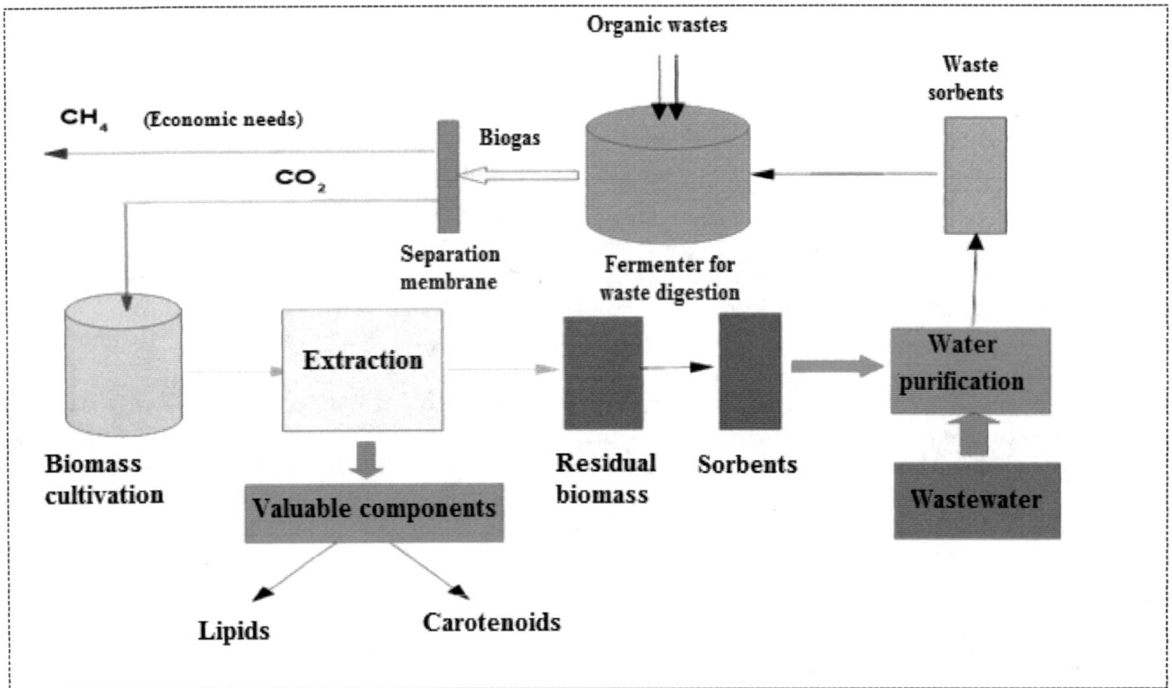

Figure 7. Schema of biomass processing of *Chlorella sorokiniana*.

4. Conclusion

As a result of the study, the following conclusions were drawn:
1. It was found that the action of Helium-Neon LR with a power of 0.3 W/m^2 with a laser ray diameter of 5 cm on the suspension of *Chlorella sorokiniana* increases the biomass growth by 15 times in 7 days.
2. It was determined that the tolerable exposure time for LR with a power of 0.3 W/m^2 and a laser ray diameter of 5 cm per suspension of *Chlorella Sorokiniana* provides an optimal biomass gain is 15 sec.

Acknowledgements
This work was performed within the implementation of Federal Targeted Programme for Research and Development in Priority Areas of Development of the Russian Scientific and Technological Complex for 2014-2020, the project «Development and implementation of innovative biotechnologies for treatment of microalgae *Chlorella sorokiniana* and duckweed *Lemna minor*» (Agreement № 14.587.21.0038). The unique identifier of the project RFMEFI58717X0038

References

[1] Piligaev AV, Sorokina KN, Parmon VN Production of high-energy biomass in heterotrophic cultivation of microalgae in water purification processes // Vestnik Novosibirsk State University. Series: Biology, clinical medicine. 2015. Vol. 13, no. 4. P. 19-26.

[2] Influence of ultraviolet on the development of duckweed Lemna and extraction of copper from contaminated sewage Olshanskaya LN, Sobgayda NA // *in the collection: promising polymeric composite materials. Alternative technologies. Recycling. Application. Ecology ("composite-2016") reports of the international conference "Composite-2016"* (dedicated to the 60th anniversary of the Engels Technological Institute (branch) of the SSTU named after Gagarin Yu.A.). Editor Panova L... 2016. P. 404-408.
[3] Effect of a magnetic field on the extraction of heavy metals from wastewater containing duckweed/Ol'shanskaya L.N, Sobgaida N.A., Tarushkina Yu.A., Stoyanov A.V.//Chemical and Petrolium Development. 2008. P. 44. N° 7-8. Pp. 475-479.
[4] The influence of the magnetic field on the extraction of heavy metals by duckweed / Olshanskaya LN, Sobgayda NA, Stoyanov AV, Kuleshova ML // *News of Higher Educational Institutions. Series: chemistry and chemical technology.* 2010. P. 53. № 9. P. 87-91.
[5] Kuklev, Yu.I. The ecological ecology: Textbook. allowance / Yu.I. Kuklev. M.: Higher School, 2001.-357 p.
[6] Reggie, *J. Industrial use of lasers* / J. Reggie. M.: The World, 1987. - 236 p.
[7] Priezzhev, A.V. Azerne diagnostics in biology and medicine / A.V. Priezzhev, V.V. Tuchin, L.P. Shubochkin. M.: Science, 1989.-321s.
[8] Belsky, A.I. Quantum processing of laser radiation in a magnetic field in the technology of annual growing of tree seedlings of fruit crops / A.I. Belsky // Plant growing. - 2007. - №2. Pp. 15-19.
[9] Patent RF RU No. 2240663 Method of industrial cultivation of agricultural crops using laser radiation. Zhurba PS, Zhurba TP, Zhurba EP 2003.03.11
[10] Alekseev, Yu.V. Some Aspects of BioPhotometry with LED Light in the Near-IR Range / Yu.V. Alekseev, M.E. Sokolov, T.V. Degrave // Laser and Information Technologies in Medicine of the XXI Century: Materials of the Intern. Conf.: at 2 pm - St. Petersburg, 2001. - Part II. - P. 462 - 463.
[11] Fain, S. Biological effect of laser radiation / S., ayn, E. Klein; trans. with English - World, 1968. - 336 p.
[12] Gamaleya, N.F. Some questions of biodynamics and bioelectronics of the organism in the norm and pathology, biostimulation by laser radiation / NF. Gamaleya // Materials of Intern. Conf. Krasnodar: KPI. - 1972. - P. 286 - 289.
[13] Engineering ecology: a textbook / ed. prof. V.T. Medvedeva. - M.: Gapdariki, 2002. - 687 p.
[14] Kaplan, M. A. Biological action of a laser of low intensity of a near infrared spectrum / M.A. Kaplan // Radiation Biology. Radioecology. - 1999. - T.39, No.6. - P. 683-691.
[15] Kaplan, M. A. Biological action of a laser of low intensity of a near infrared spectrum / M.A. Kaplan // Radiation Biology. Radioecology. - 1999. - T.39, No.6. - P. 683-691.
[16] Combined action of electromagnetic radiation of optical and millimeter bands on the growth of unicellular ones TI Belaya, LD Gapochka, MG Gapochka and dr // Moscow *Vestnik* university. Ser. 3. Physics. astronomy. 1994. Vol. 35, No. 4

Analysis of the insolation regime of the territory for various types of residential development

Tatiana Germanova [a], Natalya Bessonova [b]
Tyumen Industrial University, Engineering Economic Institute, Lunacharsky str. 2, 625 001 Tyumen, Russia

E-mail: [a] ecogtv@mail.ru, [b] sheneps@mail.ru

Abstract. The characteristics of insolation, favorable and unfavorable effects are analyzed. The requirements for insolation in Russia and the subjects of the Federation are reflected. The methods of investigation of insolation in the program complex "Citys: Solaris" are reflected. The calculation and analysis of insolation of the residential areas of the city of Tyumen.

1.Introduction

At present, the increase in density and number of storeys in the planning and reconstruction of residential buildings is at a rapid pace, which in turn should maintain a comfortable and safe living conditions. One of the most important environmental and hygienic factors in town planning [1], providing environmental improvement and comfort in urban spaces and buildings is insolation. According to [2], the provision of regulatory parameters of illumination and insolation is an indicator of the consumer quality of housing. Compliance with sanitary rules is mandatory for citizens, individual entrepreneurs and legal entities involved in the design, construction, reconstruction and operation of facilities in Russia.

The erection of new buildings and structures on the territory of existing buildings should not worsen the lighting performance. The magnitude of gaps between buildings is mainly determined based on the conditions for ensuring the standard values of insolation and illumination, as well as in accordance with the requirements of fire safety. Thus, it is mandatory to follow the norms of sanitary rules and norms, building codes and regulations, as well as territorial building codes that determine the required level of insolation and natural lighting.Insolation is an important health factor and should be used in all residential and public buildings and in the residential area. Optimal effectiveness of insolation - its general health, psychophysiological, bactericidal and thermal action is achieved by providing a daily continuous 3-4-hour exposure to direct sunlight in rooms and territories.

Local territorial construction norms (TSN) can be adopted on the territory of the constituent entities of the Russian Federation. In the territory of other subjects of the Russian Federation such documents lose their legal force and can only be taken into consideration, and be considered as recommendations. An example is "Natural, artificial and combined lighting. MGSN2.06-99 "and" Standards of outdoor lighting of urban and rural settlements. TSN 23-330-2002 KhMAO.

2.Methodology

Learning of regulation and standard technical documentation makes it possible to sum up the main insolation requirements:

- buildings should be designed so as to provide sufficient duration of insolation aiming at creating safe living conditions regardless of term (Federal Act №384);
- standard duration of continuous insolation for accommodations is set for certain calendar periods depending on types of dwellings, functional specification of buildings and latitude;
- standard duration of insolation should be ensured in one room in one- or two-bedroom apartments, and in at least two rooms in three and more-bedroom apartments (SRaN 54.13330.2011);
- standard duration of insolation on the territories of playgrounds, recreation grounds, playgrounds of pre-school facilities should be no less than 3 hours on the 50% of the lot independent of latitude (SRaN 2.2.1/2.1.1.1076-01)

So, providing standard insolation of constructions is realized through the following:

1. Location and orientation of dwellings and public buildings should provide continuous insolation required by sanitary standards.
2. Spacing of buildings is determined by the light conditions and according to number of stories in the opposite building but no less than 2.5 of the height.
3. Insolation routing does not take into consideration some raised parts of buildings.

Location and orientation of dwellings and public buildings (excluding pre-school institutions, comprehensive schools, boarding schools) should provide continuous insolation of accommodations and areas: to the north of 58^0 of latitude north of the equator – no less than 3 hours a day for the period from the 22nd of April till the 22nd of August; to the south of 58^0 of latitude north of the equator – no less than 2 hours a day for the period from the 22nd of March till the 22nd of September.

The insolation regime of territories and premises mainly depends on the climatic region of the locality of the locality and can vary according to the regulated insolation time both in the large and in the smaller direction. This is due to the assessment of the effect of the established time of action of the sun's rays on the microclimate of the premises. So, when the established norm is exceeded, the overheating of premises is often observed, which in turn creates discomfort to the population. Most often this occurs in one- and two-bedroom apartments with the arrangement of windows on one side of the building. However, such layouts of apartments in the greater part suffer from a lack of sunlight. Such a violation of the insolation rate is observed in windows directed to the north, north-west, north-east and west directions. In such directions, insolation of living quarters is much lower than the established norm or it is intermittent, which can not have the desired health-improving effect for the microclimate of the room.

Year-round shadowing of facades of buildings and residential areas is not allowed. Semiannual shades (from the 22nd of September till the 22nd of March) should not exceed in general 10% of undeveloped residential areas, healthcare centers in localities to the south of 58^0 of latitude north of the equator, and 20% to the north of 58^0 of latitude north of the equator

Single intermittency of insolation in domestic premises is permissible while housing with 9 and more-storied blocks. However, this can be conditioned only by half an hour increase of overall length of insolation during the daytime for each zone correspondingly.

Favorable effect of insolation on sanitary and hygienic atmosphere in rooms is achieved through a bacterial and biological quality of UV light to kill microorganisms or to slow down their multiplication. Hygienic effect of insolation is important for the purpose of letting in some certain daily amount of UV sunlight during some continuous period of time. biological quality of UV light to kill microorganisms or to slow down their multiplication. Hygienic effect of insolation is important for the purpose of letting in some certain daily amount of UV sunlight during some continuous period of time. Getting inside solar radiation forms some sanitary and hygienic conditions which can be either favorable or unfavorable [2,3].

Unfavorable effect is connected with overheating of rooms in summertime. The heat getting inside the rooms with solar radiation through light transmissive and lightproof screens causes air temperature rise inside.

Actual insolation is always different from the theoretical (presumable) one and can be defined only while observing it. Actual insolation depends on housing orientation and layout, aperture of windows, location of rooms, balconies and recessed balconies.

Duration of permanent and interruptive insolation of buildings and territories can be calculated with the help of program complex "City`s Solaris". It allows to make calculations for any geographic values and for a given date. The program provides opportunities to create, edit, copy objects, change their spatial altitude, and present objects as 3D models.

Calculations of insolation examination can be visualized in diagrams, display of shadows, as well as in drafts with design parameters of insolation of constructions and areas.

IOP Conf. Series: Earth and Environmental Science **115** (2018) 012002 doi:10.1088/1755-1315/115/1/012002

3.Results

Tyumen and its green zone are situated in the southern part of West Siberian lowland plain. The climate in Tyumen is not quite favorable for air pollutants dispersion. It is characterized by low annual total insolation. Accumulated temperatures are also relatively low above 10°C. More detailed information can be found in [4].

A study was made of the territories represented by different types of residential buildings. The first option of development is a 17-storey building with single-entrance houses, which are located on an area of 164,161 m^2. The second option is presented by residential buildings with a height of 9 and 10 floors of various shapes located on an area of 134974 m^2.

Calculation of insolation was carried out with the help of the program "Sitis Solaris" on the date set by the standards - this is March 22 on the day of the equinox. The estimated time was chosen in view of the territory's designation, namely the most probable time for walking with children at 11:00 and 16:00(Fig. 1, 2).

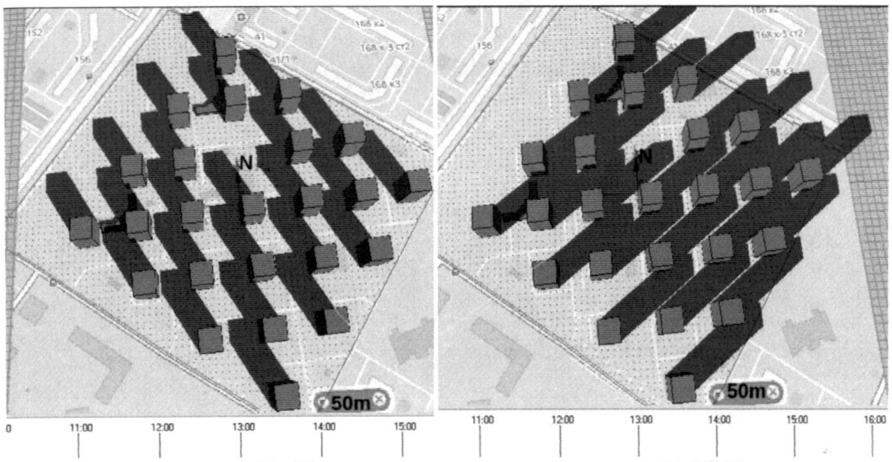

a) 11:00 b) 16:00

Fig. 1. Model shading the territory of residential development № 1, March 22.

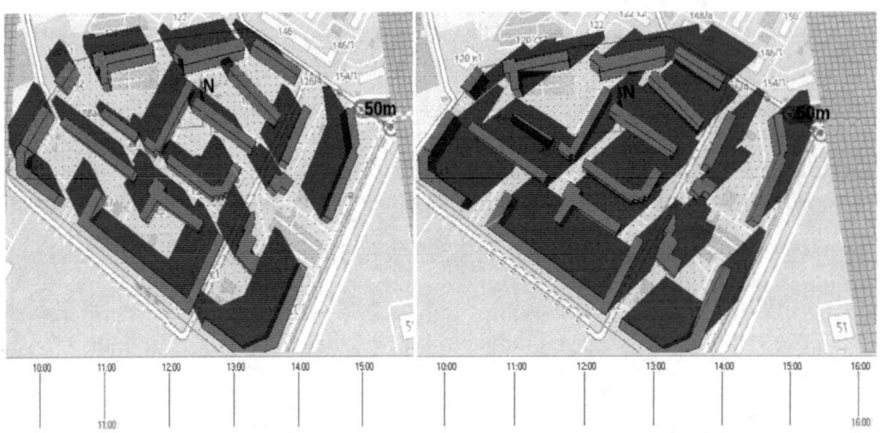

a) 11:00 b) 16:00

Fig. 2. Model shading the territory of residential development № 2, March 22.

Also, the calculation of the insolation of the territory No. 2 was made in the case of a change in the direction of development (Fig. 3, 4).

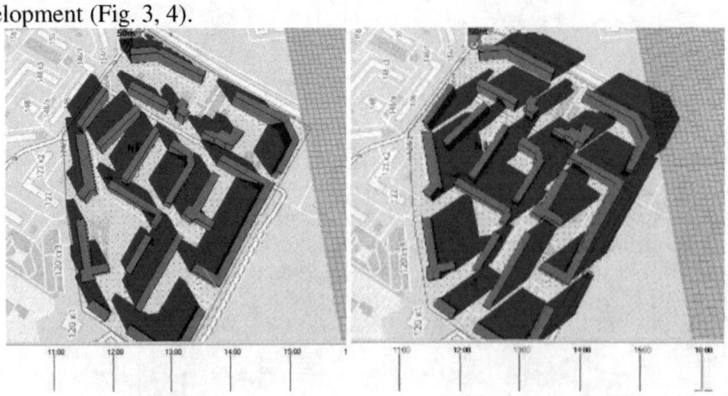

a) 11:00 b) 16:00

Fig. 3. Model shading the territory of residential development number 2 when the direction of the north is changed to 90^0, March 22

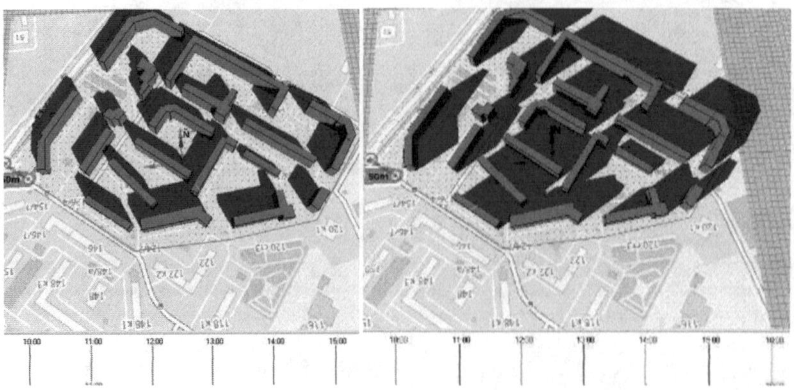

a) 11:00 b) 16:00

Fig. 4. Model shading the territory of residential development number 2 when the direction of the north is changed to 180^0, March 22

Based on the results of the calculation, one can draw a conclusion about the light comfort of the territory and the adjacent construction. In the construction presented in option number 1, the shading of the facades is the same for all residential buildings. With the variant of development of territory No. 2, shading in the interior corners of buildings is almost constant regardless of the direction of development, which is an unfavorable factor [5,6].

4. Discussion

The requirements for insolation of premises in different climatic regions are not the same. In the southern regions, where there is an excess of solar radiation during a long hot period, insolation of premises in the summer months is a negative factor. In these areas insolation is favorable mainly in the relatively cold period of the year. In the central zone of the Russian Federation, insolation of the premises of residential buildings, children's and medical institutions from March to September is considered mandatory [4], in the rest of the period - desirable. In the northern regions, where there is an acute shortage of solar radiation, it is necessary to make full use of the favorable effect of insolation of premises.

Conditions for year-round shading are calculated on the day of the summer solstice (June 22), and half-yearly on the days of the spring and autumn equinox (March 22 and September 22).

Irradiation of premises with solar beams (insolation) and natural lighting are important components of the microclimate of the apartment.

For the organization of literate natural lighting and insolation of buildings, the calculation technique is very important, it allows correctly to determine the gaps between buildings under natural lighting conditions [7,8], which is especially important in connection with the mass building in the cities of the Tyumen region. Thus, it can be concluded that a competent calculation of insolation is necessary in the planning of the territory.

5. Conclusion

As a result of the carried out researches of insolation of territories presented by two variants of building in the city of Tyumen it is evident that the most appropriate architectural solution is building No. 1, regardless of its orientation to the north. In the case of building No. 2 - the insolation regime of the territory is observed in the actual location of the building in the afternoon of the day. If you change the orientation to the north for 180^0 building number 2 - in the afternoon, the insolation regime is observed, but in the afternoon the area is almost completely obscured. This may serve as a favorable factor in the summer, but in the calculated winter period affects negatively.

Similar studies of the territory at the stage of the project make it possible to give the necessary recommendations on the direction of development in a particular case, as well as on a possible change in the designation of the territory.

Accounting for the period of insolation in the construction of the city's territory will positively affect the health and working capacity of the population, which reduces the cost of hospital sheets and treatment of adults and children.

References

[1] Eléonore Loiseau, Guillaume Junqua, Philippe Roux, Véronique Bellon-Maurel Environmental assessment of a territory: An overview of existing tools and methods Journal of Environmental Management, Volume 112, 15 December 2012, Pages 213-225
[2] Bakharev D.V., Orlova L.N. On regulations and calculation of insolation // Illumination engineering. 2006. № 1, Pages 18-27
[3] Ioan Sarbu, Calin Sebarchievici Chapter 2 - Solar Radiation Solar Heating and Cooling Systems, 2017, Pages 13-28
[4] Bessonova N.S., Germanova T.V.. Scientific and Technical Gazette Volga. 2013. № 6. Pages. 153-156
[5] Emilie Nault, Peter Moonen, Emmanuel Rey, Marilyne Andersen Predictive models for assessing the passive solar and daylight potential of neighborhood designs: A comparative proof-of-concept study Building and Environment, In Press, Accepted Manuscript, Available online 31 January 2017
[6] Building and Environment, In Press, Accepted Manuscript, Available online 31 January 2017
[7] Assessment of insolation at a playground Germanova, T., Bessonova, N., Davydova, I. MATEC Web of Conferences2017.
[8] Bessonova N.S., Germanova T.V. Bulletin of Perm National Research Polytechnic University. Applied Ecology. Urbanity. Number 1, 2014. (13). Pp 124-128.

Use of the method of biosphere compatibility for the assessment of environmental protection methods

Sergey Vorobyov

Oryol State Universiti, Oryol, Komsomolskya Street 95, Russia 302026

E-mail: vorser323@rambler.ru

Abstract. The article is devoted to the question of using the indicator of biosphere compatibility for assessing the effectiveness of environmental protection methods. The indicator of biosphere compatibility was proposed by the vice-president of RAASN (Russian Academy of Architecture and Building Sciences), Doctor of Technical Sciences, Professor V.I. Ilyichev. This indicator is allows not only qualitatively but also quantitatively to assess the degree of development of urban urban areas, from the standpoint of preserving the biosphere in urban ecosystems while realizing the city's main functions. The integral indicator of biosphere compatibility is allows us to assess not only the current ecological situation in the territory under consideration, but also to plan the forecast of its changes for building the new construction projects, or for reconstructing existing ones. The indicator of biosphere compatibility, which is a mathematical expression of the tripartite balance (technosphere, biosphere and population of this area), is allows us to quantify the degree of effectiveness of different method of protecting the environment for choose the most effective for these conditions.

1. Introduction.

The problem of preservation of the environment is based on the provisions of the concept of sustainable development, which served as the basis for developing strategies for the protection of the environment in most countries [1, 2, 3, 4, 5], including Russia. Despite the enormous positive role of sustainable development in international cooperation on the protection of the environment (agreement on limiting greenhouse gas emissions) and urban development [2], still not existed approaches to transition to biosphere compatibility city. In this regard, of particular interest is the concept of biosphere compatibility offered by Vice President of Russian Academy of Architecture and Construction Science, Doctor of Technical Sciences, Professor V.I. Ilyichev. [6]. The concept of biosphere compatibility is means the balance of biosphere and technosphere, for evaluation of this balance we can use as indicators of environmental pollution traditional values of industrial emissions compared to maximum permissible concentration, green areas, the intensity of a cleaner environment by recreational areas and other indicators of the anthropogenic impact on the biosphere. The general concept of hierarchy and relationships represented in the matrix of the transformation of the city in biosphere compatibility's city (Table 1) [7].

Table 1. Matrix development of urban areas based on the balance of homeostasis Biotechnosphere.

1. The harmonious coexictence of urban areas and the natural environment	4. Regalatory support maintaining the Balance of Biotechnosphere	7. Functions of the city, providing development of urban areas
2. Analysis of the interaction of the components of urban ecosystemes (emissions technosphere, biological objects, people)	5. Knowledge management as a basis biosphere compatibility processes in urban ecosystems	8. Reliability – program for the development of urban settlements based on the concept of biosphere compatibility
3. Balance Biotechnosphere – Triple drmographic, technosphiric biological indicators of urban ecosystems	6. Progress, a comparison of the planned parameters of urban ecosystems with actual	9. Knowledge – new approaches to the urban environment and the development of urban areas

In this approach, the numerical value of the biospheric compatibility is defined as follows [7]:

$$\eta = \sum_{n} \sum_{i} (D_{in} \cdot \xi^{*}_{in} \cdot \Theta_{in} - A_{in} \cdot \gamma_{in} \cdot m_{in}) \quad (1)$$

where the right side of this formula is a relative index of pure environment; left - a relative indicator of pollution from the technosphere with maximum concentrations that allow the development of urban area.

D_{in} - The ratio of the required area of the biosphere wich necessary to neutralize the pollution from the technosphere is based on the i-th source polution in the n-th function of the city; ξ^{*}_{in}- Uniformity coefficient of the biosphere, taking into account the different intensities of pollutants; Θ_{in} - the number of sources, pollution from which can be absorbed by the biosphere, in the zone of influence of the i-th source in the implementation of the n-th function of city; A_{in}- The amount of pollution from the i-th source in the implementation of the n-th function of the city for the territory of distribution of pollution; γ_{in} - The coefficient of reduction of pollution parameters from each source; m_{in} - The relative number of sources, pollution from which must be absorbed by the biosphere in the zone of influence of the i-th source in the implementation of the n-th function of the city.

For calculation of the ratio of the area of the biosphere in the region or district of the village needed to neutralize contaminants from (D_{in}) we had the formula [7]:

$$D_{in} = \{(V_{in} / \Theta_{in}) / k_{in}\} / S_{all} \quad (2)$$

where V_{in} - volume of contaminants from the i-th source for implementing the n-th function of city kg/year; k_{in} - the volume one pollutant wich absorb by 1 m^2 biosphere, kg/year; S_{all} - total area (m^2).

The value of parameter pollution from the i-th source in the implementation of the n-th function of the city (A_{in}) is calculated by the formula [7]:

$$A_{in} = (S_{con} / \Theta_{in}) / S_{all} \quad (3)$$

where S_{con} - the area of contamination from the i-th pollutant in the implementation of the n-th function of the city, m^2.

The numerical value of the integral indicator of biosphere compatibility varies in the limit from 0 (complete suppression of the biosphere by the technosphere) to 1 (no emissions of the technosphere and natural development of the biosphere).

With the help of this technique, it is possible to compare quantitatively the different territories of urbanized territories (neighborhoods, city districts, separate cities and settlements) according to the level of development of both the human potential and the development index of the biosphere (as the most important factor providing comfortable and safe living conditions for the population) for planning the development of these territories [7].

2. Results and discussions

The development of urban areas in modern conditions is impossible without the application of the methods of protect the main environments of urban ecosystems (atmosphere, lithosphere, hydrosphere, biosphere) and the population of urbanized areas from the effects of various types of pollution. It is very important to choose the method that will most effectively eliminate this type of pollution in this conditions. The effectiveness of a different methods and engineering solutions for environmental protection today is estimated by the economic criterion, which allows calculating the amount of costs for their implementation, as well as saving resources and money from their implementation [3]. Also, the ability of the method to reduce the degree of this type of pollution to the level of MPC. This approach allows us to quantify the parameters of the technosphere when we choosing a method or engineering solution. But the parameters of the development of the biosphere in this case (for example, the ability to accumulate pollution) are estimated only approximately (qualitatively) [8].

So, the methodology for calculating the integral index of biosphere compatibility is universal and allow not only qualitatively but also quantitative assessment of the effectiveness of various methods of environmental protection [9, 10].

As an example, we have analyzed the land territory Orel city as a typical regional center of Central Federal District of Russia [8]. The area of the Orel city is 1,452,000 m^2, industrial and communal storage areas - 16% [9]. The amount of generated waste is 64,000 t/year from 320,000 inhabitants, or 19,200 m^3/year from each resident ($V_{i=1}$). The area of the landfill for solid household waste (including a 500-meter sanitary protection zone) is 1,040,000 m^2, or 3.25 m^2 per inhabitant. The volume one pollutant wich absorb by 1 m^2 biosphere (k_{in}) is 0.024 kg/year. So we can calculate the ratio of the required area of the biosphere wich necessary to neutralize the pollution from the technosphere is based on the i-th source polution in the n-th function of the city (D_{in}) by formula 2.

$$D_{i=1} = \{(V_{in} / \Theta_{in}) / k_{in} / S_{all} = \{(19200 / 320000) / 0.024\} / 3.25 = 0.76 \quad (4)$$

The area of contamination from the i-th pollutant in the implementation of the n-th function of the city is 124,000 m^2. So we can calculate the amount of pollution from the i-th source in the implementation of the n-th function of the city for the territory of distribution of pollution (A_{in}) by formula 3.

$$A_{in} = (124000 / 320000) / 3.25 = 0.12 \quad (5)$$

Waste from the whole city is buried in the landfill area therefore $m_{i=1}$=320000=100%=1. So the value of the indicator of biosphere compatibility for the Orel city is:

$$\eta = \sum_{n=1} \sum_{i=1} (D_{in} \cdot \xi_{in} \cdot \Theta_{in}) - (A_{in} \cdot \gamma_{in} \cdot m_{in}) = (0.76 \cdot 1 \cdot 1) - (0.12 \cdot 1 \cdot 1) = 0.64 \quad (6)$$

Values of the indicator of biosphere compatibility shows that the environment is pressing by a human. Innovative methods of waste disposal reduce the human impact on the environment. The level of effectiveness of different methods of waste disposal can be assessed by using the indicator of biosphere compatibility. As an example, the construction of a waste processing plant is planned in the city of Orel. Part of the solid household waste will be recycled (48%) and the rest will be pyrolyzed in the gasifier to produce combustible gases. The gasifier is a shaft furnace with a working diameter of 1500 mm and a working part height of 5000 mm, equipped with devices for automatic fuel loading and unloading of ash residue and a control system. To implement this method, an inert heat-resistant material is loaded into the gasifier together with the fuel, which accumulates the heat of the exhaust gas and passes to the afterburner zone and transfers heat to the incoming air. The gasification process has a high efficiency (up to 95%), and the use of high temperatures (up to 1200^0 C) and the process of binding pollutants with an ash residue can significantly reduce the concentration of pollutants in emissions to the atmosphere [7].

The territory of the plant will be located in 4 km from the Orel city in an easterly direction from the leeward side to the prevailing winds. The area of the plant with a sanitary protection zone is 800,000 m^2, or 2.5 m^2 per inhabitant (S_{all}), k_{in} - is 0.03 kg/year.

Figure 1. Scheme of waste disposal plant

The ratio of the required area of the biosphere wich necessary to neutralize the pollution from the technosphere is based on the i-th source polution in the n-th function of the city *(D_{in})* is:

$$D_{i=1} = \{(V_{in} / \Theta_{in}) / k_{in} / S_{all} = \{(19200 / 320000) / 0.03\} / 2.5 = 0.8 \quad (7)$$

The value of the indicator of biosphere compatibility for Orel city in the application of the waste processing plant is:

$$\eta = \sum_{n=1}\sum_{i=1}(D_{in}\cdot\xi_{in}\cdot\Theta_{in}) - (A_{in}\cdot\gamma_{in}\cdot m_{in}) = (0.8\cdot 1\cdot 1) - (0.08\cdot 1\cdot 1) = 0.72 \quad (8)$$

3. Conclusions

Comparison of these two values of η (0.64 and 0.72) shows that the utilization of the method of disposal of solid waste at the waste processing plant is more effective than traditional disposal at the landfill site by 22%.

Thus, the methodology for calculating the indicator of biosphere compatibility allows us to evaluate the effectiveness of different methods for improving the urban environment and choose the best one.

Refereences

[1] Noor Farihah Mohd Noor, «Environmental sustainable devepment: the» – Jornal of International Scientific Publications: Ecology&Safety. V. 11. 2017 pp. 101-110.
[2] Daniela Palaščáková, Gabriela Koľveková, «Alternatives in wast management development in Slovakia compared to Germany ». Jornal of International Scientific Publications: Ecology&Safety. V. 11. 2017 pp. 130-141.
[3] Ruth D., Craig R. See, John L. Campbell «Current Practices in Reporting Uncertainty in Ecosystem Ecology» - Ecosystems. V. 20. 2017 pp. 1-11.
[4] Hua Shi, Matthew Rigge, Collin G. Homer. «Historical Cover Trends in a Sagebrush Steppe Ecosystem from 1985 to 2013: Links with Climate, Disturbance, and Management» Ecosystems. V. 20. 2017 pp. 1-17.
[5] Alain Paquette, Jordi Vayreda, Lluís Coll, Christian Messier, Javier Retana. «Climate Change Could Negate Positive Tree Diversity Effects on Forest Productivity: A Study Across Five Climate Types in Spain and Canada». Ecosystems. V. 20. 2017 pp. 1-17.
[6] V. Ilyichev, «About the dynamic formation of the urban livelihood system compatible with the biosphere» – Applied mechanics and materials. V. 725-726, 2015, pp. 1224-1230.
[7] V. Ilyichev, «Estimation of indicators of ecological safety in civil engineering» - PROCEDIA ENGINEERING. V. 117 2015, pp. 126-131.
[8] S. Vorobev «Evaluation of urban environment by bioindication methods from position the concept of urban settlements biosphere compatibility» - The strategies of Modern Science Development: Proceedings of the V International scientific-practical conference (Yelm, WA, USA, 26-27 December 2013). – Yelm, WA, USA: Science Book Publishing House, 2013. pp.149-152.
[9] S. Vorobev, «Application Biosphere Compatibility Concept To Evaluate The Quality Of Urban Environment By Bioindication Methods» IOP Conference Series: «Earth and Environmental Science» Volume 50, 2017, pp. 1-8.
[10] S. Vorobev, « Distribution of Heavy Metals in Soil of Urban Ecosystems by the Example of the Oryol City» IOP Conference Series: «Earth and Environmental Science» Volume 66, 2017, doi:10.1088/1742-6596/755/1/01100.

Synthesis and Physicochemical Properties of $SiO_2 - Ta_2O_5$ Thin-Film Systems and Powders

A G Malchik [a], S V Litovkin[b], P V Rodionov[c], S O Kryuchkova[d]

Yurga Institute of Technology (branch) of National Research Tomsk Polytechnic University,
652055, Russia, Yurga, 26, Leningradskaya st.

E-mail: [a]ale-malchik@yandex.ru, [b]protoniy@yandex.ru, [c] rodik-1972@yandex.ru, [d] lizakup@bk.ru

Abstract. Within the research, $SiO_2 - Ta_2O_5$ films from film-forming solutions based on tetraethoxysilane and tantalum chloride (V) are synthesized. Physicochemical patterns and specifics of producing the films are determined by means of IR spectroscopy and thermogravimetric analysis. The processes of $SiO_2 - Ta_2O_5$ films formation are studied. Interrelation between $SiO_2 - Ta_2O_5$ composition, concentration of incoming components and their physicochemical properties are revealed.

1. Introduction

Many branches of economy use modern ultraviolet technologies such as ultraviolet drying devices, disinfection plants for air, surfaces and water, climatic chambers, weather modification machines, various irradiators for medicine, cosmetology equipment (lamps for hairdressing, manicure, solariums, etc.). Ultraviolet radiation is widely used for industrial and household purposes among which one can find identification of special signs, control of securities, medical, health and cosmetic purposes, etc. [1,2,3].

Excessive ultraviolet radiation can cause a number of diseases and health disorders. These are primarily eye damages including cataracts or clouding of the eye lens, inflammation of cornea and mucous membranes. UV radiation can also lead to skin diseases: inflammatory skin erubescence or erythema, acceleration of skin aging, allergic reactions to UV radiation, skin tumors including skin cancer [4,5].

In this regard, there is a need in various protective technologies to reduce or limit the UV spectrum. Applying a special film on light sources to adjust the UV radiation is one of such technologies. The applied coating is transparent to the required radiation wavelength absorbing the rest parts of the spectrum and thereby reducing the radiation effect.

2. Methodology

To obtain thin-film systems with a given composition, we used film-forming solutions (POR) made from dried ethanol, tetraethoxysilane and tantalum chloride (V). The dried ethanol was used to prevent $TaCl_5$ hydrolysis. The films were produced on silicon and glass substrates. Two production methods were used: centrifugation at a speed of 3000 to 5000 rpm, and drawing at a speed of 1 to 5 mm/s. The samples were incubated at a temperature of 333 K for 30 minutes. The complete filming was carried out at a temperature of 873K for an hour. The refractive index and film thickness were measured at 5 points of each sample using the LEF-3M and "SE400advanced" laser ellipsometers. Phase composition of the films was studied using a DRON-3M diffractometer (using CuKα radiation and a shooting speed of 2 deg/min).

IOP Conf. Series: Earth and Environmental Science **115** (2018) 012004 doi:10.1088/1755-1315/115/1/012004

3. Results and discussion

The essence of the oxide film formation is that the chemical composition of initial film-forming compounds undergoes a number of changes: first at the solution stage, then at the moment of film formation on the surface of the substrate and finally during thermal processing. To establish technological conditions for film synthesis, we have studied the main stages of simple oxides and their double oxides formation. Heat treatment reduces concentration of crystal lattice defects, crystal structure becomes thermodynamically more stable which provides stable physical and chemical properties. By changing heat treatment modes and conditions, we can shape the structure and therefore the properties of a product. This problem cannot be solved without a thorough research into kinetics of heat treatment processes [6,7,8].

Thermal gravimetric and IR spectroscopic analysis were carried out for hydrolysed powders dried by thermostating at 333 K. Fig. 1 and 2 show thermogravimetric curves for simple oxides formation.

On the DTA curve (Fig. 1), two endothermic effects are recorded with maxima at 383K and 653K. They are caused by water evaporation from the surface, these processes involving mass loss, the TG curve . The endothermic effect at 653K is caused by evaporation of water generated by condensation of silanol groups when siloxane bonds are being formed. The exothermic effect at 863K is characteristic to combustion of alcohol and high-molecular hydrocarbons to form SiO_2.

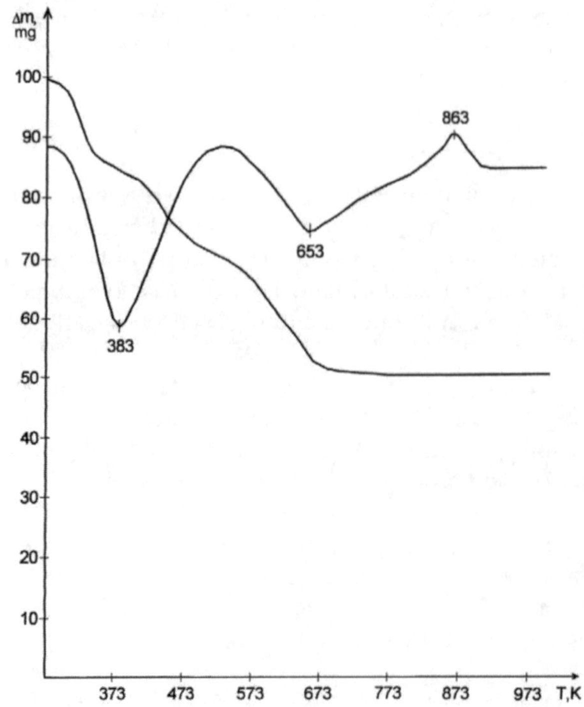

Figure 1. Decomposition derivatogram of film-forming solution dried at 333K.

The DTA curve (Fig. 2) shows that Ta_2O_5 formation also proceeds in three stages. The first stage is involves evaporation of physically adsorbed water and alcohol. At the second stage (exothermic effect at 523K) $TaCl_5$ is oxidized with air oxygen by the equation:

$$TaCl_5 + O_2 \rightarrow TaO_2Cl_3 + Cl_2$$

At the third stage, combustion of oxychlorides and alkoxy derivatives of tantalum takes place to form Ta_2O_5 oxide.

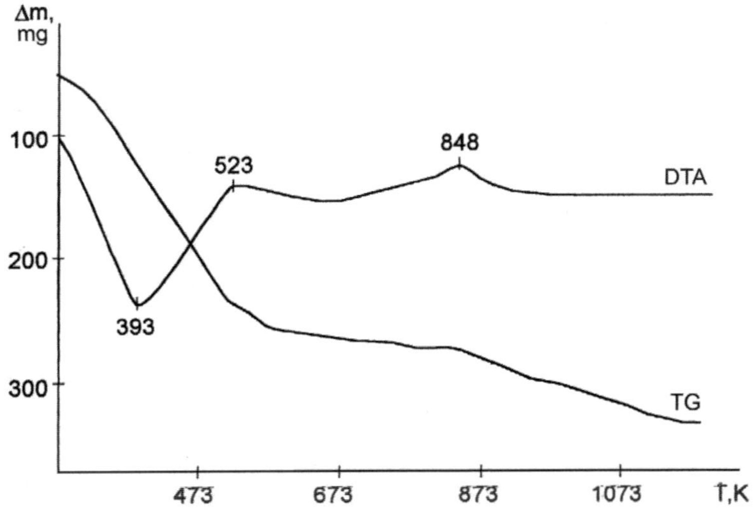

Figure 2. Derivatogram of $TaCl_5$ alcohol solution dried at 333 K

The IR spectra data (Table 1) prove that at 800-840 K there are valence and deformation vibrations associated with ethoxy groups. At higher temperatures, these vibrations do not occur.

The physicochemical processes of double oxides formation are complex. When $SiO_2 - Ta_2O_5$ are formed, the DTA curve shows an effect with a maximum at 393K which coincides with SiO_2thermal formation effect and corresponds to the processes of polycondensation and physically adsorbed water removal (Fig. 3). The process of $SiO_2 - Ta_2O_5$ complex oxides formation takes place in four stages The exothermic effect at 523K corresponds to $TaCl_5$ oxidation. The processes of bulky ethoxy groups oxidation and $SiO_2 - Ta_2O_5$ oxides formation take place at stages 3 and 4.

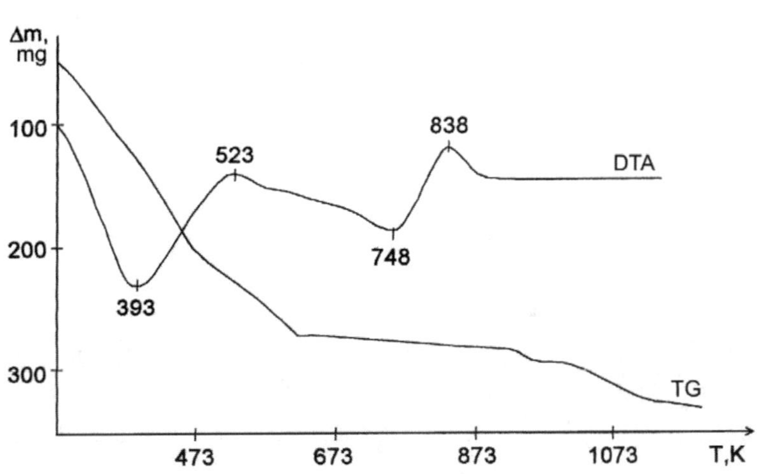

Figure 3. Derivatogram of Ta_2O_5formation from film-forming solutions with $TaCl_5$

Table 1. Infra-red spectrum bands of powders obtained from film-forming solutions at various annealing temperatures

Film-forming solution composition; T,K	Vibration type					
	H – OH	Valence $-OC_2H_5$	H\ O H/	Deformation $-OC_2H_5$	Si – O – Si	O\ Si; O/
$TaCl_5$ + C_2H_5OH						
T=373	3600	2900	1640	1200	-	400
T=773	-	2900	-	1100	-	400
T=848	-	2900	-	1100	-	400
T=900	-	-	-	-	-	400
ТЭОС +$TaCl_5$						
T=373	3560	-	1640	1000	600-800	460-400
T=600	3600	2900	1640	1000	600	460-400
T=633	-	2900	-	1100	600	460-400
ТЭОС + C_2H_5OH						
T=373	3660	2935	1640	1095	600-800	460
T=473	1380	2935	1640	1100	800	460
T=773	-	-	-	1100	800	460

Process activation energies are calculated for each stage, the data are summarized in Table 2. The activation energy of the first process, E = 41.3 kJ/mol, agrees with these values for condensation of hydroxyl-containing organosilicon compounds [9,10,11].

Table 2. Kinetic parameters for f SiO_2, Ta_2O_5 and $SiO_2 - Ta_2O_5$ powders production (according to thermogravimetric and DTA methods)

Formation stage	Powder SiO_2			Powder Ta_2O_5			Powder $SiO_2 - Ta_2O_5$		
	T interval, K	Degree of transformation, %	E_a, kJ/mol						
	1	2	3	1	2	3	1	2	3
1	298-473	33.0	41.4	298-473	51.7	44.1	298-473	50.0	40.5
2	473-623	29.0	51.8	473-573	15.0	32.1	473-623	34.5	32.6
3	623-973	37.5	68.5	573-923	33.3	107.8	623-763	3.4	60.9
4							763-973	12.0	117.4

The research data prove that it is possible to control the film synthesis by changing concentration of incoming components and process conditions. Obtaining films of a certain thickness and with given physical and chemical characteristics can be provided only by using high quality film-forming substances and properly prepared solutions and by following the processing methods.

Based on the results of the research into physical and chemical properties of film-forming solutions, we have developed a technology for preparing solutions to produce homogeneous films of $SiO_2 - Ta_2O_5$ systems.

Coatings produced from these solutions can form complete glassy films which have a number of valuable qualities. The surface holding power of the films as well as their optical and electrophysical properties make their practical use efficient. The research showed that the obtained films have good adhesion to various substrates and are nonporous.

Optical, electrophysical and adhesion properties of $SiO_2 - Ta_2O_5$ films have been studied (Table 3). The refractive index changes linearly from 1.45 (for SiO_2) to 1.98 (for Ta_2O_5) (Fig. 4), which indicates there are no chemical compounds in this system and a mechanical mixture of oxides is formed similar to bulk samples state diagram [12,13].

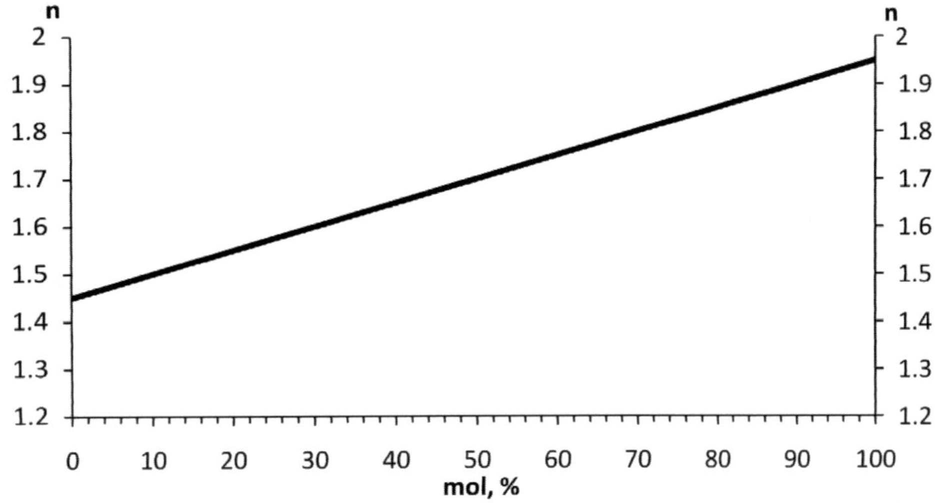

Figure 4. Dependence of $SiO_2 - Ta_2O_5$ film refractive index on Ta_2O_5 content

Table 3. Physical and chemical properties of $SiO_2 - Ta_2O_5$ films

Properties	Ta_2O_5 content in films, %								
	10	20	30	40	50	60	70	80	90
Refractive index, n	1.51	1.61	1.63	1.63	1.69	1.72	1.75	1.83	1.91
Dielectric permittivity, ε	4.6	5.1	7.4	9.2	11.0	12.5	14.3	15.1	16.5
Adhesive power, F, kg/mm^2	0.93	0.95	0.91	0.96	0.97	0.94	0.96	0.94	0.95

Electrophysical properties show that SiO_2-Ta_2O_5 films are dielectrics (Table 3) which have a lower dielectric constant but a smaller loss angle (tg = 10-3) in comparison with pure Ta_2O_5. Small *tg* values speak for either a defect-free film structure or a structure with very uniform defectiveness.

A coating method can be assessed an efficient one only if it provides obtaining layers of different materials with a wide range of physical properties to meet engineering requirements. Film optical properties (refractive index and the absorption coefficient) are the most significant of their physical properties. The problem of regulating and filtrating optical radiation in the ultraviolet spectrum is important for various fields of technology. Using light-redistributive UV-radiation coatings for high-intensity light sources is of particular interest [14,15].

Figure 5 shows optical transmittance spectra of SiO_2 and Ta_2O_5 films. Since optical properties of films depend on their molecular structure, coatings produced from multicomponent solutions have different spectral absorption characteristics from respective one-component coatings. Figure 6 shows transmittance spectra of SiO_2-Ta_2O_5 films of various compositions.

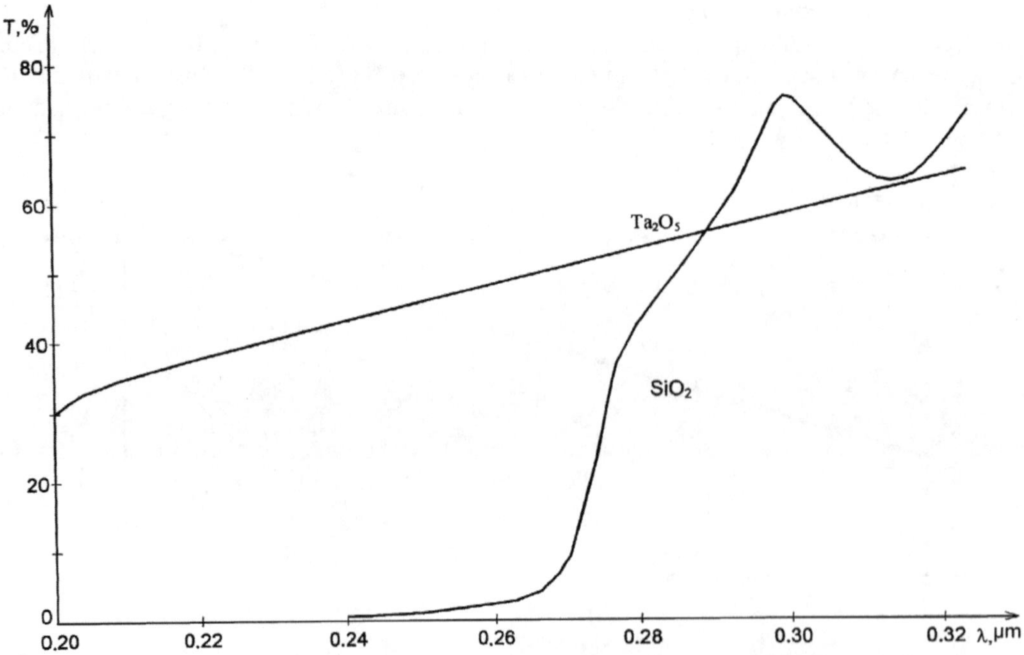

Figure 5. Spectral transmittance of SiO_2 and Ta_2O_5 base films

As can be seen from the optical studies results, when the oxide composition changes from pure SiO_2 to pure Ta_2O_5, the absorption wavelength is in the range of 160 to 300 nm. The research prove that it is possible to control λgr depending on the oxide.

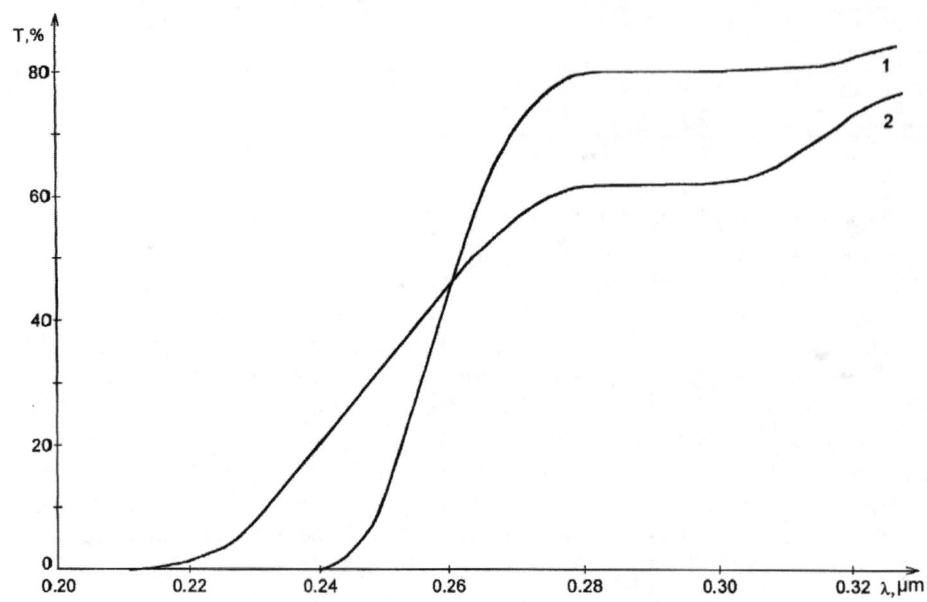

Figure 6. Spectral transmittance of $SiO_2 – Ta_2O_5$ films
1 - 40% Ta_2O_5 concentration; 2 - 20% Ta_2O_5 concentration

Since the operating lamp temperature reaches 770-970K, the thermal stability of the films was evaluated at 670, .870 and 970K. Analysis of transmission spectra showed that λgp position and transmittance do not change at increasing annealing temperature and transmittance index at maximum annealing temperature decreases no more than 10% of the initial one. The results of the study showed that $SiO_2 - Ta_2O_5$ films are of practical interest for correcting ultraviolet spectrum radiated by light sources. Tests of gas-discharge light sources coated with $SiO_2 - Ta_2O_5$ has shown that a film allows reducing ozone generation to 80%. The coating is thermally stabile and does not change the light transmittance index during a light source operation. It is proved that by changing the composition of $SiO_2 - Ta_2O_5$ film it is possible to control the UV radiation transmittance boundary of a gas discharge lamp in the range of 220-300 nm. As a result ultraviolet radiation with a predominantly erythemic, bactericidal or highly inhibitory effect can be generated.

4. Conclusions

A sequence of main formation stages of Ta_2O_5 as well as $SiO_2 - Ta_2O_5$ from film-forming solutions was determined by means of IR spectroscopy and thermogravimetry. Optimal methods of thin-film synthesis were proposed. They consist in preliminary heat treatment at T = 333K and subsequent annealing at T = 873K. Structure and physicochemical properties of the obtained films are analysed. The refractive index for $SiO_2 - Ta_2O_5$ films varies from 1.45 to 1.98. The developed thin-film materials can be used in the production of zero-gap UV radiators with bactericidal and erythemic effect, as well as dielectric coatings of high thermochemical stability.

References

[1] Biotesting as a Method of Evaluating Waste Hazard in Metallic Mineral Mining Lugovaya, Y.R., Orlova, K.N., Litovkin, S.V., Malchik, A. G., Gaydamak, M.A. 2016 IOP Conference Series: Materials Science and Engineering 127 (1), 012026

[2] Spirin, E.K., Lugovtsova, N.Yu., Khomich, O.I. Special features of VP-1Ap anion exchanger behavior under deep silicifi cation // Russian Journal of Applied Chemistry. - 2012. Vol. 85, Is. 6, P. 921-925.

[3] Timofeeva, S.S., Khamidullina, Y.A., Davydkina, O.A., Lugovtsova, N.Y. Technology of Water Purification with Chlorinated Derivatives and Assessment of Risk Associated with Human Exposure to These Substances // IOP Conference Series: Materials Science and Engineering. - 2016. Vol. 127 - №1, Article number 012025. – p. 1-7.

[4] Il'yaschenko D.P., Chinakhov D.A., Danilov V.I., Schlyakhova G.V., Gotovschik Y.M. Increasing Strength and Operational Reliability of Fixed Joints of Tubes by MMA Welding. IOP Conf. Series: Materials Science and Engineering. 91 (2015) 012007. doi:10.1088/1757-899X/91/1/012007. http://iopscience.iop.org/1757-899X/91/1

[5] Il'yaschenko D.P., Chinakhov D.A., Danilov V.I., Schlyakhova G.V., Gotovschik Y.M. Physical Nature of the Processes in Forming Structures, Phase and Chemical Compositions of Medium-Carbon Steel Welds. IOP Conf. Series: Materials Science and Engineering. 91 (2015) 012006. doi:10.1088/1757-899X/91/1/012006 http://iopscience.iop.org/1757-899X/91/1

[6] Ievlev V.M. Structural transformations in thin films. - Moscow: Metallurgy. - 1998. – 132p.

[7] Physicochemical Study of Ta_2O_5 and SiO_2-Ta_2O_5 Film Formation from Film-Forming Solutions Gryaznov, R. V., Borilo, L. P., Kozik, V.V., Mal'chik, A.G. 2001 Russian Journal of Applied Chemistry, 74 (1), p. 18-20

[8] Physicochemical investigation of the formation of SiO_2 powders and films from film-forming solutions Mal'chik, A.G., Borilo, L.P., Kozik, V.V. 1996 Russian Journal of Applied Chemistry 69 (2), p. 198-201

[9] Voronkov M.G., Mileshkevich V.P., Yuzhelevsky Yu.A. Siloxane bond. - Novosibirsk: Science. - 1976. – 413p.

[10] Toropov N.A., Barzakovskiy V.P. State diagrams of silicate systems. Reference book. Issue 2. Metal - oxygen compounds of silicate systems. - L .: Science. - 1969.

[11] Borilo L.P. Thin-film inorganic nanosystems / ed. V.V. Kozik. - Tomsk: Izd-vo TSUU, 2003. – 134p.
[12] Synthesis and properties of films in the SiO_2-Bi_2O_3 system. Borilo L.P., Kuznetsova S.A., Kozik V.V., Zabolotskaya A.V., Mal'chik A.G. Russian Journal of Inorganic Chemistry. 2014. T. 59. № 10. P. 1065-1068.
[13] Synthis and Phisical and Chemical; Properties of SiO_2 - B_2O_3 and SiO_2 - P_2O_5 Thin Film Systems and Powders Mal'Chik, A.G., Litovkin, S.V., Seregin, V.I., Rodionov, P.V., Kryuchkova, S.O. 2016 IOP Conference Series: Materials Science and Engineering 142 (1), 012066
[14] Sergeev A.N., Serebrennikov V.V. Material science of refractory oxides films. - Tomsk: Tomsk University, 1981. – 145p.
[15] J.Pout, K.Tu, J.Meyer. Thin films. Mutual diffusion and reaction / trans. from English. Ed. V.F. Kiseleva. - Moscow: Mir. - 1982. – 382p.

Ecological Safety of the Internal Space of the Cattle-Breeding Facility (Cowshed)

A A Potseluev, I V Nazarov[a], T N Tolstoukhova, M V Kostenko

Azov-Black Sea Engineering Institute FSBEE HE «Donskoy GAU» in Zernograd, ul. Lenin, 21, the town of Zernograd, Russia, 347740

E-mail: [a] niv671@rambler.ru

Abstract. The article emphasizes the importance of observing the ecology of the internal air-space. The factors affecting the state of the air in the internal space of the cattle-breeding facility (cowshed) are revealed. Technical and technological solutions providing for a reduction in the airspace contamination of the livestock facility are proposed. The results of investigations of a technological operation for treating skin integuments of cows with activated water are disclosed, as well as the constructive solution of a heat and power unit that ensures a change in the hydrogen index of the treated water. The justification of the efficiency of the proposed technical and technological solutions is given.

1. Introduction

Efficiently selected technological equipment in the cattle-breeding premises, in this case it is a barn, allows improving the quality of the performance of technological processes for servicing animals, increasing the productivity and comfort of the maintenance personnel and, as a consequence, increasing the quantity and quality of the main product (milk, meat) and secondary raw materials (hide, hair, manure).

However, failure to perform certain operations of the mechanized technological process or poor performance of these processes increase the risk of disturbance of the ecological balance within the cattle-breeding premises. In the context of the technological processes of servicing animals, the technological equipment used inside the cattle-breeding premises also affects the ecological background of the internal space of the cattle-breeding premises, in particular the cowshed. The use of mobile means in the process of distribution of fodder increases the gas contamination and the dustiness of the space (distribution of dry coarse fodder) and stationary means which as a result leads to the contamination of the interior space of the premises, stabling equipment and the surface of the skin of animals (cows). It contributes indirectly to contamination of the final product, in our case it is milk.

A number of researchers have established that the dust particles of dry fodder, as well as the ones of the introduced litter within the technological process of the maintenance of the cows, contain a certain kind of microorganisms (bacteria, molds and others), which increases the risk of contamination of the final milk product, as well as the sickness rate of the maintenance personnel [1, 2, 3]. In this connection, it becomes necessary to perform a technological operation of processing fodder at the stage of fodder intake into the zone of its distribution along the aft line and the inclusion of a periodic operation of wetting the skin of the cows with activated water (according to the content of the pH-value) into a technological operation for treating the skin integuments of cows.

2. Results and discussion.

To reduce the dustiness of the air space from the dust particles of the fodder, we propose to use heat treatment of coarse fodder with warm water at a temperature of 20...30 ° C with a periodically valid qualitative pH indicator (alkaline or acidic medium) during the process of its distribution [4].

IOP Conf. Series: Earth and Environmental Science **115** (2018) 012005 doi:10.1088/1755-1315/115/1/012005

Depending on the used production line of fodder distribution, it is advisable to perform point-by-point processing of coarse fodder in the feeding zone of a stationary fodder distributor or continuously along the fodder distribution line (mobile distribution) Figure 1.

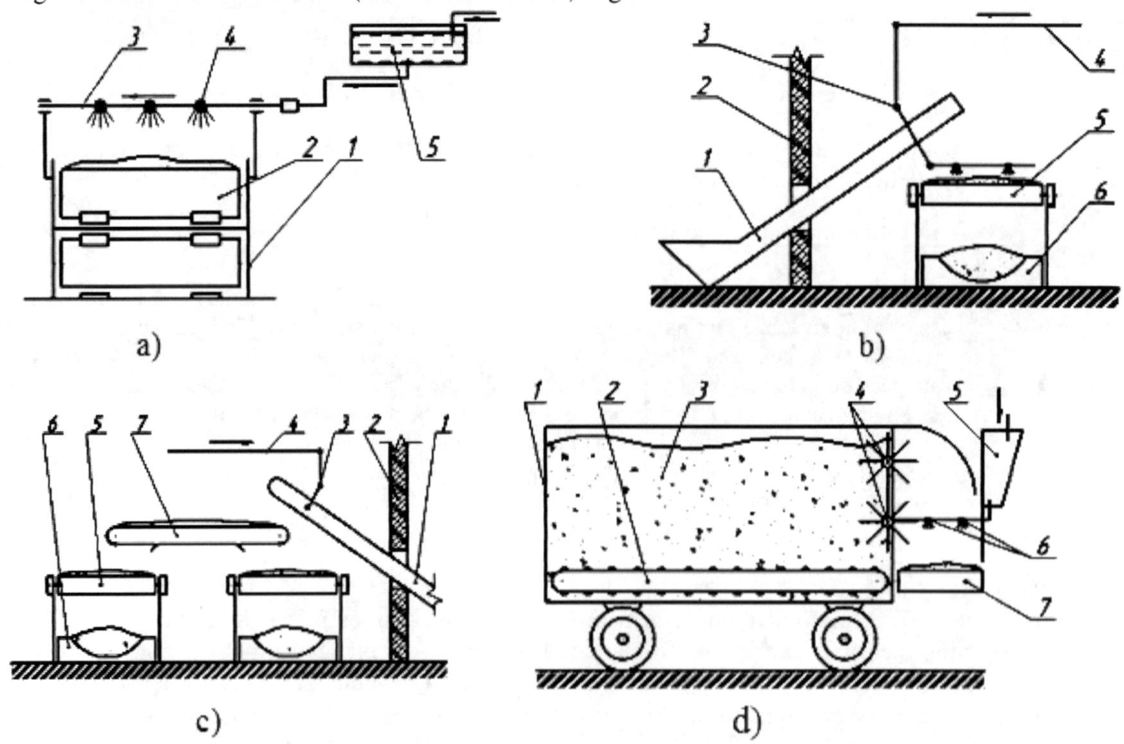

Figure 1. Schemes of organizational options for processing coarse fodder:
a) distribution of fodder to a stationary fodder distributor: 1 – the body of the fodder distributor inside the feeding bowl; 2 – movable operating element of the fodder didtributor (scraper plate or tape); 3 – distribution perforated pipeline; 4 – pulverizing hole; 5 – water storage; b) distribution of fodder to the fodder distributor above the feeding bowl: 1 – inclined conveyor; 2 – wall of the cowshed; 3 – perforated pipeline for wetting fodder with water; 4 – water supply line; 5 – fodder distributor; 6 – feeding bowl; c) distribution of fodder to the fodder distributor above the feeding bowl with reverse loading: 1 – powered incline; 2 – wall of the cowshed; 3 – a perforated pipeline for wetting the fodder water; 4 – water supply line; 5 – fodder distributor; 6 – feeding bowl; 7 – powered reverse; d) mobile fodder distributor: 1 – bunker of the fodder distributor; 2 – longitudinal conveyor; 3 – monolith of the fodder; 4 – beater; 5 – storage capacity of water; 6 – perforated pipeline for wetting fodder with water; 7 – transverse fodder conveyor

In order to improve the protective properties and the metabolism in the animal body (cow), to increase productivity and reduce contamination of the product in the process of milking cows, technological equipment has been developed and is still being developed to care for the skin of animals [5, 6, 7, 8]. In the technological environment, special spring brushes or rotating brushes with an electric drive are used to care for the skin. At the feedlots in the USA special equipment "Chesalo" is used to take care of the cattle. In some cases during the technological process, vacuum cleaners of various designs were used, as well as mobile cattle washing appliances. In the design solutions of livestock buildings, a separate room is sometimes provided for cleaning, washing and drying of animals. However, for the most part they do not correspond to the complexity of performing technological operations, they do not exclude the possibility of contaminating the air of the environment of the cattle-breeding premises with structural components of the skin, they are also low in effectiveness of reducing the bacterial contamination of the skin and do not exclude the possibility of injury to the maintenance personnel by the animal on the leash (cows) [9].

In order to preserve the ecological safety of the internal space of the cattle-breeding premises, reduce the risk of sickness of maintenance personnel and injure them while caring for the animal (cow), and reduce the harmful effects of bacterial contamination of the animal's skin on the quality of the products, we have developed a system for the sanitary treatment of the skin of the cows on a leash Figure 2 [10].

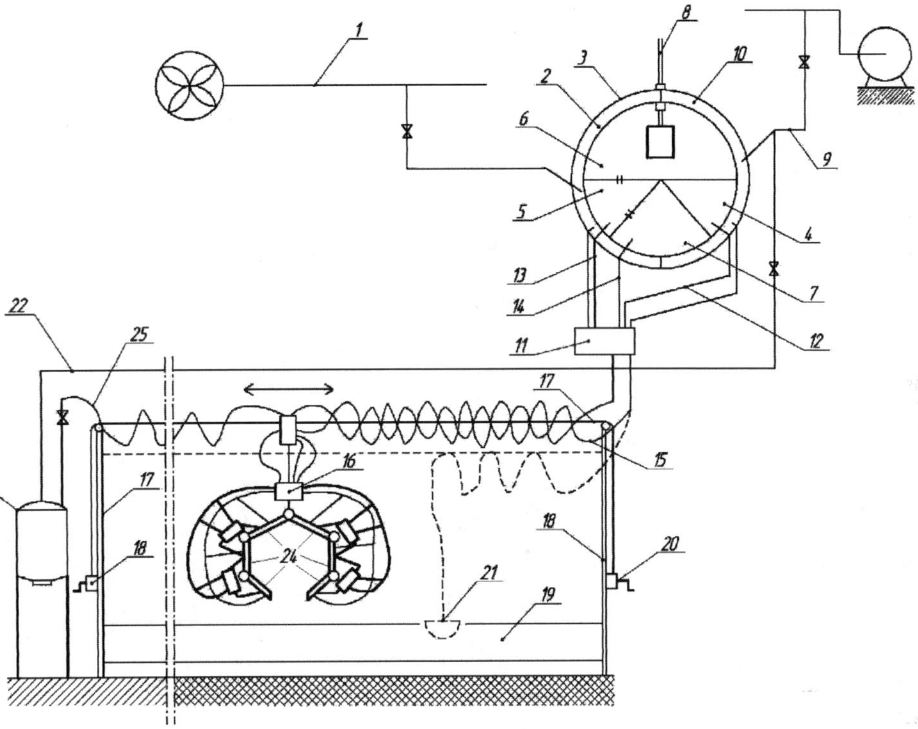

Figure 2. Scheme of the sanitary treatment system for the skin of the cows on the leash:
1 – high pressure air supply pipeline; 2 – chamber of internal air pressure accumulation; 3 – water heating tank; 4 – chamber for disinfecting solution; 5 – water chamber with a lower pH relative to the normal one; 6 – water chamber for needs of automatic drinking; 7 – water chamber with increased pH; 8 – supply pipeline; 9 – pipeline of vacuum supply; 10 – chamber of vacuum accumulation; 11 – distributor of processing agents; 12, 13, 14 – trunk pipelines for delivering disinfectant solution and water with different pH levels; 15 – working pipeline for the supply of processing agents; 16 – two-section device for treating the skin; 17 – flexible suspension; 18 – supports; 19 – fodder table; 20 – winch; 21– device for washing the udder of the cows; 22 – a vacuum tube; 23 – a tank for collecting contaminants and hair; 24 – working organs for treating the skin; 25 – transport pipeline for the transport of contamination

The proposed system of sanitary treatment of the skin of the cows provides for a full range of technological operations, including mechanical processing with vacuum removal of contaminants and hair during the shedding of hair into storage tanks, the wetting and the drying of the skin, periodic sanitization.

One of the features of this system is the possibility of treating the skin with activated water with a selected (justified) hydrogen index (pH).

To justify the feasibility of this technological operation, studies were carried out to determine the species composition of the bacterial contamination of the skin and the effect of activated water with a different pH value on its vital activity. The research was based on taking washouts from patches of the

IOP Conf. Series: Earth and Environmental Science **115** (2018) 012005 doi:10.1088/1755-1315/115/1/012005

skin of the cows before their treatment and after their treatment with a solution (activated water). The material was taken by sterile swabs, mounted into sterile tubes with a content of 0.5 ml. sterile saline solution. Before sowing the resulting material, a series of tenfold dilutions were carried out, followed by seeding on Petri dishes with nutrient agar to determine the total microbial count (TMC) and with blood agar for the growth of the microorganisms. Aerobic and facultatively anaerobic microflora were subjected to the analysis. As a result, it was found that the microflora of the skin of the cow on the leash consists of the cocco flora and microorganisms p. Corynebacterium (corynebacterium). Along with this, there are spore and mold microorganisms (transient bacteria coming from outside) on the skin Figure 3.

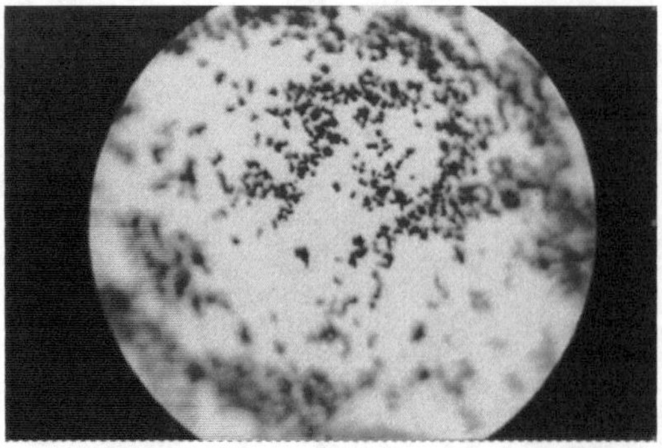
cocco flora

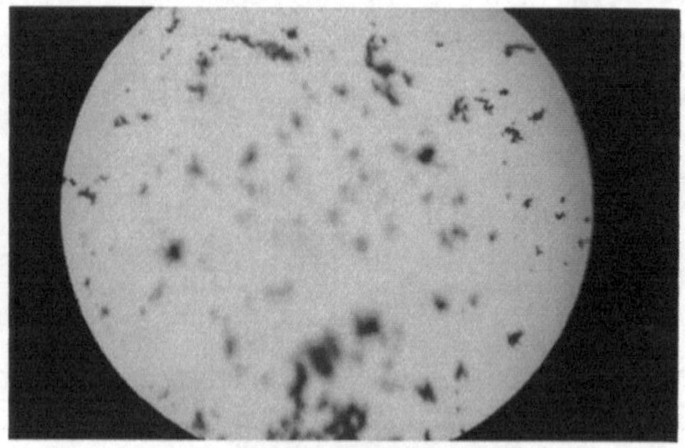
corynebacterium

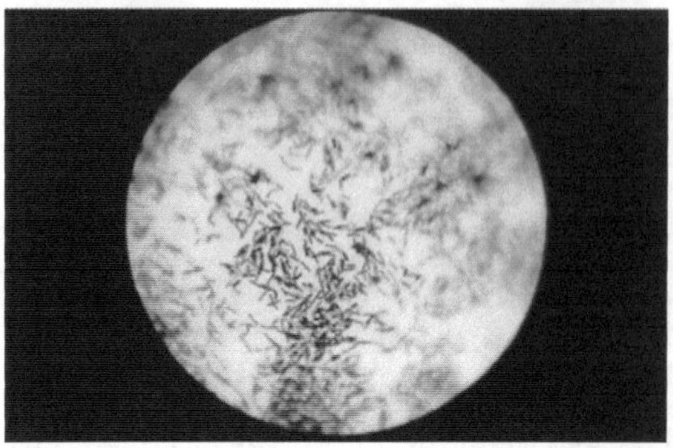
cryptogamous microflora

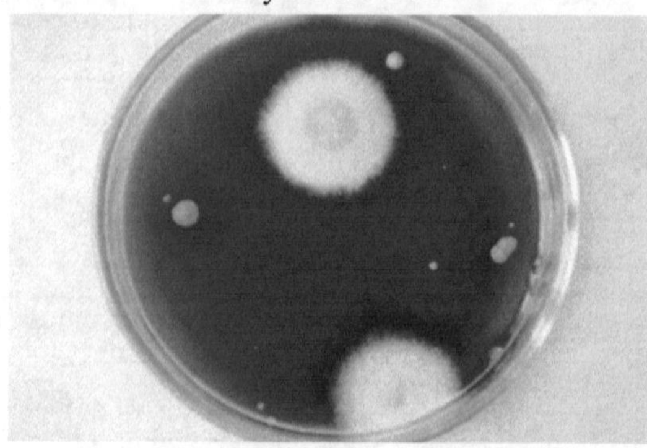
mould

Figure 3. Fragments of sowing the microflora of the skin of the cow

Studies have also found that the use of activated water in the treatment of the skin reduces the total microbial count (TMC) and suppresses the growth of spore microflora.

The process of water activation is provided in Sections 5 and 7 of Storage Tank 3 of the system Figure 2 using a heat and power unit Figure 4.

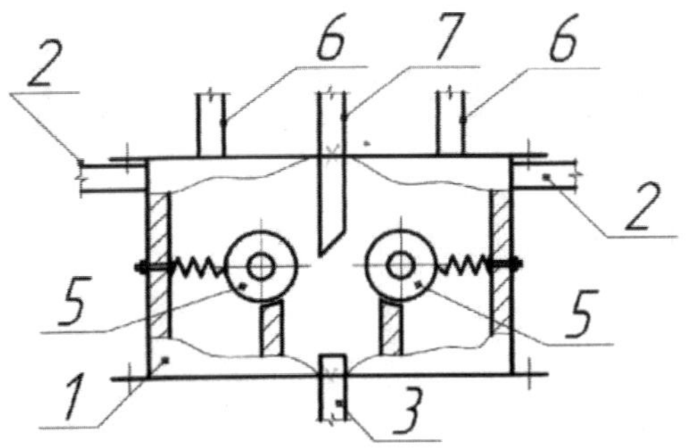

a)

b)

Figure 4. The construction scheme of the heat and power unit (a) and its full-scale sample (b): 1 – the case of the block; 2 – circulation pipes; 3 – an inlet branch pipe; 4 – adjusting screws; 5 – electrodes; 6 – gas pipes; 7 – division wall

The investigations have established the time and temperature regime of the block, as well as the changes in the physical and chemical characteristics of water during its activation, which corresponds to the technological process of treating the skin of cattle (cows) Figures 5, 6.

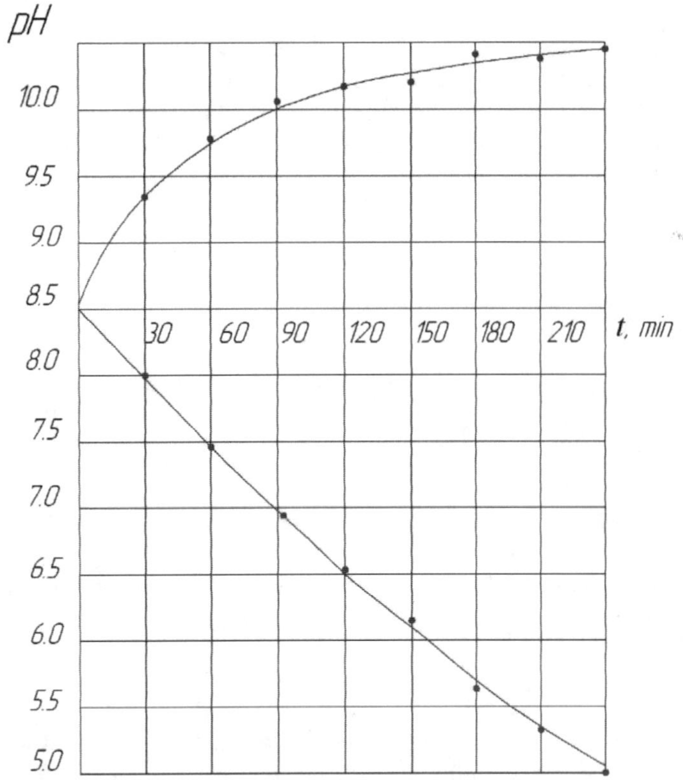

Figure 5. pH change depending on the duration of water treatment

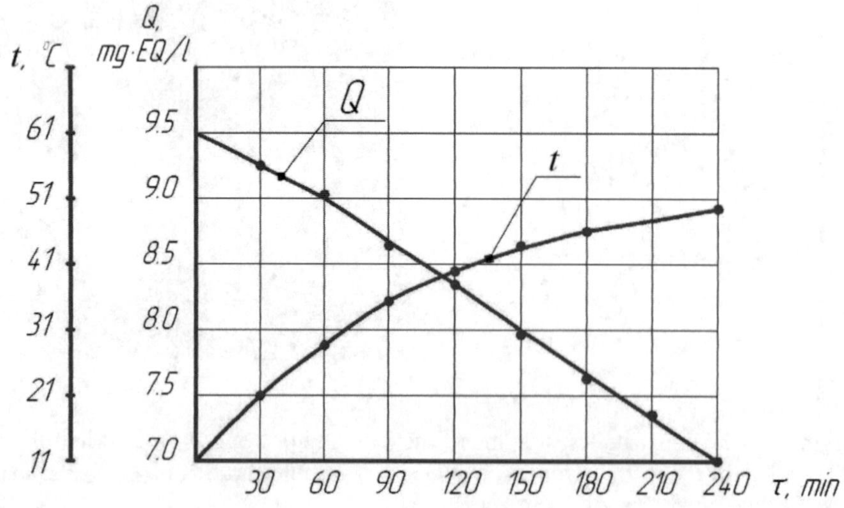

Figure 6. Change in temperature t and water hardness Q, depending on the treatment time

3. Conclusions

1. Ecological safety of the internal space of the cattle-breeding premises (cowshed) depends on the strict performance of zootechnical and sanitary requirements for performing the technological process for servicing animals and in particular the process of sanitary treatment of the skin of animals (cows);

2. The proposed systems for processing coarse fodder and skin of cows on the leash, while observing the technological requirements for processes, allow to reduce the ecological safety of the internal space of the cattle-breeding premises (cowshed) for maintenance personnel.

3. The proposed system and the technology of treating the skin of the cows on the leash allow to reduce the level of pollution of the internal air environment, the risk of diseases of the maintenance personnel (pneumonia, sepsis, endocarditis, meningitis, wound suppuration, bacteriemia and others) caused by the microflora of the skin of the animal, as well as the maintenance personnel' injury risk through contact with animals;

4. It was found that activated water with hydrogen pH in the ranges 5.8...6.0 and 8.6...9.0 is the most effective in the treatment of the skin.

5. The implementation of the technological operation for the treatment of the skin with activated water helps to reduce the total microbial count (TMC) by 11.1 ... 12.8 times, while the growth of spore microflora being suppressed completely.

References

[1] Bogdanov V.M. Microbiology of milk and dairy products / V.M. Bogdanov. Moscow: Pishchepromizdat, 1946. 147 p.

[2] Tomescu V. Zoonosy / V. Tomescu, I. Gavrile, D. Gavrile / Moscow: Kolos, 1982. 304 p.

[3] Water supply resource-saving systems of technological processes for servicing cattle: Diss. .... doct. tech. sciences. Zernograd, 2011.

[4] Agricultural machinery. Product catalog, 2007. LLC Science and Production Enterprise, Izhevsk.

[5] Fedorenko V.F. Innovative technology for livestock breeding / V.F. Fedorenko / Moscow: Rosinformagrotech, 2013. 205 p.

[6] Patent of the Russian Federation No. 2438301 M.A. K 13/00 Animal cleaning device / Vasilyeva T.N., Vasilyev V.G., Vasiliev V.V. Publik. 10.01. 2012.

[7] Nigmatov L.G. The design development and the justification of the parameters of the device for mechanical cleaning of the skin of cattle. Diss. ...cand. tech. sciences. Orenburg, 2015.

[8] Khlopko Yu. A. The development of a multifunctional device for mechanical cleaning of the skin of cattle / Yu.A. Khlopko, L.G. Nigmatov, A.A. Panin, I.V. Gerasimenko // Bulletin of VNIIMZh, 2014 / 4
[9] Baranov Yu.N. Methodology of increasing the safety of livestock workers through engineering technical and organizational measures: Diss .... doct. tech. sciences. S. Peterburg, 2010.
[10] Urban V.P. Workshop on episiotomy and infectious diseases with veterinary sanitation / V.P. Urban, M.A. Safin, A.A. Sidorchuk. Moscow: Kolos, 2002.

Direction of the Rational Use of Water at the Facilities for Growing Poultry

A A Potseluev, I V Nazarov [a], A K Porotkova, N V Volovikova

Azov-Black Sea Engineering Institute FSBEE HE «Donskoy GAU» in Zernograd, ul. Lenina, 21, the town of Zernograd, Russia, 347740

E-mail: [a] niv671@rambler.ru

Abstract. The article notes the effect of water use in the technological process of automatic drinking agricultural poultry on the quality and the quantity of outputs. At the same time, the requirements to the quality of the used water, the regimes of its consumption by the poultry and the role of mechanization of the process of automatic drinking in the rational use of the water resource, the processing and the reuse of contaminated wastes are disclosed. Within the framework of this concept, we propose constructively technological solutions of systems and means of automatic drinking agricultural poultry, providing the rational use of water as one of the important products of vital activity of agricultural poultry.

1. Introduction

At agricultural facilities for the production of poultry products, the main consumption of water and the formation of contaminated wastes occur during the technological process of drinking poultry. When carrying out some production studies in the poultry-house on 366 hen heads, it was found that the daily water consumption for the researched object is 2244 m^3. However, it should be noted that the quality and quantity of produced products is not only affected by the amount of consumed water, but also by its quality. At the same time, the amount of used water and the factor of its rational use directly affect the cost of produced products (eggs, meat).

The quality of water used for the needs of automatic drinking poultry and the mode of its use must meet the zootechnical requirements and we should take into account the recommendations of the scientists in this regard.

The researchers [1, 2, 3, 4, 5, 6] have established that, depending on the species of the poultry and their sex and age, the consumption of water relative to the consumed fodder varies over a wide range and can be accounted for by the water consumption coefficient of irregularity in the approximate range of 1.4 ... 2.5. It should be noted that the demand for water increases to 30% when using granulated fodder. Also, the demand for water increases depending on the content of salt, meal, molasses, fiber and protein in the fodder. The deviation of the temperature regime from the comfort temperature (up to 30 °C) can cause a surge in water consumption in 2.5...3 times relative to the stable regime of water consumption. However, despite the importance of the considered factors, the maintenance of the required water temperature regime in the systems and the means of automatic drinking, the ensuring of a necessary water quality, the decrease of the nonproduction water costs and labor costs for servicing the systems and the means of automatic drinking are the directions characterizing the level of the technological effectiveness of the process of automatic drinking and the resource saving.

2. Results and discussion

The analysis of zootechnical requirements to the process of automatic drinking hens and turkeys, constructive solutions of the systems and the means of automatic drinking at the level of their production, the research and the patent developments [7, 8, 9, 10] showed that:

1. The developed and recommended systems and means of automatic drinking chickens and turkeys do not provide a comprehensive technical and technological solution that would ensure the compliance with zootechnical requirements for the process of automatic drinking and rational use of water.

2. The comprehensive solution of the unit providing the intake and reservation of water, its treatment in view of maintaining the temperature and the physical and mechanical parameters, has not been developed for poultry farming systems. The issues of collecting contaminated water, its treatment and secondary use, as well as the parallel drug feeding of poultry haven't been solved either.

3. In the group means of automatic drinking, the question of the universality depending on the product drunk by the poultry (water with different physicochemical properties) has not been solved.

4. In the group means of automatic drinking and the lines of group drinking, the problems of the self-purification of the drinking bowls, the withdrawal of contaminated water, its preliminary filtration, bactericidal treatment and transportation to a storage tank has not been solved.

With this in mind, we have developed a scheme for a multifunctional poultry self-drinking system Figure 1, a scheme of the technological line for distributing water by automatic drinking bowls with their periodic cleaning from contamination, and a design and technological solution for a group automatic drinking with mechanical removal of contaminants and the withdrawal of contaminated water (waste water).

One of the basic elements of the automatic drinking system for poultry is the technological line for distributing water by automatic drinking bowls Figure 2.

The working process of this automatic drinking line is as follows. The water prepared in the storage tank according to the temperature index and the hydrogen index (pH) flows through the piping system into the valve-float device, and enters the regulating tank and then flows along the distribution pipeline in the regime of communicating vessels to the automatic drinking bowls for poultry. To clean periodically the drinking bowls from contamination and to change the drinking water , the cleaning unit is also periodically switched on. In this case, the vacuumed pulsator provides the reciprocating movement of the cleaning device and the purification of the drinking bowls from contamination through the flexible rod and the shock absorber. At the end of the working cycle, the technological line is brought into operation by means of the distribution pipeline valve (the drinking bowls are filled with clean water).

One of the basic elements of the system of the automatic drinking is the means of automatic drinking (automatic drinking bowls). The poultry are directly in contact with the automatic drinking bowls. Therefore, the constructive solution of the drinking bowls should ensure the rapid adaptation of the poultry to the drinking bowls, exclude the cases of injure to the poultry, maintain the designated temperature regime of the drinking water and its quality, their multifunctionality and versatility, optimize the water consumption, as well as reduce labor costs for their maintenance.

On the basis of the analysis of zootechnical requirements to the process of the withdrawal of water from automatic drinking bowls and the known design solutions of the means of the automatic drinking [11, 12, 13, 14, 15, 16], we found that:

1. The designed and recommended for use means of drinking the poultry, particularly turkeys, do not provide a comprehensive technical solution that ensures the compliance with zootechnical requirements for the process of the automatic drinking;

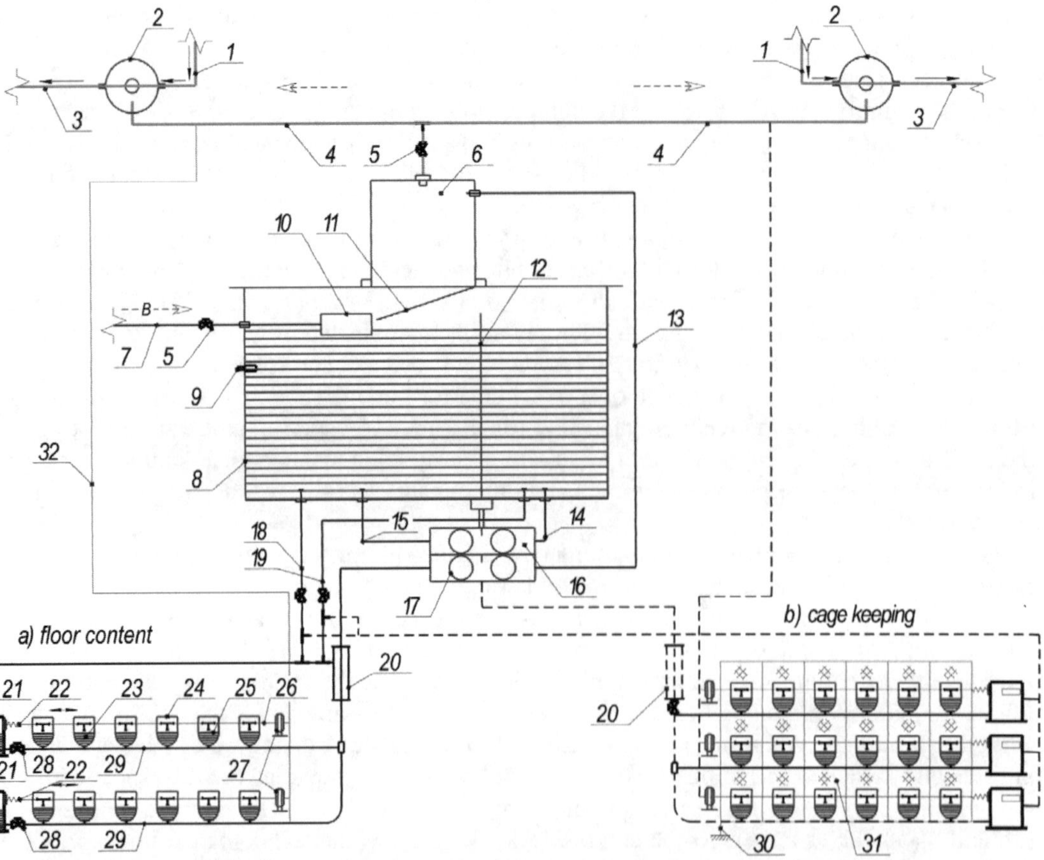

Figure 1. Scheme of a multifunctional automatic drinking system for poultry:
1 – air intake duct; 2 – ventilation fan of the poultry house; 3 – air supply duct to the poultry house; 4 – a vacuum line; 5, 28 – electric valves; 6 – accumulator of circulating water; 7 – water supply pipeline; 8 – water storage tank; 9 – temperature sensor; 10 – valve-float device; 11 – the circulating water dump valve; 12 – partition of the storage tank; 13 – pipeline of reused, circulating water; 14, 15 – pipelines of waters divided in accordance with their pH; 16 – electrode block for water treatment; 17 – electrode block for treating circulating water; 18, 19 – water supply pipelines with different pH values; 20 – block of filtration for circulating water; 21 – tank of the water dispensor; 22 – shock absorber; 23 – drinking bowl; 24 – guide bracket; 25 – cleaner; 26 – rod; 27 - pulsator; 29 – a pipeline of water supply to drinking bowls; 30 – cage battery; 31 – a cage (section for poultry).

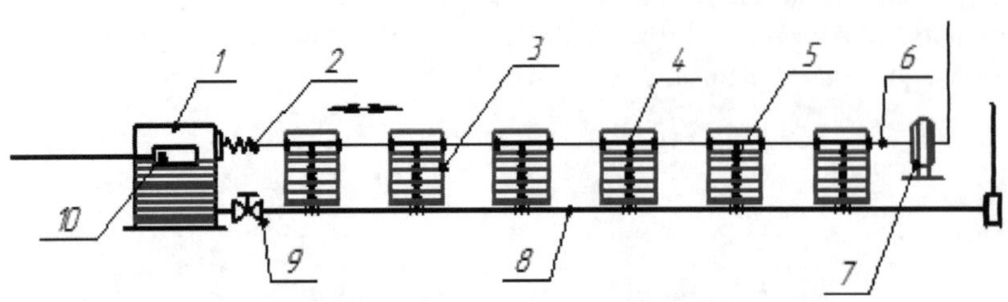

Figure 2. The scheme of the technological line of water distribution to the drinking bowls with periodic cleaning and the withdrawal of contaminated water:
1 – regulating tank; 2 – shock absorber; 3 – automatic drinking bowl; 4 – suspension slider, cleaning device; 5 – the actuator of the cleaning device; 6 – flexible rod; 7 – pulsator; 8 – distribution pipeline; 9 – valve (electric valve), 10 – valve-float device

2. Speaking about the means of the automatic drinking for poultry, particularly turkeys, a constructive solution ensuring the access of the poultry to an alternative product in terms of its physico-chemical properties (a different value of pH water; water, and medicinal solution) has not been developed yet.

3. Speaking about the group means of the automatic drinking, the issue of their self-cleaning, pre-filtration and the withdrawal of contaminated water has not been solved.

4. Speaking about the group means of the automatic drinking, the issue of resource saving for the drive operation of cleaning devices has not been solved.

To eliminate these shortcomings, we have developed a constructive solution for the group automatic drinking for turkeys and chickens Figure 3.

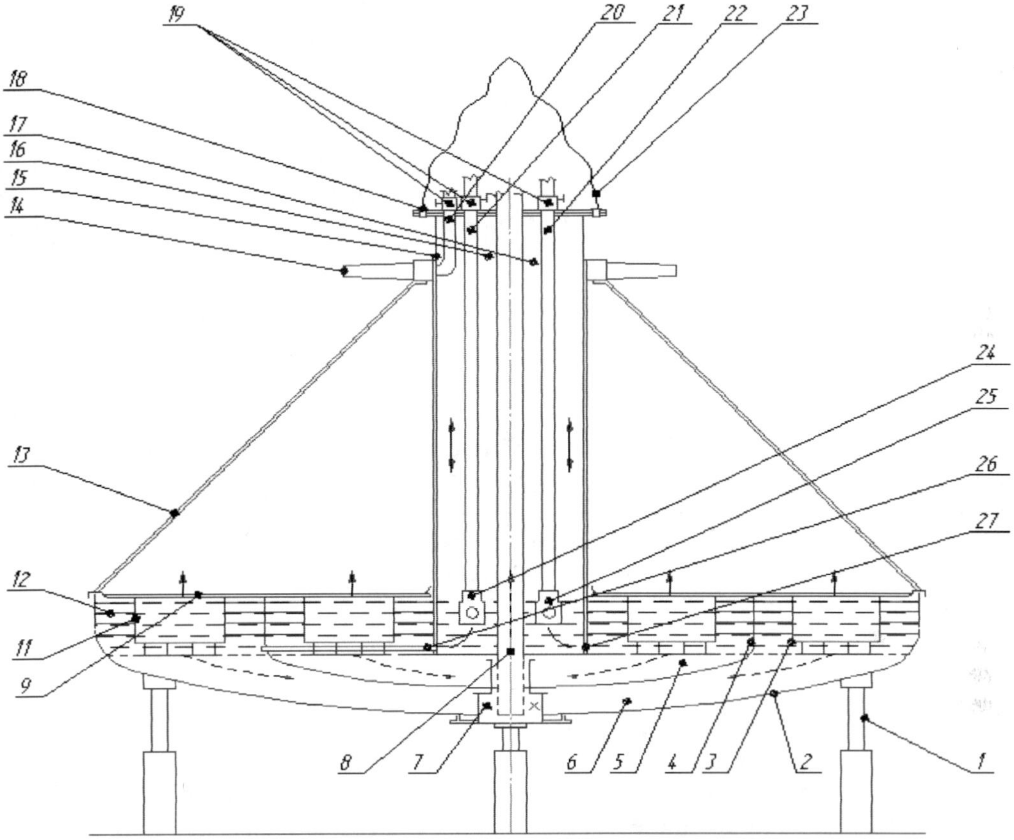

Figure 3. Scheme of the group drinking for the poultry (turkeys, chickens):
1 – removable, transformable supports; 2 – body of an automatic drinking bowl; 3; 4 – separate containers with a perforated bottom; 5; 6 – pollutant discharge chambers; 7 – cascade filter; 8 – pipeline for the withdrawal of contaminated water; 9 – replaceable lid; 10 – water-filled windows; 11 – water drinking glasses; 12 – brush working parts; 13 – spacers; 14 – jet spraying working organ; 15– water supply pipe; 16; 17 – isolated cavities; 18 – lid; 19 – the lock of the pressure pipeline of air supply; 20 – jet spraying working element; 21 – pressure water supply of the drinking water with a standard pH value; 22 – pressure pipe of the drinking water supply with a reduced pH value or medicinal solutions; 23 – suspension (traction) of the drinking bowl; 24; 25 – valve mechanisms for supplying drinking water with a standard pH value and supplying drinking water with a reduced pH or medicinal solutions; 26 – metering channels of water supply; 27 – metering inlets for supplying water or solution to a separate container

IOP Conf. Series: Earth and Environmental Science **115** (2018) 012006 doi:10.1088/1755-1315/115/1/012006

When developing the design of the drinking bowl, the task was to improve the quality of the process of the automatic drinking for the poultry, reduce the nonproduction consumption of water and energy to maintain a given temperature regime of drinking water and labor costs for its technological maintenance.

To achieve this technical and technological result, a two-channel supply of drinking water with different pH values or drinking water and medicinal products is provided in one of the two isolated containers of the drinking bowl closed with removable lids with water-filled windows for the maintenance of the livestock and water drinking glasses with brush working parts for the periodic cleaning drinking bowls from contamination, which are connected through the spacers with the jet organ, ensuring their periodic rotation together with the lid. At the same time, the presence of perforation in the bottoms of the drinking bowls, pollutant removal chambers, a cascade-type detachable filter and a contaminated water discharge pipeline make it possible to reduce labor costs for sanitation of the drinking bowls and provide the treatment and secondary use of purified water.

3. Conclusions

1. The proposed technical and technological solutions due to the stability of the process of the automatic drinking in terms of the quality of water preparation and its temperature regime make it possible to reduce the irregularity of the water consumption by the poultry depending on some external factors and thereby make it possible to reduce the total water consumption per unit of produced products;

2. The availability of filtration units for nonproduction water (contaminated effluents) and its electrothermal treatment in the proposed systems and means of the automatic drinking allow it to be reused in the structure of technological processes of servicing poultry, which ensures its rational use with respect to the final product (egg, meat, feather);

3. The presence of the water activating unit in the system of the automatic drinking (division according to the hydrogen index, pH) and the use of periodically acidified water additionally reduces the incidence and the mortality of the poultry, which is especially characteristic when growing turkeys in the first temporary technological cycle;

4. In the proposed development an additional positive effect is the reduction of labor expenditures (the use of manual labor) in the process of the maintenance of the means of the automatic drinking (the cleaning and the washing of drinking bowls) and, as a consequence, the reduction of irreversible water losses.

References

[1] Kavtarashvili, A.Sh. Water quality is the most important condition for the health and the productivity of the poultry / A.Sh. Kavtarashvili - M .: GNU VNITNP Rosselkhozakademiya; 2013.

[2] Kornilova V.A. Efficiency of growing turkeys of heavy and medium crosses in their different ways of keeping / V.A. Kornilov – Diss. of Cand. of Agricultural Sciences, Samara, 2001.

[3] Maevska Teresa. Turkeys are not chickens / Teresa. Maevska – Faculty of Poultry, UVM Olshtyn.

[4] Ageev A.A. Efficiency of ways of growing young turkeys in the conditions of the peasant farm economy «Mars» in Zelenodolsk district of the Republic of Tatarstan. / A.A. Ageev, V.I. Shilova – FSBEE of the Higher Professional Education of the KSAVM.

[5] Shevchenko A.V. History of the turkey farming / A.V. Shevchenko – Poultry farming, 5 - LLC Intervesa, 2010.

[6] Fisinin V.I. Development of the poultry farming / V.I. Fisinin / Moscow: Kolos / Poultry farming # 2. – 2009.

[7] Standards of the technological design for poultry enterprises. Scientific and Technical Advance of Agroindustrial Complex 1.10.05.001-01 M .: Ministry of Agriculture of the Russian Federation, 2001.

[8] Requirements for the drinking water: State Standard P 51232-98, State Standard 2874-82, State Standard 2761-84.

[9] Potseluev A.A. Water supply of agricultural facilities / A.A. Potseluev – Zernograd: Publishing House of FSEE HPO Azov-Black Sea State Agroengineering Academy, 2005.

[10] Technologies and Equipment for Poultry Farming: Reference Book / V.T. Sklyar, A.V. Sklyar, T.N. Kuzmina, V.A. Gusev. – Moscow: FSBSE "Rosinformagrotekh". 2014. 188 p.
[11] Means of Mechanization for the Production and the Processing of Agricultural Products in Small Forms of Management. / Catalog – Moscow: FSBSE "Rosinformagrotech, 2008.
[12] Drinking machine. Patent No. 2573329 dated December 17, 2015.
[13] Drinking bowl for poultry and small animals. Patent No. 2242121, December 20, 2004.
[14] Drinking bowl for poultry and small animals Patent No. 1367929, January 23, 1988.
[15] Drinking bowl for poultry. AC # 738564, Newsletter No. 2, 1986.
[16] Drinking bowl for poultry. AC # 321238, Newsletter No. 35, 1971.

Environmental Hazard of Some Types of Expanded Polystyrene

R Fediuk[1, a], A Pak[1], V Ginevskiy[1], N Stoyushko, N Gladkova

[1]School of Engineering, Far Eastern Federal University, 8, Sukhanova St, Vladivostok, 690950, Russian Federation

E-mail: [a] fedyuk.rs@dvfu.ru

Abstract. Chromatography of five different samples of expanded polystyrene is given. The ecological danger of low-quality polystyrene foam for living creatures is proved. There is a need to develop a new high-density concrete structure that protects the interior area from toxic fumes of expanded polystyrene.

1. Introduction

Despite the wide range of expanded polystyrene (EP), which corresponds to Russian sanitary and hygienic standards, in the Far East, non-certified foreign expanded polystyrene was often used, as well as a number of locally produced materials . In this article, it examines the most commonly used in the Russian constructions polystyrene foam of five types, different grades and densities (Table 1).

Table 1. Samples of the tested polystyrene foam

No.	Type of expanded polystyrene	Manufacturer	Density, кg/m^3
1	Europlex-35	"Europlast" (Vladivostok)	30-35
2	Penoplex	"Penoplex" (St. Petersburg)	25-35
3	PSB-S 15	"Primstroitel" (Владивосток)	up to 10
4	PSB-S 25	"Europlast" (Vladivostok)	15-25
5	PSB-S 35	"Europlast" (Vladivostok)	25-35

Due to high sensitivity and automation, the chromatograph without any sample preparation allows analyzing the content of harmful substances in the air in a wide range of concentrations: from the MPC in the atmosphere to industrial emissions, as well as in emergency situations [1-3]. Chromatograph is a unique means of express analysis, it is designed to work, both in laboratory and in field conditions, directly on the object under study [4-5]. Contains its own means of electrical and gas power [6-10]. The results of the analysis, comments to them and the chromatograms themselves are automatically documented in the computer's memory [11-14]. In addition to determining the concentration of substances, the chromatograph automatically identifies them [15-18].

To clarify the results obtained, the samples were also investigated using a Shimatsu QP-5050A chromatography-mass spectrometer. A flexible quadrupole gas chromatography-mass spectrometer along with electron impact (EI) ionization and a positive chemical ionization (PCI) system for high-accuracy molecular weight determination also has a negative chemical ionization (NCI) system for highly sensitive analyzes of electrophilic compounds. It has a computer-optimized ion source and a system of magnetic lenses [4-6, 19-21].

2. Results and Discussion

The obtained results of studies of the quantitative and qualitative composition of toxic substances released during the destruction of expanded polystyrene are graphically depicted in Fig. 1-3, and the characteristics are presented in Tables. 2-4.

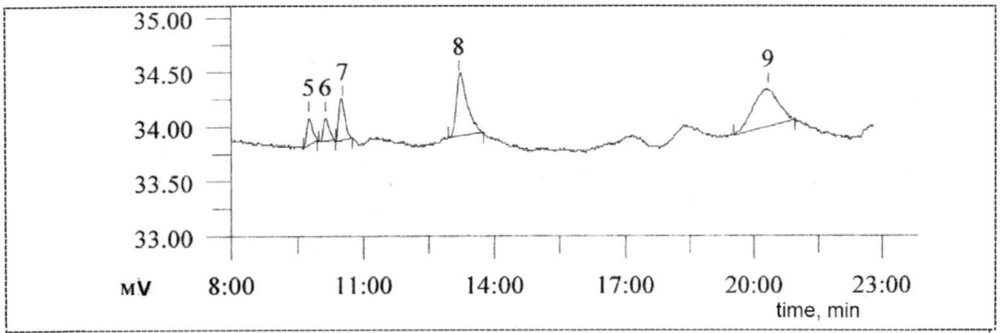

Figure 1. Chromatogram of evaporation of EP No. 1:
5-ethylbenzene, 6-p, m-Xylene, 7-undetermined, 8-o-Xylene, 9-styrene

The sample of expanded polystyrene Europlex-35 releases 60 times the maximum permissible concentration (MPC) of styrene [22-24]. There is also an excess of MPC of propyl alcohol and ethylbenzene (Table 2).

Table 2. The results of the tested EP No. 1

No.	The substances to be determined	Concentration of the substance in the sample, mg / m^3	MPC, mg / m^3
1	Styrene	0.12	0.002
2	Propyl alcohol	1.3	0.3
3	Isopropyl alcohol	0.4	0.6
4	Ethylbenzene	0.025	0.02
5	Xylenes (sum of isomers)	0.05	0.1
6	Benzaldehyde	0.01	0.04
7	Phenol	undetermined	0.003
8	Formaldehyde	undetermined	0.01

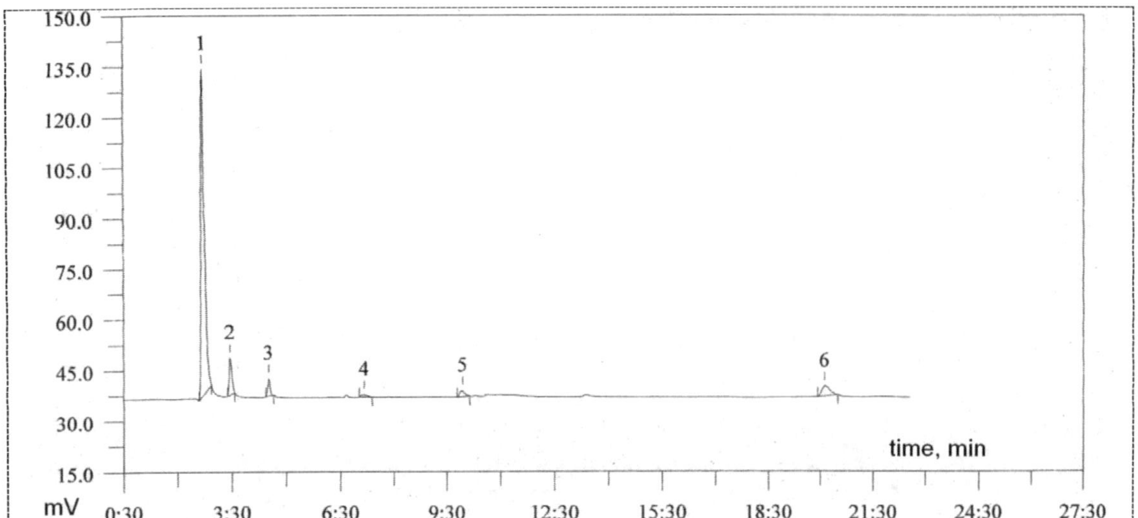

Figure 2. Chromatogram of evaporation of EP No. 2:
1 - vinyl chloride, 2 - octene, 3 - methyl acrylate, 4 - not detected, 5 - ethylbenzene, 6-styrene

During the testing of the EP Penoplex sample, the highest level of toxicity values of the material with excess of the MPC of styrene is observed more than 200 times, as well as a significant excess of the MPC of ethylbenzene, xylenes, benzaldehyde.

Table 3. The results of the tested EP No. 2

No.	The substances to be determined	Concentration of the substance in the sample, mg / m³	MPC, mg / m³
1	Styrene	0.41	0.002
2	Hexane	5.1	60
3	Acetone	0.2	0.35
4	Ethylbenzene	0.3	0.02
5	Xylenes (sum of isomers)	0.2	0.1
6	Benzaldehyde	0.08	0.04
7	Phenol	undetermined	0.003
8	Formaldehyde	undetermined	0.01

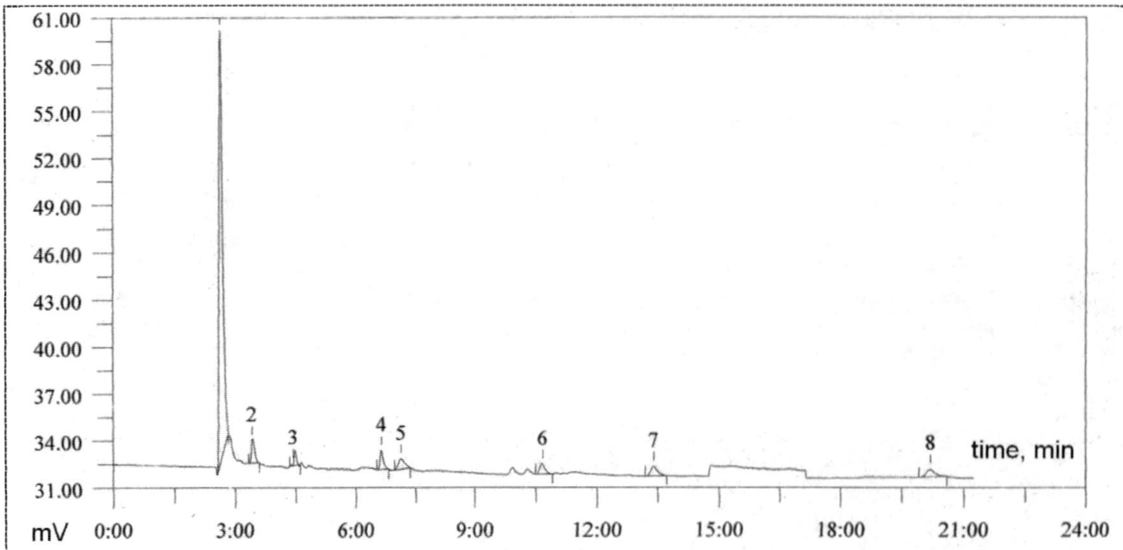

Figure 3. Chromatogram of evaporation of EP No. 3:
1-butane, 2-acetone, 3-ethyl alcohol, 4-toluene, 5-not detected, 6-butyl alcohol, 7-o-Xylene, 8-styrene

According to the test results presented in Table. 4, in the PSB-S15 brand sample, in contrast to the other samples considered, in addition to the presence of styrene, there are also toxic fumes such as acetone phenone and formaldehyde.

Similar results were obtained for samples No. 4 and No. 5. The results of the tests revealed an excess of the maximum permissible concentrations of all types of expanded polystyrene considered, which indicates their high toxicity and the environmental dangers of their use. It should be noted a significant difference in the chemical composition of fumes of toxic substances, which implies a violation of technological operations in the production of expanded polystyrene.

IOP Conf. Series: Earth and Environmental Science **115** (2018) 012007 doi:10.1088/1755-1315/115/1/012007

Table 4. The results of the tested EP No. 3

No.	The substances to be determined	Concentration of the substance in the sample, mg / m^3	MPC, mg / m^3
1	Styrene	0.06	0.002
2	Ethanol	0.5	5
3	Acetone	0.03	0.35
4	Toluene	0.04	0.6
5	Xylenes (sum of isomers)	0.03	0.1
6	Benzaldehyde	0.01	0.04
7	Acetophenone	0.07	0.003
8	Phenol	undetermined	0.003
9	Formaldehyde	0.03	0.01

For further investigations, samples No. 1, No. 2 and No. 3 (from Table 1), coated with a protective layer of a 20mm cement-sand mortar, were used in the work. The content of toxic substances released during the destruction of expanded polystyrene, taking into account this protection is shown in Table. 5.

Table 5. The results of tests of expanded polystyrene coated with a 20mm layer cement-sand mortar

The substances to be determined	Concentration of the substance in the sample, mg / m^3			MPC, mg / m^3
	Europlex-35	Penoplex	PSB-S15	
Styrene	0.04	0.08	0.005	0.002
Propyl alcohol	0.05	undetermined	undetermined	0.3
Isopropyl alcohol	0.03	undetermined	undetermined	0.6
Ethanol	undetermined	undetermined	0,007	5
Toluene	undetermined	undetermined	undetermined	0.6
Xylenes (sum of isomers)	undetermined	0,006	undetermined	0.1
Benzaldehyde	undetermined	undetermined	undetermined	0.04
Hexane	undetermined	0,03	undetermined	60
Ethylbenzene	undetermined	0,005	undetermined	0.02
Acetone	undetermined	0,006	undetermined	0.35

At the next stage of the study, expanded polystyrene samples were covered with a protective layer of fine-grained concrete with a thickness of 40mm. According to the results presented in Table. 6 it is noted that the concentration of toxic styrene in the EP Europlex-35 sample when it is coated with fine-grained concrete 40 mm thick is reduced by a factor of 2. For the EP Penoplex sample, the concentration of toxic styrene is 1.5 times more than the allowable, and in the PSB-S15 sample there is no toxic styrene at all.

Table 6. The results of tests of expanded polystyrene coated with a 40mm layer fine-grained concrete

The substances to be determined	Concentration of the substance in the sample, mg / m^3			MPC, mg / m^3
	Europlex-35	Penoplex	PSB-S15	
Styrene	0.001	0.003	undetermined	0.002
Propyl alcohol	0.05	undetermined	undetermined	0.3
Isopropyl alcohol	0.03	undetermined	undetermined	0.6
Ethanol	undetermined	undetermined	undetermined	5
Toluene	undetermined	undetermined	undetermined	0.6
Xylenes (sum of isomers)	undetermined	undetermined	undetermined	0.1
Benzaldehyde	undetermined	undetermined	undetermined	0.04
Hexane	undetermined	undetermined	undetermined	60
Ethylbenzene	undetermined	undetermined	undetermined	0.02
Acetone	undetermined	undetermined	undetermined	0.35

To study the influence of the thickness of the protective layer on the concentration of toxic fumes, a sample of the EP Penoplex was tested more detail. It was covered with a heavy concrete layer of 20, 40 and 60 mm . The data obtained (Fig. 4) demonstrate that after coating of EP with a concrete layer of 60 mm thickness, exceeding the MPC of toxic substances in the air was not detected.

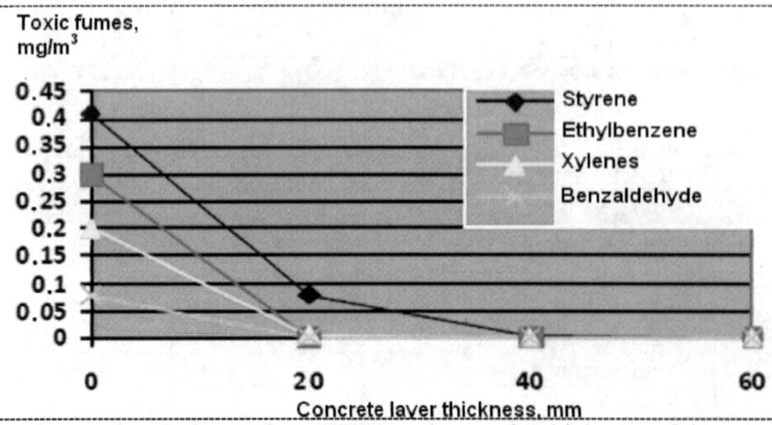

Figure 4. Dependence of toxic fumes from EP Penoplex on the thickness of the protective concrete layer

However, with an increase in the ambient temperature to 40 ° C, the intensification of evaporation of styrene is again observed (Table 7).

Table 7. The results of tests of expanded polystyrene coated with a 60mm layer fine-grained concrete (at 40°C)

The substances to be determined	Concentration of the substance in the sample, mg / m^3			MPC, mg / m^3
	Europlex-35	Penoplex	PSB-S15	
Styrene	0.004	0.002	undetermined	0.002
Propyl alcohol	0.03	undetermined	undetermined	0.3
Isopropyl alcohol	0.01	undetermined	undetermined	0.6
Ethanol	undetermined	undetermined	0.002	5
Toluene	undetermined	undetermined	undetermined	0.6
Xylenes (sum of isomers)	undetermined	0.004	undetermined	0.1
Benzaldehyde	undetermined	undetermined	undetermined	0.04
Hexane	undetermined	0.01	undetermined	60
Ethylbenzene	undetermined	0.001	undetermined	0.02
Acetone	undetermined	0.002	undetermined	0.35

It is known that among the substances contained in construction expanded polystyrene, styrene possesses the greatest degree of cumulative activity. This raises the need to amend the current regulations - reducing the MPC of EP in housing construction by approximately 600 times.

3. Conclusions

Thus, carrying out the analysis of toxicity, there is an environmental hazard for biological individuals in the use of certain types of expanded polystyrene. This calls for the development of a new high-density concrete structure that protects the interior from toxic vapor of expanded polystyrene.

4. References

[1] Peschard A, Govin A, Grosseau P, Guilhot B and Guyonnet R 2004 *Cem. Concr. Res.* **34** 2153–2158.

[2] Sazonov T V, Fershalov Y Y, Fershalov M Y, Fershalov A Y, Ibragimov D I 2014 *Applied Mechanics and Materials* **635-637** 155-158

[3] Fediuk R S 2016 *IOP Conf. Series: Materials Science and Engineering* **116** 012020

[4] Chihaoui R, Khelafi H, Senhadji Y, and Mouli M 2016 *J of Adhesion Science and Technology* **30** 1930-1944.

[5] Fediuk R S and Yushin A M 2016 *IOP Conf. Series: Materials Science and Engineering* **116** 012021

[6] Pelin G, Pelin C-E, Stefan A, Dinca I., Ficai A, Andronescu E, Trusca R 2016 *Bulletin of Materials Science* **39(3)** 769-775

[7] Cai J, Hu X, Xiao B, Zhou Y, Wei W 2017 *International Journal of Heat and Mass Transfer* **105** 623-637

[8] Glagolev E, Suleimanova L, and Lesovik V. 2016 *International Journal of Environmental and Science Education* **11(18)** 12383-12389.

[9] Ibragimov R 2016 *ZKG International*. **69 (6)** 34-39

[10] Zagorodnjuk L H, Lesovik V S, Volodchenko A A and Yerofeyev V T 2016 *International Journal of Pharmacy and Technology* **8(3)** 15146-15155

[11] Ibragimov R A and Pimenov S I 2016 *Magazine of Civil Engineering* **62 (2)** 3-12

[12] Ranjbar N et al. 2016 / *Journal of Cleaner Production* **112** 723-730.

[13] Fediuk R S 2016 *IOP Conf. Series: Materials Science and Engineering* **125** 012019.

[14] Buyantuev S L, Urkhanova L A, Lkhasaranov S A, Stebenkova Y Y, Khmelev A B and Kondratenko A S 2016 *IOP Conf. Series: Materials Science and Engineering* **168 (1)** 012046.

[15] Buyantuev S L, Urkhanova L A, Kondratenko A S, Shishulkin S Y, Lkhasaranov S A and Khmelev A B 2016 *IOP Conf. Series: Materials Science and Engineering* **168 (1)** 012048.

[16] Semenov A P, Smirnyagina N N, Urkhanova L A, Kanakin S V, Lkhasaranov S A, Semenova I A, Tsyrenov B O, Dasheev D E and Khaltarov Z M. 2016 *IOP Conf. Series: Materials Science and Engineering* **168 (1)** 012059.
[17] Fediuk R S and Yushin A M 2015 *IOP Conf. Series: Materials Science and Engineering* **93** 012070
[18] Koniorczyk M, Gawin D and Schrefler B A 2015 *Computer Methods in Applied Mechanics and Engineering* **297** 38-61.
[19] Marcin K 2015 *Int J of Heat and Mass Transfer* **81** 260-271.
[20] Liu Z, Zhang Y and Jiang Q 2014 *Constr. Build. Mater.* **53** 26-31.
[21] Fediuk R S, Smoliakov A K, Timokhin R A, Stoyushko N Yu and Gladkova N A 2017 *IOP Conf. Series: Materials Science and Engineering* **66** 012026.
[22] Liu Z, Zhang Y, Sun G, Jiang Q and Zhang W 2012 *J of Civil, Architectural and Environmental Engineering* **34** 148-153.
[23] Schmidt M, Pöllmann H, Egersdörfer A and Göske J, Winter S 2010 *32nd International Conference on Cement Microscopy* 86-118.
[24] Quintard R 2015 *Transfers in porous media Special Topics & Reviews – An International Journal*, **6(2)** 91-108.

IOP Conf. Series: Earth and Environmental Science **115** (2018) 012008 doi:10.1088/1755-1315/115/1/012008

Power Consumption Optimization in Tooth Gears Processing

N Kanatnikov, G Harlamov, P Kanatnikova, A Pashmentova

Orel State University, Russia, Orel, Naugorskoe shosse, 29

E-mail: NKanatnikov@yandex.ru

Abstract. The paper reviews the issue of optimization of technological process of tooth gears production of the power consumption criteria. The authors dwell on the indices used for cutting process estimation by the consumed energy criteria and their applicability in the analysis of the toothed wheel production process. The inventors proposed a method for optimization of power consumptions based on the spatial modeling of cutting pattern. The article is aimed at solving the problem of effective source management in order to achieve economical and ecological effect during the mechanical processing of toothed gears. The research was supported by Russian Science Foundation (project No. 17-79-10316).

1. Introduction

In the current conditions of global competition and modern state of ecology, it is hard to overestimate the impact of power engineering on the all fields of man's activity. Consequently, there is a problem of effective energy source management in order to achieve both economic and ecological effects. The distinguishing characteristic of engineering industry is a high level of power consumption of manufactured items. The share of the energetic component in the manufacturing cost of production made on the machine-building enterprises amounted to 25% [1, 2, 3]. In Russian's machine-building enterprises, 450-600 kilowatt-hours of electric energy are spent for production of 1 ton of facing. Analogous production made in the industrialized Europe countries consumes is 1.5-2 times smaller amount of energy [4]. Main of the reason for it is that the energy intensity and metal removal criteria's are applied along with the major criteria (cost value, accuracy, manufacturing capability) during the developing the technological processes in the Germany, France and other countries [5, 6, 7].

The major part of shaping parts (more than 2/3 of the total items) account for gear wheels, profile and other parts of complex shape, and the accumulated park of technological equipment does not allow to reduce the production operating time. Moreover, the park is constructed in 80-yes, in the era of cheap energy. Processing of the surfaces can take up to 70% of normal time of item manufacturing.

Hence, increasing of energy efficiency of technological processes of tooth wheel machining by the optimization of energy costs is the urgent task of modern machinery production.

The aim of this work is to develop a method of determining the optimal parameters of the technological system (process conditions and design of the cutting tool) in which the desired quality of surfaces, cutting tool durability and productivity will be ensured by the minimum necessary amount of energy.

2. Optimization of energy cost in the processing of gear wheels

The problem of increasing of energy efficiency in general can be solved by reducing the energy losses in the cutting area, mechanical or electrical part of the equipment.

In the article, the option of the management of the energy efficiency of the process. Since the mechanical work performed by the cutting tool during each working stroke and determined the regularities of chip formation affects the load losses in transmission and determines the total amount of energy consumed by the machine engine.

There are a number of parameters, by which it is possible to evaluate the cutting process from the point of view of supply energy, in particular the specific energy of cutting process. The energy intensity of the cutting process characterizes numerically the amount of energy expended by the cutting tool to the separation the unit volume of the shear layer in the chip form (a preliminary stage processing), or for the formation of a unit area of new item surface. The advantages of this index include its simple determination by both theoretical and experimental methods.

However, the energy intensity index is proportional to the cutting power, but the power is supposed to be constant during the whole operating stroke, which is rare and becomes true only for stationary cutting.

It is possible to use nondimensional parameter K, which is an energetic performance factor of cutting process, as integral characteristics of the efficiency of machining proposed:

$$K = \frac{\Delta w V}{n_C A_C} = \frac{\Delta w V}{n_C \int_0^{\tau_c} N(\tau) d\tau} \qquad (1)$$

where Δw – energy intensity of processing material; V – the volume of treated material being exposed; n_c –the number of cycles of cutting power $N_{(\tau)}$ changing during the working cycle of the tool; A_c – cutting job during τ_C one cycle of output variation.

Duration τ_c of one cycle of the cutting power change refers to the time period of the working stroke or period of time during which there is a complete one-time change in cutting power at the transient mode.

Parameter K meets requirements established for criteria of optimality of technological processes. This indicator expresses the efficiency of the cutting process and thus has a physical meaning. It clearly described in a mathematical form. This form can be reduced to the objective function of the form K→max; indicators of properties of processed and tool materials, geometric parameters of the tool and the shear layer, the elements of the cutting mode, the kinematics of the machining process are the arguments of the function.

However, if it is necessary to evaluate the energy efficiency of cutting processes of gear machining (gear shaping, gear milling, gear planing, etc.), expression 1 can not be used, since at each cycle of loading the tool can remove chips of various configurations, therefore, the cutting job will change. Expression 1 for description of efficiency of gear wheels processing will take the following form:

$$K = \frac{\Delta w V}{\sum_{i=1}^{n_C} A_{Ci}}. \qquad (2)$$

Cutting job can be expressed through processing modes and workpiece parameters:

$$K = \frac{60 \cdot 1020 b \Delta w V}{\upsilon^2 \sum_{i=1}^{n_c} P_{Zi}} \qquad (3)$$

where b – gear face width; υ – cutting speed; P_{Zi} – axial cutting force.

3. Cutting force definition

The empirical approach is currently widely used to determine the cutting forces. The approach is based on the "Mechanistic Force Modelling". Method essentially consists in determining the specific forces per unit length of cutting blade and then summing the forces by all cutting edges involved in cutting.

Power for the processes of cutting by the tool with custom profile can be calculated using the method, when several edges are involved in the process of the chip removal. R. V. Anisimov calculated by this method the force produced by shaping the inner noninvolute gears [8]. According to this approach L. Berglind, D. Plakhotnik developed a model to calculate the cutting forces occurring during machining of an arbitrary configuration on the five-axis milling centers [9].

A disadvantageous feature of this approach is that the model has to be calibrated in order to refine adjustment factors in case of any changes (use of new processing modes, using a new cutting tool, etc.).

Calibration is implemented by means of complex experimental studies. The forecast error will be less than 5% [10] after calibration of the model.

Calculation of the specific forces is made first.

According to [8] specific forces in tooth gears processing are defined as follows:

in the range of cut layer thickness from 0.01 to 0.5 mm

$$\Delta P_z = \left(1.7 \cdot 10^2 a^{0.9} + 4.3 - 9(\gamma + \Delta\gamma)\right)(\alpha - \Delta\alpha)^{-0.07} 9.8 K_{Pz} \tag{4}$$

in the range of cut layer thickness from 0 to 0.01 mm

$$\Delta P_z = (57 + 2210 a) K_{Pz} \tag{5}$$

where a – thickness of cutting layer; γ – front angle of the instrument; Δγ – kinematic change of the front angle; α – back angle of the tool; Δα – kinematic change of the back angle;

K_{Pz} and K_{Py} – generalized correction factors which are equal to the product of the coefficients K_M, K_{CCF}, K_h, and K_V. These coefficients take into account the following factors: K_M – material; K_{CCF} – coolant-cutting fluid; K_h – tool wear; K_V – cutting speed.

Summing up the works of specific forces occurring on the elementary segment of cutting blade, we will get:

$$P_z = \sum \Delta P_z \Delta l K_{CC} \tag{6}$$

where K_{CC} – the factor of the chip complexity.

4. Determination of kinematic cutting parameters

Spatial mathematical model of the process of the blade machining is used for determining the kinematic changes of the cut angles and the geometry of the chip [8]. It includes a description of the motion of each elementary area of the cutting edge; moreover, the model allows to make the necessary calculations.

The process of machining is presented in the form of dependencies:

$$\begin{aligned} X &= f(\varphi(V), \varphi(S), \varphi(l), t, u); \\ Y &= f(\varphi(V), \varphi(S), \varphi(l), t, u); \\ Z &= f(\varphi(V), \varphi(S), \varphi(l), t, u), \end{aligned} \tag{7}$$

where φ(V) – cutting speed parameter; φ(S) – carrying parameter; φ(l) – cutting blade parameter; t – time; u – control parameter.

The proposed mathematical spatial of cutting pattern mapping allows to calculate parameters of the machining process based on the use of vector analysis and numerical methods. The scheme presented in fig. 1.

The kinematic change of the back angle in the direction of the greatest movement of the flow is determined by the following formula:

$$\Delta\alpha = \operatorname{arctg} \frac{\sqrt{X_F^2 + Y_F^2 + Z_F^2}}{\sqrt{X_V^2 + Y_V^2 + Z_V^2}}, \tag{8}$$

where X_F, Y_F, Z_F – the coordinates of the velocity vector in the direction of the handling; X_V, Y_V, Z_V – the coordinates of the vector of cutting movement.

The thickness of the cutting layer:

$$a = \frac{X_a X_F + Y_a Y_F + Z_a Z_F}{\sqrt{X_a^2 + Y_a^2 + Z_a^2}} \tag{9}$$

where X_a, Y_a, Z_a – are the coordinates of the vector in the direction of which the thickness of the shear layer changes.

IOP Conf. Series: Earth and Environmental Science **115** (2018) 012008 doi:10.1088/1755-1315/115/1/012008

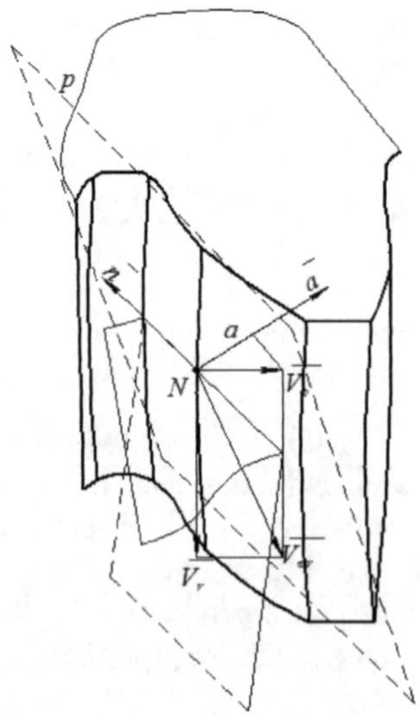

Figure 1. The scheme of determining the thickness of layer cut at bevel gear processing

The kinematic change of the front angle is determined by the following formula:

$$\Delta\gamma = arctg\frac{a}{\sqrt{X_V^{\,2}+Y_V^{\,2}+Z_V^{\,2}}} \tag{10}$$

Then, it is necessary to compare the indicators of "energy efficiency of the cutting process" for different types of gears profile processing of spur bevel wheel (ms=6 mm, b=30 mm). We consider cutting in two passes, first (rough) is the method of cutting a trapezoidal cutter, the second (finishing) is performed by generating a standard finishing cutter shaping. Another option is cutting in one pass by the special stepped gear planing cutters, which is shown in fig. 2.

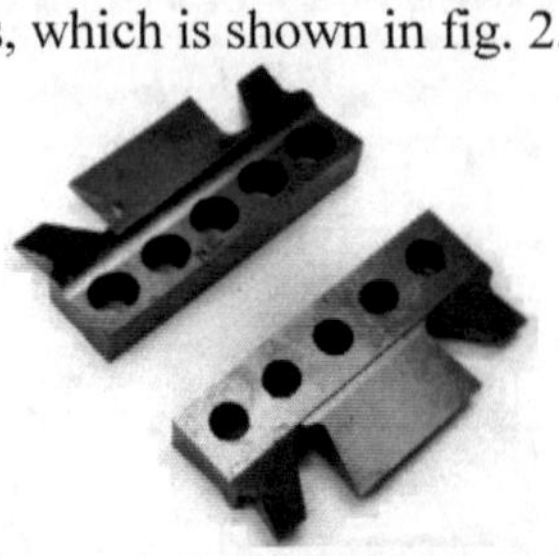

Figure 2. Stepped gear cutting tools

Possible options of processing modes are shown in the table 1. The options are limited by the minimum required parameters of quality and capabilities of common gear planing machines (5A250П, 5С286П, etc.)

Table 1 – Comparison "energy efficiency of the cutting process" for different technological solutions for the machining of gear profiles of the spur bevel wheel

Processing mode		In two cuts		In one strike by cutter with differentiated cutting scheme
		Roughing (penetration)	Finishing (motion)	
1	t_z, minute per tooth	0.56	0.15	0.75
2	υ, meters per minute	15	20	13
3	t_{mach}, minute	22		13.7
4	$K_1/K_{pro.}$	1		1.38

Line 4 of table 1 presents a comparison of energy criterion of cut. The value of K for processing into two passes was used as K_1. Analysis of the results of the analytical calculation shows that the standard processing option has a greater energy efficiency, however, is considered to be less effective in performance.

5. Conclusions

The achievement of environmental benefits is closely connected to the energy savings, as the use of any kind of energy resource is accompanied by emissions. The economic effect is achieved due to the reduction of the energy intensity of manufactured products, thereby the cost of production reduces.

The results showed that depending on the cutting conditions, the amount of energy spent on processing may considerably vary even when using the same processing method and tool. The use of the proposed methodology for assessing the efficiency of the cutting process at the stage of technological preparation allows to determine a rational processing method, the parameters of the technological system, processing modes and tool design, which are required for processing the detail with the necessary accuracy with minimal energy consumption.

Acknowledgements
The research was supported by Russian Science Foundation (project No. 17-79-10316).

References

[1] Hu L. et al. Sequencing the features to minimise the non-cutting energy consumption in machining considering the change of spindle rotation speed //Energy. 2017. V. 139. pp. 935-946.
[2] Li L. et al. An integrated approach of process planning and cutting parameter optimization for Energy-aware CNC Machining //Journal of Cleaner Production. 2017.
[3] Shin S. J., Woo J., Rachuri S. Energy efficiency of milling machining: Component modeling and online optimization of cutting parameters //Journal of Cleaner Production. 2017. V. 161. pp. 12-29.
[4] Stepanov Y. S., Barsukov G. V., Bishutin S. G. Technological Fundamentals for Efficiency Control of Hydroabrasive Cutting //Procedia Engineering. 2016. V. 150. pp. 717-725.
[5] Wang B. et al. Proper selection of cutting parameters and cutting tool angle to lower the specific cutting energy during high speed machining of 7050-T7451 aluminum alloy //Journal of Cleaner Production. 2016. V. 129. pp. 292-304.
[6] Wójcicki J., Leonesio M., Bianchi G. Integrated energy analysis of cutting process and spindle subsystem in a turning machine //Journal of Cleaner Production. 2018. V. 170. pp. 1459-1472.
[7] Deng Z. et al. Optimization of process parameters for minimum energy consumption based on cutting specific energy consumption //Journal of Cleaner Production. 2017. V. 166. pp. 1407-1414.
[8] Tarapanov A. et al. Improving Efficiency of Gear Shaping of Wheels with Internal Non-involute Gears //IOP Conference Series: Materials Science and Engineering. IOP Publishing, 2016. V. 127. 1. 012052.
[9] Berglind L., Plakhotnik D., Ozturk E. Discrete Cutting Force Model for 5-Axis Milling with Arbitrary Engagement and Feed Direction //Procedia CIRP. 2017. V. 58. pp. 445-450.
[10] Altintas Y., Merdol S. D. Virtual high performance milling //CIRP Annals-Manufacturing Technology. 2007. V. 56. 1. pp. 81-84.

The Composition and Characteristics of the Dendroflora in the Transformed Conditions of the Middle Reaches of the River Khoper

M V Larionov[1, a], N V Larionov[2], I S Siraeva[3] and A S Ermolenko[3]

[1]Balashov Institute (branch) National Research Saratov State University named after N.G.Chernyshevsky, 412300, Russia, Balashov, K. Marx str., 9.
[2]School No. 2 urban district Kotelniki, 140054, Russia, Moscow region, Kotelniki, Novaya str., 39.
[3]National research Saratov state University named after N. G. Chernyshevsky, 410012, Russia, Saratov, Astrakhanskaya str., 83.

E-mail: [a] m.larionow2014@yandex.ru

Abstract. Green plantings of the cities perform a number of important functions which provide ecological needs of cities population for a healthy surrounding medium, and his needs for communication with wildlife. The comparative analysis of species of wood plants of the towns of the Voronezh region (Borisoglebsk and Novokhopersk) was carried out. 110 types (to Borisoglebsk) and 91 species (to Novokhopersk) species of trees and bushes are registered. The systematic structure which showed a dominance of angiospermous plants that in general is characteristic of the urbanized territories. The range of geographical elements is presented by three types (European-West Asian – 35% and 40%, European – 30% and 26%, North -American – 14% and 21%). The prevailing life forms of the cities are trees, they make 56% and 60% respectively. Researchers showed, in the cities of east part of the Voronezh region many types of various origin can grow that helps to solve problems in gardening, afforestation and an agrarian and forest melioration, in creation of dendrology nurseries for the purpose of maintaining a genetic variety and the subsequent introduction of wood plants and bushes in culture.

1. Introduction

The problem of the conservation of biological diversity in anthropogenically transformed territories is especially significant, especially in connection with the processes of an urbanization, agglomeration which are followed by negative technogenic and chemical and mechanical impacts on landscapes and ecosystems [1-13].

Prikhopyorye's average falls into regions with a high level of anthropogenic transformation of a environment which reasons the urbanization, the developed agrarian complex and also the amplifying industrial and technogenic and transport load. In this territory the largest level of anthropotechnogenic transformation characterizes city settlements [8].

It is known that in settlements, especially in suburbs and especially in the cities of a plant test a potent press [3, 7, 11] from an economic complex (direct anthropogenic influence) and in the course of biotic interactions of various views-vselentsev with native organisms (the mediated anthropogenic influence on city ecosystems), in particular, at naturalization of various Advents and competitive replacement of local types by them. Including therefore the specific structure and structure of city plantings undergoes essential changes.

The research of features of trees and bushes in the conditions of the modern city settlements represents a relevant and significant scientific and practical task. It is known that in town planning, at arrangement of various functional zones green plantings of the corresponding target categories have to be created. City plantings, first of all, wood and shrubby, perform a number of the useful functions, including carry out a regulation of a microclimate, create a biological barrier on the way of an emission

IOP Conf. Series: Earth and Environmental Science **115** (2018) 012009 doi:10.1088/1755-1315/115/1/012009

and migration in a surrounding medium of pollutants and considerably provide ecological equilibrium between natural (organisms, the soil, water objects) and anthropogenic components urbosisty [3, 7].

The composition and structure of wood and shrubby vegetation within settlements in the east of the Voronezh region are studied not enough therefore this direction of researches is represented rather relevant.

The purpose of this work was carrying out the comparative analysis of wood plants of the Average Prikhopyorya (on an example urbosistem the Voronezh region) by results of the executed field researches.

Now the studying dendroflora of the urbanized areas of the Voronezh region continues. According to characteristic accounts the abstract of wood plants in the towns of Borisoglebsk and Novokhopersk located in east part of the Voronezh region is made.

2. Materials and methods

The basis of work is made by the materials of field researches of wood and shrubby vegetation of the towns of Borisoglebsk and Novokhopersk collected from 2009 to 2017 by method of route account [2]. All wood plants and bushes from structure of green plantings of different target categories (the common, restricted and express use) growing in city conditions and on suburban sites were considered. Work is documented by a herbarium from the introduced and wild-growing species of wood plants. The registered species of plants were differentiated also in life forms [6].

3. Results of study

The towns of Borisoglebsk and Novokhopersk chosen as areas of field researches are located in north-east and east parts of the Voronezh region, are the regional centers, motor transportation clusters. Local climatic conditions (moderate and continental climate, mainly slabovskholmlenny type of a relief, distribution of steppe, meadow and steppe, inundated meadow and forest vegetation) are characteristic of all Average Prikhopyorya.

Common information about above the called areas is provided in table 1.

Table 1. Total characteristic of Borisoglebsk, Novokhopersk

City	Borisoglebsk	Novokhopersk
Area	49 sq.km	14 sq.km
Population	63,000 residents	14000 residents
Location	it is located on the left-hand bank river of the Raven, near its confluence with the Khopyor River	it is located on the right bank river Khopyor (Don inflow at the height of 50-70 m over river level)
Education time	the city is founded in 1698	it is mentioned in the chronicle since the 17th century
Building	the majority of both cities is built up with private one-storey houses with gardens and kitchen gardens	
Features	in the south borders on the Tellermanovsky forest area, other part of the city is surrounded with agrotsenoza	the East and the north-east of the city borders on the Khopyor reserve, other part of neighborhood of the city adjoins to agrotsenoza

Systematic structure. During the researches it is established that 110 types, from them 28 families and 59 sorts (tab. 2) enter into structure of trees and bushes of Borisoglebsk.

Table 2. Systematic structure dendroflora of Borisoglebsk

Department	Class	Family	Sort	Species
Pinophyta	1	3	6	12
Magnoliophyta	1	25	53	100
Total	2	28	59	112

Dendroflora of Novokhopersk is presented by 91 species, to the falling into 23 families and, respectively, to 49 childbirth that it is reflected in table 3.

Table 3. Systematic structure dendroflora of Novokhopersk

Department	Class	Family	Sort	Species
Pinophyta	1	2	5	8
Magnoliophyta	1	21	44	83
Total	2	23	49	91

The systematic analysis of wood plants of areas of researches revealed that its basis is made by representatives of angiospermous plants (*Magnoliophyta*) – 89% in Borisoglebsk, and in Novokhopersk their quantity reaches 91%. Thus, the quantitative ratio of the considered group angiospermous approximately equal in both settlements. At the same time the share of representatives of gymnospermous (*Pinophyta*) is rather small: in Borisoglebsk it makes 11%, in Novokhopersk – 9%.

The ratio of departments and classes shows empery of representatives of angiospermous plants. Such ratio in general is characteristic of these urbanized territories. Increase in number of department of *Pinophyta* pays an attention that it is bound to a larger variety cultivated in the territory of the cities gymnospermous, mainly from among introduced species.

Further on table 4 the range of the leading families of the considered plants is presented.

Table 4. Range of the leading families dendroflora of Borisoglebsk, Novokhopersk

p/o	Family	The place in a range	Number of sorts			
			Borisoglebsk		Novokhopersk	
			Absolute quantity	Share as a percents, %	Absolute quantity	Share as a percents, %
1	*Rosaceae*	1	25	33	18	34
2	*Salicaceae*	2	10	13	7	13
3	*Caprifoliaceae*	3	9	12	5	9
4	*Pinaceae*	4	8	11	5	9
5	*Grossulariaceae*	5-7	5	6	4	8
6	*Oleaceae*	5-7	5	6	4	8
7	*Aceraceae*	5-7	5	6	4	8
8	*Ulmaceae*	8	4	5	3	5.5
9	*Betulaceae*	9-10	3	4	3	5.5
10	*Cupressaceae*	9-10	3	4	1	2
	Total:		77	100	54	100

As appears from table 4, the leading position in structure of wood plants and bushes of Borisoglebsk and Novokhopersk occupies the *Rosaceae* family numbering 25 sorts and 18 sorts respectively that makes 33% and 34% of total number of the registered representatives of the studied group of plants for each of these areas. The second place is taken by the *Salicaceae* family (13%). These two families take the leading positions that in general is characteristic of regions of Eastern Europe with a temperate climate. Such ratio is explained by historically developed introduction due to enrichment of wood and shrubby vegetation decorative and fruit and berry cultures. *Rosaceae* – one of large families of the flowering plants presented in areas of researches. Besides, his representatives are widespread almost in all areas of the globe where can grow floral, but after all their main part is concentrated in a moderate zone of the northern hemisphere. They meet in the most various vegetable communities and though do not play the dominating role, are, nevertheless, one of the major for us families of plants. The made family taking the leading place reckons the greatest number of fruit and berry, medicinal and vitamin-containing cultures that is rather fully used in the national economy, but often in private enterprises by means of gardening, truck farming, cultivation of numerous front gardens, hotbeds and flower beds. Representatives of the *Salicaceae* family originally were native types which then were widely cultivated that, by the way, is observed and presently in view of the fact that plants of the specified family have high decorative qualities and are allocated with vitality in culture. They frosr-are also wind-resistant, photophilous.

Also it should be noted that the role in structure of a city dendroflora in the conditions of the Average Prikhopyorya of the *Pinaceae* family differing in a small amount of sort, but a large number of species is rather big. Species of this systematic group are highly decorative and in too time have high qualities of wood. Therefore they are often used as in gardening, forestry, and in other branches of the national economy: construction, housing and communal services, by production of stationery (various products from wood), the pharmaceutical industry.

Biomorphological structure. On the developed I.G. Serebryakov [6] classification, a range the ecomorph of wood and shrubby plants of the towns of Borisoglebsk and Novokhopersk is created by three types. The prevailing life forms are trees which make 56% and 60% respectively of total number of the registered species.

These are such widespread representatives as *Populus bolleana* Lauche, *Salix caprea* L., *Quercus robur* L., *Pinus sylvestris* L., *Malus domestica* Borkh., *Cerasus vulgaris Mill.* and others. Bushes are presented by such types as *Rubus idaeus* L., *Prunus domestica* L., *Ribes nigrum* L., *Ribes aureum Pursh, Rosa canina* L. Most poorly in structure an ecomorph of wood and shrubby plants lianas – only 4% are presented: *Vitis vinifera* L., *Parthenocissus inserta (*Kern.) Fritsch, *Schisandra chinensis* (Turcz.) Baill.

Composition of geographical elements. Plants which form this or that phytocenosis in the concrete region differ in features of their geographical origin, that is in geographical structure. Having analysed similarities and distinctions of areas of different types from among the researches considered in areas, it is possible to allocate the following geographical elements dendroflor that, actually, is reflected in table 5.

This table shows that a basis of structure of wood and shrubby plantings of Borisoglebsk, Novokhopersk species from structure of an European-West Asian geoelement (35%, 37%) make. These species are widespread mainly in moderately warm zone of Europe and the Western Asia. The continentality of climate increasing in east direction and also its intensive aridization acts as the major factor limiting their distribution. Among representatives of this group it is possible to call *Rosa canina* L., *Prunus spinosa* L., *Prunus domestica* L., *Padus avium* Mill., *Quercus robur* L., *Tilia cordata* Mill.

Among geographical elements of the European group of which 30% and 26% respectively for each of the considered urbanized areas are the share it should be noted the European species taking the third place (*Salix caprea* L., *Salix fragilis* L., *Acer platanoides* L., *Juniperus communis* L., *Sorbus aucuparia* L.).

Table 5. Geostructure dendroflora of Borisoglebsk, Novokhopersk

Groups and geocells types of geographic elements	Borisoglebsk		Novokhopersk	
	Quantity of species	Share as a percents, %	Quantity of species	Share as a percents, %
European-Asian group	38	35	37	41
Euroasian	10	9	15	17
The European-West Asian	24	22	20	22
The European-South West Asian	3	3	1	1
European-Caucasian	1	1	1	1
South East European and Asian group	3	3	2	2
European group	33	30	24	26
European	7	6	6	7
East European	4	4	2	2
South East European	6	5	4	4
Western European	13	12	10	11
Mediterranean	3	3	2	2
Asian group	12	11	10	11
Central Asian	5	5	4	4
The South West Asian	1	1	1	1
The East Asian	6	5	5	6
North -American group	16	14	11	12
Cultural group	8	7	7	8

Also the third place together with the European species was divided by natives of North America (respectively, on 14% and 12%). Emergence of these plants is caused by an introduction in the past. Plants of *Picea pungens* Engelm., *Acer negundo* L., *Ribes aureum, Thuja occidentalis* L. – representatives of this group. Introduced species of *Acer negundo* L. Prikhopyorya including within the Saratov region, next to it, is very widespread in city and natural conditions of the Voronezh region and across all territory.

Species Asian by origin make 11% of total of the registered representatives of wood and shrubby vegetation and take the fourth place. Treat them, in particular *Armeniaca vulgaris* Lam., *Rosa rugosa* Thunb.

The species known only in culture growing in the explored territories have also great economic value. First of all, they are used as sources of food (*Malus domestica* Borkh., *Prunus domestica* L., *Cerasus vulgaris* Mill.) and also as raw materials for production of diverse medicamentous tools.

4. Conclusion

The executed researches established rather high variety dendroflora of the considered urbanized territories that in general it is comparable to a specific variety of the residential district of the large city, such as Voronezh, and even exceeds that [7].

On the basis of the analysis of the executed accounting of objects of researches the following conclusions are received:

1. The systematic analysis of wood plants showed that its basis is made by angiospermous plants (*Magnoliophyta*) – 89% in Borisoglebsk, in Novokhopersk – 91%. The share of gymnospermous (*Pinophyta*) is rather small: in Borisoglebsk – 11%, in Novokhopersk – 9%.
2. In a range of families *Rosaceae* (33% and 34%) dominates, takes the second place this. *Salicaceae* (13%) that shows features of formation of wood and shrubby plantings of Borisoglebsk and Novokhopersk owing to mass cultivation decorative (*Salicaceae, Rosaceae)* and to a large extent fruit and berry cultures (*Rosaceae*).
3. Among an ecomorph of areas of researches there are 56% and 60% of total number of species it is the share of wood plants which make a basis of city green plantings and, respectively, an ecological framework of these territories.
4. Dendroflora has the dominating European-West Asian range of geographical elements (35% and 41%). The group, the second for number, is formed by the European species (30% and 26%).
5. A variety of geoelements is bound to formation of city plantings from structure of the native and introduced species of plants that in total helps to develop and to successfully solve problems of improvement of city residential districts, first of all, by means of green construction that as a result creates a possibility of achievement of effect of ecological comfort of the urban environment for locals.

So in the explored urbanized territories many species of wood plants and bushes, and various origin grow that represents an opportunity for a further introduction of representatives of these groups of organisms according to their bioecological features and further target use. It promotes optimum scheduling and the subsequent realization of actions for local and complex environmental engineering of urban and suburban areas in the conditions of the Voronezh region: gardening, afforestation, recultivation, agrarian and forest melioration.

References

[1] Bochkova I Yu and Hokhlacheva Yu A 2015 Assessment of influence of the invoice on decorative qualities of flower plants in the system of gardening of the city *Forestry Bulletin* 19 (5) 102-106.
[2] Voronov A G 1973 Geobotany. 384.
[3] Gavrilin I I and Runova E M 2012 Some features of a gas-absorbing capacity of trees in an urboekosistema of Bratsk *Forestry Bulletin* 1 (84) 135-138.
[4] Gilyarov A M 2011 Inevitable threats to biological diversity *Nature* 9 3-12.
[5] Fifth national report "Preservation of a biodiversity in the Russian Federation" Moscow Ministry of Natural Resources and Environmental Protection of the Russian Federation publishing house 2015 124.
[6] Serebryakov I G 1964 Life forms of the higher plants and their studying *Field geobotany*. Moscow-Leningrad III 146-205.
[7] Tregubov O V *et al* A specific variety and a condition of plantings of the green space in the northern residential district of the city Voronezh//the Timber magazine. 2014. T. 4. No. 3 (15). Page 61-76.
[8] Lubimov V B, Larionov M V *et al* 2016 Prospects Of Employing The Ecological Method Of Plant Introduction While Establishing The Man-Made Ecosystems Of Different Designated Use *Research Journal of Pharmaceutical. Biological and Chemical Sciences* 7 (4) 1481-1486.

[9] McGill B J 2010 Towards a unification of unified theories of biodiversity Ecology Letters 13(5) 627-642.
[10] Niemelä J and Breuste J H 2011 Urban Ecology: Patterns, Processes, and Applications. Chippenham: Oxford University Press 392.
[11] Rehder A. 1949 Manual of cultivated trees and shrubs New York 725.
[12] Firsov G A *et al* 1994. A comparison of the assortment of broad-leaved trees and shrubs used in City Planting in NW Russia and NE. Sweden *Röbäcksdalen meddelar* Umeå 2 25.
[13] Walker B H 1992 Biodiversity and ecological redundancy Conservation Biology 6 18-23.

Solid-Lubricant, Polymer – Polymeric and Functionalized Fiber– and Powder Reinforced Composites of Ultra-High Molecular Weight Polyethylene

S V Panin[1,2,a], V O Alexenko[1,2], D G Buslovich[1,2], Nguyen Duc Anh[2], Huang Qitao[2]

[1] Institute of Strength Physics and Materials Sciences SB RAS, 634055, Tomsk, Russia
[2] National Research Tomsk Polytechnic University, 634050, Tomsk, Russia

E-mail: [a] svp@ispms.tsc.ru

Abstract. Mechanical and tribotechnical characteristics of solid-lubricant and polymer-polymeric composites of UHMWPE were studied for the sake of design extrudable, wear-resistant, self-lubricant polymer mixtures for Additive Manufacturing (AM). Tribotechnical properties of UHMWPE blends with the optimized content of solid lubricant fillers (polytetrafluoroethylene, calcium stearate, molybdenum disulphide, colloidal graphite, boron nitride) were studied under dry sliding friction at different velocities (V = 0.3 and 0.5 m/s) and loads (P = 60 and 140 N). Also, in order to increase strength and wear-resistance of UHMWPE composites they were reinforced with wollastonite microfibers and aluminum metahydroxide AlO (OH) microparticles preliminary treated (functionalized) in polyorganosiloxane. The comparison on measured mechanical and tribotechnical properties are given with interpretation of the mechanisms of observed phenomenon.

1. Introduction

Ultra-high molecular weight polyethylene (UHMWPE) possesses high enough strength properties as well as the low friction coefficient, high wear- and chemical resistance in aggressive environment. For this reason, it is applied for manufacturing friction units of machine parts and mechanisms. Also, it has found wide application in medicine for fabricating components of orthopedic implants (artificial joints) [1–3]. Extrudable high-strength and solid lubricant nano- and microcomposites based on ultra-high molecular weight (UHMWPE) matrix are intensively developed recently [1–6]. However, due to the chemical inertness of the UHMWPE–matrix, at loading of various additives (mostly, mineral fillers) no chemical bonds are formed at the interphase boundaries. This is manifested in extremely low interphase adhesion. In this context, it is of importance to study addition of micron size fillers into UHMWPE with functionalization of the former using coupling agents (compatibilizers/adhesion promoters) in order to improve the mechanical and tribotechnical characteristics of composites.

Besides the adhesion problem, because of the long polymer chains (molecular weight of 4.0×10^6 c.u.), UHMWPE possesses extremely low melt flow index (MFI–0.06 g/10 min) that limits its processing by traditional methods for structural plastics such as injection moulding, screw extrusion, etc. In so doing, the issues of improving the processability (extrudability) of composites based on the high molecular weight matrix, in particular UHMWPE, are remained unresolved.

In order to develop extrudable powder composites for further application in additive manufacturing (AM) at fabricating the complex shape components of the friction units for mechanical engineering and medicine, the flowability and tribomechanical characteristics of polymer-polymeric mixtures of UHMWPE with the block–copolymers of linear low-density polyethylene with polypropylene (PP-b-LLDPE) as well as the cross-linked polyethylene (PEX-b) were studied. The ultimate aim was the further development the composites with a multiply increased wear resistance and satisfactory melt flowability (MFI) [7–11]. The use of extrudable composites in AM technologies allows one to design high-strength and wear-resistant products of complex shape, including friction units for the machines operating under the harsh environment and loading conditions [7–12].

2. Results and Discussion

UHMWPE powder (GUR-2122, Ticona) with a molecular weight of 4.0 × 106 c.u., cross-linked polyethylene PEX-b, block copolymers of PE with polypropylene (PP-b-LLDPE) particle size ~100–200 μm, grafted high density polyethylene HDPE-g-VTMS (particle size 100–300 μm), were used for the specimen preparation; as well as solid lubricant powders of: (i) polytetrafluoroethylene (PTFE, ∅14 μm), (ii) molybdenum disulphide (MoS2, ∅1–7 μm), (iii) colloidal graphite (CG ∅1–4 μm), iv) calcium stearate (CS, ∅1–7 μm) and v) hexagonal boron nitride (NB, ∅10 μm).

Wollastonite microfibers (∅ 10 μm, aspect ratio 5 : 1) (GEOKOM CJSC, Kaluga region), and micropowder of aluminum metahydroxide AlO(OH) (∅ 2–5 μm) (modifying additive TS-104, Techno-Service Ltd., St. Petersburg) were used for specimen preparation. To modify the filler surfaces, the Penta-1006 liquid reagent (, PENTA-91 ltd. Russia) was used. The latter represents a composite based on poly-[oxy(dimethylsilylene), $[R_2SiO]n$ where R is the oxygen-containing high-molecular-weight organosilicon compounds] [13].

Functionalization of fillers was carried out by their treatment in water-ethanol (pH 4.5-5.5) solutions with added Penta-1006 modifier at ambient temperature with continuous mixing for 5 min. The content of the Penta-1006 reagent was 1 wt % related the filler weight (or 0.2 wt % over the total weight of the powder mixture). The suspension was then dried in oven at the temperature of $T = 120°C$.

The bulk specimens of polymer composites were fabricated by compression moulding at the specific pressure of 10 MPa and the sintering temperature $T = 200°C$ with subsequent cooling rate of 3–4 °C/min. Powders of UHMWPE and fillers (PP-b-LLDPE, PEX-b and nanoparticles) were mixed in a planetary ball mill MP/0.5*4 with preliminary dispersion of the components in an ultrasonic bath in alcohol suspension.

The wear resistance of the specimens under the dry sliding friction was determined by a "block-on-ring" scheme with the 2070 SMT-1 friction testing machine (following the ASTM G99 and DIN 50324). The loading applied to a specimen fixed in a holder was $P = 60$ and 140 N while the sliding velocities were equal to $V = 0.3$ and 0.5 m/s. The size of the specimens was 15.8 × 6.4 × 10.0 mm^3. The diameter of the counterbody made of the bearing steel (HRC 60) steel was 35 mm.

The roughness of the wear track surface as well its volume (being proportional to wear factor) of the specimens was examined by the Step IQ (KLA-Tencor) profilometer.

The degree of crystallinity was determined with the use of SDT Q600 combined analyzer. IR-spectra were recorded with the help of NIKOLET 5700 (Thermo Fisher Scientific, USA) spectrometer. The structural studies were carried out with the use of LEO EVO 50 (Carl Zeiss, Germany) scanning electron microscope at the accelerating voltage of 20 kV. So, the rupture surface were prepared by mechanical bending fracture of the notched specimens after the preliminarily exposure in liquid nitrogen.

Table 1 show the data on mechanical properties of the polymer-polymeric mixtures of UHMWPE with the fillers with the optimized content [7–10] in terms of the extrudability degree (proportional to the MFI) and the minimum decrease in the mechanical characteristics of the mixtures (yield point $\sigma_{0.2}$, ultimate strength σ_U, value of elongation at break ε). The revealed dependencies of the mechanical and tribotechnical properties of polymer mixtures versus the filler weight fraction carried out in [7–11] have shown that for all the fillers under study (Table 1), the optimum filler loading makes 5–10 wt %. For this reason, the composites with the optimum compound of the polymer mixture were chosen to study the effect of the sliding velocity and load on the wear resistance.

The physical and mechanical properties of UHMWPE composites loaded with 20 wt % of the wollastonite and 20 wt % aluminum metahydroxide are given in Table 2. It follows from the table that the density and hardness of UHMWPE composites increase with enlarging weight fraction of the microfibers and microparticles. We suggest that initiation of the hydrolysis process would not affect values of these parameters.

Table 1. Physical-Mechanical and tribotechnical properties of UHMWPE and its composites

Filler content, wt %	Density, g/cm³	Shore hardness *D*	Elastic modulus *E*, MPa	Yield point $\sigma_{0.2}$, MPa	Ultimate strength σ_U, MPa	Elong. at break ε, %	Extrud. temp./pressure, °C/MPa	MFI, g/10 min
UHMWPE	0.928	55.6 ± 0.2	405 ± 17	22.1 ± 0.4	36 ± 1.6	482 ± 6	205/5.52	0.126
UHMWPE + 10% HDPE-g-VTMS	0.924	57.6 ± 0.5	398 ± 9	22.5 ± 0.9	29.0 ± 1.3	416 ± 57	205/5.52	0.291
UHMWPE + 10% PEX-b	0.931	57.4 ± 0.5	354 ± 39	22.8 ± 0.8	31.2 ± 2.9	429 ± 38	205/4.69	1.122
UHMWPE + 5% PP-b-LLDPE	0.916	56.1 ± 0.5	403 ± 25	21.1 ± 0.5	29.4 ± 1.6	383 ± 28	205/4.34	3.574

Table 2. Physical-mechanical properties of UHMWPE and its composites

Filler content, wt %	Density, g/cm3	Shore hardness D	Elastic modulus E, MPa	Yield point $\sigma_{0.2}$, MPa	Ultimate strength σ_U, MPa	Elong. at break ε, %
UHMWPE + 20 wt % Wollast	1.062	58.9 ± 0,7	501 ± 30	24.7 ± 1.1	36.8 ± 5.7	454 ± 60
UHMWPE+ 20 wt % AlO(OH)	1.07	60.23 ± 0.6	509 ± 55	21.9 ± 1.3	33.2 ± 1.9	425 ± 39
UHMWPE + 20 wt % Wollast. + 0.2% Penta 1006 (hydrolysis)	1.062	59.19 ± 0.6	620 ± 46	25.2 ± 1.3	34 ± 4	415 ± 23
UHMWPE + 20 wt % AlO(OH) + 0.2% Penta 1006 (hydrolysis)	1.064	59.4 ± 0.5	564 ± 19	23.1 ± 0.5	37.0 ± 1.5	446 ± 26

The permolecular structure typical for UHMWPE (a) and all polymer mixtures under study at the filler content up to 10 wt % (b–d) are given in Fig. 1. It is shown that with the further increase of the filling degree, a spherulitic structure is not formed, and the degree of crystallinity of the polymer-polymeric composites is substantially reduced [7, 8]. At the same time, the mechanical properties are also substantially decreased (yield strength, ultimate strength, value of elongation at break). For this reason, the criterion for choosing polymer processing modifier for UHMWPE should be the balance between satisfactory flowability of the mixture and lowest loose of strength characteristics.

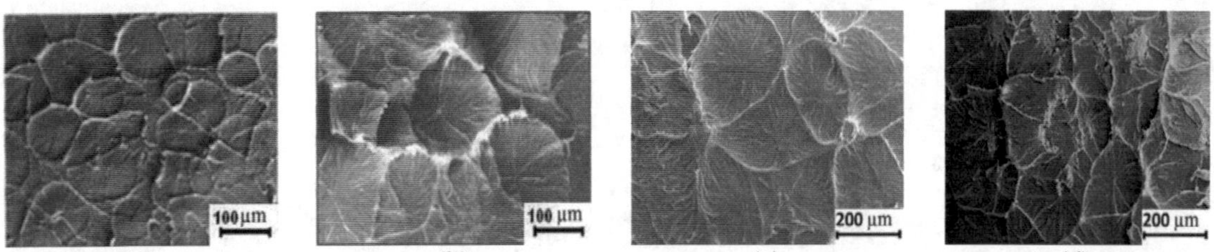

(a) (b) (c) (d)

Figure 1. SEM micrographs of permolecular structure of UHMWPE (a) and its composites "UHMWPE + 10 wt % HDPE-g-VTMS" (b), "UHMWPE + 10 wt % PEX-b" (c), "UHMWPE + 5 wt % PP-b-LLDPE" (d)

Polypropylene has the higher melt flowability (MFI = 0.6 g/10 min). However it does not possess thermodynamic affinity to UHMWPE and its loading substantially reduces the strength properties of the mixture due to lack of adhesion (Table 1). A block copolymer of the polypropylene with the linear low density polyethylene (PP-b-LLDPE) provides better adhesion to UHMWPE and maximizes the strength characteristics of the matrix (Table 1). The analysis of the IR-spectra of UHMWPE blends with the block copolymer PP-b-LLDPE showed the increased peak intensity at 1240 cm^{-1} (C-O-C), 1040 cm^{-1} (C-C-O), 1620 cm^{-1} (C = C) and 1700 cm^{-1} (C = O) for the spectra of polymer-polymeric composites which testifies for the bonding between UHMWPE and the block copolymer [8]. In other words, by adding a block copolymer containing the linear low–density polyethylene it was possible to mechanically and chemically "bond" the fillers containing polypropylene with the UHMWPE matrix. As a result, the flowability (extrudability) increased while the minimum loss of mechanical properties took place (under the conditions of the adhesion of UHMWPE with processing filler).

Analysis of the permolecular structure of powder and fiber filled UHMWPE composites shows that at loading of 20 wt % of the fillers the spherulitic structure is no longer formed. Microfibers and microparticles are evenly distributed in the matrix. The degree of crystallinity in the composites does not exceed 15 % [7, 8].

Physical and mechanical properties of polymer-polymeric mixtures of UHMWPE with solid-lubricant fillers of optimum content were determined by the authors earlier [7]. It was motivated by reaching maximum level of wear resistance and retaining the spherulitic permolecular structure of the composite (see Table 3). As follows from the table the Shore D hardness of UHMWPE composites is determined solely by the type of filler: the composite with calcium stearate possesses the lowest hardness (54.3), while the composite with molybdenum disulfide MoS_2 has the highest one (57.4). The value of ultimate strength σ_U for all composites has decreased as compared with the neat polymeric matrix. The value of elongation at break ε of the composites also decreases or increases with respect to UHMWPE depending on the type of the filler varying in the range 360–503 %. In other words, all the fillers for the ultra-high-molecular weight matrix (UHMWPE) under investigation cannot be treated as the reinforcing ones.

The degree of crystallinity of the UHMWPE composites with increasing weight fraction of all the above mentioned solid lubricant particles is decreased (Table 3). However, the type of the permolecular structure remains spherulitic (Fig. 2). It was found that various solid-lubricant fillers provided different wear resistance of the UHMWPE composites, which also depends on sliding velocity and applied load. The values of the friction coefficients for all the studied composites are presented in Table 3. It is easily seen that for most of them the friction coefficient has decreased by 60–70 %. Further studies of the UHMWPE based mixtures were focused on their wear resistance at various sliding velocities (V = 0.3 and 0.5 m/s) and loads (P = 60 and 140 N).

Table 3. Physical-mechanical properties of UHMWPE and its composites

Composition	Density ρ, g/cm^3	Shore hardness D	Ultimate strength σ_U, MPa	Elongation at break ε, %	Crystallinity χ, %	Friction coefficient f
UHMWPE	0.928	55.6±0.2	36±1.6	482±6	56.5	0.12
UHMWPE + 5 wt % PTFE	0.951	54.5±0.6	29.2±1.0	465±23.6	39.5	0.07
UHMWPE + 5 wt % CS	0.913	54.3±0.2	33.1±0.9	411±19.5	41.0	0.11
UHMWPE + 5 wt % MoS_2	0.928	57.4±0.3	28.9±1.6	500±25.3	43.6	0.8
UHMWPE + 5 wt % CG	0.953	56.8±0.3	29.7±1.5	503±25.1	42.1	0.08
UHMWPE + 5 wt % BN	0.955	55.7±0.4	24.1±0.9	360±16.4	41.2	0.08

Loading of the functionalized wollastonite also demonstrates high efficiency at high sliding velocity of $V = 0.5$ m/s (Fig. 3b). At the high applied loads the wear multiplies by the factor of 5–10 at both sliding velocities for pure UHMWPE and its composites with functionalized fillers (Fig. 3). In doing so, the composites containing the wollastonite treated by polyorganosiloxane with the initiated hydrolysis process at the high load provides the highest wear resistance among all the studied blends. This testifies for the efficiency of this technique of the filler treatment in comparison with a simple mechanical mixing of the composite components.

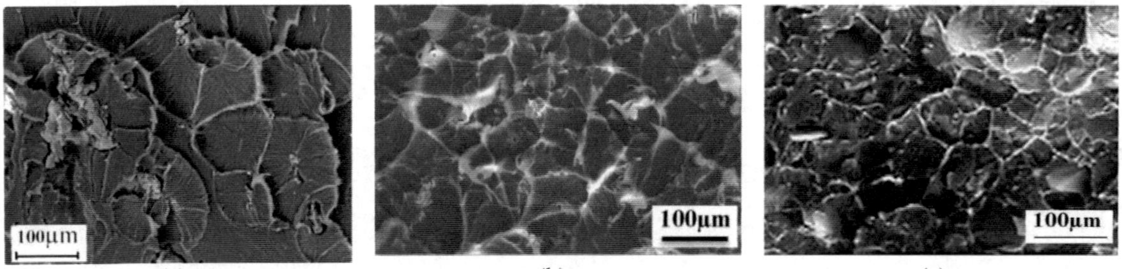

(a) (b) (c)

Figure 2. Permolecular structure of composites based UHMWPE with 5 wt % PTFE(a), 5 wt % MoS2 (b), 5 wt % CG (c)

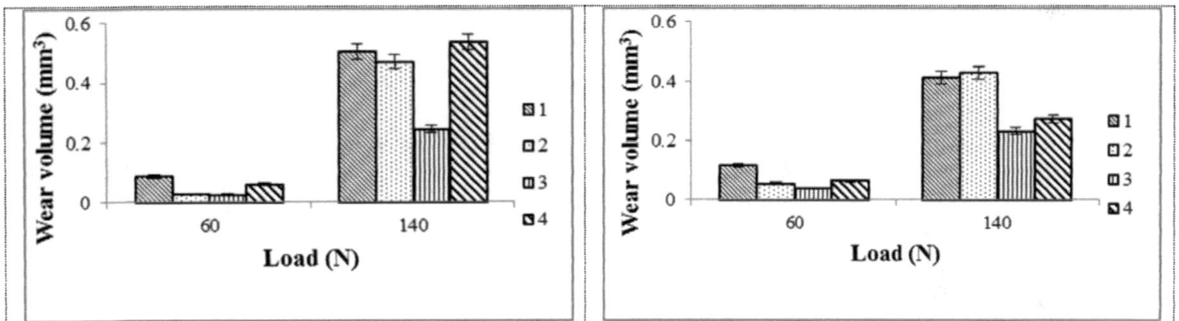

(a) (b)

Figure 3. Volumetric wear (mm^3) at sliding velocities $V = 0.3$ (a) and 0.5 m/s (b) of UHMWPE (*1*) and its composites with 20 wt % of wollastonite (*2*), UHMWPE + 20 wt % of wollastonite + 0.2 wt % Penta 1006 with hydrolysis (*3*) and UHMWPE + 20 wt % of wollastonite + 0.2 wt % Penta 1006 without hydrolysis (*4*) at the loads $P = 60$ and 140 N

When UHMWPE is filled with the functionalized aluminum metahydroxide particles with initiating the hydrolysis, wear decreases both at moderate and high sliding velocities ($V = 0.3$ and 0.5 m/s) at the load of $P = 60$ N (Fig. 4). With increasing the load up to $P = 140$ N the wear of UHMWPE and its composites also multiply increases at both sliding velocities.

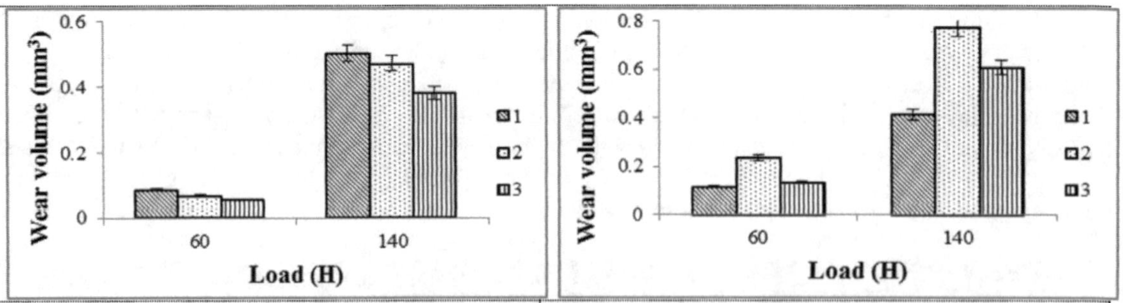

(a) (b)

Figure 4. Volumetric wear (mm^3) at the sliding velocities V = 0.3 (a) and 0.5 m/s (b) of the UHMWPE (*1*) and its composites UHMWPE + 20 wt % AlO (OH) (*2*), UHMWPE + 20 wt % AlO (OH) + 0.2 wt % Penta 1006 with hydrolysis (*3*) at loads P=60 and 140 N

The volumetric wear diagram for all the composites studied is shown in Fig. 5, at the moderate sliding velocity (V = 0.3 m/s) and loads of P = 60 and 140 N. It is seen from the diagram that at low load P = 60 N the wear resistance of the composites increases by 2–3.5 times. The temperature at the counterface surface makes T = 27–30°C. Note, that the most efficient solid lubricant filler for the UHMWPE under these testing conditions is polytetrafluoroethylene.

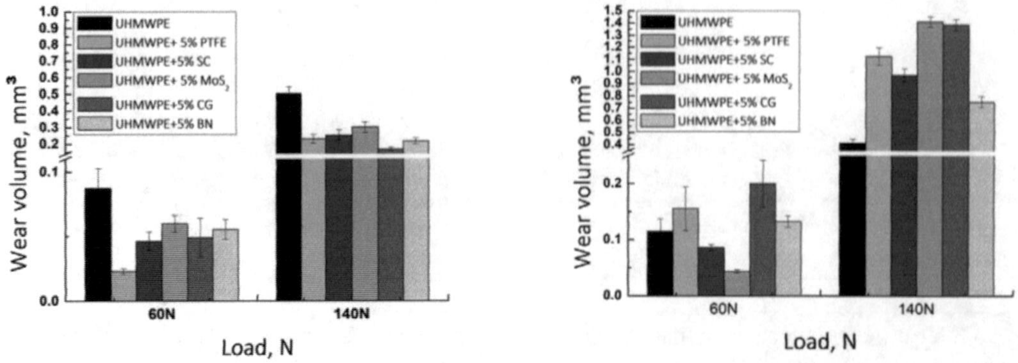

(a) (b)

Figure 5. Volumetric wear of UHMWPE and its composites at the steady-state wearing stage. Sliding velocity 0.3 (a) and 0.5 m/s(b), and loads 60 and 140 N

At the load of P = 140 N the wear resistance of the UHMWPE and its composites decreases more than twice. In doing so, the temperature at the counterface surface (that should be equal to one at wear track specimen surface) increases from 27 to 39 °C.

At higher sliding speed (V = 0.5 m/s) and moderate load (P = 60 N) the wear of the neat UHMWPE as well as its composites is twice higher than that at the sliding velocity of V = 0.3 m/s (Fig. 5b). At these loading conditions the molybdenum disulphide provides the most efficient wear resistance. This means that it can be used as solid lubricant filler for fabricating the products, which operate in friction units at moderate loads and high sliding velocities.

Finally, at high sliding velocity and load (up to P = 140 N) a multiple increase in the wear of the UHMWPE and all its composites takes place (by the factor of 5–10). In doing so, the temperature of the friction surface rises up to T = 64°C.

3. Conclusions

The physical-mechanical and tribotechnical properties of the polymer-polymeric, solid lubricant and functionalized fiber and powder reinforced composites based on UHMWPE at different sliding velocities (V = 0.3 and 0.5 m/s) and loads (P = 60 and 140 N) under dry sliding friction have been studied. It is shown that the treatment of the fibrous and powder fillers by the polyorganosiloxane with initiating the hydrolysis process increases the mechanical characteristics of composites based on the high molecular weight PE matrix by 10 %. The wear resistance of the microcomposites increases twofold at sliding velocities V = 0.3 m/s and 0.5 m/s at the moderate load (P = 60 N) as compared to the neat UHMWPE. When the load is increased up to P = 140 N, the wear resistance of the composites is sharply reduced. For this reason, micro-composites UHMWPE based with the functionalized fibrous and powder fillers can be recommended for operation in friction units operating in a wide range of sliding velocities (up to V = 0.5 m/s) and at the moderate loads up to P = 60 N. The technique for compatibilizing mineral fillers (fibrous and powder) with the high molecular weight PE matrix by the treatment in the polyorganosiloxane can be applied as an efficient approach for increasing the wear resistance of highly filled UHMWPE composites.

The wear resistance of solid lubricant UHMWPE composites at the moderate sliding velocity (V = 0.3 m/s) and load (P = 60 N) increases by 2–3 times in comparison with pure UHMWPE. However, when increasing the load up to P = 140 N the wear resistance of UHMWPE and its composites is reduced twice. At high sliding velocity and load (up to P = 140 N) the multiple increase in wear of the UHMWPE and all its composites (by the factor of 5–10) occurs.

Acknowledgments

The study was conducted at Tomsk Polytechnic University within the framework of Tomsk Polytechnic University Competitiveness Enhancement Program grant, Project Number TPU CEP_IHTP_73\2017. Authours acknowledge support from Fundamental Research Program of Russian State Academies of Sciences for 2013–2020 as well as RFBR projects No. 16-48-700192.

References

[1] B. J. Briscoe and S. K. Sinha, in Tribological Applications of Polymers and Composites: Past, Present and Future Prospects (Elsevier, 2008), pp. 1–14.
[2] S.M. Kurtz, The UHMWPE Handbook: Ultra-High Molecular Weight Polyethylene in Total Joint Replacement (Academic Press, 2004).
[3] M.C. Galetz, T. Blar, H. Ruckdaschel, K. W. Sandler, and V. Alstadt, J. Appl. Polym. Sci. 104, 4173–4181 (2007).
[4] O. V. Gogoleva P. N. Petrova, S.N. Popov, and A. A. Okhlopkova, Friction Wear 36(4), 301–305 (2015). https://doi.org/10.3103/S1068366615040054
[5] Y. Khalil, A. Kowalski, and N. Hopkinson, Manufactur. Rev. 15(3), 1–9 (2016).
[6] A. P. Krasnov, A. E. Said-Galiev, O. V. Aphonicheva, et al., Friction Wear 28(3), 292–299 (2007). doi 10.3103/S1068366607030099
[7] S. V. Panin, L. A. Kornienko, V. O. Alexenko, and L. R. Ivanova, Key Eng. Mater. 712, 155–160 (2016). doi 10.4028/www.scientific.net/KEM.712.155
[8] Y. Khalil, A. Kowalski, and N. Hopkinson, Manufact. Rev. 3(15), 1–9 (2016).
[9] E. D. Prut and A. N. Zelenetsky, Usp. Khim. 1, 72–87 (2001).
[10] A. P. Krasnov, E. E. Said-Galiev, O. V. Afonicheva, et al., Frict. Wear 28(3), 288–295 (2007).
[11] S. V. Panin, L. A. Kornienko, S. Wannasri, et al., Frict. Wear 31(5), 353–360 (2010).
[12] S. V. Panin, L. A. Kornienko, and S. T. Nguen, et al., Frict. Wear 36(6), 652□661 (2015).
[13]http://itwpenta.ru/area/proizvodstvo_himicheskoj_produkcii1/pererabotka_termoplastov_poliolefinov_i_pvh/modifikator_penta-1006

IOP Conf. Series: Earth and Environmental Science **115** (2018) 012011 doi:10.1088/1755-1315/115/1/012011

Increasing fuel efficiency of passenger cars with the "climate-control" system as a method of improving the ecological safety of the urban infrastructure

L N Burakova [a], A D Burakova [b], O D Burakova and V O Dovbysh

Tyumen Industrial University, str. Volodarskogo 38, Tyumen, 625000 Russia

E-mail: [a] burakova.1992@mail.ru, [b] burakova.96@bk.ru

Abstract. Motor vehicle should provide safety and a high ecological standard of living for the population. One of the methods to improve the ecological friendliness of motor vehicles in particular passenger cars (cars), which are considered in this article, is the growth of their fuel economy. It is established that fuel consumption and the amount of specific emissions of harmful substances with exhaust gases of cars when using the "climate control" system depend on the effective ambient temperature, the color of the opaque car body elements, the power of the car engine and the interior volume. However, the simplest controlled factor is the color of the opaque car body elements, which is characterized by the coefficient of light reflection. In the course of experimental studies, we established the dependences of a change in fuel consumption and a share of reducing emissions of harmful substances with exhaust gases of passenger cars with the "climate control" system on the coefficient of light reflection. A method has been developed to reduce fuel consumption and the amount of specific emissions of harmful substances with the exhaust gases of passenger cars when using the "climate control" system, which involves painting the vehicle roof white and allows reducing fuel consumption by 5.5-10.3% and the amount of specific emissions of harmful substances by 0.37-1.13% (CO) and 0.47-1.08% (CH).

1. Introduction

Motor vehicles are an integral part of urban infrastructure, which should meet the requirements of safety and ecological friendliness of vital activity. The modern concept of its improvement is the direction of management decisions to increase the ecological responsibility through the introduction of new technologies [1,2,3,4]. According to the Transport Strategy of the Russian Federation until 2030, the ecological nature of the transport system plays an important role in the socio-economic development of the country, so reducing the harmful impact of transport on the environment is one of its main tasks [5]. GOST R 52033-2003 "Cars with gasoline engines. Emissions of pollutants with exhaust gases. Norms and methods of control in assessing the technical condition (Amend. No. 1 of 01.07.2012)" indicates values of the content of carbon monoxide and hydrocarbons in exhaust gases.

In the European Union, there is a program that reduces fuel consumption and the negative impact of vehicles on the environment through the formation of behavioral rules of the driver. However, motor vehicles operation specifics in the Russian Federation necessitate the adaptation of these rules to the existing conditions [6,7,8,9,10,11,12]. It is low density of the street and road network, rapid growth in the number of vehicles, separation of pedestrian and traffic flows at intersections. This reduces the throughput and increases the average idle time of vehicles. As a result of the analysis of previous studies, it has been established that for the last 30 years the idle time of cars with a running engine has increased by 18% and in modern conditions is 30 ... 35% [13,14,15,16].

The Russian Federation territory location also influences the operation of motor vehicles. For example, the number of car-days in the operation of cars at temperatures above +20 ° C is more than 51%. Deviation of operating conditions from standard affects the motor vehicles by changing the thermal state of the elements of their design. In particular, in the warm period of the year, there is an increased temperature of the ambient air (above + 20 ° C) and an increase in the amount of solar radiation entering the Earth's surface. This causes an increase in the amount of heat entering the interior of the car, hence increasing the air temperature inside the vehicle. According to the norms of POT RM-008-99 "Inter-industry rules for labor protection" in the operation of industrial vehicles, the air temperature at the driver's workplace during the warm season should be 20-25°C or no more than 3°C above the average outside air temperature at 13.00 of the hottest month, but not more than 28°C. To create the required microclimate in the car, "climate control" systems are applied, which are the auxiliary equipment of the car. The inclusion of this in the work leads to an increase in the proportion of the car engine power spent on the work of the auxiliary equipment, which causes an increase in the consumption of fuel by the vehicle. The greatest impact is observed during the idle time of the car with the running engine; it is caused by the absence of streams of air cooling the interior during movement, and the minimum value of the effective capacity of the car engine.

The increase in fuel consumption of cars with the use of "climate control" during the idle time with the engine running increases the amount of exhaust gases [19,20,21,22]. Their composition is 95 ... 98% products of complete combustion, unused oxygen and nitrogen of the air, and 2 ... 5% - several hundred components that determine the degree of a harmful impact of the vehicle on the environment. However, in the products of complete combustion, there are chemical compounds that have a negative effect on the ecological state of the atmosphere, for example, carbon dioxide, which contributes to the development of the greenhouse effect. Reducing the concentration of this compound in the exhaust gas can be achieved by reducing the fuel consumption of cars.

Thus, the purpose of this study is to improve the fuel economy and environmental friendliness of passenger cars when using the "climate control" system during idle time with the engine running.

2. Methodology

The ecological nature of the transport system can be characterized as the ability of cars to carry out transport work or transportation of passengers with the minimum possible emissions of harmful substances in strictly defined conditions. Vedenyapin G.V., Govorushchenko N.Ya., Gutarevich Yu.F., Evtin P.V., Gelagin O.I., Zvono V.A., Magaril E.R. Reznik L.G. studied the problems of reducing the toxicity of exhaust gases and increasing fuel economy. Fuel consumption and the content of harmful substances in the exhaust gases of cars depend on many factors, in particular, on incorporating the auxiliary equipment and operating conditions.

An increase in the ambient air temperature leads to an increase in the number of heat gain to the car interior, therefore, the microclimate inside the vehicle changes [23]. Assessment of the impact of operating conditions on the microclimate in the vehicles was carried out by Palutin Yu.I., Kormin A.M., Semikin S.N., Shelyakin A.S. and others [24]. The given works proposed a set of natural and climatic factors of the conditions of vehicle operation to evaluate the effective temperature of the ambient air with the help of a complex index, the numerical value of which corresponds to the true temperature of the still and saturated air. However, the effective temperature of the ambient air is the determining factor only when heat is supplied to the car interior by heat transfer.

When assessing the heat balance of the vehicle interior, it should be taken into account that heat gain from external sources is realized not only by heat transfer, but also by thermal radiation. The amount of heat entering the passenger car interior through the opaque car body elements with the help of thermal radiation depends on the reflectivity of the paintwork. The color characterizes the reflectivity of the opaque body elements. For example, an absolutely white body has the coefficient of light reflection equal to 1, and absolutely black (soot) -0. However, it is impossible to achieve absolutely white or black colors when painting, therefore, as a rule, cars with a white color of the opaque body elements have the coefficient of light reflection equal to 0.95, and with a black color - 0.05. Thus, the amount of heat entering the interior of the car depends on the effective ambient temperature and the coefficient of light reflection of the opaque body elements and influences the microclimate inside the vehicle. An increase in the amount of heat input causes an increase in the temperature of the air inside the car; in order to reduce and maintain it, within the limits corresponding to comfort conditions, a "climate control" system is used. The main characteristics of its operation affecting the fuel consumption of a car are the cooling

IOP Conf. Series: Earth and Environmental Science **115** (2018) 012011 doi:10.1088/1755-1315/115/1/012011

capacity and power consumption which depend not only on the amount of heat entering the interior of the car, but also on the interior volume. An increase in the interior volume, as well as the amount of incoming heat lead to an increase in the power consumption of the "climate control" system. It is established that the increase in the fuel consumption of a car when the "climate control" system is switched on is determined by an increase in the proportion of the power take-off of the automobile engine for the operation of this auxiliary equipment, which is characterized by the ratio of the cooled volume of the interior to the car engine power. This value is characterized by the specific climatic power parameter, which determines the part of the power of the automobile engine spent for cooling the unit volume of the car interior.

The fuel consumption of a passenger car when using the "climate control" system that affects the amount of specific emissions of harmful substances with exhaust gases depends on the effective ambient temperature, the coefficient of light reflection of the opaque car body elements and the specific climatic power parameter (the ratio of the passenger car interior volume to its engine power).

2.1. *Equipment and devices used in studies*

The experiment involved cars of three makes - Ford Focus, Toyota Corolla, and Mitsubishi Lacer. The light reflection coefficient was measured with a thermal imager. Illumination of the site was assessed using a luxmeter. The effective ambient temperature was taken into account by measuring the three factors: ambient air temperature and humidity - by the Assmann psychrometer and wind velocity - by the Fuss anemometer. The air temperature in the car was measured with mercury and digital thermometers. Fuel consumption was measured using a BT-ECU CAN adapter connected to the diagnostic socket of the car with an adapter and a computer program Check-Engine. To monitor the volume content of harmful substances in the exhaust gases of engines, a gas analyzer GIAM-29M-1 was used. The resulting volume concentrations were converted into specific CO and CH emissions.

2.2. *Methodology of research*

Passenger cars participating in the experiment consisted of a series of pairs of cars (black, silver, gray and white), one of which had an original roof color and the other was given a white color with the help of a vinyl film. Initially, the area of the opaque car body elements, in particular the roof, was determined.

Then, the vehicles were installed on a uniformly illuminated platform without shading. At the time of the maximum warm-up of the car with the "climate control" system switched off, the air temperature in the car was measured.

Then, the "climate control" system was switched on. When the constant thermal conditions were achieved, the air temperature in the car was measured and checked for compliance with its set value in the "climate control" system. The coefficient of light reflection for opaque body elements of different colors was determined with an illumination of 60000 lux (illumination at noon on a cloudless summer day) using a thermal imager.

Fuel consumption data of a passenger car equipped with a "climate control" system were measured during the idle time at a steady coolant temperature in the operating range of 80-100°C from the vehicle ECU using the BT-ECU CAN adapter and the Check- Engine computer program. Together with the measurement of fuel consumption, carbon dioxide and hydrocarbons in the exhaust gases were measured with a gas analyzer.

3. Results and discussion

Analyzing the state of the matter, it has been revealed that the factors that have the greatest effect on the fuel consumption of a passenger car using the "climate control" system are the effective ambient temperature, the power of the car engine, the interior volume and the color of the opaque car body elements (the light reflection coefficient). Of the presented factors only the power of the car engine, the interior volume and the color of the opaque car body elements are controllable, but the change in the first two factors is difficult. Thus, one of the ways to reduce fuel consumption of a car with the use of the "climate control" system is to impart a color that is characterized by the largest coefficient of light reflection (white color). A change in the increase in fuel consumption of a Ford Focus passenger car using the "climate control" system during the idle time with the engine running from the coefficient of light reflection of the opaque body elements is shown in Figure 1.

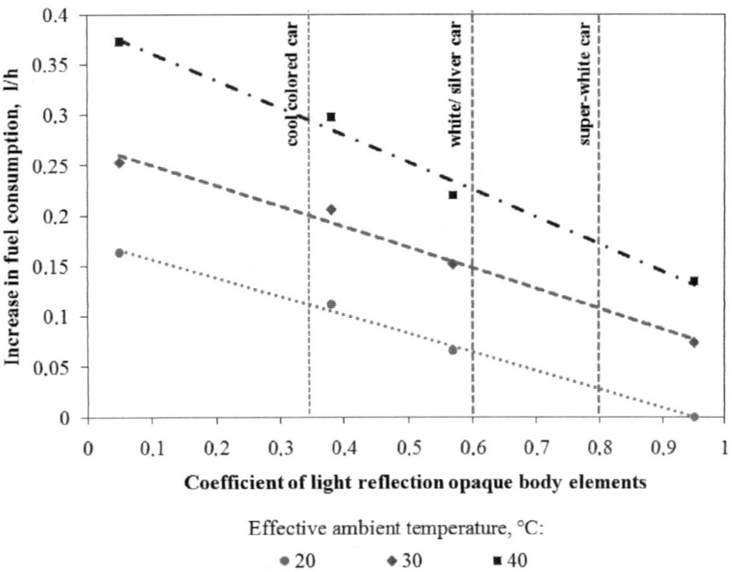

Figure 1. A change in the increase in fuel consumption of a Ford Focus passenger car using the "climate control" system during the idle time with the engine running from the coefficient of light reflection of the opaque body elements.

An increase in the fuel consumption of a car with the use of the "climate control" system during the idle time with the engine running becomes greater with an increase in the effective ambient temperature and a change in the color of the opaque body elements from light to dark characterized by a smaller coefficient of light reflection. This causes an increase in specific emissions of harmful substances with exhaust gases into the atmosphere, which is shown in Figures 2 and 3.

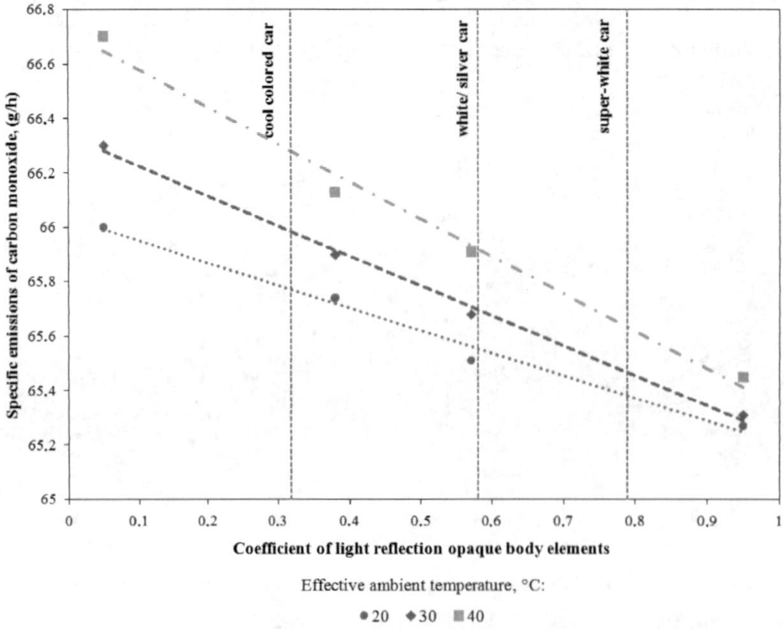

Figure 2. Reduction in specific emissions of carbon monoxide with exhaust gases of a passenger car when using the "climate control" system during the idle time with the engine running from the coefficient of light reflection of the opaque body elements.

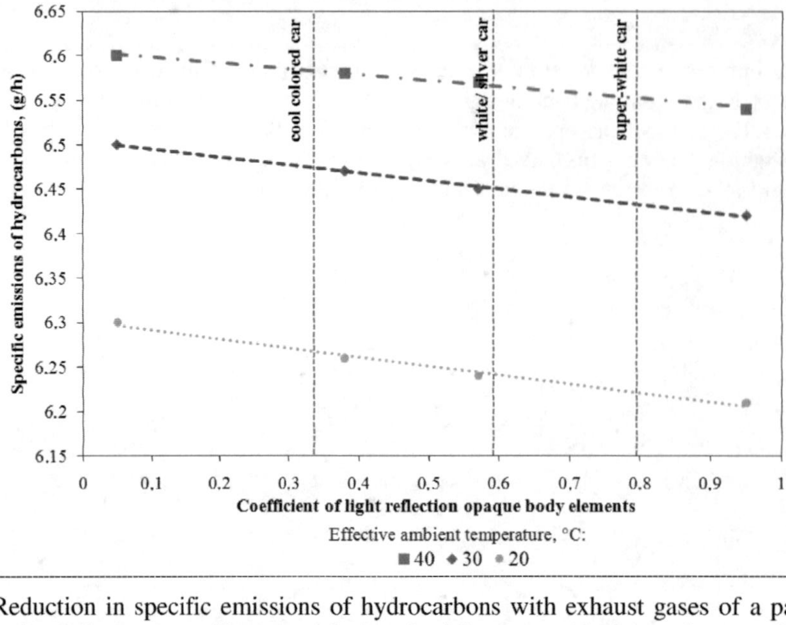

Figure 3. Reduction in specific emissions of hydrocarbons with exhaust gases of a passenger car when using the "climate control" system during the idle time with the engine running from the coefficient of light reflection of the opaque elements.

An increase in the coefficient of light reflection leads to a decrease in specific emissions of carbon monoxide and hydrocarbons with exhaust gases. As a result of the analysis of the received data it is established that the roof of the car equipped with the "climate control" system should be white to reduce fuel consumption and the amount of specific emissions of harmful substances with the exhaust gases. In the course of experimental studies, results are obtained that attest to the feasibility of introducing the developed method. Experimental data on the effect of the white color of the roof on the change in fuel consumption and the amount of specific emissions of harmful substances with the exhaust gases are presented in Figures 4 and 5.

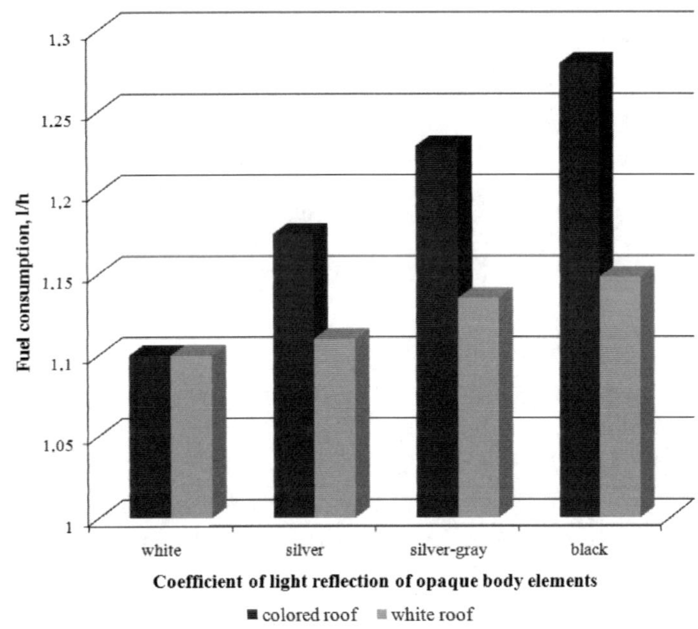

Figure 4. Change in the fuel consumption of a car equipped with the "climate control" system with the introduction of the developed method.

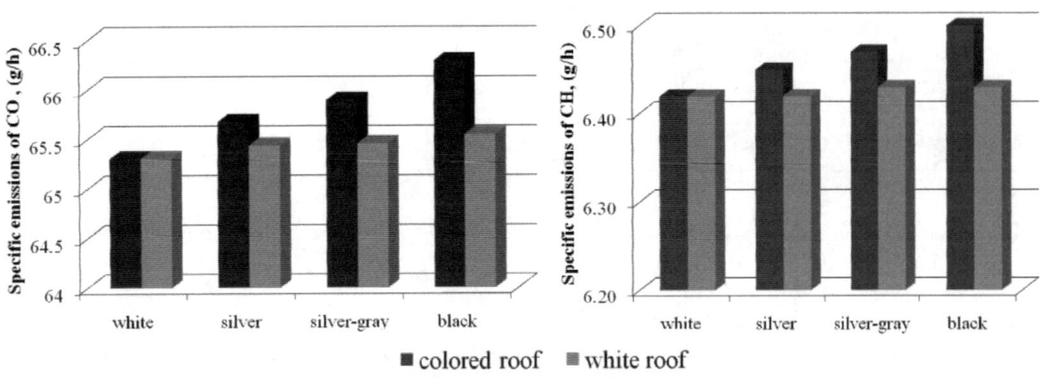

Figure 5. Decrease in specific emissions of carbon monoxide and hydrocarbons with exhaust gases of cars equipped with the "climate control" system.

Application of the developed method allowed reducing the fuel consumption of a passenger car using the "climate control" system by 5.5-10.3%. Making the roof of a silver car white led to the achievement of fuel consumption corresponding to that of a white car.

Reducing the fuel consumption of a car with the introduction of the developed method allowed reducing the amount of specific carbon monoxide emissions by 0.37-1.13% and hydrocarbons by 0.47-1.08% with the exhaust gases.

4. Conclusions

In the course of experimental studies, it has been established that when the color of the opaque body elements changes from light to dark (reducing the coefficient of light reflection), fuel consumption and concentration of carbon monoxide and hydrocarbons in the exhaust gases of passenger cars increase when using the "climate control" system.

The area of the opaque body elements and their internal arrangement are analyzed. Based on the results of the preliminary stage of the experiment, a method has been developed for reducing the fuel consumption of a passenger car using the "climate control" system. It is to give a white color to the roof of the vehicle.

According to the obtained results of the experiment, it is established that the application of the developed method allows reducing the fuel consumption of a car by 5.5-10.3% and the amount of specific carbon monoxide emissions by 0.37-1.13%, hydrocarbons by 0.47-1.08% with the exhaust gases.

References

[1] Anisimov I, Magaril E, Magaril R, Chikishev E, Chainikov D, Gavaev A, ErtmanS, Ertman Yu and Ivanov A 2016 Improving vehicle adaptability to the operating conditions of "smart" cities in the northern regions *E3S Web of Conf. Rev.* **6** 02003

[2] Panuccio P, Amodeo L, D'agostino P, Lamari D and Scattarreggia T 2015 Urban Regeneration and smart sity according to EU strategies: an urban distribution center in city ogistics *WIT Transactions on The Built Environment* vol 146 (UK: WIT Press) pp 313-324

[3] Monfaredzadeh T and Berardi U 2014 How can cities lead the way towards a sustainable, competitve and smart future? *WIT Transactions on Ecology and the Environment* vol 191 (UK: WIT Press) pp 1063-74

[4] Russo F, Rindone C and Panuccio P 2014 The process of smart city definition at an EU level *WIT Transactions on Ecology and the Environment* vol 191 (UK: WIT Press) pp 979-989

[5] Government of the Russian Federation 2014 *Transport Strategy of th Russian Federation until 2030* (Moscow: Government of the Russian Federation)

[6] Chikishev E, Chainikov D and Anisimov I 2017 Increasing the use of natural gas on motor transport by an efficient location of the NGV RS *IOP Conf. Ser.: Earth Environ. Sci. Rev.* **50** 012004

[7] Chikishev E, Chikisheva A, Anisimov I and Chainikov D 2017 Natural gas use on mimbuses, engaged in the carriage of passengers and baggage on the regular routes, as a measure for decrease I harmful environment effects *IOP Conf. Ser.: Earth Environ. Sci. Rev.* **50** 012008

[8] Chainikov D, Chikishev E, Anisimov I and Gavaev A 2016 Influence temperature on the CO_2 emitted with exhaust gases of gasoline vehicles *IOP Conf. Ser.: Mater. Sci. Eng. Rev.* **142** 012109

[9] Chikishev E, Ivanov A, Anisimov I and Chainikov D 2016 Prospects of and problems in using natural gas for motor transport in russia *IOP Conf. Ser.: Mater. Sci. Eng. Rev.* **142** 012110

[10] Ertman J, Ertman S, Anisimov I, Chainikov D and Chikishev Y 2015 Complex characteristic estimation of vehicle adaptability for low temperature operation conditions *Research Journal of Pharmaceutical, Biological and Chemical Sciences (Electronic Materials* vol 6*)* (India: RJPBCS) pp 1761-70

[11] Anisimov I, Ivanov A, Chekishev E, Chainikov D and Reznik L 2014 Assesment of gas cylinder

vehicles adaptabilty for operation at low ambient temperature conditions *WIT Transactions on Ecology and the Enviroment* vol 190 (UK: WIT Press) pp 685-695
[12] Ertman S, Ertman J and Zaharov D 2016 Raising of operating a motor vehicle effects on enviroment in winter *Innovative Technologies in Engineering Rev.* **142** 12119
[13] Cloke J and Layfield R 1996 The environmental impacts of traffic management schemes *WIT Transactions on The Built Environment* vol 26 (UK: WIT Press) p 11
[14] Ntzeremes P, Kirytopoulos K and Tatsiopoulos I 2016 Management of infrastructure's safety under the influence of normative provisions *WIT Transactions on The Built Environment* vol 164 (UK: WIT Press) pp 51-59
[15] Langeland A 2015 Sustainable transport in cities: learnin from best and worst practice? *WIT Transactions on The Built Environment* vol 146 (UK: WIT Press) pp 93-103
[16] Langer D and Mattrish G Future trends and driving forces in transport and logistics *WIT Transactions on the Built Environment* vol 75 (UK: WIT Press) p 10
[17] Galderisi A 2014 Adapting cities for a changing climate: an integrated approach for sustainable urban development *WIT Transactions on Ecology and the Environment* vol 191 (UK: WIT Press) pp 549-560
[18] Alisov B and Poltaraus B 1974 *Climatology* (Moscow: Publishing House of Moscow University)
[19] Jacyna M and Merkisz J 2015 Estimations in real urban traffic conditions s a deteminant of shaping sustainable urban development *WIT Transactions on The Built Environment* vol 146 (UK: WIT Press) pp 219-231
[20] Gong R and Waring P 1996 Investigation of emission behaviour in an urban driving cycle with a remote sensing system *WIT Transactions on The Built Environment* vol 26 (UK: WIT Press) p 10
[21] Tong H and Hung W 1999 Vehicular modal emission and fuel consumption factors in Hong Kong *WIT Transactions on Ecology and the Environment* vol 37 (UK: WIT Press) p 8
[22] Kazopoulo M, El-Fadel M and Kaysi I 2004 Emissions Standards Development for an I/M program *WIT Transactions on Ecology and the Environment* vol 74 (UK: WIT Press) p 10
[23] Bogoslovski V 1985 *Air conditioning and refrigeration* (Moscow: Stroiizdat)
[24] Ertman S, Ertman J and Zakharov D 2016 Raising of operating a motor vehicle effects on environment in winter *Innovative Technologies in Engineering Rev.* **142** 12119

IOP Conf. Series: Earth and Environmental Science **115** (2018) 012012 doi:10.1088/1755-1315/115/1/012012

Reducing environmental damage through the use of unmanned aerial vehicles as the best available technology

E A Fedulova, A O Akulov[a], A O Rada, T A Alabina and Ju Ju Savina

Kemerovo state university, Krasnaya st. 6, Kemerovo, Russian Federation, 650043

E-mail: [a] akuanatolij@yandex.ru

Abstract. The article examines the possibilities of using unmanned aerial vehicles as the best available technologies in the field of agriculture and mining. The object of the study is the use of unmanned aerial vehicles as the best available technology. The main areas of application of this technology are identified: agro technical operations, aerial photography of mining operations. The technology of unmanned aerial vehicles is compared with the technologies of ground agricultural machinery. The research methodology includes an expert evaluation of the unmanned aerial vehicle technology belonging to the class of the best available technologies by the criteria: the level of environmental impact, resource saving, the use of low-waste, non-waste processes, the existence of at least two objects, economic efficiency. Expert evaluations were processed using the apparatus of fuzzy sets, which make it possible to construct membership functions. This allowed us to prove that the technology of unmanned aerial vehicles belongs to a fuzzy set of the best available technologies. The results of the research show that the use of unmanned aerial vehicles provides a saving of resources, especially non-renewable combustible minerals, reduces emissions and discharges of pollutants into the atmosphere, and also reduces soil erosion. Unmanned aerial vehicles should be included in the national directories of the best available technologies for the mining industry and agriculture.

1. Introduction

In industrially developed countries, resource-type regions, to date, significant environmental damage has been accumulated, the ecological situation is tense. This strengthens the scientific and practical interest in environmentally safe and low-waste technologies. These technologies allow obtaining the same economic, production results with a sharp decrease in the load on the environment. To create environmentally friendly technologies there are scientific and technical and organizational and economic prerequisites. Scientific prerequisites are associated with the creation and development of fundamental principles for obtaining new materials, technical processes, physical-chemical and mechanical basis for the transformation of matter and energy. Organizational prerequisites are associated with increased interest in reducing environmental damage under the influence of state regulation, legislative requirements, and social responsibility. For the characterization of environmentally safe or low-waste technologies, the notion of "best available technologies" became widespread [1, 2, 3].

Under the best available technology, it is customary to understand the technology of production, determined on the basis of modern scientific achievements and the best combination of environmental protection criteria, provided that the technical feasibility of its application is available [4]. The best available technology is an economically efficient technology with minimal environmental damage. When implementing the best available technologies, the joint responsibility of producers and consumers is taken into account in the course of the production, sale and use of products [5]. The introduction of the best available technologies is the main direction to improve environmental safety, to achieve the effect of decoupling. The best available technologies are the basis for progressive environmental regulation, which ensures minimization of environmental damage. Legislation on environmental protection of the world's leading countries, including Russia, seeks to stimulate enterprises with maximum environmental damage to the transition to the best available technologies. These investment projects can be considered as projects of dual purpose or double effect (economic and environmental). However, this process has a long-term perspective.

The criteria that allow attributing technology to the best available technologies are: minimum level of negative environmental impact on the environment on a unit of production (level of emissions, discharges, disturbance of natural landscapes, etc.); resource saving, saving of non-renewable natural resources, in particular, fuel and mineral raw materials; use of low-waste or non-waste processes; existence of at least two objects on which technology can be applied; economic efficiency of use. Rapid development of research and development leads to the active emergence and use of new technologies that do not always have time to be evaluated according to the criteria of the best available technologies. Because of this, they do not fall into the basic directories and classifiers of the best available technologies that regulators use. As a result, a number of advanced technical solutions are not integrated into the national system of the best available technologies. This forms an urgent scientific and applied problem.

One of the progressive technologies that contribute to reducing environmental damage is the use of unmanned aerial vehicles [6, 7]. The very principle of unmanned aerial vehicles is not new (the lethal vehicles without the presence of a human operator on board were used in the 1930s), but their active introduction into civilian industries began in the 2000s. The main areas of application of unmanned aerial vehicles are various kinds of survey of surfaces, inventory of objects, carrying out of territorial measurements, etc. The application of this technology is developing in agriculture, mine surveying, logistics, as well as insurance and entertainment.

But the main advantages of the technology of unmanned aerial vehicles in the literature are higher economic efficiency, high availability, the possibility of obtaining large amounts of data, cheaper formation of information bases per unit of information [6, 7, 8]. Indeed, in comparison with classical (manned) aircraft, unmanned aerial vehicles are not only much cheaper, but also satisfy a wider range of needs, give new opportunities to producers and consumers [9, 10]. However, unmanned small aircraft has many features of the best available technology in terms of resource intensity, reducing environmental damage, economic efficiency. At the same time, the theoretical and applied aspects of the use of unmanned aerial vehicles as the best available technologies in various types of economic activity have not been studied to date. Therefore, the purpose of this study is to justify the rationality of using unmanned aerial vehicles as the best available technology for forming the ecological foundations of progressive technological schemes. The area of research is the scope of unmanned aerial vehicles in particular, agriculture and mining.

2. Results and Discussion

To develop classifications and directions for using the best available technologies, the main method is the expert method. This is due to the complexity, poor structuring of information on environmental damage and the economic efficiency of technology. The method of expert assessments allows us to formalize the verbal statements of experts in natural language. At present, special working groups are used to compile directories of the best available technologies, which in fact mean an expert method. To assess the technologies of unmanned aerial vehicles for compliance with the criteria of the best available technology by the "snowball" method, an expert group was formed from among specialists in agriculture, the coal industry, fundamental science and public administration. The required number of experts is calculated by the formula:

$$N = \frac{1}{2}\left(\frac{3}{\alpha} + 5\right) \tag{1}$$

where α is an admissible error, N is the number of the expert group.

At $\alpha = 0.05$, the number of experts N will be 33 people. A group with such a number was formed. To process the data obtained, we used the method of constructing the membership functions associated with the theory of fuzzy sets. This theory is based on a multivalued logic with an arbitrary number of truth values. It is known that the usual (clear) subset A of the set X can be associated with the membership function (2):

$$\chi_A(x)=\begin{cases}1, if\ x\in A\\ 0, in\ otherwise\end{cases} \tag{2}$$

In the theory of fuzzy sets, the possibility of working with judgments in natural language, which does not have clear formalized characteristics, is considered. A fuzzy set is understood to mean the set (3):

$$A=\{x,\mu_A(x)|x\in X\} \tag{3}$$

where X is a universal set; $\mu A\ (x)$ is the membership function characterizing the degree of belonging of the element x to the fuzzy set A.

Therefore, a fuzzy set is a collection of ordered pairs, including elements x of the universal set X and corresponding membership functions $\mu A\ (x)$. The membership functions are mathematical functions that can be piecewise linear, Z- and S-shaped, U-shaped. In the study, the S-shaped membership function (spline function), which is given by expression (4), is used to process expert estimates.

$$f_{s1}(x;a,b)=\begin{cases}0, & x<a\\ \frac{1}{2}+\frac{1}{2}cos\left(\frac{x-b}{b-a}\pi\right), & a\leq x\leq b\\ 1, & x>b\end{cases} \tag{4}$$

The criteria for assigning technology to the best available class, according to the order of the Ministry of Industry and Trade of Russia of March 31, 2015, No. 665 «On approval of methodological recommendations for the definition of technology as BAT», are: the level of environmental impact, resource saving, use of low-waste, non-waste processes, the existence of at least two objects, economic efficiency. The last criterion also reflects the productivity of the technology, taking into account the ratio of the production result and the costs of the enterprise.

By processing expert assessments of the five criteria for classifying technologies as the best available technologies, the following data were obtained (Table 1, 2, Figure 1).

Table 1. Data of the expert assessment of the compliance of unmanned aerial vehicle technology with the criteria of the best available technology (number of expert answers in the interval)

Evaluation Criteria	Number of responses in the interval				
	does not meet the criteria	weakly matched	average criterion	meets the criterion	fully meets the criterion
Level of environmental impact	2	0	7	19	5
Ensuring resource conservation	1	2	5	21	4
The use of low-waste, non-waste processes	0	4	3	20	6
The existence of at least two objects	3	2	0	23	5
Economic efficiency	2	2	2	18	9

Table 2. Accumulated response rate

Evaluation Criteria	The accumulated response frequency in the interval				
	does not meet the criteria	weakly matched	average criterion	meets the criterion	fully meets the criterion
Level of environmental impact	0.06	0.06	0.27	0.85	1.00
Ensuring resource conservation	0.03	0.09	0.24	0.88	1.00
The use of low-waste, non-waste processes	0.09	0.15	0.15	0.85	1.00
The existence of at least two objects	0.09	0.15	0.15	0.85	1.00
Economic efficiency	0.06	0.12	0.18	0.73	1.00

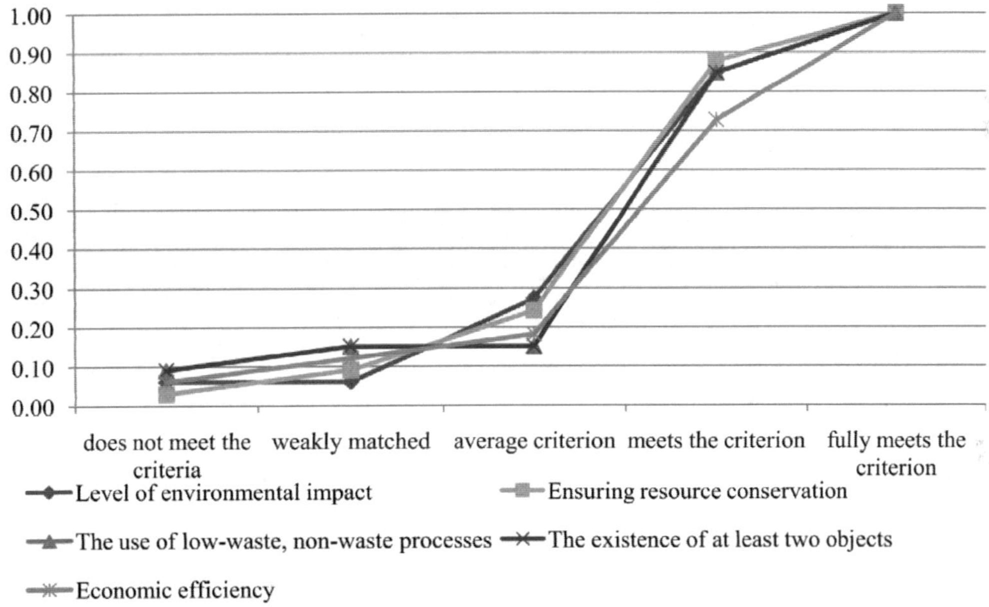

Figure 1. The functions of unmanned aerial vehicle technology belonging to the set of the best available technologies by all criteria

The experimental data on the expert assessment of the unmanned aerial vehicle technology belonging to the best available technologies show their compliance by all criteria. The form of the membership functions is such that the area of the right subspace considerably exceeds the area of the left subspace. On most criteria, the type of membership graphs is very similar, since most peer evaluations are assigned to intervals 4 and 5 (compliance and full compliance with the criteria of the best available technologies). And the most significant is the membership function according to the criterion of "economic efficiency". Here, the cumulative response rate in the intervals 1-3 is only 0.18; 82% of assessments refer to the relevance of the technology to this criterion and its full compliance. Therefore, the assessment shows that unmanned aerial vehicles are the best available technologies for a number of economic activities, in particular, agriculture and mining.

Consider the basic spheres of application of the technology of unmanned aerial vehicles as the best available technology. To this end, we will focus on those economic activities where this technology

IOP Conf. Series: Earth and Environmental Science **115** (2018) 012012 doi:10.1088/1755-1315/115/1/012012

provides a reduction in environmental damage - agriculture and mining. These fields of application retain their positions in the economy, but require the solution of environmental problems.

To select the best available technology, the following technologies were compared:

1. Technologies of agro technical operations (soil cultivation, planting of crops, fertilizers, chemicals) using ground and air equipment, respectively. The result of the application of technology is the implementation of agro technical operations required in crop production.

2. Technologies of aerial photography of mining operations using ground and air equipment, respectively. The result of applying the technology is obtaining information about the state of the "mountain environment - technology" system, which is necessary for making engineering decisions.

Let us dwell on the potential of the corresponding technologies in more detail.

Unmanned aerial vehicles in agriculture are used to monitor the growth and development of plants, the state of agricultural landings (instead of satellite imagery); carrying out agro technical measures (sowing, sprinkling, application of nutrients) [11, 12]. Technologies of the first direction practically do not involve ecological damage and are not considered in this research. Replacing unmanned aircraft with traditional agricultural equipment (tractors, cultivators, etc.), by contrast, deserves close attention as the best available technology. The principle of operation of an unmanned aerial vehicle in this case is that instead of ground equipment with an internal combustion engine, air technology, usually with an electric motor, is used (although some large unmanned aerial vehicles can also be equipped with an internal combustion engine).

The ecological effect will be caused by a number of factors. First, a smaller mass and lack of environmental resistance for an unmanned aerial vehicle lead to less need for kinetic energy for movement. This helps to save energy in principle. Further, even less powerful electric motor in comparison with the internal combustion engine practically does not produce harmful emissions. More precisely, the production of electrical energy, as a rule, is accompanied by environmental damage, but it is much lower in comparison with an autonomous internal combustion engine. Replacing fuel cycle engines (petrol, diesel) with electric motors is a worldwide trend, reflecting the development of environmentally friendly technologies. Although this practice seems to move emissions and discharges of harmful substances into the early stages of the technological chain (generation of electrical energy at thermal, nuclear, hydraulic stations), the total ecological damage is reduced for a set of economies of scale.

Secondly, moving through the air instead of moving along the ground ensures that there is no pressure on the fertile soil layer, which also sharply reduces environmental damage. Soil compaction by ground agricultural machinery is one of the main causes of soil erosion, dispersion and loss of the fertile layer. And even the use of progressive no-till technology of cultivation of the land with the exception of the overthrow of the formation does not completely exclude the compaction of soil by the movers of agricultural machinery. The transition from the movement of agricultural units on the ground to travel by air is a fundamental breakthrough solution to the problem.

Thirdly, the qualitative growth of information on the development of plants can significantly reduce the introduction in the agro-industrial landscape of unusual, in a certain sense, foreign substances – fertilizers, especially pesticides. Traditional agro technical methods of fertilizing soils, destroying pests, fighting diseases are based on unified methods of introducing certain chemicals with minimal feedback. Survey of crops with the help of unmanned aerial vehicles will allow to accurately determining the need for nutrients, the degree of infection with pests in various areas of agricultural land. As a result, only a minimum amount of new substances is introduced into the agrarian landscape.

At the same time, with the use of unmanned aerial vehicles, a new environmental problem is emerging – the recycling and recycling of batteries, which are usually used as an energy carrier for the device. However, modern hybrid batteries with the possibility of multiple recharging have less weight and more efficiency, so the amount of utilization will most likely not exceed the need for recycling of standard low-antimony and calcium batteries (they are usually used in agricultural machinery). In addition, in agricultural machinery (tractor), diesel engines are most often used, which require the use of batteries with larger capacity and starting current. Therefore, the mass of such batteries is much higher. If, on the other hand, the unmanned aircraft motor is powered by solar panels, it is possible to obtain a significant lengthening of the battery life in the absence of self-discharge.

Consequently, the use of unmanned aerial vehicles in agriculture fully meets the criteria for the best available technology: the absence or minimum level of waste, the use of less hazardous substances, the reduction or elimination of harmful emissions and discharges. Obtaining accurate estimates of changes in the ecological effect and environmental load will additionally justify the environmental effectiveness of the further introduction of unmanned aerial vehicles in agriculture. Due to the lower cost of one hour of operation, a higher economic efficiency of the technological operations is achieved.

In the mining industry (extraction of ore, nonmetallic minerals), unmanned aerial vehicles are used for geological exploration (mapping, scanning, reconnaissance, other functions, for example, sampling); operational planning of mountain preparatory and mining operations (design, monitoring and mapping, surveying measurements); monitoring of mining and geological conditions (behavior of the mountain environment and technological scheme); identification of risks and environmental threats (violations of technology, the environment, the presence of emissions). There is a practice of using unmanned aerial vehicles for direct carrying out individual mining operations (destruction of space, collapse of rocks, clearing of passages) [13, 14]. In this type of economic activity, unmanned aerial vehicles rather automate those works that were previously performed by specialists manually. Therefore, the potential for the formation of the best available technologies is estimated here below. Nevertheless, in some cases, unmanned aerial vehicles can also contribute to reducing the environmental load.

The first direction of reducing environmental damage is the exclusion of aerial photography and observations using classical aviation (helicopters). It is known that manned aviation assumes a significant consumption of combustive-lubricating materials (kerosene, gasoline), obtained from non-renewable raw materials. Their burning leads to emissions of harmful substances into the atmosphere. For a 30-minute runway, the engine of the most popular MI-8 helicopter emits 340 grams of carbon monoxide and up to 1,000 grams of nitrogen oxides (nitrogen oxide, nitrogen dioxide). When using unmanned aerial vehicles, these emissions are excluded. Environmental damage is reduced to the problem of utilizing fully used batteries.

The second direction is the reduction of environmental damage – control of spontaneous combustion of waste rock piles, stored masses of coal and other combustible minerals. Traditional methods of control are characterized by inadequate speed and precision. Spontaneous combustion of coal, slate, carbonaceous rocks, sulphide ores, peat leads to significant emissions of nitrogen oxides, hydrogen sulfide, methane, etc., which causes great harm to the health of the population. Due to demolition of precipitation products by atmospheric precipitation, additional contamination of soil and water occurs, large deformations of the land cover. Accelerating the monitoring of clusters of substances prone to spontaneous combustion and increasing the accuracy of this monitoring ensures faster measures to extinguish endogenous fires and reduce emissions. Thus, for the mining industry, the use of unmanned aerial vehicles contributes to resource saving and to reducing the negative impact on the environment. Consequently, this technology at the industry level meets the criteria of the best available technology. Unmanned aerial vehicles can be included in the lists of the best available technologies for relevant economic activities.

3. Conclusion

The development of the best available technologies in the country's economy requires the analysis, evaluation and integration of new prospective objects, processes and devices into national classifiers. The conducted evaluation of the technology of unmanned aerial vehicles proves their compliance with the criteria of reference to the best available technologies. The use of unmanned aerial vehicles ensures the saving of resources, especially non-renewable combustible minerals, reduces emissions and discharges of pollutants into the atmosphere, and also reduces soil erosion due to compaction, and contributes to the timely extinguishing of foci of spontaneous combustion. Consequently, unmanned aerial vehicles can be included in national guides for the best available technologies for mining and agriculture and receive appropriate support. The use of unmanned aircraft technology in these areas provides a significant reduction in environmental damage.

Further development of the technology of unmanned aerial vehicles, according to the authors, requires comprehensive technical, ecological, economic research to determine technological schemes throughout the production cycle. The basis of these surveys should be the principle of minimal environmental damage throughout the entire technological cycle (rather than local sub-optimization in a separate link). This principle is well known in production management, logistics, supply chains; it is focused on the search for a common optimum. Extending the principle of minimum costs (damage) to the analysis of the best available technologies will allow identifying those technologies that reduce overall damage, analyze emissions and discharges at all stages of the process chain.

In particular, considering the damage from various options for the use of machinery in agriculture, it is necessary to take into account the costs for the production of all types of energy carriers (fuels and lubricants, electric power, batteries, etc.), their transportation, periodicity of use, utilization, e. Otherwise, the problem of finding local optima in certain types of economic activity arises with the growth of cumulative environmental damage. Undoubtedly, the traditional problems of development of unmanned aircraft, in particular, legal, provision of physical and information security, organization of insurance protection require further research.

References

[1] Mazlova E A and Eremina N V 2016 Environmental regulations based on best available technologies in oil and gas industry *Oil Industry Journal* **5** 98

[2] Milone D, Peria G, Pitruzzella S and Rizzoa G 2015 Are the Best Available Technologies the only viable for energy interventions in historical buildings? *Energy and Buildings* **95** 39–46

[3] Loyon L, Burton C H and Misselbrook T 2016 Best available technology for European livestock farms: Availability, effectiveness and uptake *J. of Environ Manage* **166** 1–11

[4] Martus E 2017 Contested policymaking in Russia: industry, environment, and the "best available technology" debate *Post-Soviet Affairs* **33** 276–97

[5] Berzosa Á, Barandica J M and Fernández-Sánchez G 2014 A new proposal for greenhouse gas emissions responsibility allocation: Best available technologies approach *Integrated Environmental Assessment and Management* **10** 95–101

[6] Carrio A, Sampedro C, Rodriguez-Ramos A and Campoy P 2017 A Review of Deep Learning Methods and Applications for Unmanned Aerial Vehicles *J. of Sensors* **1** 1–13

[7] Anderson K and Gaston K J 2013 Lightweight unmanned aerial vehicles will revolutionize spatial ecology *Frontiers in Ecology and the Environment* **11** 138–46

[8] Chabot D, Dillon C, Ahmed O and Shemrock A 2017 Object-based analysis of UAS imagery to map emergent and submerged invasive aquatic vegetation: a case study *J. of Unmanned Vehicle Systems* **5** 27–33

[9] Gallacher D 2016 Drones to manage the urban environment: Risks, rewards, alternatives *J. of Unmanned Vehicle Systems* **4** 115–24

[10] Kirk B 2016 Business opportunities in automated vehicles *J. of Unmanned Vehicle Systems* **4** 4–6

[11] Xiang H and Tian L 2011. Development of a low-cost agricultural remote sensing system based on an autonomous unmanned aerial vehicle (UAV) *Biosystems engineering* **108** 174–90

[12] Torres-Sánchez J, López-Granados F and Peña J M 2015 An automatic object-based method for optimal thresholding in UAV images: Application for vegetation detection in herbaceous crops *Computers and Electronics in Agriculture* **114** 43–52

[13] Lee S and Yosoon C 2016 Reviews of unmanned aerial vehicle (drone) technology trends and its applications in the mining industry *Geosystem Engineering* **19** 197–204

[14] Jangwon S and Yosoon C 2017 Mapping hazardous mining-induced sinkhole subsidence using unmanned aerial vehicle (drone) photogrammetry *Environmental Earth Sciences* **76** 144

Investigation of the Temperature Sensors Accuracy in the Temperature Monitoring System for the Welded Joints of the Industrial Power Supply Main Busways

N V Grivennaya, A V Bazhenov, G A Bondareva, S V Malygin, A A Knyaginin

Technological Institute of Service (branch) Don State Technical University, Stavropol, Russian Federation

E-mail: katrinastenton@yandex.ru

Abstract. The article is devoted to the substantiation of the technical solution of the remote monitoring system for the temperature changes of main and branch busways of power supply to industrial enterprises of increased environmental danger. When monitoring the temperature of trunk buses of AC mains up to 1000 V, heated by an electric current, errors occur due to various factors. Studies have been carried out to evaluate the effect of temperature of surrounding objects (including neighboring busbars) on the accuracy of temperature measurements. Conclusions are made about the possibility of using alternative versions of temperature sensors as the basis of the monitoring system.

1. Introduction

In modern electrical installations of industrial enterprises, electric power networks and, in particular, AC and DC mains up to 1000V made by busways, gradually replace other types of electrical networks made by cables or uninsulated bars laid on insulators. In many cases, the transition from radial power circuits to busbar trunk systems has led to a wide application of sectional busways. The use of bus trunking allows increasing the service life of electrical networks and significantly reducing their breakdown rate [1].

Damage to the mainline leads to simultaneous disconnection of all the power collectors, feeding on it. This is particularly undesirable when many separate large consumers that are involved in a continuous technological process feed on the main line. This also applies to the enterprises classified as environmentally hazardous.

The peculiarity of the main busways installation technology is the connection of busbar sections of different lengths. So, when installing a 100 m long network, the number of connections reaches 30-40 pcs. The connection of the busbar sections is carried out by the argon-arc welding method. The technology of joints of busway sections allows isolation of the welded joint with insulation varnished cloth. The location of the welded joint is most susceptible to thermal heating, especially in case of loose connection between the busway sections. Besides, there is often need to violate the manufacturer's recommendations (for technological reasons) and welding is carried out by a single longitudinal seam. It leads to a decrease in the quality of the ohmic contact (increase in transient resistance) between the connected sections of the busway [2].

To prevent violations of technological processes and operating modes as a result of possible accidents on the electrical busbars of enterprises, it is necessary to carry out continuous (in real-time mode) or periodic monitoring of temperature changes, processing and storing (documenting) the results and predicting the changes in the busway condition. The proposed monitoring system can be constructed similarly to those presented in [3,4].

2. Results and Discussion

The remote monitoring system of temperature changes of trunk and branch busways must ensure the fulfillment of the following requirements and characteristics:

- accuracy of temperature measurement for each welded joint ± 1.0° C;

- operating range of the busbar trunking current is not more than 4000 A;
- compensation of cross impact of the neighboring buses temperature sensors;
- the maximum permissible temperature for the welded joint is 90° C.

To develop recommendations for the implementation of a system for remote monitoring of temperature changes in trunk and branch busways, it is necessary:

- to study the dependence of the temperature produced by the sensors on the current, flowing through the busbar trunking;
- to perform research to assess the influence of the temperature of surrounding objects (including neighboring busbars) on the accuracy of measurements;
- to study the effect of insulation varnished cloth on the accuracy of measuring the temperature of the busway.

To theoretically substantiate the processes of heating and cooling of a homogeneous conductor, the mathematical modeling of the dependence of the bus heating temperature on the current strength under conditions of heat transfer has been performed.

$$T = \frac{I^2 \cdot R_a}{G \cdot c} t \quad (1)$$

where T – overheating (exceeding the temperature of the conductor in relation to the environment), C; I – the effective value of the current passing through the conductor, A; R_a – resistance of the conductor at alternating current, Ohm; G – weight of current-carrying conductor, kg; c – the specific heat capacity of the conductor material, $J \cdot s/kg \cdot \deg$ [5,6].

From the preliminary analysis of the composition and characteristics of the monitoring system, it is reasonable to use alternative versions of temperature sensors, namely the non-contact infrared sensors of the MLX90614 family and DS18S2 universal digital sensors [7, 8]. This is necessary to ensure the contact temperature control of the welded joint spot and subsequent comparative analysis of contact and non-contact monitoring results.

As a source of current, a load measuring kit with a current regulator was used. The kit provides current intensity regulation, as well as setting the preset duration of the current flow and measuring the tripping time of the release with the display of information on the graphical LCD. The kits are used to test AC circuit breakers in mains up to 1000V with an industrial frequency of 50 Hz.

The range of regulation and measurement of the test current, A: from 20 to 2000. The resulted error of measurement of current strength, %, is no more than 5. The range of setting and measuring the duration of the current flow in short-time mode up to 99 s. To cool the source of current, a heat gun fan was used.

For the experiment, we used a two-package bus track SHMA-4 4000A (Figure 1).

The connection of the busway sections has been made through the bolt tightening method. Three infrared sensors MLX90614 have been placed above the connection points of three buses, belonging to one package. The IR sensors have been fixed on a board in one common housing. The data obtained in the course of the experiment on the current strength I and the temperature measured by the contact sensor on one bus bar of the busway, confirm the results calculated using the model (1). The average error value is 0.210°C (Figure 2). For aluminum busways SHMA-4 4000A, the time constant is 17.5 minutes.

When location of the infrared sensors above the bus connection points, in the chart area receiving the thermal radiation sensor fall into a surrounding bus For the sensors used, the radiation reception diagram is a 90^0–opening cone. The current was supplied to only one bus. The remaining two bus bars of the package were heated only by heat transfer inside the box. The DS18B20 universal sensor was attached directly to the surface of the bus to which the test current was applied. The data have been collected using an Asus personal computer. The I^2C-bus controller for the MLX90614 sensors' scanning and the 1-Ware-bus controller for the DS18B20 sensor scanning are implemented on Arduino Uno boards. The software provides accurate temperature measurement of the sensor DS18B20 0.1° C and sensors MLX90614 1°C.

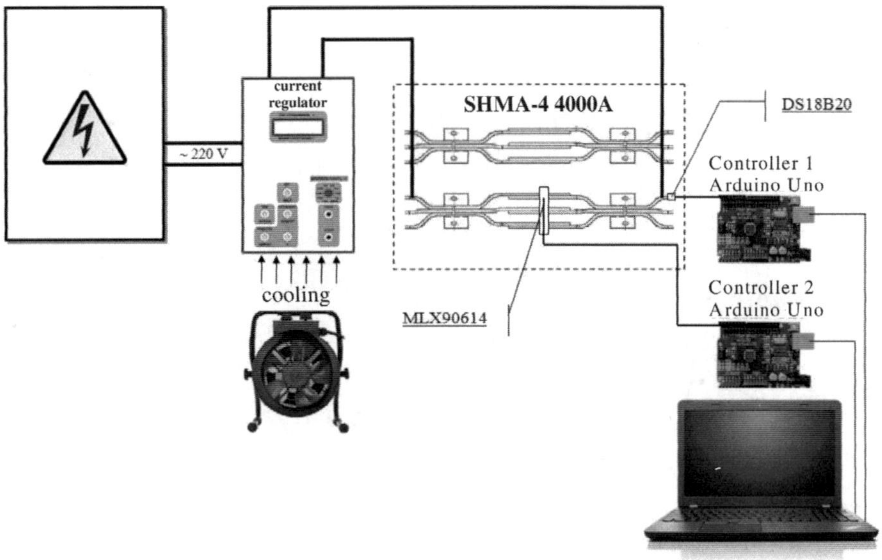

Figure 1. Scheme of the experiment on the influence of the current intensity on the busway temperature

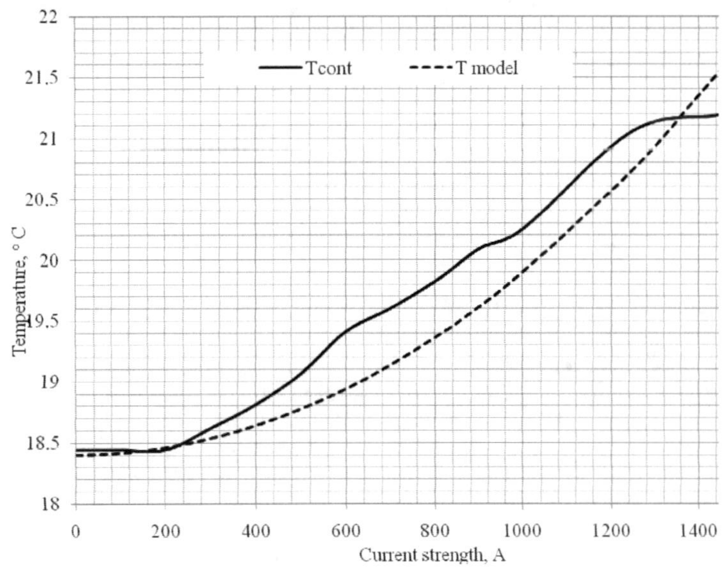

Figure 2. Heating curves of the busbar trunking system (experimental data calculated using the model (1)).

IOP Conf. Series: Earth and Environmental Science **115** (2018) 012013 doi:10.1088/1755-1315/115/1/012013

In the course of the experiment, it was found that the readings of the contact sensor and the infrared noncontact sensor are distinguished by a constant displacement value which is proportional to the area of the bus bar, getting into the infrared sensor directional pattern (Figure 3).

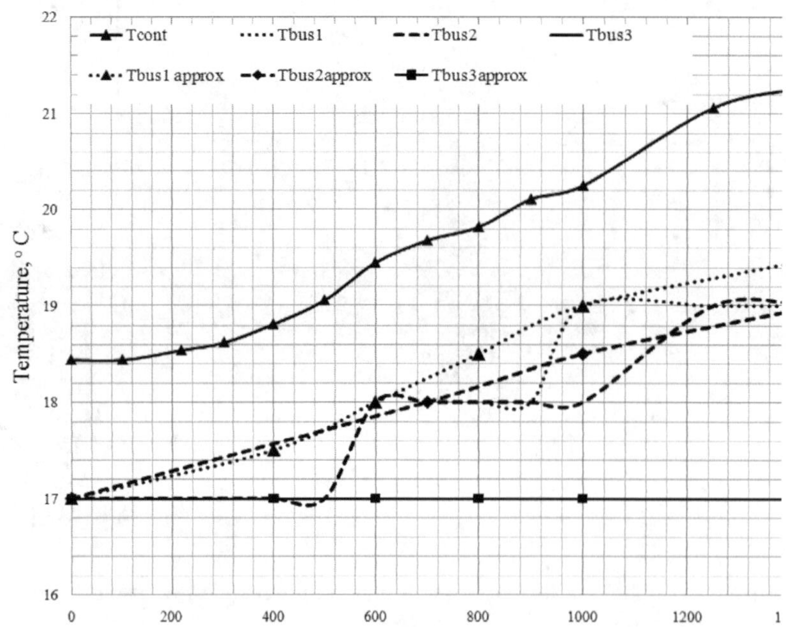

Figure 3. The graph of the dependence of the heating bus bars 1-3 (Tbus1, Tbus2, Tbus3) on the current (I) flowing through the busway 1.

To reduce the influence of radiation from foreign objects (including neighboring busbar trunking systems) falling into the radiation reception diagram, on the accuracy of temperature measurement during the research, one of the infrared sensors was placed in a metal tube. By heating the metal tube, the sensor showed the significantly higher temperature at first, since it took the total radiation of a closely located tube and a heated bus bar. Since a thin-walled tube was used, its cooling rate exceeded significantly the cooling rate of the bus, so after a certain time, the sensor readings in the tube became less than the readings of the open sensors. This is due to the fact that a narrow sector in the sensor's radiation diagram was cut out by the tube. It reduced the intensity of the received radiation and increased the error in the temperature measuring.

To reduce the influence of radiation from foreign objects was held experiment with using the polypropylene plates as limiters for infrared sensors' directional pattern. Using of plastic stoppers makes it possible to avoid the influence of spurious emission. The results are shown in Figure 4.

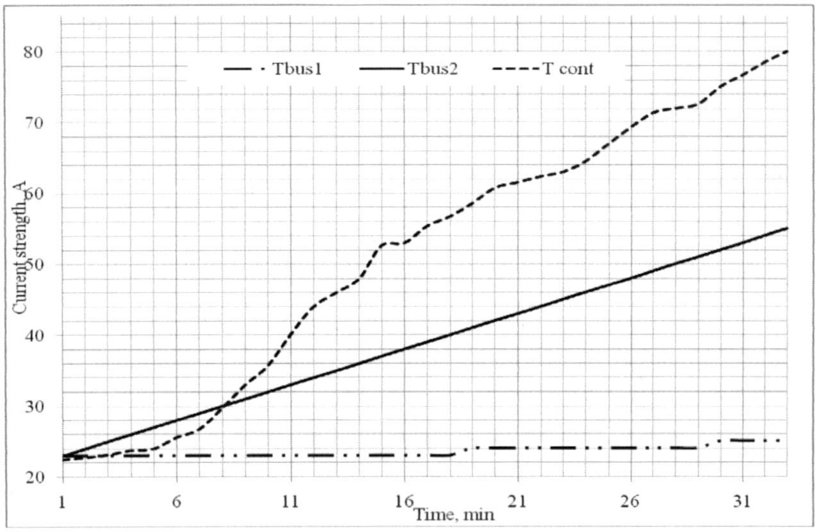

Figure 4. Influence of the heat source limiters on the MLX90614 sensor reading on the busbar 1.

In addition to the errors caused by the mutual influence of the conductors and the width of the directional pattern of the non-contact sensors MLX90614, the error results from the electrical insulation of the bus bars [9]. The electrical insulation of the welded joint of busways is made of varnished cloth. In addition to insulating properties, the varnished cloth has pronounced thermal insulation properties. While using the varnished cloth on the surface of the radiation source, a "greenhouse" effect is produced, caused by a decrease in thermal convection with the surrounding air. With the increased number of varnished cloth layers, the temperature measured by the contact sensor DS18B20 grows almost linearly. The results of the measurements are shown in Figure. 5.

The temperature monitoring of the busway during all the conducted studies was carried out with the help of the controller's Automated Workstation program (AWS program). In the program, data come from digital thermometers (contact sensor DS18B20, infrared sensors MLX90614) through the Arduino Uno platform, built on the ATmega328 microcontroller. In the AWS program, a mnemonic scheme is developed that represents the graphic scheme of a remote object and is intended to display its state at a certain point in time (Figure 6). The mnemonic diagram shows the busway where the current temperature is displayed at the junction of the sections.

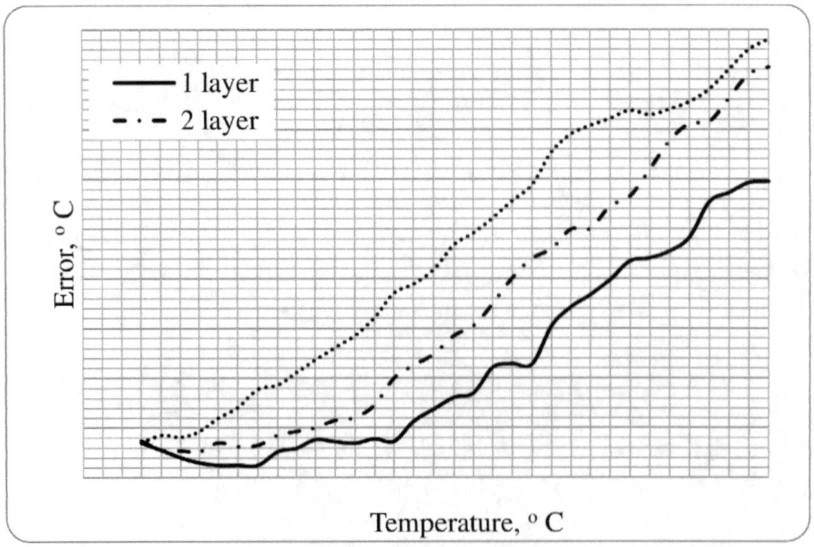

Figure 5. Error in measuring the temperature using the varnished cloth.

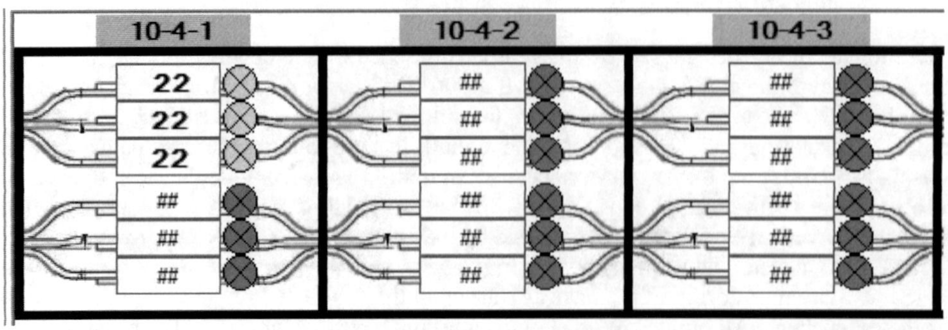

Figure 6. Fragment of the mnemonic diagram "Controller's automated workstation".

Thus, all the planned studies were carried out that allow making a conclusion about the advisability of using the MLX90614 infrared sensors and the DS18S2 universal sensors in the temperature monitoring system for the busways of SHMA-4 4000A type. All the results of the experiments were statistically processed [10].

Conclusions

Analyzing the results of the research, we can draw the following conclusions:

- non-contact infrared sensors of the MLX90614 family of the Microelectronic Integrated System (Melexis) have sufficient accuracy and operating range to control the temperature of the busways;
- to reduce the mutual influence of single buses on the measurement accuracy when using sensors, similar to the MLX90614 and with a 90-degree beamwidth, flat-plate restraints should be used (when using more expensive sensors with a 5-degree beamwidth, you do not need such limiters);
- the number of MLX90614-type sensors per one serial information bus should not exceed 127;
- when using universal DS18S20 sensors from Dallas Semiconductor, the software calibration of the results of measurements is required. The calibration must take into

account the contact density between the sensor and bus bar and the number of the varnished cloth layers. The advantage of the DS18S20 sensors is their low price (by an order less than MLX90614);

- the number of sensors used simultaneously on the same data bus of DS18S20 sensors is not limited.

References

[1] Van Hertem D and Delimar M 6 2013 High Voltage Direct Current electric power transmission systems *In Woodhead Publishing Series.* pp 143-173

[2] Bayliss C and Hardy B 26 2012 High Voltage Direct Current Transmission *In Transmission and Distribution Electrical Engineering (Fourth Edition)*, (Oxford, Newnes). pp 1027-1057

[3] Rongjun X, Shuping J and Pai X 2017 Long-term temperature monitoring of tunnel in high-cold and high-altitude area using distributed temperature monitoring system *In Measurement.* **V95** pp 456-464

[4] Bazhenov A, Grivennaya N, Malygin S, Bondareva G et al. 2017 Main Pipelines Corrosion Monitoring Device, IOP Conference Series-Earth and Environmental Science. **V50** 012032

[5] 2016 Protections for Electronic Systems, *In Power Electronics Applied to Industrial Systems and Transports* ed by Nicolas Patin (Elsevier). pp 75-127

[6] Murty P.S.R. 2017 11 Short Circuit Analysis, *In Power Systems Analysis (Second Edition)*, (Boston, Butterworth-Heinemann). pp 277-312

[7] Nenarokomov A, Alifanov O, Budnik S and Netelev A 2016 Research and development of heat flux sensor for ablative thermal protection of spacecrafts *In International Journal of Heat and Mass Transfer.* **V97** pp 990-1000

[8] Galar D and Kumar U 2017 1 - Sensors and Data Acquisition In eMaintenance (Academic Press). pp 1-72

[9] Bazhenov A, Grivennaya N, Malygin S, Knyaginin A and Kolchintsev Y 2017 Evaluation of the accuracy of the temperature measurement heated by electric current conductor contact and non-contact sensors (Stavropol, Stavrolit). pp 134-138

[10] Sheldon M. 2017 Introductory Statistics (Fourth edition) (Oxford, Academic Press). pp 381–432

The differentiated assessment of damage to economy of subjects of the Siberian Federal District from road and transport accident rate

Artur I Petrov[1, a] and Vera A Svistunova[1] and Daria A Petrova[2]

[1] Tyumen Industrial University, 72 Melnikaite str., Tyumen, 625000, Russian Federation
[2] Ural Federal University named after the first President of Russia B. N. Yeltsin, 16a Chapaev str., Ekaterinburg, 620002, Russian Federation

E-mail: [a] ArtIgPetrov@yandex.ru

Abstract. The results of assessment of damage from the road accident rate in subjects of the Siberian Federal District (SFD) of the Russian Federation are presented in the article. The thesis about spatial differentiation of the Gross Regional Product (GRP) losses in different regions of the country because of people's death and injuries in the road accidents (RA) and due to formations of property and ecological damage was chosen as a working hypothesis. The calculations, carried out for 12 subjects of the SFD, confirmed this idea. The range of calculated values of economic damage from road accident rate (in % of GRP) was from 1.3 (Tomsk region) to 12.6 (Republic of Tyva) in 2015. In article the attempt to explain the received result by heterogeneous development of economics in various Russian regions is made. The consequence of it is a heterogeneous quality of people's life and quite various perception of life value by inhabitants of different regions that influences their life safety level.

1. Introduction.

According to the Federal State Statistics Service of the Russian Federation (hereinafter - Rosstat) [1] in the first half of the year 2017 from the external reasons in our country have left life of 69,3 thousand people, including 8,2 thousand people have died in road accident. The calculation shows that according to 2017 about 12 % of all premature deaths in our country is the proportion of people killed in an accident.

The struggle to reduce this indicator has long been among the priorities of the Government of Russia. This is evidenced by the implementation of the First (in the period 2006 ... 2011) and the Second (2012 ... 2020) Federal Target Programs for Road Safety.

The results of this work are unambiguously positive in the whole country, however, in different regions and cities the dynamics of the decrease in indicators of road and transport accident rate (absolute and relative) are not the same [2]. This is evidenced by the data of other authors [3, 4]. If countrywide the share of deaths from all types of transport accidents among those who died from external causes in 2014 was about 20 %, then in the Leningrad, Novgorod, Kaluga regions and the Republic of Karachaevo-Cherkessia this indicator reaches 32...35 % , and in the Republic of Tuva it is equal to 50 % [5].

Within the framework of this article, an attempt is made to assess the damage caused by road and transport accident rate to the economy of the Siberian Federal District (SFD) of Russia. At the same time, the priority task is to make it differentiated in relation to various subjects of the SFD.

2. The studied problem

Today Russia does not have an unambiguous understanding of the cost of human life and the damage caused by the death and injury of a person in a traffic accident (road accident). Those not numerous works [3, 4], which are performed in this direction, only state need of quantitative estimation of cost of life and offer only the general approaches to the solution of this task. In this case, you can use different methodologies (the theory of human capital, the methodology of estimating the average life, methods of insurance companies in solving the problem [3, 4, 6, 7, 8]. It is also important that any of these methodologies uses the notion of "Loss of gross domestic product", and the GDP index, in spite of an unambiguous quantitative estimate by Rosstat, in different accounting systems [7, 8] is also varies greatly.

Taking into account the considerable ambiguity of both the concept of the "Average cost of living" of a person (ACL) and the constant correction of GDP (due to inflation and reassessment of value added), it is proposed to solve the problem of estimating the damage from road and transport accident rate to the economy of the SFD from the perspective of differentiating this indicator in different regions of the District. A certain error in the calculations will be present, but this error will be systemic and practically identical for all regions.

3. Formulation of the problem.

It is necessary to estimate quantitatively damages to the economy of the subjects of the Siberian Federal District (SFD), formed upon the death and wounds of people in road accidents in the respective territories. The corresponding damage will be assessed on the basis of data of the Federal Target Program for traffic safety for the period 2012...2020 and Rosstat's data on the GDP of the Russian Federation [9, 10]. In [9] it is stated that the amount of social and economic damage from road accidents and their consequences for 2004...2011 is estimated at 8188.3 billion rubles. The actual total for 2004...2011 Russia's GDP in current prices is estimated at 281161.5 billion rubles [10]. Thus, the economic losses as a whole from road and transport accident rate in the country make up about 2.9 % of GDP. It is obvious, that in some years this indicator varies in a wide range. So, according to the Federal Target Program for traffic safety for the period 2006...2011 this indicator in Russia at the beginning of the 2000th was 2.2...2.6 % of GDP. In recent years, in 2015...2017 on the fact of a very significant decrease in indicators of the road and transport accident rate in recent years (in 2016, 20.308 thousand people died in an accident, compared to 27.025 thousand people in 2013), it can be considered that GDP losses due to road accidents decreased proportionally, approximately by 25 % and amount to today (2015...2017) about 2.0...2.1 % of GDP.

Problems solved in this article.

1. Calculation of the total damage to the economy of the SFD from the road accident.
2. Differentiated calculation of direct damage (formed by loss and decline in quality of the human capital) and external or indirect damage (formed by material losses in road accidents) to subjects of the SFD in absolute dimension.
3. Estimation of the relative magnitude of gross regional product (GRP) losses in the road accidents.
4. Explanation of the results.

4. Initial data

Let's take the magnitude of the economic damage from the road and transport accident rate is approximately 2 % of GDP in Russia in 2015, which one according to Rosstat [10] was 64997.04 billion rubles in 2011 prices. Conditionally, the damage from an accident in Russia in 2015 can be estimated at 1.3 trillion rubles. According to the worldwide approaches [3, 4, 6], the division of this damage into direct and indirect can be estimated on the basis of the ratio 57 : 43 (Fig. 1), i.e. about 57 % fall on direct damage (or 741 billion rubles) and 43 % fall on external damage (or 559 billion rubles). The immediate damage is the entire amount of loss associated with the death and wound of people in the road accident. Indirect (external) damage includes property and environmental damage.

In 2015, in Russia in the road accident killed 23.114 thousand people and 231.197 thousand people were wounded (with varying degrees of severity of health consequences).

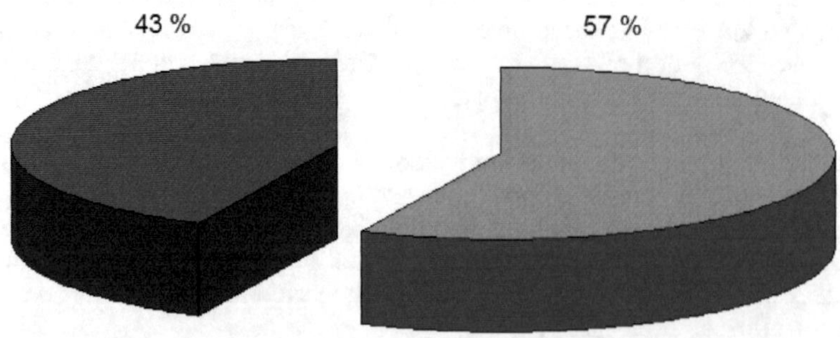

Figure 1. Differentiation of damages from the road accidents in the world practice [3, 4, 6]

5. Method of assessing damages from death and injuries in the accident

To solve the problem of a differentiated calculation of the immediate damage (generated by the loss and degradation of human capital) and external damage to the subjects of the Siberian Federal District it is necessary to quantify the "Average cost of living" (ACL) and "Average cost of injuries" (ACI) parties to the accident. The task of determining values for ACL and ACI will decide on the reverse (1).

$$\begin{aligned}&\text{Tne Number of Fatalities in Road Accidents} \cdot \text{ACL} + \\ &+ \text{The Number of Injured in Road Accinents} \cdot \text{ACI} = \text{Immediate Economic Damage}\end{aligned} \tag{1}$$

$$23114 \cdot \text{ACL} + 231197 \cdot \text{ACI} = 741 \text{ billion rubles}$$

To solve this equation, it is necessary to determine the numerical relationship between the values of the ACL and ACI. In [11], the following data on the relationship between the severity of road accidents in various outcomes for victims (Table 1) are given.

Table 1. Ratio of severity of road accidents in different outcomes for victims [11]

Type of road accident outcome	Values of gravity ratios of outcomes of road accidents	
	F. Reingold	P. Fisher
Material damage	1	1
Injury in an accident	5	2
The fatal outcome of an accident	130	40
Ratio of the severity of road accidents $\frac{\text{Injury}}{\text{Fatal outcome}}$	$\frac{5}{130} = 0.038$	$\frac{2}{40} = 0.050$

In the article [12] the data of 2013 are given, on the basis of which it is possible to determine this ratio as 0.235/11.25 = 0.02.

In [7], the analogous ratio is defined as 0.045.

Let's assume that the ratio of the severity of the outcomes of the road accident $\frac{\text{Injury}}{\text{Fatal outcome}}$ is about 0.04 , i.e. the economic damage from the ACI is about 25 times lower than the damage from the death of a person ACL.

Assuming that the $ACL = 25 \cdot ACI$, we write:

$$23114 \cdot (25 \cdot ACI) + 231197 \cdot ACI = 741 \text{ billion rubles}$$

$$577850 \cdot ACI + 231197 \cdot ACI = 741 \text{ billion rubles}$$

$$809047 \cdot ACI = 741 \text{ billion rubles}$$

$$ACI = 741 \text{ billion rubles} / 809047 = 0.915 \text{ million rubles}$$

$$ACL = (25 \cdot ACI) = 22.875 \text{ million rubles}$$

Let's check the correctness of the reasoning. Estimated damage from deaths and wounds of people in road accidents, taking into account the values of the ACL and ACW in the Russian Federation in 2015 amounted to:

$$23114 \cdot 22.875 \text{ mln. rbl} + 231197 \cdot 0.915 \text{ mln. rbl} = 740.8 \text{ billion rubles}$$

It should be noted that the assessment of the ACL is 22.875 million rub. and the ACI equal to 0.915 million rub., it is very conditional and averaged, because the quality of human capital is highly differentiated depending on age, social status, education, actual and potential contribution to the country's economic success. However, in view of the large-scale consideration of the problem, this reasoning fades into the background and can be withdrawn from consideration. Further calculations for each region of the SFD were carried out according to the following formulas (2) and (3).

$$\text{Absolute Immediate Damage}_{\text{region}} (Rubles) = \\ = (\text{Tne Number of Fatalities in RA} \cdot \text{ACL})_{\text{region}} + (\text{The Number of Injured in RA} \cdot \text{ACI})_{\text{region}} \quad (2)$$

$$\text{Relative Immediate Damage}_{\text{region}} (\% \; GRP) = \\ = 100\,\% \cdot \frac{(\text{The Number of Fatalities in RA} \cdot \text{ACL})_{\text{region}} + (\text{The Number of Injured in RA} \cdot \text{ACI})_{\text{region}}}{\text{GRP}_{\text{region}}} \quad (3)$$

6. Results and discussion

6.1. The differentiated for the SFD subjects calculation of direct (due to death and wounding of people) damage from the road accident

Further calculations are based on taking into account the statistics of deaths and wounds of people in road accidents in various regions of the Siberian Federal District in 2015.

Data on GRP in the regions of the SFD [13] and the calculated magnitude of direct damage are presented in Table 2.

The damage from death and wound to people in the road accident in the SFD in 2015 amounted to $103257.4 \text{ mln. rbl} / 741000 \text{ mln. rubl.} = 13.94 \%$ of the Russian result.

The share of the GRP of the SFD in the total Russian GDP in 2015 was $6751.9 \text{ bln. rbl} / 64997.04 \text{ bln. rubl.} = 10.38 \%$

Comparison of the shares formed by the SFD in the Russian GDP (10.38 %) and damage from deaths and wounds in road accidents (13.94 %) allows us to conclude that the problem of the negative impact of road and transport accident rate in the SFD is somewhat more acute than in Russia as a whole.

Table 2. Results of calculation of direct damage (due to death and wounding to people) from road accidents in the subjects of the SFD in 2015

Subject of the Siberian Federal District	Gross regional product (GRP) in 2015, mln. rubles [13]	Statistics of victims in road accidents (2015), people [15]		Damage (losses of GRP) due to death and wound of people (2015)	
		Fatalities in an RA	Injured in an RA	mln. rubles	% GRP
Altai Republic	41776.8	58	487	1772.35	4.24
The Republic of Buryatia	204156.2	163	1676	5262.16	2.58
Tyva Republic	47287.3	120	722	3405.63	7.20
The Republic of Khakassia	171663.9	116	1157	3712.15	2.16
Altai region	492138.9	313	4449	11230.71	2.28
Transbaikal region	248847.6	213	1834	6550.48	2.63
Krasnoyarsk region	1618166.0	567	5659	18148.11	1.12
Irkutsk region	1013542.3	479	4049	14661.96	1.45
Kemerovo Region	842618.9	435	4324	13907.08	1.65
Novosibirsk region	980850.5	360	3453	11394.49	1.16
Omsk Region	617184.4	259	4183	9752.07	1.58
Tomsk Region	473693.1	108	1040	3460.16	0.73
Total for the SFD	6751925.8	3191	33033	103257.3	1.53

6.2. The differentiated for the SFD subjects calculation of external (indirect) damage from the road accident

The external (property and environmental) damage from the road accident in the RF in 2015 amounted to 559 billion rubles. This value is calculated from the ratio of the shares between external and direct damages (43 : 57), generally accepted in the world practice [3, 4, 6]. Table 3 shows the results of calculating the external and total damage from the road accident in 2015, differentiated by the regions of the SFD.

Table 3. The results of calculating the external (property and environmental) and total damage from road accidents in the subjects of the SFD in 2015

Subject of the Siberian Federal District	External damage (losses of GRP) from the road accident in 2015		Total damage (losses of GRP) from the road accident in 2015	
	mln. rubles	% GRP	mln. rubles	% GRP
Altai Republic	1337.06	3.20	3109.42	7.44
The Republic of Buryatia	3969.78	1.94	9231.94	4.52
Tyva Republic	2569.21	5.43	5974.84	12.64
The Republic of Khakassia	2800.45	1.63	6512.60	3.79
Altai region	8472.45	1.72	19703.16	4.00
Transbaikal region	4941.68	1.99	11492.17	4.62
Krasnoyarsk region	13690.93	0.85	31839.04	1.97
Irkutsk region	11060.98	1.09	25722.94	2.54
Kemerovo Region	10491.50	1.25	24398.59	2.90
Novosibirsk region	8596.01	0.88	19990.5	2.04
Omsk Region	7356.96	1.19	17109.03	2.77
Tomsk Region	2610.35	0.55	6070.51	1.28
Total for the SFD	77897.37	1.15	181154.8	2.68

Graphically, the results of calculating the total damage from the road accident (in % of GRP region) are shown in Fig. 2.

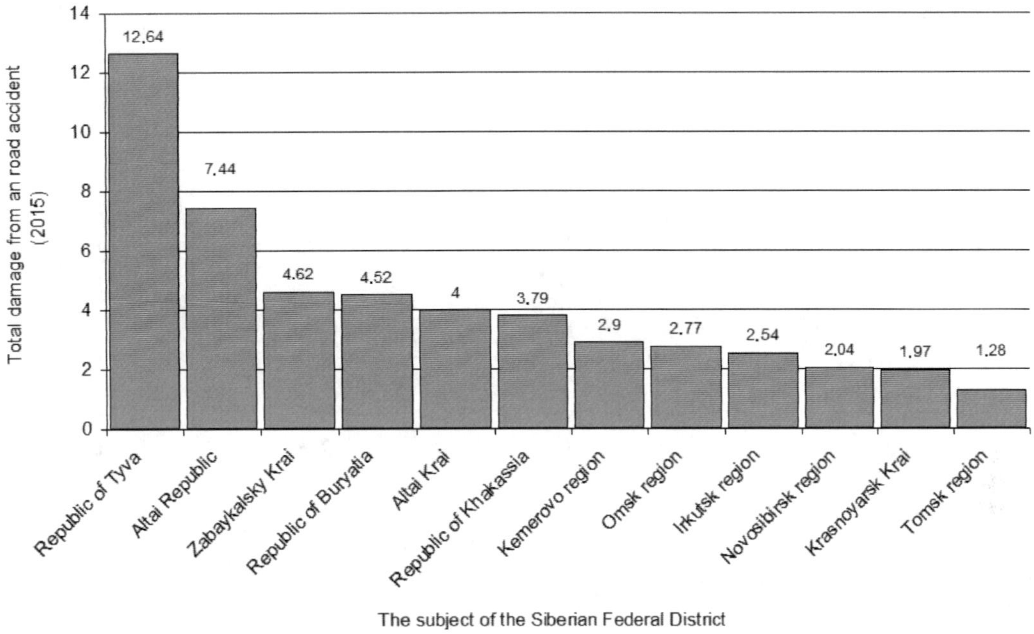

Figure 2. Total damage from road accident (in % of GRP region) in the subjects of the Siberian Federal District in 2015

6.3. Discussion of the results of damage assessment from road accident

Comparing the value of total damage from accidents in the SFD (2.68 % GRP) in comparison with the initial figure for calculations (2.00 % of GDP for the Russian Federation as a whole) allows to conclude that the situation in the Siberian Federal District in the field of road safety even on a national background.

It should be noted that in some regions of the SFD the situation in the field of road safety is simply catastrophic. Thus, the total damage from road accident in the Republic of Tyva reaches 12.5 % of GRP. This result correlates well with the data of [5]. The situation in the Altai Republic is not any better (7.44 % of GRP). Even in 7 regions, the damage exceeds the all-Russian level. And only in three regions (the Novosibirsk Region, the Krasnoyarsk Territory and the Tomsk Region), damages from road accidents are comparable with the national level or lower.

7. Explanation of the results

It is known [14], that the road and transport accident rate largely depends on the quality of life of the population. The cause and effect relationship between the quality of life of the population of the region and its deviant manifestations has been known for a long time. Incidents in transport are only part of the general situation in the region. In order to illustrate this relationship, this article presents regression models describing the statistical relationship between the regional automotive *Ua* and the human development index *(HDI)* as representative indicators of the socio-economic development of the regions and the estimated total damage from the road accident $\Sigma Damage$. The data necessary for constructing the regression models were obtained: for regional motorization in [15], and for the HDI in [16]. The data obtained in [15, 16] are presented in Table 4.

IOP Conf. Series: Earth and Environmental Science **115** (2018) 012014 doi:10.1088/1755-1315/115/1/012014

Table 4. Total damage from road accidents and representative characteristics socio-economic development in the regions of the SFD

Subject of the Siberian Federal District of the Russian Federation	Indicator values		
	Total (direct + external) damage from road accident, 2015, % GRP	Motorization of the population, 2015, passenger vehicles, units / 1000 people [15]	Human Development Index, 2014, HDI [16]
Altai Republic	7.44	220	0.814
The Republic of Buryatia	4.52	196	0.825
Tyva Republic	12.64	181	0.772
The Republic of Khakassia	3.79	308	0.844
Altai region	4.00	263	0.834
Transbaikal region	4.62	239	0.821
Krasnoyarsk region	1.97	289	0.877
Irkutsk region	2.54	283	0.858
Kemerovo Region	2.90	234	0.842
Novosibirsk region	2.04	276	0.864
Omsk Region	2.77	281	0.870
Tomsk Region	1.28	281	0.880

The processing data table 4 allowed to obtain the models $\sum Damage = f(Ua)$ and $\sum Damage = f(HDI)$. In a graphical form, these models are shown in Fig. 3 and Fig. 4.

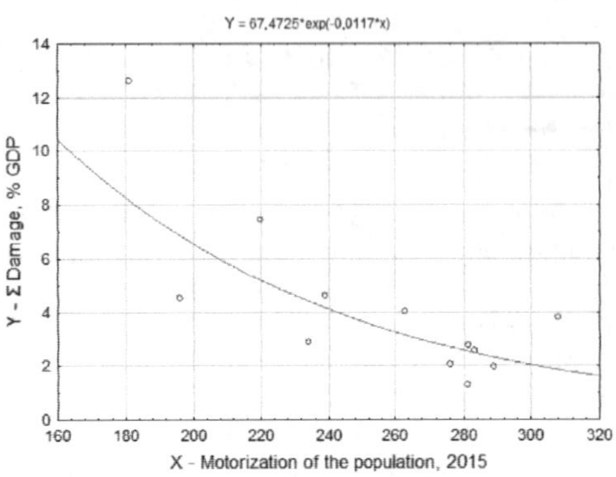

Figure 3. Model $\sum Damage = f(Ua)$

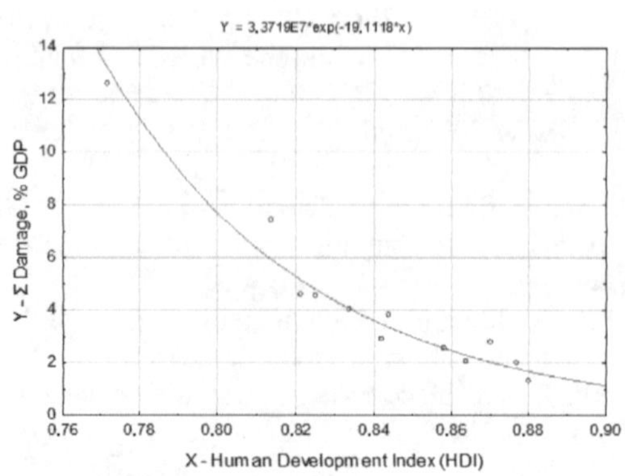

Figure 4. Model $\sum Damage = f(HDI)$

Models $\sum Damage = f(Ua)$ and $\sum Damage = f(HDI)$ for the SFD

The coefficient of determination for the model $\sum Damage = f(Ua)$ is $R^2 = 0.58$; the determination coefficient for $\sum Damage = f(HDI)$ model is $R^2 = 0.87$. The Fisher criterion F for the model $\sum Damage = f(Ua)$ is F = 13.77161. The Fisher criterion F for the $\sum Damage = f(HDI)$ model is F = 68.90596, which allows to conclude that the models are highly adequate. The standard error of approximation by the models of the experimental data is, respectively, 2.12 % and 1.16 %.

Thus, based on the presented models, it can be concluded in principle that a significant difference in the damage from the road and transport accident rate in different regions of the SFD is largely due to different quality of life of the population. This can also be judged by such an indicator as "% of the road accidents committed in a state of intoxication". According to this indicator, the Republic of Tyva is a leader not only in the SFD, but also among all 85 Russian regions, with an 22 %. Thus, it is the significant marginalization of the population of the Republic of Tyva that explains the very high level of damage from road accidents (12.64 % GRP).

8. Conclusion

The most important conclusions on the presented material are presented below.

1. Today in Russia there is no unambiguous understanding of the cost of human life and damage caused by the death and wound of a person in the road accident.
2. GDP losses due to road accidents in Russia are in 2015 ... 2017 about 2.0 ... 2.2 % of GDP (64997.04 billion rubles in 2011 prices) or about 1.3 trillion rubles in the prices of 2011.
3. Damage from an accident in Russia in 2015, conditionally determined at 1.3 trillion (741 billion rubles) and external or indirect damage (559 billion rubles).
4. The calculated values of the indicators "Average cost of living" (ACL) and "Average cost of Injuries" (ACI) in 2015 in Russia were, respectively, 22.875 mln. rub. and 0.915 mln. rub.
5. Based on the statistics of the road and transport accident rate in the subjects of the SFD, the calculated values of the ACL and ACI, the calculated values of direct and external economic damage for the regions of the District were determined.
6. The values of the total damage to the economy of the subjects of the Siberian Federal District vary in a wide range (from 1.28 % for the Tomsk region to 12.64 % for the Republic of Tyva). In general, in the SFD, the damage from the road and transport accident rate is 2.68 % GRP, which is slightly higher than for Russia as a whole (about 2.0 %).
7. Economic damage from the road and transport accident rate is a derivative of the quality of life of the population. The magnitude of damage from road accidents is determined by the influence of such particular characteristics of people's quality of life as the Human Development Index (HDI) and the provision of material goods, in particular, by means of transport. This influence is described by exponential dependencies with an inverse positive coupling.

Reference

[1] Federal state statistics service. Demographics Available from: http://www.gks.ru/wps/wcm/connect/rosstat_main/rosstat/ru/statistics/population/demography/ [Accessed 30th September 2017]

[2] Petrov A I 2015 *Features of formation of motor transportation accident rate in space and time* (Tyumen: TyumGNGU) p 254

[3] Karabchuk T S et al 2014 How to estimate life cost? *Economic sociology* **15(1)** p 89-107 Available from: http://ecsoc.hse.ru/data/2014/02/01/1328851429/ecsoc_t15_n1. pdf#page=89

[4] Karabchuk T S et al 2015 A Review of international methodologies and Russian practices for estimations of economic damage caused by death in road accidents *Ekonomicheskaya Sotsiologiya* **5** p 77-102

[5] Scherbakova E 2015 Russia: preliminary demographic results of 2014 (Part II). Rossiyksky demographic barometer Available from: http://demoscope.ru/weekly/2015/0633/barom04.php [Accessed 30th September 2017]

[6] Blincoe L J et al 2015 The Economic and Societal Impact of Motor Vehicle Crashes, 2010 (Revised) Report No. DOT HS 812 013 Available from: https://crashstats.nhtsa.dot.gov/ Api/Public/ViewPublication/812013 [Accessed 10th October 2017]

[7] Abelson P 2003 The Value of Life and Health for Public Policy *Economic Record* **79** S2-S13

[8] Abelson P 2007 Establishing a Monetary Value for Lives Saved: Issues and Controversies, Office of Best Practice Regulation Department of Finance and Deregulation Working papers in cost-benefit analysis WP 2008-02 Available from: https://www.dpmc.gov.au/sites/default/files/publications/Working_paper_2_Peter_Abelson.pdf [Accessed 10th October 2017]
[9] Federal Target Program for traffic safety for the period 2012…2020 Available from: https://rg.ru/2013/10/08/bezopas-site-dok.html [Accessed 30th September 2017]
[10] Federal state statistics service. National accounts Available from: http://www.gks.ru/wps/wcm/connect/rosstat_main/rosstat/ru/statistics/accounts/ [Accessed 30th September 2017]
[11] Sestokas V and Samoilov D 1987 *Conflict situations and traffic safety in the cities* (Moscow: Transport) p 207
[12] Kolesnikova D and Karabchuk T and Salnikova D and Fattakhov T 2016 Economics **Vol 6** p 131
[13] Federal state statistics service. Gross Regional Product Available from: http://www.gks.ru/dbscripts/cbsd/ DBinet.cgi [Accessed 30th September 2017]
[14] Petrov A I 2016 Road traffic accident rate as an indicator of the quality of life *Economic and social changes: facts, trends, forecast* **3(45)** p 154-172 Available from: http://esc.isert-ran.ru/article/1907
[15] State inspection of road traffic. Indicators of road safety Available from: http://stat.gibdd.ru/ [Accessed 30th September 2017]
[16] The human development report in the Russian Federation. The goal of sustainable development. The UN and Russia Available from: http://ac.gov.ru/files/publication/a/11068.pdf [Accessed 30th September 2017]

The phyllophagous of woody plants of genus Ulmus in protective plantings of arid zone

M N Belitskaya[1], I R Gribust [1], E E Nefed'eva [2,4, a], O S Filimonova [3], M A Golovanova [4]

[1] Federal Scientific Centre of Agroecology, Complex Meliorations and Protective Afforestation of the Russian Academy of Sciences, Volgograd, Russia
[2] Volgograd State Technical University, Volgograd, Russia
[3]Volgograd State Social-Pedagogical University
[4] Volgograd State University, Volgograd, Russia

E-mail: [a] nefedieva@rambler.ru

Abstract. Creating of long-lasting, normally developing plantings on the urbanized territories in the arid zones is difficult. The main tree species are elm (Fam. *Ulmaceae* – more than 70% of tree composition). They have a high drought and salt tolerance, rapid growth. They are widely used in the creation of artificial forest plantings in steppes and deserts. The main part of the plantations was created in the 50-60 years of 20th century. Today the aging of plants is observing. Trees of this age lose their physiological and decorative potential as well as resistance to pests under intensive anthropogenic impact. At the forest pathology research about 90 % of the plants of elm species of plantings of various types and categories were identified to be in poor condition. Leaf-eating pests play an active role in the deterioration of their sanitary state. *Xanthogaleruca luteola (*Müller, 1766), *Cladius ulmi* (Linnaeus, 1758), and *Aproceros leucopoda* (Takeuchi, 1939) have a special harmfulness as well as the regularity of the local centers among the mass phyllophagous of plants of the family. The last two species were discovered in artificial plantations for the first time. The destructive effect of those pests on certain types of elm plants is uneven.

1. Introduction

One of the most important conditions of the functional stability of agrolandscapes and urbolandscapes is the variety of biota inhabiting them [1, 2, 3, 4]. However, environmental problems which aggravated in recent decades, have led to the destruction of biocenotic relations [4, 5, 6, 7]. That has caused the disappearance of a number of organisms and sharp reduction in the number of useful biota. The widespread of herbivorous arthropods which previously had no economic value occurs now. The violation of regulatory mechanisms of ecosystem leads to the formation of a group of pests that relatively small in species, but well adapted to new ecological conditions [6, 8, 9, 10, 11]. The weakening of the natural mechanisms promoted the appearance of new aggressive biotypes of harmful species in ecosystems [5, 8, 9, 12].

The release of environmental stress is possible if the optimal biodiversity is sustained in ecosystems. This state is achieved through the creation of plantations of different functional purpose with the participation of introduced woody plants [1, 2, 9, 7, 12].

The important role in the formation of new communities play local, abundant types of dendrobionts which find optimal conditions for life here [1, 2, 6].

2. Methodics

Research was realized in the arboretum complexes, forest nurseries, forest seed plantations of Federal Scientific Centre of Agroecology, Complex Meliorations and Protective Afforestation of the Russian Academy of Sciences and in plantings of urban areas. Forest pest monitoring was conducted in the plantings of different types and environmental categories with the participation of woody plants of the families Ulmaceae. Reconnaissance surveillance was carried out with the definition of the sanitary status of planting to determine the phyllophagous. Detailed oversight was realized to determine the causes of changes in the status of forest plantations and populations of harmful insects.

Gathering of material was carried out using standard techniques (entomological mowing; visual accounting of the insects on the branches of length 0.5 running meter; 100 leaves) during the vegetation period [1, 3, 11[11], [13]13]. The species which damaged more than 50% of the leaves in the crown of woody plants were related to mass pests. Defoliation degree of plants was determined visually according to the basic methodological criteria [14].

3. Results and Discussion

The polyphagous species such as Lymantria dispar (Linnaeus, 1758), Euproctis chrysorrhoea (Linnaeus, 1758), Archips crataegana (Hübner, 1796-1799) и Agapeta hamana (Linnaeus, 1758), Geometra papilionaria (Linnaeus, 1758), Lycia hirtaria (Clerck, 1759), Alsophila aescularia (Denis & Schiffermuller, 1775), Nymphalis polychloros (Linnaeus, 1758) etc. are typical inhabitants of forest ecosystems in arid zone; they were identified to dominate among phytophagous - inhabitants of economically valuable trees and shrubs (52.9% of the species richness). A lot of those insects periodically have outbreaks and they damage the foliage in tree crowns by 70-100%.

The level of the abundance of species of specific communities of dendrobionts in the plantings of different ecological categories varies. Dendrophages which inhabit woody plants of the generic complex Ulmus L. (66 species) are characterized by high taxonomic abundance of communities.

Representatives of four environmental groups participates in the formation of specific communities of arthropods (Fig. 1). Group of pests of generative organs (1-3) and root systems (2-5 species) of plants are characterized by the lowest species abundance. The population of pests of branches and stems (24 species) a few more varied. Pests of assimilation apparatus (17-48 species, 53,1-77,4%), form the basis of the faunal abundance of the population of dendrobionts.

There are animals among them which can be occurred in plantings with high quantity. They form the centers of mass reproduction and cause severe damage.

The research identified complexes of the most numerous species of phyllophagous of elm; they are typical for different types of plantings in arid conditions.

Their composition and quantitative abundance in plantings of different natural zones are markedly different (Table 1, 2, 3). More multicast entomocomplexes support the plantings of steppe and semidesert zones. The population of phyllophagous the least various in the planting of the dry steppe zone.

Analysis of the quantitative abundance of the main species of dendrobionts demonstrated a strong dependence of this indicator on the environmental conditions of regions. During the transition from steppe to dry steppe the number of important pests in plantations was reduced on 70.5-85.2 %. but the number of pests increases 7.3 - 30.9 times with the movement into the semi-desert zone. More clearly. this tendency appears among arthropods which inhabit roadside and recreation and landscaping areas. due to specific environmental conditions (pollution of environment. microclimate. etc.). But identification of the exact causes of this phenomenon requires a detailed study.

IOP Conf. Series: Earth and Environmental Science **115** (2018) 012015 doi:10.1088/1755-1315/115/1/012015

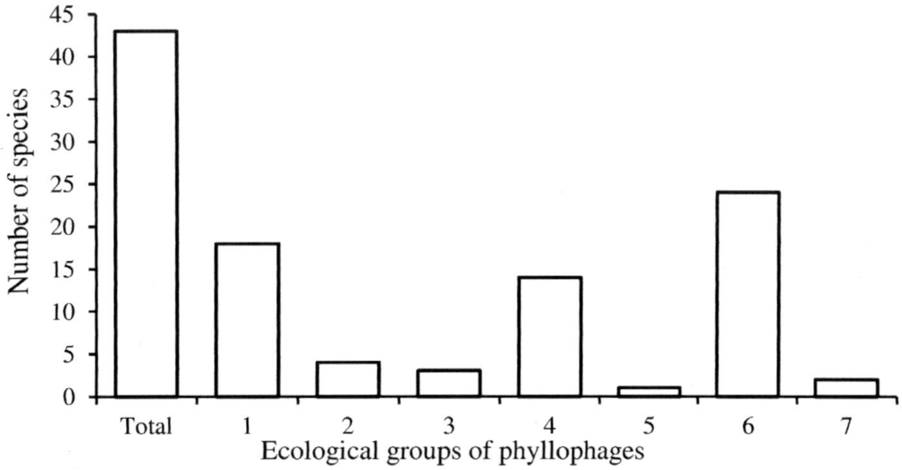

1 - biting arthropods; 2 - sucking arthropods; 3 - miners; 4 - galls; 5 - carpophages; 6 - xylophages; 7 - rhizophages

Figure 1. Structure of community of dendrobionts of plants of generic complex of *Ulmus* L.

Table 1. Traits of the composition and abundance of the main phyllophagous of elm in plantings of the steppe zone, specimens / units of account.

Family and genus of arthropods	Kind of plantings			
	Dendrological collections	Field protective	Roadside	Recreation and landscape
Cicadellidae				
Oncopsis scutellaris Fieb.	3.5±0.2	9.9±0.1	**12.7±0.2**	**28.2±0.6**
Curculionidae				
Polydrosus inustus Germ.	sporadic	**10.7±0.3**	**14.2±0.1**	**17.9±0.2**
Tenthredinidae				
Aproceros leucopoda	1.5±0.03	**68.1±0.9**	**64.2±0.5**	4.7±0.06
Cladius eradiatus Hart.	3.8±0.1	**92.6±0.4**	**74.1±0.2**	10.3±0.4
Cecidomyiidae				
Physemococis ulmi Kieffer	sporadic	**16.2±0.3**	**19.7±0.08**	7.3±0.2

A fact of significant differences in taxonomic composition of the groups of major phyllophagous in different types of plantings in different natural areas requires attention. So, common species in roadside, recreation and field protecting planting in the steppe zone are *Polydrosus inustus* (Schoenh, 1834) and *Physemocecis ulmi* (Kieffer, 1909). Here *Aproceros leucopoda* (Takeuchi, 1939) and *Cladius ulmi* (Linnaeus, 1758) [11, 12, 15-18] were discovered for the first time. Local outbreaks of those pests was noted in the Southern and Eastern districts of Samara region. Larvae of the first generation of sawflies destroyed about 85% of the leaves in the crown of trees in field protective plantings. The foliage was completely destroyed in separately standing elms and in row 3-4-row roadside plantings.

Table 2. Traits of the composition and abundance of the main phyllophagous of elm in plantings of the dry steppe zone, specimens / units of account.

Family and genus of arthropods	Kind of plantings			
	Dendrological collections	Field protective	Roadside	Recreation and landscape
Aphididae				
Tetraneura ulmi L.	2.6±0.09	**22.3±1.1**	8.9±0.1	5.8±0.3
Tortricidae				
Peronea boscana F.	sporadic	**17.6±0.2**	6.3±0.1	4.8±0.3
Notodontidae				
Dicranura ulmi Schiff. (units /1 running meter)	sporadic	**24.1±0.9**	**23.6±0.2**	-

Table 3. Traits of the composition and abundance of the main phyllophagous of elm in plantings of the semi-desert zone, specimens / units of account.

Family and genus of arthropods	Kind of plantings			
	Dendrological collections	Field protective	Roadside	Recreation and landscape
Cicadellidae				
Cicadella viridis L.	4.7±0.2	3.1±0.01	14.2±0.6	**59.5±3.1**
Aphididae				
Colopha compressa Koch.	2.7±0.6	7.1±0.03	**29.2±0.8**	5.3±0.2
Tetraneura ulmi L.	6.4±0.5	**22.3±0.9**	8.8±0.1	3.0±0.04
Chrysomelidae				
Galerucella luteola Mull.	10.7±0.5	**227.6±0.5**	**158.9±1.2**	**95.1±1.5**
Cecidomyiidae				
Janetiella lemei Kieffer	2.8±0.1	**29.5±0.3**	**32.9±0.7**	16.5±0.2
J. nervicola Kieffer	5.3±0.06	14.2±0.4	**21.2±0.3**	9.0±0.5
Eriophyidae				
Eriophyes filiformis Nal.	3.1±0.04	**23.5±0.7**	10.2±0.1	8.1±0.3
E. ulmicola Nal.	5.9±0.1	**44.9±0.3**	9.7±0.5	7.7±0.1

Sawflies caused significant damage of elms in multibreed multifunctional plantings of tight construction as well as in arboretum. The amount of damaged leaves in the crown was about 45-57%, respectively. *U. pumila* was damaged by pest significantly.

Tetraneura ulmi (Linnaeus, 1758) and *Acleris boscana* (Fabricius, 1794) are common phytophagous of woody plants of genus *Ulmus* in plantations of the dry steppe zone. They are more numerous in protective forest belts (14.8 and 11.6 specimens / accounting units correspondingly). In recreation and landscaping plantings their quantitative abundance was 65-72% and 34-46% and less. In the arboretum *A. boscana* met only in separate trees, its density was 0.2-0.5 specimens / accounting units. Settlements of *T. ulmi* were recorded in trees neighbouring elder. It populated about 85% of leaves in the lower part of the crown.

Dicraneura ulmi (Denis & Schiffermuller, 1775) dominated in forest belts and massive plantations of different ages, especially in heavily eroded areas. Local foci of the pest were located in the plantings on the right bank of the Volga along the Volga elevation. *Ulmus parvifolia* and *Ulmus laevis* were the main target of that pest. Caterpillars eat away the tissue of leaves, but they leaved intact the main and lateral veins. Damage of the assimilation apparatus was 95% at the average.

Xanthogaleruca luteola (Müller, 1766) had caused maximum damage to elm in plantations of semi-desert zone. The outbreak of that pest continues since 2013.

In these conditions, the elm-leaf beetle damages in three generations, overlapping each other. The pest populates leaves in the tree crown by groups of 5-8 individuals. The major part of the larvae in the latter stages of development lives single. Individuals of the first generation damage more than 60% of the area of the leaf lamina by the end of May. By the end of the development of the second generation almost all the leaves in the trees are destroyed.

High quantitative abundance of *X. luteola* in different types of plantations is caused by the poor condition of the trees. Plantings were created in the 50-60th years of the last century. Currently, there are more than 80% of highly weakened and drying trees.

A group of subdominant, which is presented by aphids and galls, is quite numerous in these plantings – 141.5 and 117.4 specimens / accounting units. In landscaping plantings, *Cicadella viridis* (Linnaeus, 1758) was characterized by increase in number – 59.5 specimens / accounting units.

Trophic complex of phyllophagous in plants of the dendrological collection of semi-desert region includes representatives of different ecological groups of arthropods. The quantitative abundance of individual species and whole community of that area lower on 69.7-78.8% in common than the same parameter in plantings in other natural zones.

Representatives of the genus Ulmus is among the main breeds for protective afforestation in the conditions of arid zone. They differ from other woody plants by higher drought tolerance, salt tolerance and rapid growth. However, at present a strong weakening and drying of the representatives of this kind are recorded in many plantings. Those processes are caused by activity of insects-phyllophagous.

There is an urgent necessity to develop ecological forecasting, which allows to evaluate the status of protective plantations, to provide further development of the succession process and to plan activities for enhance the biological stability of plantings. That way is realizable if information (primarily) about the peculiarities of the composition and the quantitative abundance of the most important phyllophagous in different types of elm is available. Therefore, we undertook a research of the population by pests of the most common plants of semi-desert zone, such as *U. pumila*, *U. laevis* and *U. foliacea*.

Observations have shown that the quantity of the group of dominants in different species of *Ulmus* varies greatly. Higher levels of those parameters are defined *U. pumila*. The abundance of phyllophagous on other types of elm less than there. *U. laevis* differs that in a greater degree.

Different types of leaf-eating insects respond to species of *Ulmus* as a feed tree incertainly. For example, *X. luteola* and *D. ulmi* accumulate mainly in *U. pumila*. *Physemococecis ulmi* prefers *U. grabra*. Abundance of *Eriophyes ulmicola* (Nalepa, 1898) in *U. pumila* and *U. scabra* practically did not differ, but their density in *U. laevis* decreased 1,3-1,9 times. At the same time, *E. filiformis* (Nalepa, 1898) does not show marked attachment to any species of Ulmus.

4. Conclusions

The specific complex of pests, which is based on leaf-eating insects (53.1-77.4 % of species richness of dendrobionts) is formed in the crowns of tree species of Ulmaceae.

Environmental conditions of the region affect the numerical abundance of the major species of phyllophagous. The number of important pests in plantations is reduced by 70.5-85.2% when move from the steppe to dry steppe, whereas in the direction of the semi-desert zone their number is growing at 7.3… 30.9-times.

Aproceros leucopoda Takeuchi. and *Cladius ulmi* L. first discovered in the crowns of tree species of Ulmaceae in roadside, recreational and landscape plantings as well as in field-protecting plantings.

The most abundant species were differed among the population of the insects, such as *Xanthogaleruca luteola* Müller, *Dicraneura ulmi* Den. ett Schiff., *Aproceros leucopoda* Takeuchi, and *Cladius ulmi* L.; they formed local foci in plantations of various types and categories. Their destructive effect varies at the level of 60-100% of destroyed foliage in the crowns of trees.

The abundance of phyllophagous on different types of elm is uneven. *Ulmus pumila* is the most attractive tree for pests. That species acts as a as a feed tree for *Xanthogaleruca luteola* Müller, *Dicraneura ulmi* Den. ett Schiff., *Physemocecis ulmi* Kieffer, and *Eriophyes ulmicola* Nal.

IOP Conf. Series: Earth and Environmental Science **115** (2018) 012015 doi:10.1088/1755-1315/115/1/012015

References

[1] Belickaja MN Gribust IR 2011 *Izvestija Sankt-Peterburgskoj lesotehnicheskoj akademii (Bulletin of Saint-Petersburg forestry engineering Academy)* **196** 134-139

[2] Belickaja MN Nefed'eva EE Makeev AA Zhidkoblinov SP 2015 *Vestnik Kazanskogo tehnologicheskogo universiteta (Gerald of Kazan technological university)* **18(2)** 416-419

[3] Gribust IR 2015 *Izvestija Sankt-Peterburgskoj lesotehnicheskoj akademii (Bulletin of Saint-Petersburg forestry engineering Academy)* **211** 134-139

[4] Koricheva J, Larsson S, Haukioja E Annu Rev 1998. *Entomol.* **3** 195-216.

[5] Robinson EA, Kyan GD, Newman JA 2012. *New Phytologist.* 16 p.

[6] Belickaja MN Gribust IR 2016 Vestnik agrarnoj nauki Dona (Bulletin of agricultural science of Don) **2(34)** 42-49

[7] Zvereva EL, LantraV, Kozlov MV 2010 *Oecologia.* **163** 949-960.

[8] Belickaja MN Fedotova ZA Nefed'eva EE 2016 *Paradigma (Paradigm)* **2** 207-212

[9] Belickaja MN Gribust IR Nefed'eva EE 2017 *Vestnik Novosibirskogo gosudarstvennogo agrarnogo universiteta (Bulletin of the Novosibirsk state agrarian university)* **2(43)** 41-51

[10] Karpun NN Ignatova EA Zhuravleva EN 2015 *Izvestija Sankt-Peterburgskoj lesotehnicheskoj akademii (Bulletin of Saint-Petersburg forestry engineering Academy)* **211** 189-197

[11] Mozolevskaja EG Matusevich LP 2004 Metody monitoringa sostojanija jekosistem (Methods of monitoring of the state of ecosystems, *Monitoring sostojanija lesnyh i gorodskih jekosistem (Monitoring of forest and urban ecosystems)* Moscow: MGUL p 6-9.

[12] Sorokin NS 2013 *Zashhita i karantin rastenij (Protection and quarantine of plants)* **11** 35-37

[13] Pesenko JuA 1982 *Principy i metody kolichestvennogo analiza v faunisticheskih issledovanijah (Principles and methods of quantitative analysis in faunistic studies)* (Moscow: Nauka (Science)) 284 p.

[14] Voroncov AI Mozolevskaja EG Sokolova JeS 1991 *Tehnologija zashhity lesa (Technology of protection of forest)* (Moscow: Ecology) 304 p.

[15] Gninenko JuI Gninenko MJu Rakov AG 2013 *Zashhita i karantin rastenij (Protection and quarantine of plants)* **2** 40-41

[16] Blank SM Hara H Mikulas J Csoka G Cionel C Constantineanu R Constantineanu L Roller L Altenhofer E Hufleit T Vetek G 2010 *European Jornal of Entomology* **107** 357-367.

[17] Wu Xing-yu Xin Heng 2006 *Entomotaxonomia* **28(4)** 279-280.

[18] Zandigiacomo P Cargnus E Villani A 2011 *Bulletin of Insectology. Depatment of Agroenviromental and Technologies* **64(1)** 145-149

IOP Conf. Series: Earth and Environmental Science **115** (2018) 012016 doi:10.1088/1755-1315/115/1/012016

Modelling the average velocity of propagation of the flame front in a gasoline engine with hydrogen additives

N M Smolenskaya[a], V V Smolenskii [b]

Togliatti State University,
Belorusskaya st. 14, Togliatti, Russian Federation, 445020

E-mail: [a] nata_smolenskaya@mail.ru, [b] biktor.cm@mail.ru

Abstract. The paper presents models for calculating the average velocity of propagation of the flame front, obtained from the results of experimental studies. Experimental studies were carried out on a single-cylinder gasoline engine UIT-85 with hydrogen additives up to 6% of the mass of fuel. The article shows the influence of hydrogen addition on the average velocity propagation of the flame front in the main combustion phase. The dependences of the turbulent propagation velocity of the flame front in the second combustion phase on the composition of the mixture and operating modes. The article shows the influence of the normal combustion rate on the average flame propagation velocity in the third combustion phase.

1. Introduction

The use of alternative fuels reduces toxicity without changing the design in SI Engines. Hydrogen is considered as an alternative fuel and hydrogen is intended as a substitute for fuel from petroleum products. The main reason for this is that when hydrogen is burned water is formed and there are no pollutants such as CO, CO_2, C_nH_m or solid particles. The combustion characteristics of hydrogen provide a high combustion rate and minimum ignition energy, this allows us to expand the limits of effective depletion of the mixture. Also interest in hydrogen is that hydrogen can be obtained by electrolysis of water [1,2].

For the analysis of the influence on the efficiency and toxicity of the SI Engines work process of design and operational factors and also chemical composition of fuel, it is necessary to know the mechanism of flame propagation in the combustion chamber [3, 4]. Most often in the studies the surface mechanism of the propagation of the flame front [5, 6, 7]. When a uniformly mixed mixture is burned in a turbulent flow, the fresh mixture is separated from the combustion products by a thin curved flame front. In the combustion chamber of a piston engine with an external mixture formation, the same mechanism takes place. The application of this mechanism makes it possible to identify the main combustion characteristics that determine the propagation of a flame in the combustion chamber.

The most appropriate characteristics are:

- the velocity of turbulent flame propagation, as an integral indicator of the rate of chemical reactions and gas-dynamic and thermal conditions;
- the normal flame propagation velocity, as an integral measure of the rate of chemical reactions.

Knowledge of these characteristics will allow us to assess the dynamics of combustion, which determines the toxicity and efficiency of the work process. Therefore, the aim of the work is an experimental study of the influence of the composition of the mixture, the compression ratio, the ignition timing and the rotation frequency on the average propagation velocity of the flame front in various combustion phases.

2. Experimental technique

Experimental studies were carried out on a single-cylinder UIT-85 unit. Geometric parameters of the engine: the cylinder capacity is 0.652 liters, the cylinder diameter is 85 mm, the piston stroke is 115 mm, the compression ratio is 5.9 and 7. The electric motor keeps the revolutions constant: 600±6, 900±9 min^{-1}. The uniformity of the fuel-air mixture is provided by heating the intake manifold.

Registration of the movement of the flame front inside the UIT-85 cylinder (Figure 1a) was carried out by ionization sensors. Ionization sensors are installed in a special adapter (Figure 1b). In the adapter, the ionization sensors are arranged as follows: proximity (ionization sensor 3) is located at a distance 77 mm away from the spark plug; distant (ionization sensor 5) is located at a distance 91.3 mm away from the spark plug; central (ionization sensor 4) is installed in a 4 mm hole at a distance of 9.75 mm from the plane of the combustion chamber. This location of the ionization sensors makes it possible to evaluate the propagation parameters of the flame front in the main combustion phase and in the sensor location area at different levels of turbulence [8].

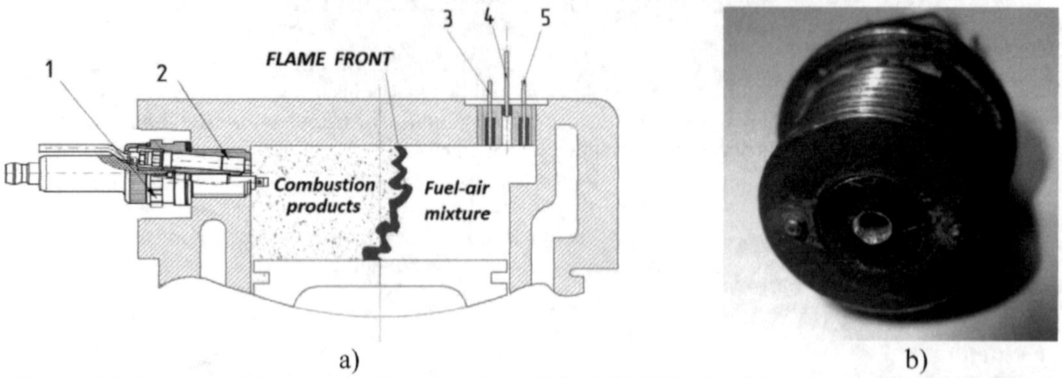

Figure 1. Diagram of the combustion chamber of the UIT-85 (a), where 1 – spark plug; 2 – pressure sensor Kistler, 3 – proximity ionization sensor; 4 – central ionization sensor; 5 – distant ionization sensor; (b) general view of the adapter with ionization sensors

The proximity ionization sensor was used to determine the average propagation velocity of the flame front in the main phase, that is from spark plug to ionization sensor 3. The central ionization sensor was used to estimate changes in the propagation velocity of the flame front in the third combustion phase. We assume that the volume formed by the cavity of the hole where the sensor is mounted will ensure that there is no effect of large-scale turbulent pulsations on combustion in the zone of the sensor. This corresponds to the combustion conditions characteristic of the third combustion phase. The turbulent propagation velocity of the flame front in the installation zone of the ionization sensors was determined from the time of the signal between the near and far ionization sensor. It can be considered as the corresponding average velocity of propagation of the flame front in the second phase of combustion.

The parameters by which the combustion process was evaluated:

– the average velocity of flame front propagation in the main phase of combustion (first and second phases):

$$U_{1+2} = \frac{77}{t_3}, \tag{1}$$

where t_3 – the time from a spark discharge to the appearance of a signal on the proximity sensor of ionization;

– the average velocity of flame front propagation in the second phase of combustion (between the proximity and distant ionization sensor):

$$U_2 = \frac{14.27}{t_5 - t_3}, \tag{2}$$

IOP Conf. Series: Earth and Environmental Science **115** (2018) 012016 doi:10.1088/1755-1315/115/1/012016

where t_5 – the time from a spark discharge to the appearance of a signal at the distant ionization sensor; t_3 – the time from a spark discharge to the appearance of a signal on the proximity sensor of ionization;

- the average velocity of flame front propagation in the third phase of combustion (between the proximity and central ionization sensor):

$$U_3 = \frac{11}{t_4 - t_3}, \tag{3}$$

where t_3 – the time from a spark discharge to the appearance of a signal on the proximity sensor of ionization; t_4 – the time from a spark discharge to the appearance of a signal on the central ionization sensor.

The investigations were carried out at the following ignition timing angles 9, 13, 16, 21, 23, 29 BTDC for each of the rotation frequencies and for each compression ratio.

The experiment was based on parallel recording of signals by a multichannel analog digital converter PCI-1712L-AE. Signals were recorded from the following sensors: ionization, induction spark sensor, optical sensor for the position of the crankshaft, with precision 0.27 degree steps of the rotation angle, pressure sensor Kistler, installed in the adapter with a spark plug, mass flow sensor Bosch. Gasoline and hydrogen were fed separately into the intake manifold for the carburetor, the mass flow rate of gas was determined from the calibrated duty ratio of the injectors. The toxicity and the coefficient of excess air (α) were determined from the gas analyzer and the lambda sensor LSU 4.9.

3. Results and Discussion

The Figure 2 shows the results of experimental studies of the average propagation velocity of the flame front in the main combustion phase. The following symbols are used in the Figure 2 and Figure 3: the compression ratio 5.9 and 7; the rotation frequency of 600 and 900 min^{-1}; 16, 21, 25 BTDC; 0, 3, 5 fraction of added hydrogen as a percentage of the mass of fuel. It can be seen from the Figure 2 that the addition of hydrogen increases the average velocity of propagation of the flame front in the main combustion phase. The nature of the effect of the addition of hydrogen depends on the change in the rotational speed. The influence of the compression ratio is compensated by an increase in the ignition timing, which increases the heat evolution in the second combustion phase.

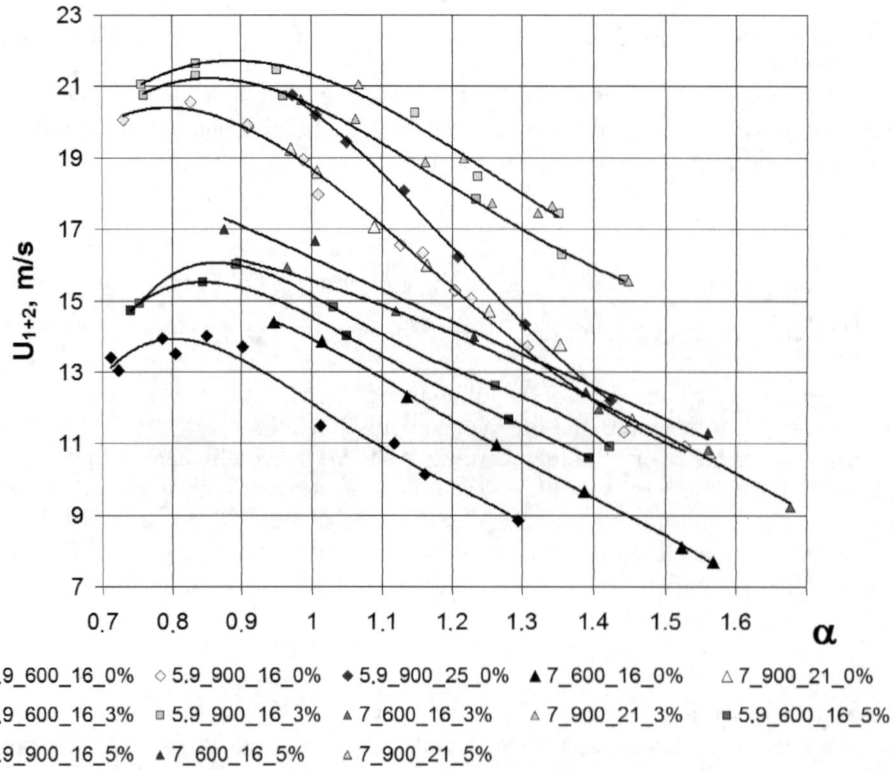

Figure 2. The influence of hydrogen addition in the gasoline-air mixture on the average propagation velocity of the flame front in the main combustion phase

The figure shows the effect of the addition of hydrogen to the gasoline-air mixture on the turbulent propagation velocity of the flame front in the second combustion phase. It can be seen from the figure that an increase in the average velocity of propagation of the flame front in the second phase of combustion occurs for all the investigated compositions of the mixture and the operating conditions of the engine. With an increase in the fraction of hydrogen, the dependence of the average velocity of propagation of the flame front in the second phase of combustion on the excess air coefficient. Because with an increase in the normal combustion rate, the flame front tends to become thinner and its propagation speed increasingly depends on the rate of involvement of the fresh mixture, which is determined by the turbulence of the flow. It can be seen from the Figure 3a, that the addition of 3 and 5% of hydrogen leads to approximately the same increase in the propagation velocity of the flame front between the ionization sensors.

The Figure 3b shows the effect of the addition of hydrogen to the benzene-air mixture on the propagation velocity of the flame front in the cavity of the central sensor hole. Where smaller turbulent pulsations and the absence of vortex motion are characteristic fuel-air mixture, which allows us to represent the motion of the flame front in a given section, as combustion in the case of small-scale turbulence. It can be seen from the Figure 3b that the addition of hydrogen in the rich mixture range from $\alpha = 0.7$ to $\alpha = 0.85$ practically does not affect the combustion rate and starting from $\alpha = 0.9$ and poorer the effect of the hydrogen additive becomes noticeable, and when the mixture is depleted its influence increases.

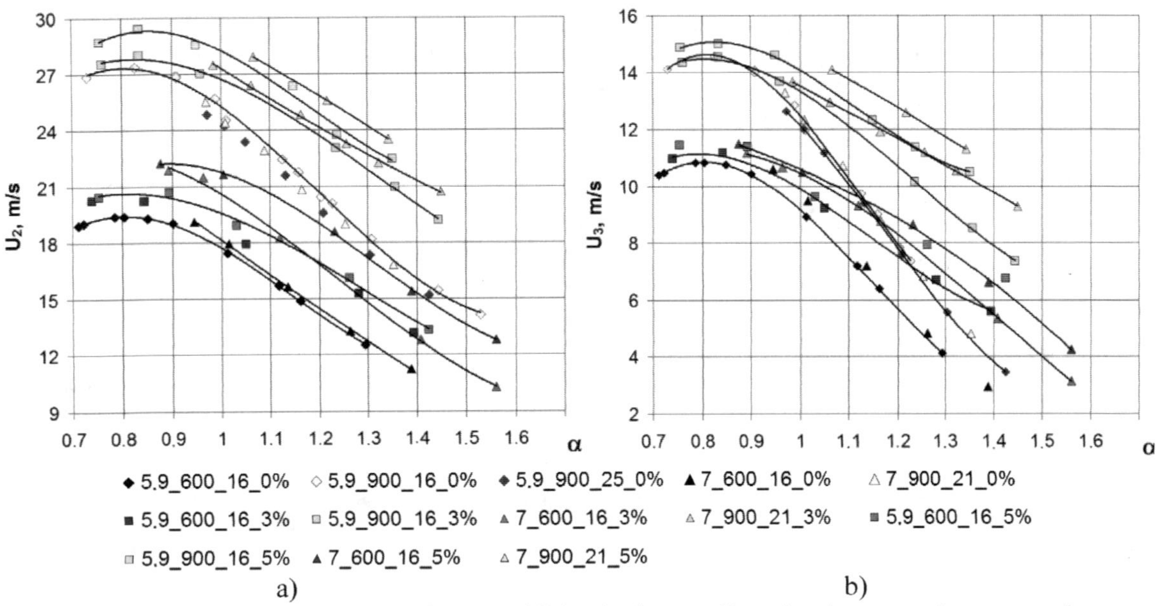

Figure 3. The influence of the hydrogen additive in the gasoline-air mixture on the propagation velocity of the flame front: (a) in the second phase of combustion; (b) in the third phase of combustion

Combustion in the main phase determines the performance indicators and is largely responsible for the toxicity of the engine. As shown by experimental studies (Figure 2), the addition of hydrogen leads to an increase in the average propagation velocity of the flame front in the main phase.

The obtained effect is ensured by the following:

- explained from the standpoint of the chain mechanism of combustion reactions of hydrocarbon mixtures in which the reaction of one hydrogen atom leads to the formation of three hydrogen atoms [6, 9]. Therefore, the chemical kinetics of combustion of hydrocarbon fuels is improved, since hydrogen is the active center of combustion.
- as shown in [6, 9, 10], in the zone of the combustion reaction the excess air factor α_{ZR} differs from α_P ahead of the flame front. This is a consequence of the difference in molecular diffusion coefficients of fuel and oxygen. For most hydrocarbon-air mixtures, including gasoline-air, $\alpha_{ZR} > \alpha_P$, and for hydrogen-air and methane-air mixtures $\alpha_{ZR} < \alpha_P$. Consequently, when hydrogen is added to the depleted gasoline-air mixture, a relative increase in hydrogen occurs in the fuel-air mixture, which leads to an increase in both the normal and turbulent rates of propagation of the flame front as compared to the depleted gasoline-air mixture of the same composition. The increase in the combustion rate makes it possible to carry out the combustion process in less time and in a smaller volume near the top dead center.

From the results obtained, it can be seen that the main effect is the addition of hydrogen on the duration of the third combustion phase. This phase is very important when working on poor mixtures, as it affects the toxicity of unburned hydrocarbons. Until now, her research has been neglected. Many researchers have shown that the combustion process with small-scale turbulence is significantly influenced by the normal combustion rate, taking into account the turbulent diffusion of the molecules of the mixture. Therefore, we present known calculation models for determining the normal propagation velocity of the tribal front at a temperature and pressure corresponding to the instant of the appearance of the signal on the central ionization sensor.

IOP Conf. Series: Earth and Environmental Science **115** (2018) 012016 doi:10.1088/1755-1315/115/1/012016

From the existing models for calculating of the laminar velocity of the flame front propagation we will choose the model of determining the laminar velocity of the flame front propagation for the mixtures of gasoline, presented in the Heywood's work [9]:

$$U_L = U_{0L} \cdot \left(\frac{T'_u}{T_0}\right)^{\alpha_t} \cdot \left(\frac{P_u}{P_0}\right)^{\beta_p}, \tag{4}$$

Where U_{0L} – the normal rate of combustion gasoline at T_0 =298 K and P_0 =0.1 MPa:

$$U_{0L} = 30.5 - 54.9 \cdot (\phi - 1.21)^2, \text{ sm/s;}$$

T'_u, P_u – the temperature and pressure for which the calculation is carried out;

α_t, β_p – power-law coefficients of the function, depending on the coefficient of excess air:

$$\alpha_t = 2.18 - 0.8 \cdot (\phi - 1),$$

$$\beta_p = -0.16 + 0.22 \cdot (\phi - 1).$$

According to the conditions of the conducted researches, combustion of gasoline with hydrogen additive up to 6%. Let us determine the laminar propagation velocity of the flame front for a hydrogen-air mixture. For this we adopt the model Iijima and Takeno, presented in the work [2, 6].

The current pressure is determined from the pressure indicator diagram, and the current temperature by the equation of state of the real gas, taking into account the compressibility.

As a result of the mathematical analysis, models for calculating the propagation velocity of the flame front in the third phase of combustion:

– when working on gasoline:

$$U_3 = \left[\left(\frac{V_\theta}{V_a}\right) \cdot \sqrt{U_{\text{mps}} \cdot U_L} + U_{\text{mps}}\right] \cdot \left(\frac{V_\theta}{V_a}\right)^{\left(1.9 \cdot \alpha^2 - 3.1 \cdot \alpha + 0.6\right)}; \tag{5}$$

– when working on gasoline with hydrogen additives:

$$U_3 = \left[\left(\frac{V_\theta}{V_a}\right) \cdot \sqrt{U_{\text{mps}} \cdot (U_L \cdot (1 - \Delta H) + U_{LH_2} \cdot \Delta H) + U_{\text{mps}} \cdot \left(1 + \alpha^{(\varepsilon + 1)} \cdot \left(\frac{H_\Sigma}{H_f} - 1\right)\right)}\right] \cdot \left(\frac{V_\theta}{V_a}\right)^{\left(1.9 \cdot \alpha^2 - 3.1 \cdot \alpha + 0.6 + 2 \cdot \Delta H\right)}; \tag{6}$$

where V_θ – the chamber volume at the time of the spark, L; V_a – the total cylinder capacity, L; U_{mps} – the average piston speed, m/s; U_L – the normal velocity of flame propagation of isooctane in the third phase of combustion, m/s; ΔH – the fraction of hydrogen additive by mass of fuel; U_{LH_2} – the normal velocity of distribution of a flame of hydrogen in the third phase of combustion, m/s; α – the coefficient of excess air; ε – the compression ratio; $H_\Sigma = \dfrac{0.145 \cdot (1 - \Delta H) + \Delta H}{(1 - \Delta H) + l_{0\text{gasoline}} \cdot \alpha \cdot (1 - \Delta H) + l_{0H_2} \cdot \Delta H \cdot \alpha + \Delta H}$ – the total mass fraction of hydrogen in the fuel-air mixture; $H_f = \dfrac{0.145}{1 + l_{0\text{gasoline}} \cdot \alpha}$ – the mass fraction of hydrogen in the gasoline-air mixture; $l_{0\text{gasoline}}$, l_{0H_2} – the theoretically necessary amount of air in kg for complete combustion of 1 kg of gasoline and hydrogen, respectively.

Analysis of the results of studies of the features of the propagation of the flame front with the addition of hydrogen in the fuel-air mixture made it possible to establish the mathematical dependences of the average rates of propagation of the flame front in the second and main phases of combustion on the propagation velocity of the flame front in the third phase and the operating conditions of the engine.

Analysis of the experimental results showed that the flame in the main combustion phase propagates along the turbulent mechanism. Therefore, for the basic model for calculating the average propagation velocity of the flame front in the main combustion phase, the formula Karlowtz (7) [11]:

$$U_T = U_l \cdot \left(1 + \sqrt{\frac{2u'}{U_l} \cdot \left[1 - \frac{U_l}{u'} \cdot \left(1 - e^{-\frac{u'}{U_l}}\right)\right]}\right), \qquad (7)$$

where U_T – the turbulent velocity of flame front propagation, U_l – the laminar flame propagation velocity, u' – the flow turbulence.

In it instead of the laminar velocity of propagation of the flame front, the average velocity of propagation of the flame front in the third phase of combustion (5) and (6), where combustion proceeds under the influence of small-scale turbulence, which is close in nature to laminar combustion [2, 5, 7, 12]. The velocity of turbulent pulsations was expressed in terms of the average piston speed, because the turbulence of the flow is proportional to the number of revolutions [5]:

$$U_{\text{mps}} = S \cdot n / 30,$$

where S – the piston stroke, n – the engine speed.

A model for calculating the average propagation velocity of the flame front in the main combustion phase:

$$U_{1+2} = U_{\text{mps}} \cdot \left(\frac{V_\theta}{V_a}\right)^{-0.325} + U_3 \cdot \left(1 + \sqrt{\frac{2 \cdot U_{\text{mps}}}{U_3} - 2 \cdot \left(1 - e^{-\frac{U_{\text{mps}}}{U_3}}\right)}\right). \qquad (8)$$

A model is obtained that takes into account the influence of the physico-chemical properties of the mixture and the effect of the regime parameters of the engine on the propagation velocity of the flame front in the second combustion phase:

$$U_2 = U_{\text{mps}} \cdot \left[(\alpha - 1) \cdot 10 \cdot \left(\frac{H_\Sigma}{H_f} - 1\right) + \left(\frac{V_\theta}{V_a}\right)^{-0.5526}\right] + U_3 \cdot \left(1 + \sqrt{\frac{2 \cdot U_{\text{mps}}}{U_3} - 2 \cdot \left(1 - e^{-\frac{U_{\text{mps}}}{U_3}}\right)}\right). \qquad (9)$$

The obtained models have good convergence (the discrepancy with the experimental results does not exceed 5%) in the following ranges of operation of a single-cylinder unit UIT-85:

– the rotation frequency $n = 600$, $n = 900$ min^{-1};
– the ignition timing from 9 to 29 BTDC;
– the coefficient of excess air from 0.7 to 1.5.

The obtained mathematical models for determining the average propagation velocities of the flame front in different combustion phases make it possible to determine the influence of the parameters of the operation of the gasoline engine and the effect of hydrogen on the rate of heat generation. This is necessary for the design of gasoline engines with improved environmental performance due to the addition of hydrogen. It is known that the concentration of NO_x in exhaust gases is largely determined by the temperature of the combustion products and the rate of lowering of the temperature upon expansion, CH are determined by the area of the layer frozen at the wall at the appropriate temperature and pressure, which are determined by the average propagation speed of the flame front in the second and third phases of combustion.

4. Conclusion

1. As a result of simulation of the combustion process in UIT-85, empirical models of average velocities of flame front propagation in the second, third and main phases of combustion for the conditions of a single-cylinder gasoline plant with a hydrogen content of up to 6% by mass of fuel.

2. The addition of hydrogen increases the rate of propagation of the turbulent flame front of the gasoline-air mixture in the second phase of combustion due to an increase in the normal combustion rate and high diffusion activity of hydrogen, this increases the burning area and the dynamics of heat generation.

References

[1] Suwanchotchoung N. Performance of a spark ignition dual-fueled engine using split-injection timing: Ph.D. Thesis.Vanderbilt University, 2003. 197 p.

[2] Verhelst S. A study of the combustion in hydrogen-fuelled internal combustion engines, PhD thesis, Gent: Gent University, 2005. 278 p.

[3] Kavtaradze R.Z. Theory of piston engines. Special chapters, Moscow, MSTU. Bauman. 2008. 712 p.

[4] Smolenskaya N.M., Smolenskii V.V., Bobrovskij I. Research of polytropic exponent changing for influence evaluation of actual mixture composition on hydrocarbons concentration decreasing on deep throttling operation, IOP Conference Series: Earth and Environmental Science 50 (2017) 012016 doi:10.1088/1755-1315/50/1/012016.

[5] Sokolik A.S. Self-ignition, flame and detonation in gases, Moscow: Academy of Sciences of the USSR, 1960. 427 p.

[6] Warnatz J., Maas U., Dibble R.W. Combustion: Physical and Chemical Fundamentals, Modeling and Simulation, Experiments, Pollutant Formation: 4th ed. Springer Berlin Heidelberg New York: 2006. 388 p.

[7] Smolenskaya N.M., Korneev N.V. Modelling of the combustion velocity in UIT-85 on sustainable alternative gas fuel, IOP Conference Series: Earth and Environmental Science 66 (2017) 012016 doi:10.1088/1755-1315/66/1/012016.

[8] Shaikin A.P., Galiev I.R. On the effect of temperature and the width of the turbulent combustion zone on the ionization detector readings, Technical Physics, 61(8) (2016) 1206-1208. DOI: 10.1134/S1063784216080247.

[9] Heywood J.B. Internal Combustion Engine Fundamentals, New York: McGraw-Hill, 1988. 931p.

[10] Kuznetsov V.R., Sabelnikov V.A. Turbulence and burning, Moscow: Science, 1986. 288 p.

[11] Hitrin L.N. Physics of Combustion and Explosion, Moscow: MGU, 1957. 442 p.

[12] Law C.K. Combustion physics, Cambridge: Cambridge university press, 2006. 722 p.

IOP Conf. Series: Earth and Environmental Science **115** (2018) 012017 doi:10.1088/1755-1315/115/1/012017

Experimental assessment of heat and mass transfer of modular nozzles of cooling towers

N A Merentsov[1, a], V N Lebedev[2], A B Golovanchikov[1], V A Balashov[1] and E E Nefed'eva[1]

[1]Volgograd State Technical University,
400131, Russia, Volgograd, Prospect Lenin, 28
[2] «LUKOIL-Engineering VolgogradNIPImorneft» branch in Volgograd
400078, Russia, Volgograd, Prospect Lenin, 96

E-mail: [a] steeple@mail.ru

Abstract. Data of experimental study of hydrodynamics, heat and mass transfer of modular nozzles of cooling towers and some comparative characteristics of the packed device with nozzles, which have wide industrial application, are given in the article.

1. Introduction

Packed heat and mass transfer devices with different configurations get application in a very wide range of technological processes in many branches of industry. So, for example, those processes in the chemical industry are the processes of absorption and rectification which are realized in packed column mass transfer apparatus Evaporative cooling of water in cooling towers sprinklers are processes in the heat-power engineering. In ecological processes it is purification of gas emissions in absorbers and scrubbers.

Named technological equipment has the distinction and the specifics of the processes which occurre there. However, the general purpose of the contact devices of the equipment is arrangement of conditions of interaction of the gas and liquid phases when it provides the greatest efficiency of that equipment processes with the least expenditure of energy. There is a wide variety of factors, chemical and thermal physical properties of the contacting flows, but the effectiveness of contact devices is mainly determined by internal geometry and embodiment of nozzles that form the hydrodynamic conditions in two-phase gas-liquid flow, as evidenced by the industrial experience and experimental studies.

Development of new perspective ways of intensification of heat-mass transfer processes in packed devices is currently an actual problem. The main purpose of this paper is to get experimental data in the field of hydrodynamics of a new modular heat and mass exchange nozzle with the function of water disinfection in electric field and comparing of its characteristics with the same characteristics of nozzles which have wide industrial application. That nozzle can get wide application for cooling of circulating water in local water supply systems for industrial enterprises.

This work is focused on experimental investigation of hydrodynamics, heat and mass transfer in new modular heat-mass exchange nozzles for cooling towers of the local water recycling systems [1]. Heat-mass exchange nozzle is made as a module for disinfection of water in electric field. That property in addition to traditional assignments, can suppress the activity of microorganisms and prevent biofilm formation on the surface of the nozzle and on the outer surfaces of the heat exchange process equipment, washed by the circulating water. Diagram of a modular heat-mass exchange in the nozzle is presented in figure 1.

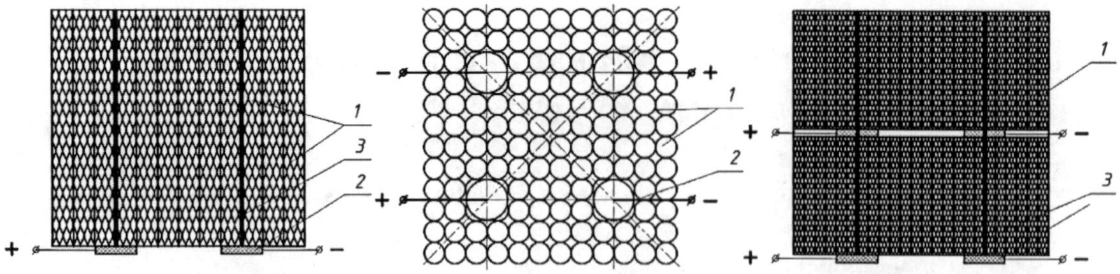

Figure 1. Modular heat-mass exchange nozzle for cooling tower [1].

Heat and mass exchange nozzle of cooling tower is made as a module which consists of layers of polymeric grid casings 1. They have cylindrical shape; they are arranged in parallel to each other in all vertical layers. Dissociating cylindrical insert 2 have the form of swirlers; they installed on the bottom surface of each module uniformly over the cross section. Each dissociating insert is made of electrically conductive material, such as sheet metal or carbon and it provided with the vertical electrode 3 with a height equal to the height of the module. Adjacent inserts 2 are attached to opposite poles of a DC source. The cooled water is intensely mixed because dissociating insert are made as swirlers. It results in the equalization of the residence time of water in the cooling tower and its disinfection by electric current. This effect suppresses bacteria and prevents biofilm formation on the surface of the polymeric grid casings 1 and intensifies heat-mass exchange processes between cooling water and air in cooling tower [1].

It should be noted that nozzle devices for evaporative cooling of water in cooling towers in accordance to their specificity, should provide a high retention capacity for fluids and have a low value of hydraulic resistance. That is, they should not maintain resistance to the upward air flow, smoothly flowing it, providing ventilation and maintaining the driving force of the process. Along with assigned tasks, the comparative characterization of the elaborated nozzles with devices that have wide industrial application is the main purpose of the research.

2. Experimental part

Experimental setup that allows to study the hydrodynamics of packed devices and heat-mass exchange in conditions of counter flow movement of air-water flow was elaborated for a wide range of experimental studies.

The setup is available for visual observation and for development of hydrodynamic and aerodynamic flow regimes of water and air through the nozzle. It allows to determine the holding capacity of hydraulic resistance of dry and irrigated nozzles and to assess the intensity of heat and mass exchange processes in packed devices [2]. Diagram and pictures which explain the construction and principle of the device for the experimentation are shown in figure 2 [2].

The device consists of a frame 1 which contains the nozzle 3. The nozzle is irrigated by water using the dispenser 4. The nozzle is blown by countercurrent air flow which arrives from a centrifugal fan 5. The frame walls are planar and transparent, it allows to observe and record the nozzle in different hydrodynamic regimes by high-speed video cameras without distortion. Cartridge system of replacement nozzles is implemented in the device; that allows to change the nozzles quickly and to ensure the reproducibility of the measured parameters in repeated experiments [2].

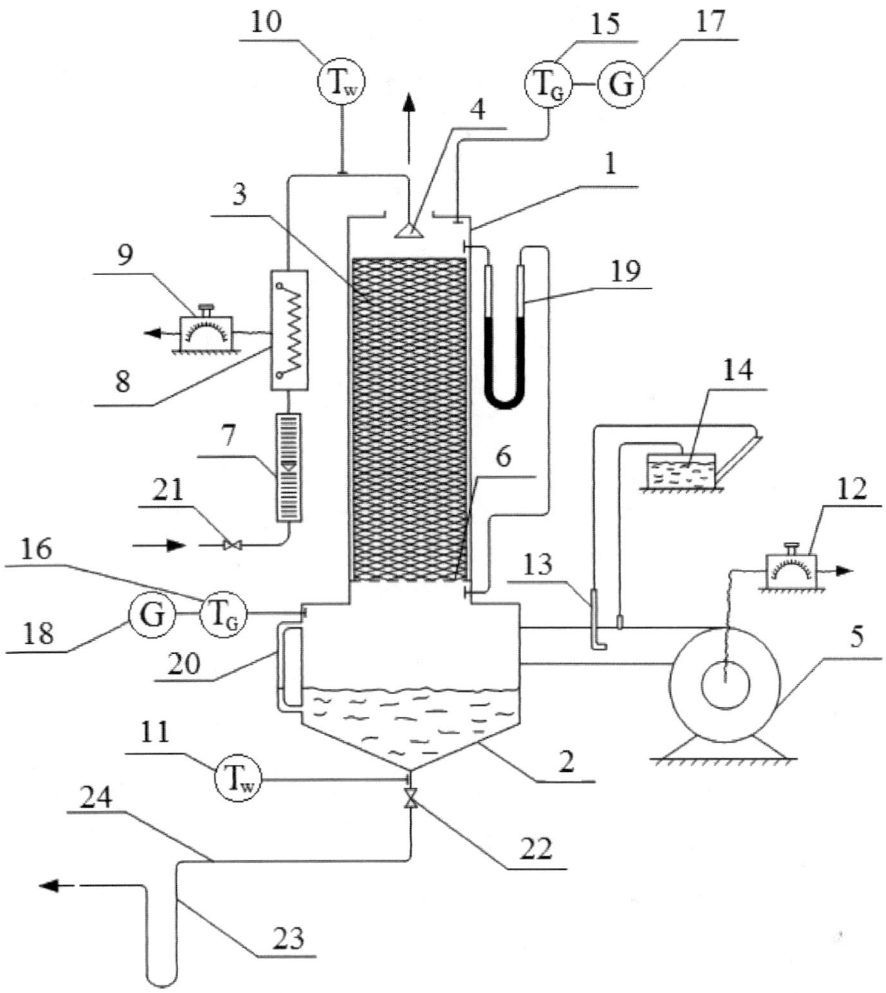

Figure 2. Diagram of the experimental setup for the study of hydrodynamics and heat and mass exchange processes in packed devices.
1 – frame; 2 – water collecting capacity; 3 – nozzle; 4 – dispenser; 5 – centrifugal fan; 6 – supporting grid; 7 – rotameter; 8 – water heater; 9 – potentiometer; 12 – frequency converter; 10, 11, 15, 18 – temperature sensor; 13 – Pitot-Prandtl tube; 14 – manometer; 16, 17 – hygrometer; 19 – differential manometer; 20 – liquid level gage; 21, 22 – valve; 23 – pipe bend; 24 – pipe for water drain to drainage

The hydraulic resistance of the nozzle without irrigation, regardless of the nature of the process, is one of the most important indicators of its work. It provides the benchmark on the possible range of processes of its industrial applications. The hydraulic resistance of the dry nozzles, in addition to industrial orientation of packing unit, is a kind of "passport" of packed devices, and the reflection of the energy cost in the realization of technological processes.

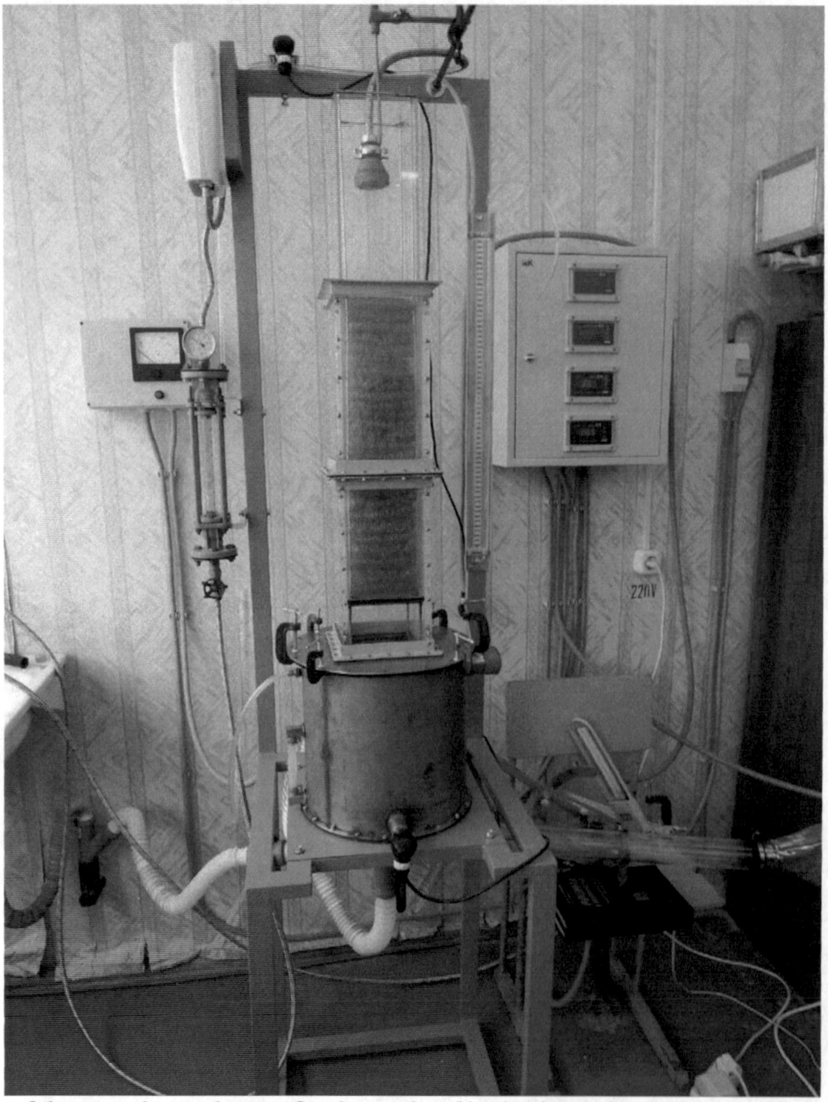

Figure 3. Photo of the experimental setup for the study of hydrodynamics and heat and mass exchange processes in packed devices.

However, the hydraulic resistance of the dry nozzles gives a not a full picture of the efficiency of the packed devices. The fact is heat and mass exchange nozzles can have completely different geometric configurations and, accordingly, they can implement fairly difficult hydraulic modes of operation. Therefore, it is necessary to explore hydraulic regimes and to conduct comparative characteristics of hydraulic resistance of irrigated nozzles.

3. Results and Discussion

The results of assessment of hydraulic resistance of dry and irrigated modular heat-mass exchange nozzle are presented in figure 4.

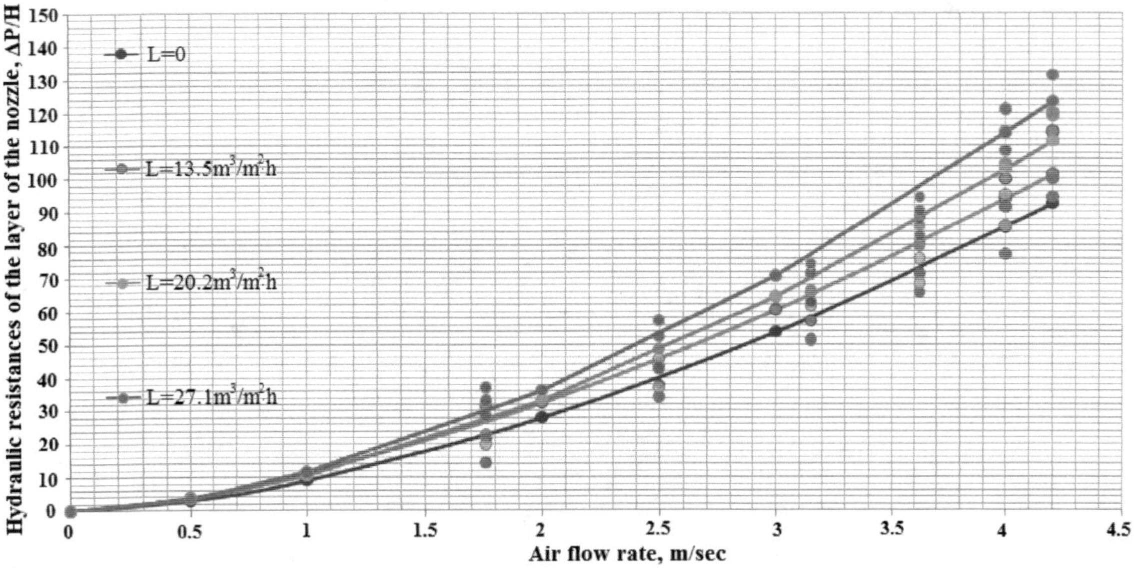

Figure 4. Diagram of dependence of hydraulic resistances of dry and irrigated modular heat-mass exchange nozzle on the air flow rate

We used criteria equation of dependence $\lambda = \frac{2}{\mathrm{Re}_m} + 2$ [3,4] to generalize the experimental data and to provide the comparative performance of the test devices and nozzles which have wide industrial application. We can compare the energy efficiency and provide the benchmark to industrial applications of absolutely different packed devices of the most difficult configurations by that criteria equation.

Linear dimensions l_1 and l_2, the values of α and β which are coefficients, respectively, of viscous and inertial components of the equation Dupuis – Forchheimer, as well as modified criteria of Reynolds Re and the corresponding coefficient of hydraulic resistance λ, were determined on the basis of experimental filtration curves for the proposed modular heat and mass exchange nozzle [1]. The data are presented in table 1 [3,4].

Table 1. The results of mathematical data processing of heat and mass exchange modular nozzle

α, m^{-2}	β, m^{-1}	Re	λ	l_{x1}, m	l_{x2}, m^2
$2.5\cdot10^5$	3.6429	0.5667	5.53	0.27451	$1.46\cdot10^{-5}$
		1.1333	3.76		
		2.2667	2,.88		
		3.4	2.59		
		4.5	2.44		
		4.76	2.42		

The results of the comparative characteristics of the packing devices are shown in figure 5.

We can conclude on the basis of comparative characteristics of kinds of nozzles, with using the generalized criterial equation of dependence of the resistance coefficient on the modified Reynolds number that a modular heat and mass exchange nozzle is in the range of very small hydraulic resistance as compared to industrial nozzles. Therefore, the proposed nozzle can be oriented to the process of evaporative cooling of water in cooling towers, as well as a to the variety of energy-efficient industrial processes [3,4].

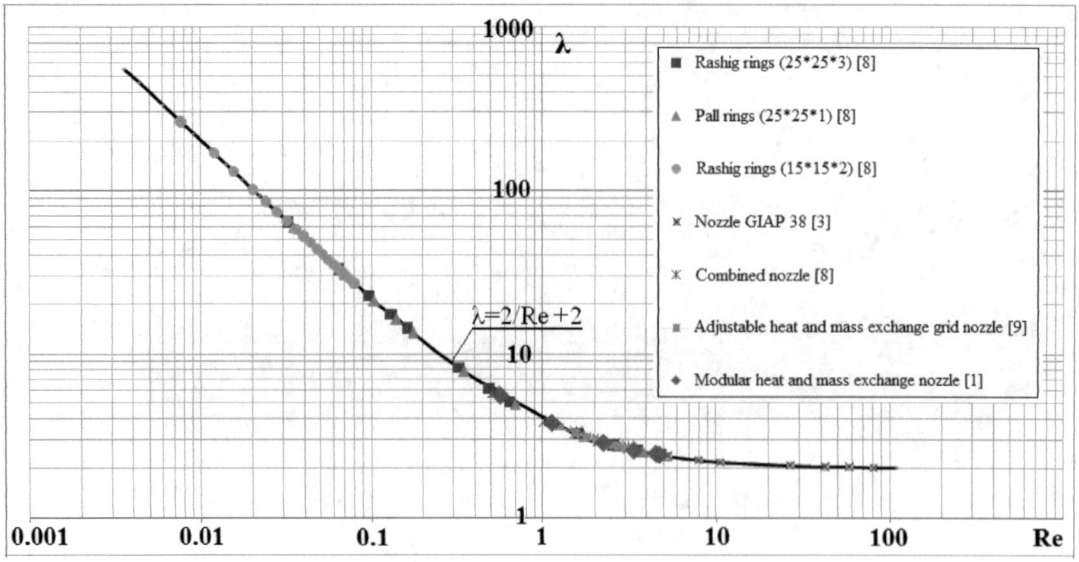

Figure 5. Diagram of dependence $\lambda = f\,(\mathrm{Re})$ for nozzles with different structure

Holding capacity is one of the main hydraulic characteristics of nozzle devices, which, to our opinion, has no enough attention. We consider it as a base for calculation methods [5,6,7] both with the coefficients of heat and mass exchange. It is the ability of the nozzle to accumulate a certain amount of liquid depending on the operation mode. That ability is a reflection of the total holding time of the liquid and the surface of the mass transfer for a range of nozzles.

The results of the research of holding capacity are demonstrated in Figure 6.

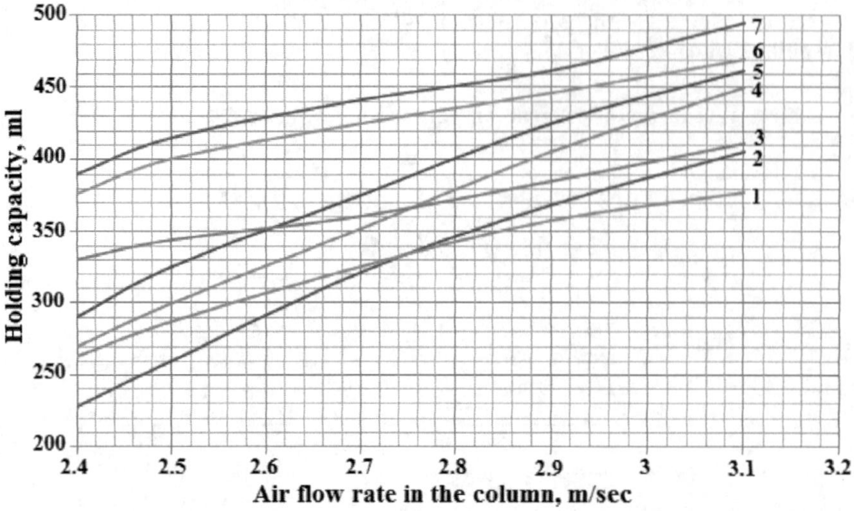

Figure 6. Diagram of dependence of holding capacity of the irrigated packing on the gas flow rate in the column:

1 – sheet nozzle; 2 – Pall rings; 3 – adjustable heat and mass exchange grid nozzle [9]; 4 – Rashig rings; 5 – nozzle with resonant effect [10]; 6 – modular heat and mass exchange nozzle for cooling towers [1]; 7 – regular block grid nozzle with an elastic liquid dispenser.

The modular heat and mass exchange nozzle is shown to have a relatively high holding capacity as compared to a number of industrial and developed nozzles which were analyzed in the experiment.

4. Conclusions

We can conclude that the modular nozzle device for evaporative cooling of circulating water in cooling towers and for devices of inertial scrubbing of gases is promising for industrial applications. That idea is based on described data of the energy-efficiency, hydraulic resistances of dry and irrigated nozzles and their high holding capacity for liquid.

Special attention, in our opinion, should be paid to the prospects of application of modular nozzle device for disinfection of industrial circulating water. That property of the nozzle could help to maintain the constancy of the surface properties of the packed devices, and partly to prevent biofouling of external surfaces of a heat transfer equipment, washed by the circulating water, by microorganisms.

References

[1] Golovanchikov AB, Sivolobova NO, Merentsov NA, Dul'kina NA, Shishljannikov VV, Dorofeeva NI 2012 *Patent* 129450 RU, IPC F 28 S 25/08 VSTU

[2] Merentsov NA, Balashov VA, Golovanchikov AB, Orljankina JaA 2012 *Izvestija Volgogradskogo gosudarstvennogo tehnicheskogo universiteta, Ser. Reologija, processy i apparaty himicheskoj tehnologii (The news of Volgograd state technical University, Ser. Rheology, processes and devices of chemical technology)* **1 (88)** 78-80

[3] Golovanchikov AB, Balashov VA, Merentsov NA 2017 *Chemical and Petroleum Engineering* **53(1-2)**10-13

[4] Merentsov NA, Balashov VA, Orljankina JaA 2013 *Izvestija Volgogradskogo gosudarstvennogo tehnicheskogo universiteta, Ser. Reologija, processy i apparaty himicheskoj tehnologii (The news of Volgograd state technical University, Ser. Rheology, processes and devices of chemical technology)* **1 (104)** 112-114

[5] Golovanchikov AB, Merentsov NA, Balashov VA 2013 *Chemical and Petroleum Engineering* **48(9-10)** 595-601

[6] Golovanchikov AB, Merentsov NA, Balashov VA, Orljankina JaA 2012 *Izvestija Volgogradskogo gosudarstvennogo tehnicheskogo universiteta, Ser. Aktual'nye problemy upravlenija, vychislitel'noj tehniki i informatiki v tehnicheskih sistemah (The news of Volgograd state technical University, Ser. Actual problems of management, computer science and Informatics in technical systems)* **10 (97)** 22-28

[7] Golovanchikov AB, Merentsov NA 2016 *Himicheskaja tehnologija (Chemical engineering)* **17(8)** 377-384

[8] Hizhnjakov IA, Rjazanov MG, H.A. Merentsov NA, Balashov VA, Golovanchikov AB 2016 *Energo- i resursosberezhenie: promyshlennost' i transport (Energy and resource saving in industry and transport)* **1 (13)** 7-11

[9] Golovanchikov AB, Vorotneva CB, Merentsov NA, Dul'kina NA, Zalipaeva OA, Sham'janova AP 2012 *Patent* 117317 RU, IPC B 01 J 19/32 VSTU

[10] Golovanchikov AB, Vasil'ev PS, Ljapkov AV, Chjorikova KV, Topilin MV, Tishhenko PO 2016 *Patent* 160198 RU, IPC B 01 J 19/00 VSTU

[11] Kagan AM, Yudina LA, Pushnov AS 2012 *Theoretical Foundations of Chemical Engineering* **46(2)** 165-171

[12] Gorodilov AA, Berengarten MG, Pushnov AS 2016 *Theoretical Foundations of Chemical Engineering* **50(3)** 325-334

[13] Gorodilov AA, Berengarten MG, Pushnov AS 2016 *Theoretical Foundations of Chemical Engineering* **50(4)** 422-429.

[14] Mitin AK, Nikolaikina NE, Pushnov AS, Zagustina NA 2016 *Chemical and Petroleum Engineering* **52(1)** 47-52

[15] Klyushenkova MI, Kuznetsova NA, Pushnov AS, Berengarten MG, Mokrousova EA 2014

Chemical and Petroleum Engineering **50(7-8)** 508-512
[16] Gorodilov AA, Pushnov AS, Berengarten MG 2014 *Chemical and Petroleum Engineering* **50(1-2)** 84-90
[17] Pushnov AS, Sokolov AS, Sidel'nikov II, Kurbatova EA, Mitrofanova EG 2014 *Chemical and Petroleum Engineering* **50(5-6)** 330-334
[18] Chizh, KV, Pushnov, AS, Berengarten MG 2014 *Chemical and Petroleum Engineering* **50(3-4)** 244-250
[19] Berengarten MG, Nevelson AO, Pushnov AS 2013 *Chemical and Petroleum Engineering* **48(11-12)** 723-729
[20] Shilin MV, Berengarten MG, Pushnov AS, Klyushenkova MI 2013 *Chemical and Petroleum Engineering* **48(9-10)** 608-614
[21] Mitin AK, Valdberg AY, Pushnov AS 2012 *Chemical and Petroleum Engineering* **48(1-2)** 50-53
[22] Tsurikova NP, Pushnov AS, Lagutkin MG, Shishov VI 2012 *Chemical and Petroleum Engineering* **48(1-2)** 3-8
[23] Kagan AM, Pushnov AS 2012 *Russian Journal of Applied Chemistry* **85(3)** 523-526
[24] Kagan AM, Yudina LA, Pushnov AS 2012 *Russian Journal of Applied Chemistry* **85(3)** 515-522
[25] Pushnov AS, Vaganov AA 2011 *Journal of Applied Chemistry* **84(9)** 1638-1641
[26] Pushnov AS, Berengarten MG, Kagan AM, Ryabushenko A 2009 *Russian Journal of Applied Chemistry* **82(4)** 723-729
[27] Tsurikova NP, Pushnov AS, Lagutkin MG 2011 *Chemical and Petroleum Engineering* **47(1)**. 97-103
[28] Petrashova EN, Lagutkin MG, Pushnov AS, Shishov VI 2011 *Chemical and Petroleum Engineering* **47(3)** 250-255
[29] Pushnov AS, Lozovaya NP, Lagutkin MG 2010 *Chemical and Petroleum Engineering* **46(1)** 3-8
[30] Kagan AM, Pushnov AS 2008 *Chemical and Petroleum Engineering* **44(3-4)** 187-192

IOP Conf. Series: Earth and Environmental Science **115** (2018) 012018 doi:10.1088/1755-1315/115/1/012018

Study of Regulatory Mechanisms of Activity of Cardiovascular System by Method of Mathematical Analysis of Heart Rhythm in Workers of Chemical Manufactures

T V Dyachkova, I A Berseneva, O A Zavaltseva, O S Mishina

State educational institution of higher professional education Moscow region State University of Humanities and Technology,
142611 Orekhovo-Zuevo, Zelyonaya street, 22

E-mail: xryylb@yandex.ru

Abstract. The article presents the results of the study of heart rate variability indices of workers engaged in the production of phenol-formaldehyde plastics and plastics at the «Karbodin» plant. 112 people aged from 20 to 50 years were studied: control group-workers with experience up to 1 year, practically healthy (n = 30), 1 group-workers with work experience up to 5 years (n = 40), 2 group workers with work experience from 5 to 10 years (n = 42). As a result of the study, violations of the functioning of the heart regulation system were revealed, depending on the length of employment. The effectiveness of the method for studying the regulation of the physiological functions of the circulatory system as well as the early diagnosis of occupational pathology has been established.

Introduction

Ecological problems, currently of global social nature, are most pronounced in the chemical industry, which is one of the leading sources of stationary emissions of harmful substances into the environment. At the same time, the air of the working area is exposed to pollutants, which is a priority environment for occupational risk assessment.

The quality of the workers' health is not only a prerequisite for high labor productivity and consequently improving welfare, but it is also a guarantee of sustainable social and economic development of the countries of the world. Specific features of the environment, technology, procurement of raw materials and production of finished products create a unique set of working conditions in the chemical industry that directly affect health. At the same time, the factors of the work environment and the characteristics of labor require an integrated approach to solving the problem of improving working condition.

Studying the features of the development of occupational diseases, using the achievements of modern science in the study of various physiological mechanisms of the vital activity of the human body, its adaptive compensatory and pathophysiological reactions, occupational pathology, occupational hygiene, physiology have made their significant contribution to understanding the general biological regularities of general pathology.

For the modern stage of development of hygienic and physiological disciplines is characterized by an isolated or combined effect of professional factors of low intensity in combination with mental and emotional stress, hypokinesia or monotonous work. This leads to the development of erased forms of occupational diseases with small clinical symptoms of their manifestation, which makes diagnosis difficult and requires in-depth knowledge of various areas of medical and biological science. The introduction of modern types of energy, chemicals, synthetic materials into the industry, intensification of labor, the growth of information loads create conditions for the development of new forms of occupational diseases.

Industry, rapid technical progress, the emergence of complex types of work have changed the habitual rhythm of life, which cannot but affect the human body, its health. These conditions have a significant impact on the working capacity and health of the worker. The unfavorable influence of the nature of labor or the production environment on human health indicates the presence of harmful production factors in the production. A harmful and dangerous production factor is a factor, the impact of which on the worker leads to significant violations of the regulation of the basic physiological functions of the organism.

Indicator of the functional state and adaptive reactions of the body can rightly be called the cardiovascular system. The functional state of the organism in the process of adaptation to environmental conditions has two limiting values: health and disease, norm and pathology. Between these values are various donorological conditions, differing in the degree of tension of regulatory systems, the degree of adaptation [1, 2]. The cardiovascular system with its neurohumoral mechanism of control and self-regulation reacts to the slightest changes in individual organs and systems and ensures the coordination of blood flow in them with hemodynamic parameters at the organism level [3].

The pathology of the cardiovascular system is a serious problem, as it is one of the most common causes of disability and deaths of a significant number of people. In Russia, mortality from circulatory system diseases has remained the highest in the world in the last two decades.

A new methodology for studying the processes of regulation of physiological functions, where the circulatory system is considered as an indicator of adaptive reactions of the whole organism, is the analysis of heart rate variability (HRV) [4, 5, 6]. The pathology of the cardiovascular system is a serious problem, as it is one of the most common causes of disability and deaths of a significant number of people. In Russia, mortality from circulatory system diseases has remained the highest in the world in the last two decades. A new methodology for studying the processes of regulation of physiological functions, where the circulatory system is considered as an indicator of adaptive reactions of the whole organism, is the analysis of heart rate variability (HRV) [4, 5, 6].

Currently, HRV analysis is recognized as the most informative non-invasive method for quantifying vegetative regulation of heart rhythm. The HRV analysis method is a nonspecific diagnosis of cardiovascular pathology with an assessment of the state of peripheral vegetative and humoral-metabolic regulation of pacemaker activity in the sinouauricular node of the heart, based on a high-resolution analysis of wave multifrequency variations in the duration of inter-cystolic intervals (the so-called RR intervals) of an electrocardiogram. Decrease in HRV values precedes haemodynamic,metabolic, energy disorders and is the earliest prognostic sign of the patient's unhappiness [7, 8, 9, 10].

The mechanisms of heart rate variability have not been fully deciphered and are still in the literature. To date, there is no accurate interpretation of the results of HRV analysis, and there is no HRT under different production conditions. Therefore, the introduction of a new method for studying the influence on the vegetative and humoral metabolic regulation of the selective effect of harmful production factors.

There remain many unresolved problems in the study of HRV, which require careful study and therefore research in this area must necessarily continue. In connection with this, **the purpose of this work is to study** the regulatory mechanisms of heart activity in workers of the «Karbodin» plant in Orekhovo-Zuyevo, Moscow Region, using the analysis of heart rate variability.

To achieve this goal, the following **tasks** must be accomplished:

1.To study statistical and index indicators of heart rate variability in the study groups of workers.

2. To study the parameters of the spectral analysis of heart rate variability in the study groups of workers.

3. Determine the effect of harmful and dangerous production factors on the state of the heart's regulatory systems in the employees of the enterprise, depending on the length of service.

IOP Conf. Series: Earth and Environmental Science **115** (2018) 012018 doi:10.1088/1755-1315/115/1/012018

Materials

112 people aged 20 to 50 years were studied, and the subjects were divided into three groups:

control group, which includes employees who have recently arrived at the enterprise (up to 1 year), without significant bad habits and almost healthy (30 people).

And two experimental groups of workers in the production of chemical plastics. The total number of 82 people. The considered groups of workers were ranked according to the duration of work at the enterprise:

- with a work experience of up to 5 years
- with a work experience of 5 to 10 years.

A 5-minute ECG recording was performed in a quiet separate room, with a constant temperature (C). The period of adaptation to the study conditions (horizontal position of the body) was 10 min. Registration of cardiointervals was carried out with the help of hardware and software complex "Varicard" with subsequent processing of the received material under a special program.

Statistical processing of data was carried out with the help of: a statistical package STATISTICA 6.

In the time domain, the statistical characteristics of the dynamic series were calculated:

• mean value of the cardio interval (R-R).

• standard deviation of the average NN intervals computed over 5-minute time intervals (low-frequency components of the HRV) - SDNN.

• rms deviation of absolute increments of cardiocycle durations, (high-frequency components of HRV) - RMSSD.

• Percentage of cardiocycles, the duration of which differs from the previous one by more than 50 milliseconds - pNN50%.

The parameters of variation pulsometry were estimated:

• mode values (Mo, ms) - the most frequently encountered in this dynamic range is the value of the cardiointerval

Amplitude mode - (AMO,%) - percentage of R-intervals that entered the modal class with a width of 50 milliseconds, characterizing the measure of the influence of the sympathetic department of the ANS [3].

The obtained data allow to determine the stress index (IN), which characterizes the degree of functional tension of the regulatory mechanisms of the circulatory system:

In the frequency range of cardiac rhythm variability in stationary intervals, cardiointervalograms were determined:

• TP - total power of all normal intervals R-R (ms^2), reflecting the total effect of all regulatory systems on the heart rhythm.

• Spectral power densities in the very low frequency range - VLF (ms^2) - the effect of thermoregulatory and peripheral motor systems and inter-system integration at the higher brain level, emotional and psychogenic effects on HRV.

• power in the low frequency range - LF (ms^2) - reflecting sympathetic and parasympathetic cardiac and vasomotor influences.

• power in the high frequency range - HF (ms^2) characterizing the parasympathetic cardiac effects caused by the respiratory waves.

• LF / HF index, reflecting the balance of sympathetic and parasympathetic influences on ANS.

Results and Discussion

Our analysis of statistical and index indicators of heart rate variability revealed that the workers of the study groups statistically significantly differ in heart rate indicators, which indicates differences in the functioning of their cardiovascular system and regulatory systems in general. The average values of the working indicators are given in Table 1 (the mean values and standard deviations are indicated).

Table 1

Statistical and index indicators of heart rate variabilityin the study groups of workers.

Indicator	Control	Group 1	Group 2
R-R , ms	876.00±126.01	928.44±234.6	698.73±65.32***
SDNN, ms	58.12±26.12	49.87±15.62	40.63±20.76***
RMSSD, ms	52.18±31.59	51.67±23.64	39.17±25.90***
pNN50, %	38.60±17.57	32.59±14.32	22.32±12.44***
Mo, ms	710.126±32.12	660±25.44	590±21.45***
AMo, %	39.1±3.41	42.4±2.92	53±3.95***
Indicator	117.26±32.12	201.52±54.05	252.98±46.25

** - differences between groups are statistically significant at p≤0.01

*** - differences between groups are statistically significant at p≤0.001

The mathematical expectation (mean value of the RR-interval) reflects the result of all regulatory influences on the heart and the circulatory system as a whole. The indicator has the smallest variability among all medical and statistical indicators, and its deviation from the individual norm usually signals an increase in the load on the circulatory system or the presence of pathological abnormalities. In our studies it was shown that the average value of the RR-interval is within the limits of the norm for workers of the first group, with an increase in the length of service from 5 to 10 years (group 2), the value of the RR-interval significantly decreases by 20.2% (p≤0.01) in comparison with the value of this indicator of the control group.

The mean square deviation (SDNN) is extremely sensitive an indicator of the state of the mechanisms of regulation of the heart rhythm. A decrease in the index is associated with an increase in sympathetic regulation, which enhances the activity of the autonomous circuit. Our studies show that the SDNN value of the workers of the first group differed insignificantly from the SDNN values of the control group and was less by 14.2% with an increase in work experience from 5 to 10 years (group 2) there is a significant decrease in the parameter compared to the control group by 30.1% (p <0.001).

The RMSSD indicator reflects the activity of the autonomous regulation loop. The higher the value of RMSSD, the more active the link of parasympathetic regulation. In the case of workers in Group 1, this was lower than in the control group, but there was no significant difference. With an increase in the length of service from 5 to 10 years (group 2), there is a significant decrease in the parameter compared to the control group by 25.1% (p <0.001).

pNN50% - gives information similar to RMSSD , but since only difference values above 50 ms are taken into account here, this index is more sensitive to high-frequency, respiratory vibrations of the heart rhythm and, therefore, better reflects the activity of the autonomous regulation loop. From the values of pNN50, one can judge the relative predominance of the parasympathetic or sympathetic link of regulation. In the case of workers in Group 1, this was lower than in the control group, but there was no significant difference. With an increase in the length of service from 5 to 10 years (group 2), there is a significant reduction in the parameter compared to the control group by 31.5% (p <0.001).

MS - indicates the dominant level of functioning of the sinus node. In the first group, the Mo value is at the lower limit of the norm and 7.4% lower than in the control group. In group 2, the value of Mo decreases reliably by 16.9% (p≤0.01).

Based on the data obtained, it can be said that most of the workers in the 2nd group are sympathicotonics, i.e. characterized by increased activity of the sympathetic department of the autonomic nervous system.

On the basis of the data obtained, it can be said that most of the workers in the 2nd group are sympathicotonics, i.e. characterized by increased activity of the sympathetic department of the autonomic nervous system.

The amplitude of the MS (mode) - reflects the effect of the stabilizing influence of the sympathetic nervous system on the cardiorhythm. The conditional norm of AMO is 32-41% (eutonium).

A decrease in AMO below 32% indicates vagotonia; an increase of more than 41% - about sympathicotonia.

The mean values of AMO in group 1 are near the upper limit of the norm, and do not differ significantly. In group 2, the AMO value increases significantly by 26.4% ($p \leq 0.001$).

The stress index (IN), proposed by R.M. Bayevsky, is an integral indicator designed to evaluate the "degree of stress of regulatory mechanisms." The higher it is, the higher the "activity of the sympathetic regulation channel," and the lower, the lower the "activity of the humoral and parasympathetic channels," that is, the increase in IN indicates on the "tension of adaptation", and its decrease indicates a stable adaptation to the effects of various environmental factors. A small load (physical or emotional) increases the IN 1.5 - 2 times. According to our data, the IN of the workers of the 1st group was 201.52 ± 54.05, which is 1.72 higher than in the control group. With an increase in the length of service from 5 to 10 years, ID was 252.98 ± 46.25, which is 1.26 ± 0.05 times higher than in the first group and 2.16 ± 0.05 times higher than in the control group.

Workers engaged in the production of phenol-formaldehyde plastics and plastics showed significant changes in the parameters of spectral analysis of heart rate variability (Table 2)

Table 2

The parameters of spectral analysis of heart rate variability in the study groups of workers

Indicator	Control	Group 1	Group 2
TP, ms^2	4196±328.21	3554.44±234.61	6814±324.12***
VLF, ms^2	1356 ±323.23	923±245.25***	1860±299.38***
LF, ms^2	1443 ±417.2	1534± 245.28	3520±299.32***
HF, ms^2	1397±523.7	1097± 232.29	1434 ±299.3***
(LF/HF)	1.03±0.85	1.36±0.36	2.45 ± 0.36 ***

** - differences between groups are statistically significant at $p \leq 0.01$

*** - differences between groups are statistically significant at $p \leq 0.001$

Thus, the total power of the spectrum (TP, ms^2) for workers in the first group differed insignificantly from the values (TP, ms^2) of the control group and was less by 15.3%. With an increase in the length of service from 5 to 10 years (group 2) TP (6814 ± 324.1 ms^2)) compared with the control group by 32.5% ($p < 0.001$).

When studying the ultra-low-frequency component of the spectrum responsible for the humoral-metabolic effects of the heart rhythm, it is noted that workers with a small amount of work experience decrease the VLF value as compared to the control group. Thus, in group 1 the index of the ultra-low-frequency component of the spectrum is 923 ± 245.2 ms^2, which is 31.9% lower than in practically healthy control group ($p < 0.001$).

In the second experimental group, the VLF (1860 ± 299.3 ms^2,) increased by 29.8% ($p < 0.001$) compared to the control group.

The LF index, which characterizes the maximum power of low frequency waves, reflecting the activity of the sympathetic centers of the medulla oblongata (cardiostimulating and vasoconstrictor) in workers of the first group increases insignificantly as compared to the control group. For example, in group 1, LF is 1534 ± 245.2 ms^2, which is 6% higher than the practically healthy control group - 1443 ± 417.2 ms^2. In the second group, the LF (3520 ± 299.3 ms^2) increases significantly compared to the control group.

The high-frequency component of the HF spectrum is associated with respiratory movements and reflects the vagal control of the heart rhythm (fluctuations in the parasympathetic part of the autonomic nervous system). In the first group, at the beginning of the labor activity there is a slight decrease in the HF index. So, in the 1st group the HF index is 1097 ± 232.2 ms^2, which is 14.3% lower than the practically healthy control group.

In workers of the second group, the HF index (1434 ± 299.3 ms^2) slightly increased compared to the same indicator of the control group (1397 ± 523.7 ms^2) by 4%.

The sympathetic-vagal index (LF / HF) conditionally characterizes the percentage contribution of sympathetic and parasympathetic influences to autonomic regulation of the heart rhythm. Its increase indicates a sympathetic regulation of the heart rate, a decrease - about the opposite effect. In the first group of workers, the value of this index was slightly shifted towards the predominance of the influence of the sympathetic department of the autonomic nervous system (1.36 ± 0.36), while in the control group this index was 1.03 ± 0.85 ($p < 0.05$).

With the increase in the length of employment in Group 2, there is a significant increase in the sympathetic-vagal balance in the direction of the predominance of sympathetic influences, which indicates an increased tension in the mechanisms of regulation of cardiac activity. For example, in group 2 the indicator (LF / HF) is 2.45 ± 0.36, which is 1.8 times higher than the indicator (LF / HF) of practically healthy control workers.

Conclusion

Our studies have revealed the negative impact of harmful factors of production on the system of regulation of cardiac activity. Workers engaged in the production of phenol-formaldehyde plastics and plastics, regardless of the length of service, there are changes in the statistical parameters of the heart rate. Thus, in both groups of workers, the reduced RR-interval, SDNN, RMSSD, pNN50%, Mo mode are noted, which reflects an increased tone of the sympathetic part of the autonomic nervous system. In workers of the 2nd group, these indicators decrease reliably, which reflects the increased tension of the mechanisms of regulation of cardiac activity. The values of the index of vegetative tension, the adequacy of regulatory processes, the vegetative index of the rhythm are significantly higher in workers of the 2nd group, which indicates a higher activity of the sympathetic department of the autonomic nervous system. As noted in the literature, the indicators SDNN and RMSSD, pNN 50% change unidirectionally, as shown in our experiment.

Our studies have revealed a negative effect of the harmful. In both groups, as the intensity of the adaptive processes increases, along with a decrease in the RR-interval, SDNN, RMSSD, pNN50%, Mo, AMo and IN indicators increase, which confirms the consistency of changes in the heart rhythm, indicates centralization of control heart rate and increased sympathetic effects.

When considering the wave characteristics of HRV in the company's employees, we come to a conclusion about the negative impact of harmful factors of production on the system of regulation of cardiac activity. Workers engaged in the production of phenol-formaldehyde plastics and plastics, regardless of length of service, observe a change in the overall power of neurohumoral regulation associated with the mobilization of functional reserves. The total power of the spectrum of neurohumoral regulation in cardiac rhythm modulation among workers in group 1, at the beginning of work, changes in the direction of decline, i.e. a decrease in the regulatory capacity of the body, with increasing length of service, activation of the lower levels of control is observed. This jump in the total power of the spectrum for workers with a work experience of 5-10 years can characterize the displacement of the pacemaker (within the zone of the sinus node).

When studying very low-frequency oscillations of the spectrum, there is a significant change associated with the work experience of workers. Workers of the first group have a decrease in humoral metabolic and cerebral ergotropic influences in modulation of the heart rhythm, as evidenced by a change in VLF. The obtained data indicate that the response of the regulation channel depends on the initial voltage of the tone of vegetative innervation of the heart.

Workers of Group 2 show a significant increase in VLF, which indicates activation of the superegmental structures of the brain, and this change indicates the transition of the regulation of the heart rate from the reflexive vegetative control level to the humoral metabolic, which is unable to provide homeostasis quickly.

According to some authors, the arrhythmia, tachycardia, diencephalic pathology, autonomic dysfunction, asthenic syndrome of vegetative dystonia, and damage to the peripheral nervous system are observed among workers in the chemical industry, as well as working with formaldehyde resins [11,12].

The magnitude of LF in the work of group 1 tends to increase, indicating an increase in the influence of the sympathetic vascular center. As the work experience increases, the activity of the sympatho-adrenal system in workers of the 2nd group increases significantly.

Sympathetic-parasympathetic equilibrium among workers of the 1st group was slightly shifted towards the predominance of the sympathetic department of the autonomic nervous system. As the work experience increased in Group 2, a reliable shift of the sympathetic balance to the predominance of sympathetic influences was observed, indicating an increased tension of the regulatory mechanisms cardiac activity.

Thus, with prolonged stays in unfavorable working conditions, persistent changes in physiological functions are possible, accompanied by a disruption in the function of the cardiovascular system [12]. In the case of an extreme force of influence or a long duration of production factors, a pronounced tension of the regulatory systems arises that is required to mobilize the body's functional reserves and to include the appropriate adaptive reactions necessary to maintain and preserve homeostasis. Mobilization of functional reserves can be considered as one of the results of regulatory systems to protect the body from adverse effects or to adapt it to new conditions of existence.

Recommendations

Based on the data obtained during the study, **it is possible to recommend**:

1. The introduction of high-resolution ECG diagnostics into practice (HRV analysis) will allow to identify early professional pathology (at the level of the donorosis state).
2. The method can be used for mass preventive examinations of the working population (as a method of additional diagnosis of cardiovascular pathology with an assessment of the state of peripheral vegetative and humoral-metabolic regulation of pacemaker activity of the sinus node of the heart).
3. The use of this methodology will allow under the conditions of production to assess the vegetative regulation of cardiac activity in healthy people and in workers with various pathologies.
4. Use of this technique will allow to predict the current functional state and state of physiological systems of the body during professional selection, and the definition of occupational fitness.
5. The obtained data can be used in the planning of preventive measures, adequate monitoring of the health status and prescription of drug therapy.
6. These materials can be used for lecture materials and sanitary and industrial research by the department of industrial ecology and labor protection.

References

[1] Bayevsky, R.M. Prediction of conditions on the verge of norm and pathology. Moscow: Medicine, 1979. 295 p.

[2] Bayevsky, R.M. Analysis of heart rate variability in space medicine // Human Physiology. 2002., June. T. 28. - 2. P. 70-82.

[3] Training manual on the physiology of the heart. Ed. Udelnova, MG, Samonina, G.E. Moscow: MSU, 1986. 79 p.

[4] Baevsky, R.M. The problem of evaluation and prediction of the functional state of the organism and its development in space medicine // Uspekhi fiziologicheskikh nauk. 2006. T.37. - 3. P. 42 57.

[5] Bayevsky, RM, Ivanov, GG, Chireikin, L.V. and others. An analysis of heart rate variability when using various electrocardiographic systems. Methodical recommendations // Herald of arrhythmology. 2001. - 24. P. 65 89.

[6] Baevsky, RM, Nikulina, GA, Funtova, II, Chernikov, AG Vegetative regulation of blood circulation. Orbital station Mir. 2000. T.2. P. 123 145.

[7] The variability of the rhythm of the heart. Ryabykina, GV, Sobolev, A.V. M .: Publishing house of Starko, 1998. 68 p.

[8] Heart rate variability. The idea of the mechanisms. S.A. Kotelnikov, A.D. Nozdrachev, M.M. Odinak, E.B. Shustov, I.Yu. Kovalenko, V.Yu. Davydenko. Physiology of humans and animals, 2002. T.28. 49 p.
[9] Heart rate variability: standards, measurements, indicators, features of the method. Zarubin, F.E. Herald of arrhythmology, 1998. - 10. 27 p.
[10] Heart rate variability: theoretical aspects and possibilities of clinical application. Bayevsky, R.M. Moscow: Nauka, 2000. 61 p.
[11] Harmful substances in industry. A reference book for chemists, engineers and doctors. L .: Chemistry, 1976.
[12] Occupational health. Navrotsky, V.K. Moscow: Medicine, 1974. 350 p.

Risk Assessment for Public Health from Transportation Noise (on the example of the city of Belgorod)

A Yu Semeykin

Belgorod State Technological University named after V.G. Shukhov,
46, Kostyukova str., Belgorod, 308012, Russia

E-mail: alexsem.n@gmail.com

Abstract. In this paper, an assessment, modelling and calculation of risks for the health of the population of the Southern district of Belgorod was carried out. The study included acoustic calculations carried out on the territory of the assessment within the boundaries of the settlement area of the rectangle with the square 6.72 km^2, where about 80000 residents live. As a result of computational and instrumental data 4 zones with different levels of potential chronic acoustic impact on the population were allocated. With the use of mathematical models of risk the trend of the formation of moderate and high risk was determined in certain areas within the boundaries of the territory. It is shown that 65% of the residents of the district live in conditions of moderate health risks and about 9% – in conditions of a high risk of developing health disorders from the effects of noise (disorders of the nervous system, hearing and the cardiovascular system).

1. Introduction

At present, the problem of the impact of transportation noise is becoming more acute for many developing urban areas in various countries of the world. It's confirmed by an increase in the number of studies in this area. One of such developing urban territories of the Russian Federation is Belgorod, located in the southern part of the Central Federal District, near the border with Ukraine. The development strategy of the city of Belgorod provides the creation of comfortable and environmentally friendly urban environment through spatial development and improvement of the street-road network [1, 2]. The growth of the territory and population of Belgorod causes an inevitable increase in the intensity of traffic flows, which are the main cause of noise pollution of the urban area and the occurrence of acoustic discomfort. Federal Service for Surveillance on Consumer Rights Protection and Human Wellbeing notes an increasing number of complaints from the public about road traffic noise [3].

Noise influences in the urban environment are currently one of the most harmful technogenic factors, along with environmental pollution, and can lead to disorders of the nervous system and cardiovascular diseases, ultimately reducing a person's life by 8 to 10 years. According to studies conducted in Europe and Asia, about 20% of residents suffer from noise, which level exceeds permissible standards and adversely affects health. According to the World Health Organization, about 40% of Europeans are affected by traffic noise, which exceeds 55 dB in daytime and at night [4, 5, 6].

In many cases, the purpose of describing the noise environment (noise regime) is to develop a forecast of the noise situation that arises from the construction of projected industrial enterprises or highways of road, rail and air transport. For quiet residential areas, the problem must be solved using appropriate calculations or by conducting studies on models [7, 8, 9, 10].

The results of full-scale studies and simulation of the noise situation are necessary to identify areas with the greatest acoustic discomfort and, accordingly, a risk to the health of residents. The basis for predicting the risk to public health from different types of noise in the urban environment is to study the dependence of «exposure – response». The studies of W. Babich, E. van Kempen, H. Ising, J. Heinrich, J. M. Fields, Laure-Anne Gille, K. Kryter etc. are devoted to the investigation of the effect of elevated levels of transport noise [5, 6, 7, 8, 12]. In studies of W. Babich, H. Miedema and C. Oudshoorn

IOP Conf. Series: Earth and Environmental Science **115** (2018) 012019 doi:10.1088/1755-1315/115/1/012019

evolutionary mathematical models describing the «dose – effect» and «exposure – response» relationships under the influence of urban traffic noise are developed [5, 11].

The purpose of this work is to predict the magnitude of the population's risk to public health (in the form of hearing impairment, as well as in the cardiovascular and nervous systems) using the example of certain areas of the city of Belgorod (Russia). The location of the city and the study area are shown in figures 1 and 2.

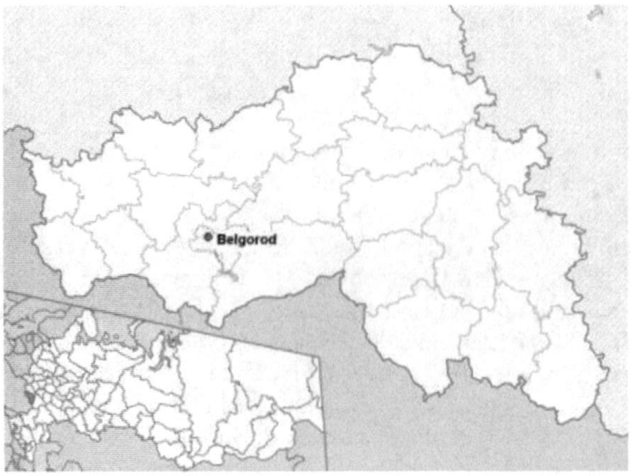

Figure 1. Location of Belgorod, Russian Federation

Figure 2. Belgorod city map and measurement site

2. Methods

2.1. Noise measurement and modelling

Investigation of the noise situation in some areas of Belgorod was carried out in accordance with the requirements of state standards GOST 20444-2014 and GOST R 53187-2008 [13, 14]. As the main noise characteristics of traffic flows, the equivalent L_{Aeq} and maximum L_{Amax} sound levels, dB(A), measured in daytime (from 7.00 to 23.00) and night time (from 23.00 to 7.00) are selected for the periods of maximum traffic intensity. For full-scale measurements, the integrating sound level meter Ecophysics-110, which meets the requirements of GOST 17187-2010, was used.

The results of field measurements of noise levels were used to simulate the noise environment and build noise maps of the territory with the help of the «Acoustics» software package. The software complex «Acoustics» allows the user to create a spatial plan of the surveyed area, taking into account the building and operating sources of noise (linear and point sources), and calculate the levels of acoustic impact anywhere in the space, build color fields and isolines of sound levels in the horizontal and vertical planes with user defined parameters [15, 16, 17, 18]. The spatial scheme of the location of objects in the investigated area was constructed on the basis of the 3D model of the city district obtained using the geographic information system 2GIS [19].

As a basic indicator of operating noise levels, the L_{den} (the A-weighted average level of sound pressure for the sum of day time periods over a year being used for the assessment of discomfort during daytime) was adopted. Indicator L_{den} can be determined according to the equation (1):

$$L_{\text{den}} = 10\lg\frac{1}{24}\left(16\cdot 10^{\frac{L_{day}}{10}} + 8\cdot 10^{\frac{L_{night}+10}{10}}\right) \quad (1)$$

where L_{day}, L_{night} – A-weighted average levels of sound pressure for the sum of periods of 16-hour day time and 8-hour night time respectively.

2.2. Risk assessment

Calculation of the predicted value of the risk of health disorders was carried out on the basis of the methodology of guidelines MR 2.1.10.0059-12 «Assessment of the health risk of the population from the effects of traffic noise». As the initial data for calculation and forecasting of the risk, noise measurements in the streets of the city and simulation of the noise situation were used. Based on the received data, noise maps were constructed, and areas of the urban territory in which the sound pressure levels exceed sanitary and hygienic standards were identified.

The calculation of risk values and the assessment of the aggregated risk of cardiovascular, nervous and hearing disorders were conducted using evolutionary mathematical models for the development of adverse effects under the influence of noise. Applied models take into account domestic and foreign data on the dynamics of these effects against the background of natural aging of the organism. Calculation of risk values was carried out by solving a system of recurrent equations [5, 11, 12, 20].

3. Results and discussion

3.1. Modelling of noise situation

An analysis of the noise situation and the assessment of the magnitude of the health risk were carried out for the urban area in the Southern District of Belgorod, as the most densely populated part of the city with a large number of streets with high traffic intensity. About 80000 people live in the district. In order to assess the level of noise pollution in the territories, we measured noise levels at various points in the districts between the streets of Shchors, Gubkin, Oktyabrskaya, Budyonny, Sportivnaya, Yesenin. The main source of noise in the territory of these districts is road transport, so to assess the noise level, we considered the complex sources of noise – traffic flows. Noise map of the Southern district of Belgorod is shown in figure 3.

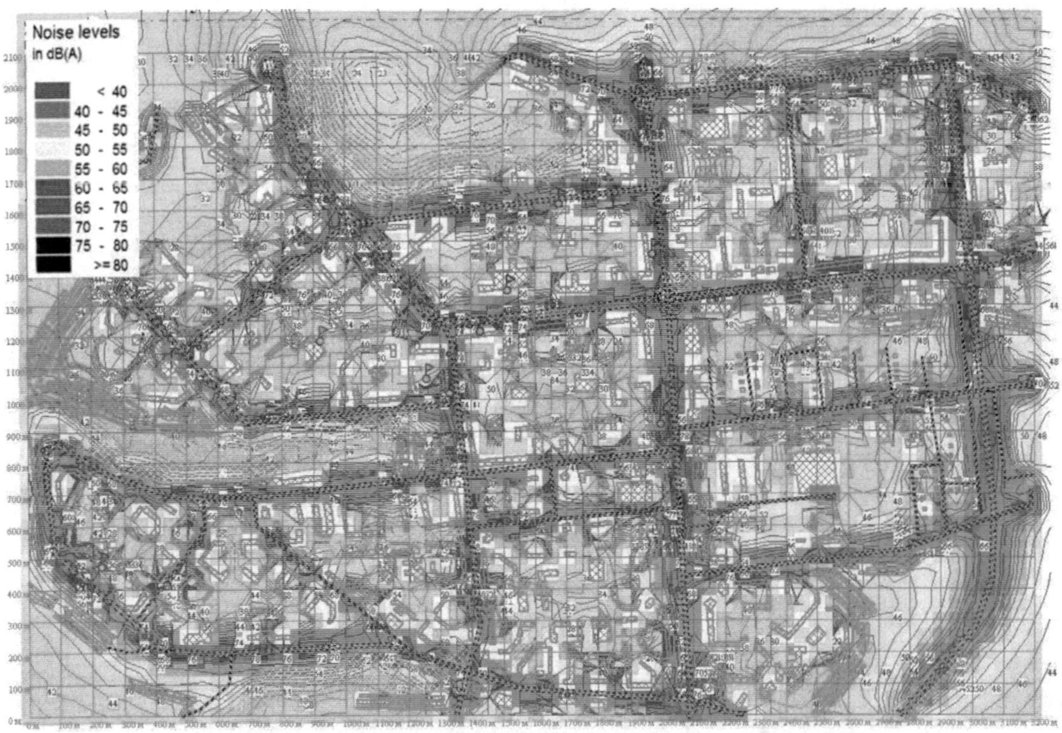

Figure 3. Noise map of the Southern district of Belgorod

Acoustic calculations were carried out for an area of 6.72 km^2 at nodal points with a grid spacing of 100×100 m. In addition, the noise level was calculated at points located on the territory of residential buildings, medical and educational institutions. It was revealed that in some areas of the city territory with a traffic intensity of more than 1000 cars per hour, the estimated sound pressure levels reached 78 dBA in the morning and evening hours. This is confirmed by the data of field measurements.

The analysis of the obtained noise map makes it possible to determine the most problematic zones of the Southern district of Belgorod by the noise factor. These include the crossroads of Gubkin and Shchors streets (exceeding 8 dBA), Gubkin and Budyonny streets (exceeding 4 dBA). All residential buildings located near the streets under study are in the zone of acoustic discomfort, which increases the risk of specific functional deviations in health due to the impact of traffic noise, which can be determined by special methods.

3.2. Results of risk assessment

The studies allowed to identify 4 zones of urban area, with average daily noise levels, respectively: zone 1 – L_{den} = 38 dB; zone 2 – L_{den} = 49 dB; zone 3 – L_{den} = 62 dB; zone 4 – L_{den} = 74 dB. Thus, about a third of the study area falls into the zone of acoustic discomfort with an increased risk of health problems.

The results of calculating the population risk to the public health under exposure of daytime average level of sound pressure 74 dBA in the Southern district of Belgorod are presented in Table 1.

Table 1. The results of the calculation of health disorders risk under exposure of daytime average level of sound pressure 74 dBA

Age, years	Aggregated risk of hearing disorders R_t^{Ahd}	Aggregated risk of cardiovascular disorders R_t^{Acd}	Aggregated risk of developing disorders of the nervous system R_t^{Ans}
1	0.023717	0.010902	0.041491
5	0.024442	0.014979	0.055423
6	0.025174	0.019255	0.070366
7	0.025914	0.023739	0.086394
9	0.026662	0.028442	0.103588
10	0.027417	0.033375	0.122034
15	0.02818	0.03855	0.141823
20	0.028951	0.04398	0.163056
25	0.029729	0.049677	0.185838
30	0.030516	0.055656	0.210285
35	0.031311	0.06193	0.236519
40	0.032114	0.068515	0.264672
45	0.032926	0.075428	0.294886
50	0.033745	0.082684	0.327313
55	0.034574	0.090302	0.362117
60	0.03541	0.0983	0.399474
70	0.036256	0.106698	0.439572

In accordance with the recommendations of MR 2.1.10.0059-12, the value of risk (R) less than 0.05 is estimated as a low risk, weakly affecting the level of health status in the study area. Measures are recommended to organize selective monitoring of noise load, planning activities that can be implemented in the long term (5 years or more). The value of R in the range 0.05 – 0.35 is estimated as moderate risk. Measures are recommended for the organization of continuous monitoring of the noise load. Measures to reduce the noise load are recommended to be developed taking into account the medium and short term (1 to 3 years). The planned revision is recommended at a frequency not less than

IOP Conf. Series: Earth and Environmental Science **115** (2018) 012019 doi:10.1088/1755-1315/115/1/012019

once every three years. Measures to reduce the noise load are recommended to be developed taking into account the medium and short term (1 to 3 years). The value of R in the range 0.35 – 0.6 is estimated as a high risk. Measures are recommended for the organization of an extended noise load monitoring program with additional studies in places and / or periods of maximum noise levels. It is necessary to develop measures to reduce the noise load for the short term within a year. It is recommended that the risk level be reviewed every year.

The calculation data show that when exposure to elevated noise levels to 74 dBA, the aggregate risk of developing nervous system disorders (nervous tension, sleep disorder, cognitive impairment, and vascular dystonia) is characterized as moderate, from the age of 7 to 50 years. Since age 55, the risk of developing disorders of the nervous system has reached a high level. The risk of hearing disorders (tinnitus, conductive sensor neural hearing loss, hearing loss caused by noise) is estimated to be insignificant up to the age of 50 years and moderate for a population aged 50 years. The aggregated risk of developing cardiovascular disease is estimated as moderate.

Table 2. The results of the calculation of the cumulative risk to public health under the exposure to transportation noise

Age, years	Cumulative risk of health disorders R_t^{Sum}			
	Zones under chronic exposure of noise level of sound pressure, dBA			
	38	49	62	74
1	0.0052571	0.05927833	0.067172	0.074426
5	0.0047394	0.06124523	0.077472	0.092314
6	0.0041945	0.06341676	0.088439	0.111219
7	0.0036208	0.06580626	0.100119	0.131196
9	0.003017	0.06843539	0.112559	0.152304
10	0.0023813	0.07132796	0.125812	0.174604
15	0.001712	0.07451009	0.139934	0.198157
20	0.0010075	0.07801046	0.154983	0.223029
25	0.002657	0.08186058	0.171023	0.249286
30	0.00515	0.08609507	0.18812	0.276995
35	0.01337	0.09075194	0.206347	0.306226
40	0.02203	0.09587301	0.225778	0.33705
45	0.03115	0.10150422	0.246497	0.369536
50	0.04074	0.10769613	0.268588	0.403757
55	0.05085	0.11450432	0.292143	0.439782
60	0.06149	0.12199001	0.317261	0.47768
70	0.0727	0.13022055	0.344045	0.517519

A space-time analysis of the magnitude of the cumulative health risk to public health under chronic exposure to noise is shown in Table 2. The results of the calculations showed that the risk of health disorders for zones with noise exposure level L_{den} = 38 is low and becomes moderate after the age of 55 years; in the zone with the value L_{den} = 49 dBA and L_{den} = 62 dBA the risk is moderate; in the zone with a value of L_{den} = 74 dBA, the risk of developing health disorders becomes high for the population aged 45 years.

The total population living in these zones of noise exposure is correspondingly: low risk for health 20800 (26%), moderate risk 52000 (65%), high risk of 7200 (9%).

4. Conclusion

The most probable consequences of the impact of traffic noise in the Southern district of Belgorod are the aggregated risk of developing disorders of the nervous system (nervous tension, sleep disorder, cognitive impairment and vascular dystonia), starting from the age of 5 years. The risk of hearing disorders (tinnitus, conductive sensor neural hearing loss, hearing loss caused by noise) is estimated as moderate for the population over the age of 50 years. The aggregated risk of developing cardiovascular disease is estimated as moderate. A space-time analysis of the magnitude of the cumulative health risk to public health under chronic exposure to noise shows that total population living in these zones of noise exposure is correspondingly: low risk for health 20800 (26%), moderate risk 52000 (65%), high risk of 7200 (9%). Thus, it requires mandatory implementation of a set of measures to reduce noise pollution in the city, as envisaged by the Concept «Comprehensive scheme of urban transport development for 2009 – 2025», which sets the task of developing a system of measures to reduce noise pollution of the urban environment by road transport in order to reduce the population, living in conditions of noise discomfort by 10%.

Acknowledgments
The work is realized in the framework of the Program of flagship university development on the base of the Belgorod State Technological University named after V.G. Shukhov, using equipment of High Technology Center at BSTU named after V.G. Shukhov.

References
[1] *Strategy of the development of Belgorod until 2025*, approved by The Council of Deputies of Belgorod from 30.01.2007 №413 (in Russian)
[2] The concept *Comprehensive scheme for the development of urban transport for 2009-2025*, approved by The Council of Deputies of Belgorod from 09.12.2008 №131 (in Russian)
[3] *The state of sanitary and epidemiological welfare of the population of the Belgorod region in 2016: report* (Belgorod: Department of Federal Service for Surveillance on Consumer Rights Protection and Human Wellbeing in the Belgorod region) 2017 p 211 (in Russian)
[4] World Health Organization 2011 *Burden of disease from environmental noise. Quantification of healthy life years lost in Europe* (Copenhagen: The WHO European Centre for Environment and Health) p 128
[5] Babisch W 2006 *Transportation noise and cardiovascular risk, review and synthesis of epidemiological studies, dose-effect curve and risk estimation* (Berlin: Umweltbundesamt) p 113
[6] Ising H, Kruppa B 2004 Health effects caused by noise: Evidence in the literature from the past 25 years *Noise and Health* **6(22)** 5–13
[7] Gille L-A, Marguis-Favre C, Morel J 2016 Testing of the European Union exposure-response relationships and annoyance equivalents model for annoyance due to transportation noises: The need of revised exposure-response relationships and annoyance equivalents model *Environment International* **94** 83–94
[8] Fields J 1993 Effect of personal and situational variables on noise annoyance in residential areas *J. Acoust. Soc. Am.* **93 (5)** 2753–63
[9] Chubirko M I, Stepkin Yu I, Seredenko OV Hygienic assessment of noise factor of the large city *Hygiene and sanitation* **94 (9)** 37–38 (in Russian)
[10] Semeykin A Yu 2015 Modeling of the noise effect of traffic flows in the urban environment by the example of certain areas of Belgorod. *Ecology and safety in the technosphere: modern problems and solutions: Proceedings of All-Russian Conference vol 1* (*Tomsk: TPU Publ.*) 303–6 (in Russian)
[11] Miedema H, Vos H 1998 Exposure-response for transportation noise *J. Acoust. Soc. Am.* **104 (6)** 3432–45
[12] Kryter K 2009 Acoustical model and theory for predicting effects of environmental noise on people *J. Acoust. Soc. Am.* **125 (6)** 3707–21
[13] GOST 20444-2014 *Noise. Traffic flows. Methods of noise characteristic determination* Moscow

2014 (in Russian)
[14] GOST R 53187-2008 *Acoustics. Noise monitoring of cities* Moscow 2008 (in Russian)
[15] Bilaşco Ş, Govor C, Roşca S Vescan I, Filip S Fodorean I GIS model for identifying urban areas vulnerable to noise pollution 2017 *Front. Earth Sci* **11 (2)**. 214-28.
[16] Cai M, Zou J, Xie J, Ma X 2015 Road traffic noise mapping in Guangzhou using GIS and GPS *Applied Acoustics* **87** 94–102.
[17] Ko J H, Chang S I, Lee B C 2011 Noise impact assessment by utilizing noise map and GIS: A case study in the city of Chungju, Republic of Korea *Applied Acoustics* **72** 544–50.
[18] Murphy E, King E A 2010 Strategic environmental noise mapping: Methodological issues concerning the implementation of the EU Environmental Noise Directive and their policy implications *Environment International* **36 (3)** 290–8
[19] Semeykin A Yu 2017Assessment of the noise situation of urban environment in certain districts of Belgorod *Bulletin of BSTU named after V.G. Shukhov* **9** 56–60 (in Russian)
[20] May I V, Koshurnikov D N, Galkina O A 2017 Space-time analysis of risk to public health under the exposure to urban noise (on the example of Perm) *Hygiene and sanitation* **96 (1)** 35–39 (in Russian)

Engineering and Computing Portal to Solve Environmental Problems

A M Gudov [a], S Y Zavozkin [b], I Y Sotnikov [c]

Kemerovo State University, Kemerovo, Russia

E-mail: [a] good@kemsu.ru, [b] shade@kemsu.ru, [c] mxtfonlife@mail.ru

Abstract. This paper describes architecture and services of the Engineering and Computing Portal, which is considered to be a complex solution that provides access to high-performance computing resources, enables to carry out computational experiments, teach parallel technologies and solve computing tasks, including technogenic safety ones.

1. Introduction

Human-made technogenic environment is known for a lot of potential and real threats that are dangerous for both the environment and human beings themselves. The role and impact of man-made factors on habitat and human-beings are necessary to be studied in order to get the system overview.

Coal mining companies can be vivid examples of man-made entities that negatively affect the environment. The development of coal industry enterprises causes the increase in water consumption that is associated with both production and subsequent coal enrichment. As a result, significant increasing of the amount of wastewater is detected, that is considered to be one of the major factors of water pollution. Quite a lot of wastewater components can accumulate in water reservoirs and inside aquatic organisms, causing irreversible effects. Particularly, implementation of highly mechanized complexes with complex hydraulic systems at mines causes large consumption of oil products in the context of mining operations, some of them remain in mine water and further pollute it. Therefore, the development of wastewater treatment methods is currently of great value. To assess the effectiveness of such methods, software tools are needed that provide an opportunity for mathematical modeling of the wastewater treatment process, since field experiments are very labor-intensive. As a result of this modeling, it is determined that one of the most cost-effective methods is wastewater treatment in spent mining workings of flooded coal mines.

Based on the information above, the important tasks are as follows: assessment of a current level of facility technogenic safety, possible threats forecasting and efficiency evaluation of their elimination methods. Specialized software and hardware are usually used to solve such problems. Often you have to deal with a variety of programs, different ways of interacting with them, various formats of input and output data at the same time. Such kind of software is not always open source and publicly available. Commercial products are known for a high license cost to use them. Moreover, computational experiments require high-performance computing resources and technologies.

Cloud computing model can be used in order to reduce the cost of computing experiments, so the payment is required only in case of software renting and computing resources usage. Taking into consideration this model, the Engineering and Computing Portal has been developed at Kemerovo State University, which provides a set of services for solving science-intensive tasks, including those related to technogenic safety [14]. Access to the following services is currently available:

– The "Onlide" service [15,16] for the development of own serial and parallel programs. Here are the main system functions: design of multi-file program projects; available initial code uploading and downloading the one created by using the system; compilation and launching on the proposed computing resources, including high-performance resources; saving program results to a user's workstation. It also has a debugging service for parallel programs, based on the automated control of the correctness approach. Users have the opportunity to independently determine the situations in the MPI program that are supposed to be detected by the system.

– Service for managing a user's file storage. Calculation results, user program outputs as well as software projects are stored in a dedicated storage for each user.

– Service for solving the problem of mine flooding that uses a multiparameter mathematical model that allows studying the processes of flow, distribution and settling of non-dissolved impurities, taking into consideration the possibility of changing the mine working form due to sediment accumulation [13]. The service enables to carry out numerical experiments with possible environmental consequences caused by using flooded underground mine workings as treatment facilities, as well as visualize the results obtained.

– The "Virtual Laboratory Practice" service [15,16], providing teaching materials devoted to the theory and technology of high-performance computing, as well as virtual laboratory workshops on parallel programming.

2. Architecture

The portal is based on the service-oriented architecture (SOA), used to build distributed systems that provide services for other systems or other services [4]. SOA simplifies the integration of new components to enhance the portal's capabilities. The SOA uses Web services technology [9] based on such standards as: WSDL [11,12] - used to describe the Web service, SOAP [5,6] - represents message format for interacting with the Web service and BPEL [10] - used to describe business processes (orchestration). Business process is a high-level Web service that combines capabilities of the existing ones.

Figure 1 shows the portal architecture. Components such as Liferay Portlet, Apache Axis2, Apache ODE, and LDAP Server will be discussed in more detail in the next section. For the other components, a brief description is given below:

– Client System is any external client system that interacts with Web services / business processes of the portal and / or with the LDAP directory.

– User Client Workstation is a user's workstation. For example, a personal computer.

– Web-browser is a web browser installed on a client's workstation; it is used to deal with the portal web-interface.

– Portal Server is a server computer that hosts the portal.

– Nginx is used as a proxy server, passing requests to other components (Apache Tomcat 7, Tornado and PHP-FPM). Static file (images, JavaScript files, etc.) requests are considered to be exceptions, because Nginx is optimized for this type of query.

– Tornado is a Web server that is used as a platform for the "Onlide" program online development service performance.

– PHP-FPM is a FastCGI process manager used to generate dynamic PHP content for the "Virtual Laboratory Practice" service, because Nginx does not have native support for generating such kind of content.

– Apache Tomcat 7 is a Web application server that provides support for servlet specifications. Here Liferay, Axis2 and ODE are deployed.

– AppGenPortlet is a portlet for generating Web interface of the portal services.

– HPCAdminPortlet is an administrative portlet for managing access to computing resources and resources accounting.

– AuthHandler is responsible for verifying user access to Web services and business processes.

– Artifact HPCWebService is a Web service for interacting with high-performance computing resources.

– StorageWebService is a Web service for interacting with a user's file storage.

– MineFloodingWebService is a Web service for interaction with the program that solves the problem of mine flooding.

– VTKUtilsWebService is a Web service for dealing with VTK format files.

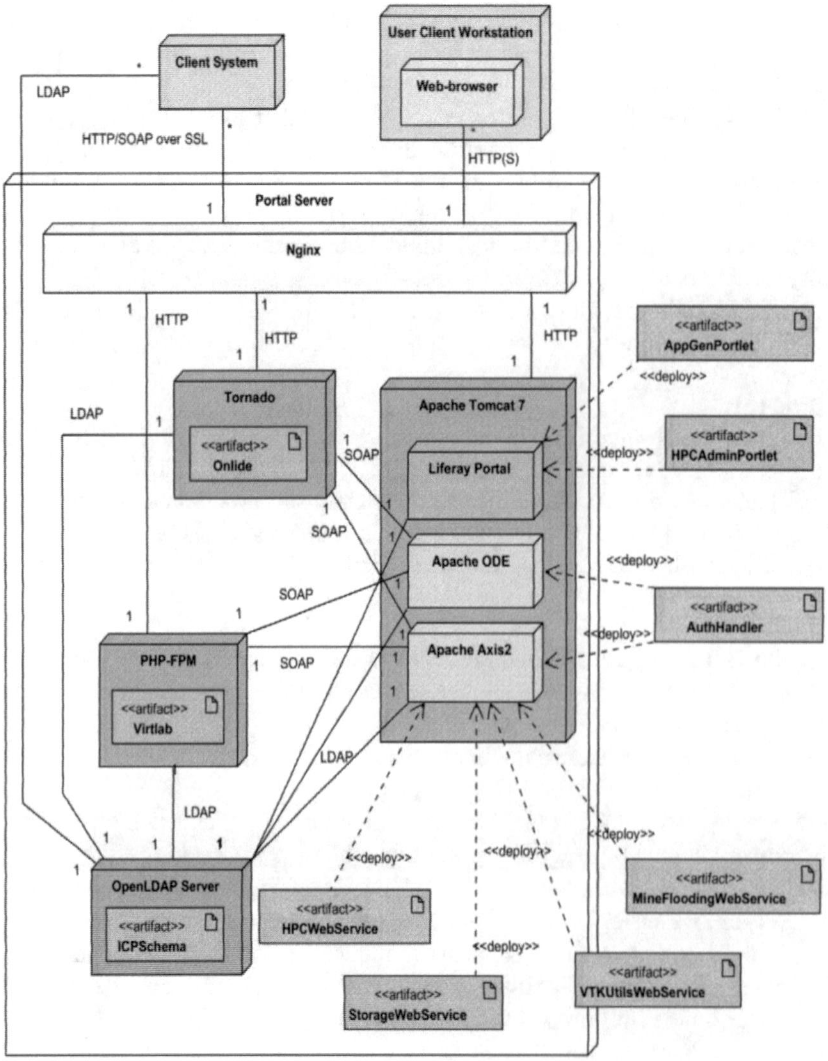

Figure 1. Portal architecture as a UML deployment diagram

3. Solutions used

Information concerning portal users, Web services, business processes, computing resources is stored in the LDAP directory [8]. OpenLDAP open solution is chosen as an implementation.

To integrate the Web-based service approach, Apache Axis2 [1,7] and Apache ODE solutions are used. Apache Axis2 represents the system of integration and management of the life cycle of Web services. The main features of this system are as follows:

– SOAP versions 1.1 and 1.2 support. This feature of Axis2 can be useful in case the client system does not support the latest version of the SOAP protocol;

– WSDL versions 1.1 and 2.0 support;

– A modular structure that provides extensibility by integrating new modules and handlers.

Apache ODE is the BPEL business process execution system.

Liferay Portal is a configurable integrated solution for building Internet portals. Liferay enables to integrate existing users from LDAP directories.

All pages of the portal are built by using portlet technology. A portlet is a Web application developed in accordance with the JSR-168 or JSR-286 specification [2,3]. They generate fragments of HTML content that are embedded in a Web page. Such kind of a page can consist of many portlets.

Liferay provides the ability to create personal pages for each user. This feature is used to create a personal user workspace. When registered, a user can be associated with a certain user group. Liferay enables to set personal site templates for each group, which are added to the user's desktop as soon as the user is added to the corresponding group. Moreover, Liferay provides roles that allow to limit user's access to portlets or ban to perform certain actions on the portal.

4. Services

The descriptions of the already implemented services are presented below.

1) Service for managing the storage of user files.

Each registered user is considered to have his own directory for storing files (results, initial programming codes, images, etc.). There is the service implemented that is used by users to interact with their storages via the portal web-interface. Fig.3 presents its interface.

Figure 2. Service for user interaction with data storage

2) Service for solving the problem of mine flooding.

The corresponding portal service is implemented for the multiparameter model [13] to solve the problem of mine flooding. This service authoring involves the following stages:

1. Create a new project or open a previously created one;
2. Create a new task or select an existing one;
3. Configure the model parameters (Figure 3), start the task and wait for it to be completed;
4. Start image generation according to the results achieved and wait for it to be completed;
5. Result visualization (Figure 4).

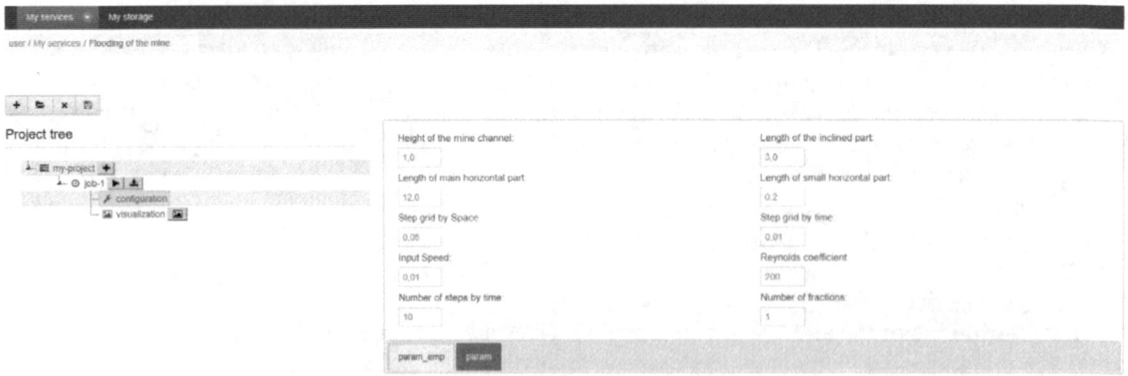

Figure 3. Service for solving the problem of mine flooding (mode of setting the model parameters)

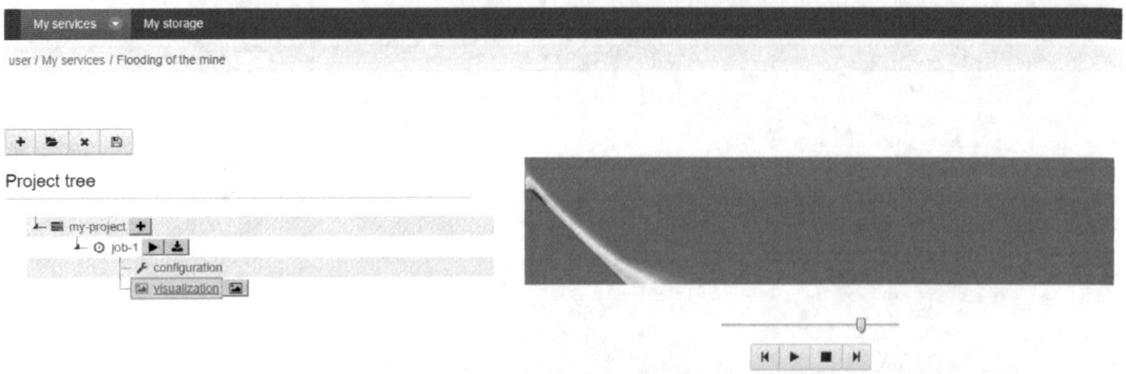

Figure 4. Service for solving the problem of mine flooding (rendering mode)

A user can finish the work with the portal and use the service later when starting a task and visualizing results. The solution results can be downloaded from the server by using the service itself, or via the previously presented service for interaction with the storage.

3) «Onlide» service.

This service is used for remote designing of parallel and sequential programs. Here is a list of available functions:

1. Designing of software projects, which can consist of many source files:
– Projects can be of different types;
– Various settings can be customized for different types of projects, such as compiler, compilation keys, plug-in libraries, etc .;
– Templates with already defined type, settings, files, etc can be used to design projects;
– Projects can be saved on a remote computing resource or client workstation, downloaded from them.
2. Project editing:
– Downloading files from a user's local workstation;
– Downloading project files to a user's local workstation;
– Creating new files;
– Deleting project files;
– Viewing and editing project files. Various editors and display methods are available for different file extensions.
3. Compiling and starting a project;

4. Using and creating plug-ins for enhancing functionality of the development environment.

5. Conclusion

The engineering and computing portal is implemented at Kemerovo State University. It is used for computational experiments, teaching theory and technology for high-performance computing. This comprehensive solution is available for a wide range of engineers, students, graduate students and researchers. Portal services as well as computing resources are provided to users on a lease basis.

In the future, it is planned to expand the list of computing services. The CFD Phoenics solution in the form of services is under active integration process.

References

[1] Jayasinghe D. Apache Axis2 Web Services. D. Jayasinghe, A. Azeez; Packt Publishing; 2 edition, 2011. 308 pages
[2] JSR 168: Portlet Specification 1.0. Available at: https://www.jcp.org/en/jsr/detail?id=168
[3] JSR 286: Portlet Specification 2.0. Available at: https://www.jcp.org/en/jsr/detail?id=286
[4] Patterns - Service-Oriented Architecture and Web Services. Available at: http://www.redbooks.ibm.com/redbooks/pdfs/sg246303.pdf
[5] Simple Object Access Protocol (SOAP) 1.1. Available at: https://www.w3.org/TR/2000/NOTE-SOAP-20000508/
[6] SOAP Version 1.2 Part 1: Messaging Framework (Second Edition). Available at: https://www.w3.org/TR/soap12/
[7] Tong K. Developing Web Services with Apache CXF and Axis2. Tong K. – TipTec Development, 2010. JSR 168: Portlet Specification 1.0. Available at: https://www.jcp.org/en/jsr/detail?id=168
[8] Understanding LDAP Design and Implementation. Available at: http://www.redbooks.ibm.com/redbooks/pdfs/sg244986.pdf
[9] W3C. Web Service Architecture. Available at: http://www.w3.org/TR/ws-arch
[10] Web Services Business Process Execution Language Version 2.0. Available at: http://docs.oasis-open.org/wsbpel/2.0/OS/wsbpel-v2.0-OS.html
[11] Web Services Description Language (WSDL) 1.1. Available at: https://www.w3.org/TR/wsdl
[12] Web Services Description Language (WSDL) Version 2.0 Part 1: Core Language. Available at: https://www.w3.org/TR/wsdl20/
[13] A.M. Gudov, S.Y. Zavozkin, I.V. Grigorieva, L.V. Bondareva, V.A. Perminov High Technology Software Web-Tools to Solve Environmental Problems of Coal Region. Proceedings of the XIV International Terpogov Conference "Information Technologies and Mathematical Modeling (ITMM-2015)", Tomsk: Tom., 2015. Part 2. P. 120-124
[14] Goudov, A., Perminov, V., Filatov, Y., Un, L.H., Zavozkin, S., Grigorieva, I., Sotnikov, I. High technology software web-Tools to solve environmental problems of coal region (2017) CEUR Workshop Proceedings, 1839, pp. 61-73.
[15] Sotnikov I. Y. Web-system Paralarea for e-learning and program development in high-performance computing. Proceedings of the XVI All-Russian Conference of Young Scientists in Mathematical Modeling. Krasnoyarsk, Russia. October 28-30, 2015. Novosibirsk: ICT SB RAS, 2015. P. 91-92
[16] Sotnikov I. Y. Development of the Paralarea system for studying the theory and technologies of high-performance computing and its Onlide subsystem for program development. Collected materials of the All-Russian Scientific and Practical Conference "Information and Telecommunication Systems and Technologies", October 16-17. 2015, Kemerovo. FGBOU HPE "Kuzbass. State. Tech. Un-t them. TF Gorbachev »; Rare: Trofimov I.E. (Editor's note) [and others]. Kemerovo, 2015

Study of Water-Oil Emulsion Breaking by Stabilized Solution Consisting of Anionic Surface Acting Agent – Soda Ash – Polymer (ASP)

S V Kulichkov [a], **E G Avtomonov** [b], **L V Andreeva**[c], **S F Solomennik**[d], **A V Nikitina** [e]

Far Eastern Federal University,8 Sukhanova St., Vladivostok, Russia, 690090

E-mail: [a] seku230@yandex.ru, [b] avtomonov@inbox.ru, [c] 708738@gmail.com, [d]sfsol@yandex.ru, [e] rabota_oos@mail.ru

Abstract. The paper provides a laboratory research of breaking natural water-oil emulsions: - by non-stabilized ASP; by stabilized ASP; by mixture of stabilized and non-stabilized ASP in different proportions and production of refinery water of the required quality with the use of IronGuard 2495 as flocculant. Oil-in-water emulsion is stable. Classic methods are not suitable for residual water treatment: sediment gravity flow; filtration; centrifuge test. Microemulsion formed after ASP application has low boundary tension and high pH. It contributes to transfer of oil phase into a water one, forming oil-in-water emulsion. Alkaline condition has adverse effect on demulsifying ability of agents, flocculation and boundary tension. For breaking of water-oil emulsion at EBU before the interchanger water or water-oil emulsion from the wells that were not APS-treated in ratio of 1:9 shall be delivered. Residual water after EBU must be prepared in water tanks by dilution in great volume.

1. Introduction

Mature assets are characterized by a high water-cut and low recover coefficient. It leads to cost increase for oil production and processing.

ASP water flooding is a method of enhanced oil recovery. It is based on the idea of injection into formation of the solution consisting of surface active agent, soda ash (caustic soda) and polymer. This method allows producing oil that is left in the formation after utilization of the traditional flooding method.

After ASP application the products of flowing wells are characterized by presence of stable microemulsion.

Chemical method of enhanced oil recovery of the formation is based upon utilization of SAA solutions, including solutions of polymers, caustics and mixtures of chemical agents (ASP) in injection wells. In such case the products of flowing wells become caustic (high value of pH medium >9-10), they are characterized by super low value of boundary tension tending to zero and contain hard-breaking microemulsion.

After injection of agents increasing oil recovery of the formation developed, multiple emulsion is delivered to initial water separation units and oil treatment units. Such emulsion is hard to break and consists of a mixture of oil-in-water, water-in-oil and microemulsion. Such emulsions are accumulated on the interface boundary in bulk-capacity tanks for oil, water preparation and are the reason for upset conditions in initial water separation units and oil treatment units.

Zapadno-Salymskoye oilfield is located in the Khanty-Mansijsk Autonomous District (120 km to the south-west of Surgut and 30 km to the west of Salym settlement). It was put in operation in 2005.

In terms of gravity the oils of Zapadno-Salymskoye oilfield are referred to middle oils – 0.876 g/cm^3. Asphaltens content is 0.85-1.87 percent, pitches – 8.4-9.8 percent, paraffins – 1.53-1.81 percent, paraffin melting temperature is 63-64°C.

In terms of ratio of paraffin and asphaltic resinous components of oil they are classified as paraffin.

Temperature of the formations being developed is 100-110°C, temperature of water-oil emulsion at well mouth is 70°C. Gas-oil ratio is 45 m^3/t.

Produced waters are of low mineral content, their density comprises 1.008-1.011 g/cm^3, pH equals to 7.1-7.2.

After ASP application water-oil emulsion produced of 95 percent water-cut represents a stable microdispersed oil-in-water emulsion, and its boundary tension tends to $\sigma \rightarrow 0$ (pH=9.9).

Produced water-oil emulsion after ASP application is characterized by high water-cut of 90-95 percent, pH=9.88, water density of 1.013 g/cm^3 and super low values of boundary tension.

Applied ASP solution consists of the following components:

- Enordet surfactant (by SHELL) – 0.43 %;
- Isobutanol – 2.0 %;
- Soda – 2.0 %;
- Sodium Chloride – 1.1 %;
- FLOPAAM polymer (by SNF) – 0.25 %
- Soda (2%) forming a part of ASP contributes to $CaCO_3$ sedimentation of pump elements.

To determine the possible ways to solve these problems the following laboratory test were performed:

- water-oil emulsion breakdown:
  - by non-stabilized ASP;
  - by stabilized ASP;
  - be mixture of stabilized and non-stabilized by ASP in different proportions;
- water-oil emulsion breakdown and production of refinery water of required quality with the use of IronGuard 2495 as flocculant.

Pilot tests for breakdown of water-oil emulsion treated by ASP solution were performed on pilot plant Emulsion Breaking Unit (EBU).

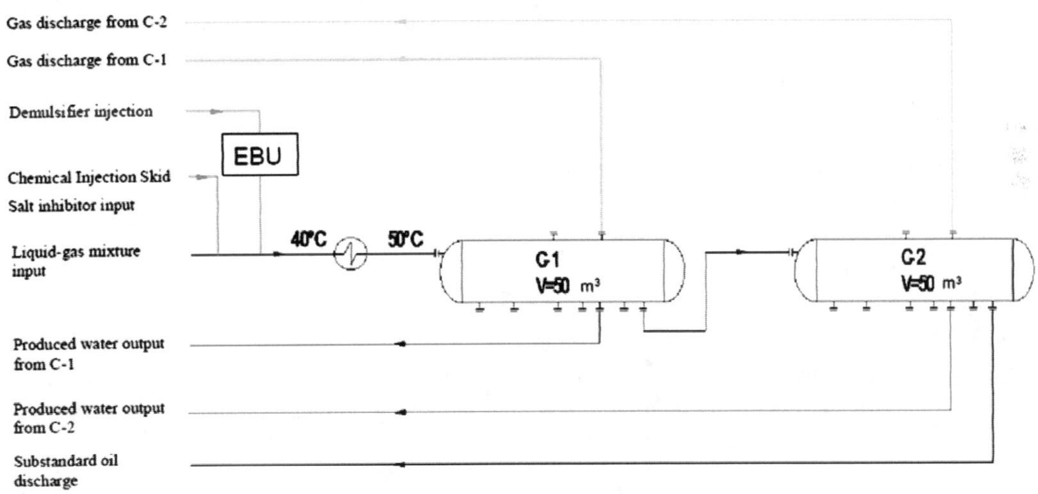

Figure 1. EBU basic diagram.

Water-oil emulsion (water-cut – 95 %, temperature – 15°C) is delivered to EBU.

Oil demulsifier Emulsotron R2601A (specific flow – 50 g/t) and salt inhibitor GYPTRON R4601-C (specific flow – 50 g/m^3) are dosed at EBU input. Liquid treated by demulsifier and salt inhibitor arrives to heat interchanger where it is heated to 45 – 50°C, and then directed to two consecutively tied separation plants C-1,2 (V=50 m^3 each).

In 75-percent-full separation plants of the first and second stage gas withdrawal takes place and discharge of produced waters. Hold-on time of liquid in separation plants is within 4 – 4.5 hours at 50 °C.

Originally there was almost no water discharge into C-1 due to high content of oil products in refinery water (low value of boundary tension). The more water was discharged, them more oil products were in refinery water– from 1,500 mg/l to 7,258 mg/l.

After first-stage separation in C-1 partially dry and separated liquid arrives to the last stage separation at C-2 where gas is separated and produced water is discharged. Water content in oil after C-2 separation plant is from 8 % to 25 %, chlorides content – 3,125–5,130 mg/l; oil products in refinery water – up to 190,390 mg/l (1.5%).

After EBU oil with different degree of water-cut (%) and chlorides content (mg/l) as well as refinery water were delivered to pre-treatment tanks of oil treatment unit.

Pilot tests for breakdown of water-oil emulsion performed at EBU showed that this scheme and applied chemical agents didn't allow producing oil and water of required quality.

To enhance oil recovery of the formation the laboratory by SHELL selected ASP solution which is capable of forming a stable microemulsion with oil in formation conditions at high temperatures, pressure, high pH and super low boundary tension.

Laboratory tests for water-oil emulsions breaking were carried out by well-known bottle test in accordance with the following technique:

- emulsion sample is taken from the well;
- upon presence, free water is separated from emulsion;
- emulsion sample is carefully mixed;
- water content in emulsion is determined;
- emulsion is poured into graded catchers;
- catchers with emulsion are thermostated in water-bath, in this case at 70 °C and 50 °C;
- demulsifier in commodity form is dosed into the catchers with the help of microliter syringe;
- emulsion treated by demulsifier is manually mixed within three minutes;
- catchers with emulsion treated by demulsifier are placed in water-bath;
- sedimentation temperature for emulsion from Zapadno-Salymskoye oilfield is 70 °C and 50 °C;
- sedimentation time is fro 5 to 60 minutes.

The volume of separated water was recorded at set time intervals upon achievement of the preset temperature in catchers.

Content of water separated from water-oil emulsion was calculated according to the formula:

$$W_{\text{sep.}} = \frac{V_i}{V_{\text{orig.}}} \cdot 100 \qquad (1)$$

Where V_i - the amount of separated water in a given period of time, ml;
$V_{orig.}$ – amount of water in original water-oil emulsion, ml;
$W_{sep.}$ – amount of water separated from emulsion, %.

Content of residual water in oil emulsion was calculated according to the formula:

$$W_{\text{resid. water}} = \frac{W_{\text{orig.}} \cdot 0.01 \cdot V_{\text{samp.}} - V_{\text{i}}}{V_{\text{samp.}} - V_i} \cdot 100 \qquad (2)$$

Where V_i is the amount of separated water in a given period of time, ml;
$V_{samp.}$ – volume of sample taken for de-emulsification, ml;
$W_{orig.}$ – original water-cut of emulsion sample, %;
$W_{resid.\ water}$ – residual water content in oil phase, %

Process efficiency was estimated by the speed of water separation from broken emulsion.

2. Results and discussion

The study of the mixture of water-oil emulsions of wells that were treated and non-treated by ASP showed that:

- mixture of emulsions breaks down better at 70 °C;
- favorable ratio is 9:1 - between emulsions that were treated and non-treated by ASP;
- residual water content in broken emulsion comprises up to 10%;
- with any emulsion ratios separated water contains a lot of oil products 560 mg/lat emulsions ratio of 20:1;
- application of flocculant IronGuard-2495 has not achieved satisfactory results;

The photos of emulsion sample under the microscope with 100 g/ м^3 of IronGuard-2495 and 50 g/t of Emulsotron R2601A are represented in figure 2.

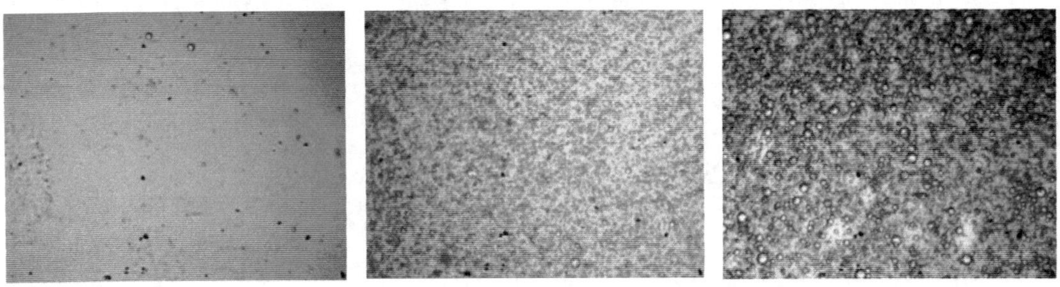

a b c

Figure 2. Emulsion sample under the microscope (600x).
Oil layer (a), interlayer (b), water phase (c)

Examining of emulsion treated by Emulsotron R2601A under the microscope DINO AM-451 (600x) showed that oil layer contains small amount of water, interlayer is a complex emulsion, and water phase is characterized by high oil products content.

Table 1. The results of laboratory researches of breaking of water-oil emulsions in Zapadno-Salymskoye oilfield

Sample no.	Agent, g/t	Water-cut, %	Amount of separated water in % over the time, min.						Residual water content in % over the time, min.						Note
			5	10	15	20	30	40	5	10	15	20	30	40	
1	Intex-720, 50 g/t	40	63	95	99	99	99	99	20	3.2	0.8	0.8	0.8	0.8	Sharply defined interface, visually clean
2	Emulsotron R2601A, 50 g/t	40	88	95	99	99	99	99	8	3.2	0.8	0.8	0.8	0.8	Sharply defined interface, dirty water
3	Without agent	40	0	0	0	0	0	0	40	40	40	40	40	40	No boundary interface, emulsion was not broken

IOP Conf. Series: Earth and Environmental Science **115** (2018) 012021 doi:10.1088/1755-1315/115/1/012021

Water-oil emulsion of the wells that were not ASP treated breaks by 99% over 15 minutes, having sharply defined interface and clean water. In case of thorough mixing emulsion is kneaded.

3. Conclusions

On the basis of laboratory research and results of pilot tests on Emulsion Breaking Unit (EBU) the following results were obtained:

1. Water-oil emulsion of the flowing wells that were not ASP treated breaks over 10-15 minutes at 70°C with specific flow of demulsifier of 50-70 g/t of oil. Residual water content in oil is less than 5 percent. Refinery water is visually clean with sharply defined interface.
2. Microemulsion (multiple emulsion) formed after ASP application has low boundary tension, high pH. It contributes to transfer of oil phase into a water one, forming oil-in-water emulsion. Residual water is characterized by high oil products content.
3. According to the data obtained in the course of the laboratory research on dilution of ASP-treated emulsion it is found that favorable ratio is 9:1. Residual water content in oil phase is less than 10 percent.

Dosing of demulsifiers at EBU input has not achieved satisfactory results. In liquid arriving for treatment on EBU boundary tension has the value close to zero. Additional dosing of the agent decreasing boundary tension has an adverse effect both on oil and residual water quality.

Oil-in-water emulsion is stable. Classic method are not suitable for residual water treatment:

- sediment gravity flow;
- filtration (filtration through Blue ribbon paper filter with pore size 3-5 microns.
- centrifuge test. (The test was performed at 170° revolutions, temperature – 70°C during 40 minutes).

Alkaline condition (pH=10-11) has adverse effect on demulsifying ability of agents, flocculation and boundary tension.

For breaking of water-oil emulsion at EBU before interchanger water or water-oil emulsion from the well that were not APS-treated in ratio of 1:9 shall be delivered.

Residual water after EBU must be prepared in water tanks by dilution in great volume.

There is no necessity in demulsifier dosing at EBU input.

All produced water shall be completely discharged at EBU.

References

[1] Konoryshkina I.Y., Tsutskareva N.V. Development of effective separation of compositions based on surfactants of domestic production for oil treatment light type. // Perm: Publishing house of the Perm state technical university.2000. P.78-82.

[2] Galyautdinov A.A., Basimova R.A., Rahimov H.H. и др. The new demulsifiers for dehydration and desalting of crude oil. // Refining and petrochemicals. 2003. No. 10. P.73-75.

[3] Tronov V.P. The destruction of water-oil emulsion under the influence of the surfactant. // Chemistry and technology of fuels and oils. 1982. No 12. P.24-26.

[4] Usheva N.V., Moyzes O.E., Kuzmenko E.A., Kim S.F., Khlebnikova E.S., Gizatullina S.N., Filippova T.V. Analysis of technological conditions influence on efficiency of oilfield treatment (Article number 012047) // IOP Conference Series: Earth and Environmental Science. – 2015 – Vol. 27. – P. 1-5

[5] Mikula R.J.; Munoz V.A. Characterization of Demulsifiers in Surfactants, Fundamentals and Applications in the Petroleum Industry, Schramm, L.L. (Ed.), Cambridge University Press: Cambridge, UK, 2000, pp. 51–78.

[6] Smith V.H.; Arnold K.E. Crude Oil Emulsions in Petroleum Engineering Handbook, Bradley H.B. (Ed.), Society of Petroleum Engineers: Richardson, TX, 1992

[7] Pozdnyshev G.N., Stabilization and destruction of the oil emulsion. -M.: Nedra, 1982, 221 p.

[8] Buhidma A. and Pal R. Flow Measurement of Two-phase Oil-in-water Emulsions using Wedge Meters and Segmental Orifice Meters // Chem. Eng. J., 1996 v N 63., P. 59-64.

[9] Pal R. Techniques for Measuring Composition (Oil and Water Content) of Emulsions // Colloids & Surfaces, 1994. - N 84. v P. 141-193.

[10] Sokolov A. G., Shabaev E. F., Vladimirov Yu. D. current status and ways to improve dewater oil // VNIIOENG, Oilfield business.1984. - 12 (84). 56 p.

[11] Sergienko N. D. the Research, development and implementation of process preparation of oil for processing oil-water resistant vysokoporodnyh emulsions with a high content of mechanical impurities. // Diss. M.: VNII NP. 2005. 154 p.

[12] R. Z. Sakhabutdinov, F. R. Gubaidulin and others. Features the formation and breaking of water-oil emulsion at a later stage the development of oil fields. - .M: VNIIOENG publ., 2005. - 324 p.

[13] Surface phenomena and surfactants: a Handbook. Ed. by A. A. Abramson and E. D. Shchukin, L.: Chemistry, 1984. 392 p.

[14] Khutoryanskaya F. M., Somov, V. E., V. P. Gaskin etc. Develop and the introduction of oil-soluble demulsifier "Hercules 1017". // SB. scientific. "Kirishinefteorgsintez" LLC NIF "Inzhener-SERVIS, VNII NP ". Moscow: Tsniiteneftekhim, 2005. Pp. 116-145.

[15] Khutoryanskaya, F. M., Antonov, T. A., Potapochkina N. N. etc. New oil-soluble demulsifier domestic production - "Hercules 1603". Development, laboratory and pilot testing. // SB. scientific. "Kirishinefteorgsintez" LLC NIF "Inzhener-SERVIS, VNII NP ". Moscow: Tsniiteneftekhim, 2005. Pp. 146-151.

[16] James G. Speight, Baki Oum. Petroleum refining processes. New York: Marcel Dekker inch. 2002. 695 pages.

[17] Jean Pierre Wauquier. Le raffinage du petrole.Tome 2,Procedes de separation.Paris: Technip, 1998, 627 pages.

[18] GOST R 51858-2002 Crude petroleum. General specifications Standards publishing house, 2002. Standards publishing house, 2002.

Reduction of Harmful Emissions During Start and Warming Up of the Engine

N Volkov, D Chainikov

Motor transport operation, Industrial University of Tyumen, 625039, Tyumen, Russian Federation

E-mail: 9044955509@mail.ru

Abstract: The question of decrease in harmful emissions when idling of a truck engine in the conditions of low temperatures is considered. The implementation of the thermogenerator for a power supply of electrical elements is offered in a design of the self-powered heater. The principle of the device operation is based on a thermoelectric effect at which there is heat absorption and thermo-EMF emergence. In a consequence of this process electricity is produced. The exhaust gases of the self-powered heater are the source of the absorbed heat and act as fuel for the thermogenerator. It allows developing energy for a power supply of electrical elements of the heater. It gives the chance not to start the engine for warming up during the long parking, thereby reducing harmful emissions.

1. Introduction

The number of the devices consuming the electric power in the vehicle rapidly grows. It concerns the increasing use in vehicles of navigation aids and video fixing, charging of portable mobile devices, complications of electronic systems of vehicles, etc. For trucks installation of the self-powered heater is very relevant for the heating of engine and salon during the parking. Installation of the heater also increases the number of the electric power consumers[1,2]. Additional consumers increase fuel usage during the trip and amount of the harmful substances (HS) emitted with exhaust gases (EG). Also the volume of harmful emissions in from trucks increases during the winter period because of long warming up before a trip to the working temperature of the engine and on the parking before loading/unloading, warm-up period depends on ambient temperature. Standard time of warming up is provided in the figure 1.

Structure of vehicle HS emitted with EG is multiunit. The greatest attention in the world [3] and in Russia in recent years is given to CO_2. So, according to the Transport strategy of the Russian Federation [4] the volume of CO_2 specific emissions on the motor transport has to be reduced on 22 percent.

For decrease in negative impact of vehicles on the environment it is offered to use thermoelectric modules (TEM), having implemented in a design of the self-powered heater (SPH). Their operation principle is based on a thermoelectric effect at which there is heat absorption and and thermo-EMF emergence. In a consequence of this process the electricity capable to supply the heater electrical elements is produced.

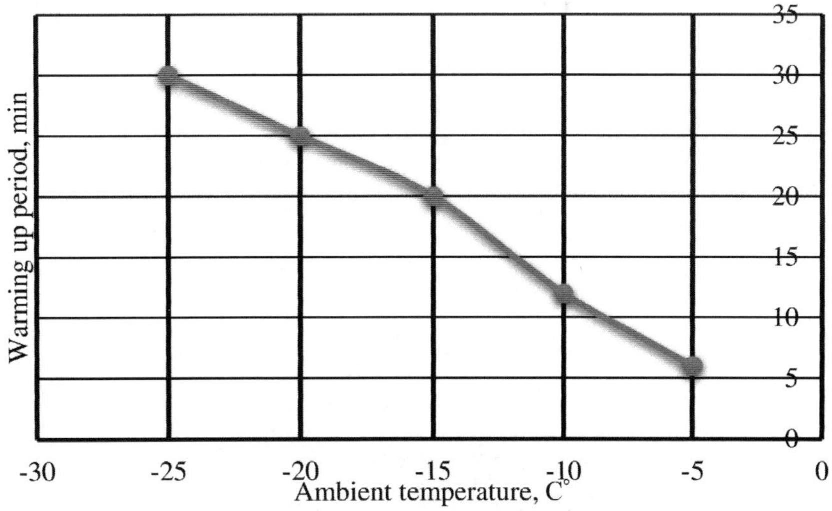

Figure 1. Warm-up period on not heated parking

TEM is a set of the thermocouples electrically connected, as a rule, in series (figure 2). In standard TEM thermocouples are located between two flat ceramic plates. Plates are made of oxide or nitride of aluminum. When a direct current passes through the TEM between its sides, a temperature difference forms. One side (cold) is cooled, and other (hot) heats up. If the efficient heat dissipation is provided from the hot side of TEM [5], for example, by means of a radiator, then on the cold side it is possible to receive temperature which will be lower than the ambient temperature by tens of degrees. The degree of cooling will be proportional to current level. In the presence of different temperatures on the different sides of TEM, it produces electricity.

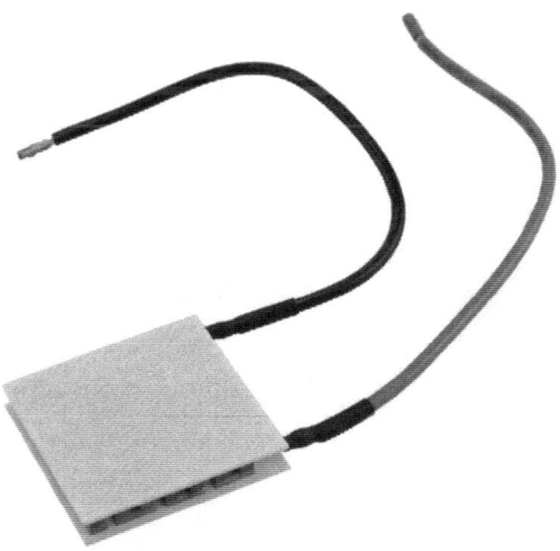

Figure 2. Thermoelectric module (TEM)

SPH is devices which serve for maintenance of optimum temperature condition in salon and preheat the engine before starting (figure 3). Their main feature is that they can function without the started engine. There are several modes of operation: standard ventilation and suction of air mix outside as well as heating the coolant. The external thermostat or the built-in timer is responsible for monitoring the operation. Heaters provide comfortable conditions for passengers, long-haul drivers, operators of the special equipment. Structurally the device is suitable for any vehicle and doesn't depend on dimensions and assignment. The SPH is able to provide with warm air, as well as the small-capacity passenger car, and the huge truck and also is capable to warm the engine up to the working temperature by heating the coolant[5,6]. By means of the special control board it is possible to control the air temperature and a flow of its direction. In addition, the devices are divided according to the power supply principle. Electricity (12 and 24 volts), gas, diesel fuel or gasoline can be used as power supply [7]. Also there are models that heating cooling liquid of the vehicle which in turn heats the vehicle's interior. For a power supply of electrical elements of the heater it is offered to use TEM.

Figure 3. The principle of the self-powered heater operation

2. Methodology

It is offered to place TEM in the area with different temperatures (on one side of the module temperature higher, on the other is less) where it can produce the electric power. This feature allows to use TEM on surfaces with different temperatures for receiving the electric power [8,9]. Temperature of the EG in the heater is about 270 °C, this value is significantly higher than the ambient. Similar indicators allow to achieve a significant temperatures difference on the device. Implementation TEM in a design of the SPH, will allow using the heat of the EG and further to transform the existing temperatures difference to the electric power for the vehicle [10,11]. Thereby reducing harmful emissions, as there is no need to recharge the vehicle's standard battery (SB) without which the device can't function. Figure 4 shows the location of the TEM in the heater.

In an operating mode the TEM (Figure 4 № 17) will transform the heat of EG to the electric power for a power supply of the heater electrical elements. This makes the device autonomous and independent of the vehicle's on-board network. Table 1 describes the operating mode of the heater when it starts and the power sources are specified, depending on the cycle of operation. At the moment of start, the heater begins to consume current from the SB, this process takes 5 minutes, it is caused by the fact that at start of the heater, electricity consumption sharply increases (approximately three times), this energy is spent for heating of a heating plug [12,13]. Further the device passes into the warming up mode and switchs to TEM from which the current to the heater is supplied. All remained operating time of the heater, consumption of current comes from TEM without the use of SB.

To determine the amount of CO_2 produced, an experiment was performed using the gas analyzer GARO AVG-4 (Figure 5). There are channels in the gas analyser for measuring the engine speed and the oil temperature [14,15]. The maximum measurement error for CO_2 is ±0.5%. The operating principle of the CO_2 volume fraction sensor is optical absorption.

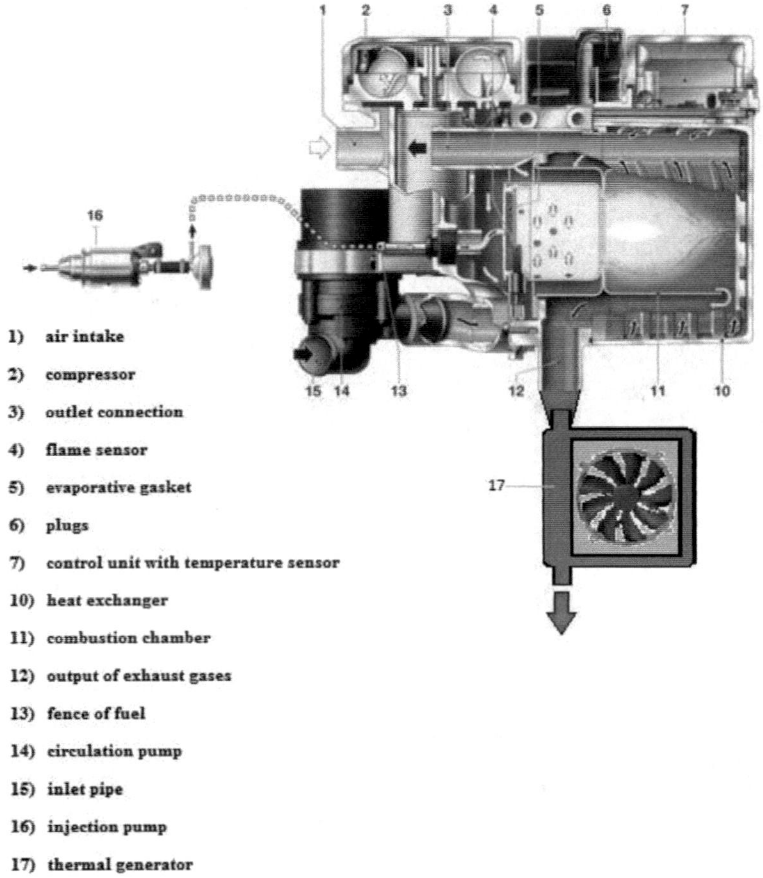

Figure 4. TEM location in the SPH

The instrument readings were recorded after steady-state values in the range of 15–30 seconds. The measurements were made on vehicles of brands: Freightliner, Volvo, MAZ. As SPH the Planar of 3 kW were used (Table 2).

Table 1. TEM operating mode

Power consumer			Power source	
Current intensity(A)	Electric potential (V)	Operation period(min)	Vehicle SB 55 Ah 12V	TEM 5A 12V
12	12	3	+	-
3.8	12	20	2 min SB+TEM	2 min SB+TEM
2.4	12	7	-	+
0.7	12	5	-	+

Table 2. Planar heater characteristics

Heat producing capability, kW	1 - 3
Fuel consumption, l/h	0.12 – 0.37
Heater input, W	10 – 42
Amount of heated air, m^3/h	70 - 120
Fuel type	Diesel fuel
Operating supply voltage, V	12

IOP Conf. Series: Earth and Environmental Science **115** (2018) 012022 doi:10.1088/1755-1315/115/1/012022

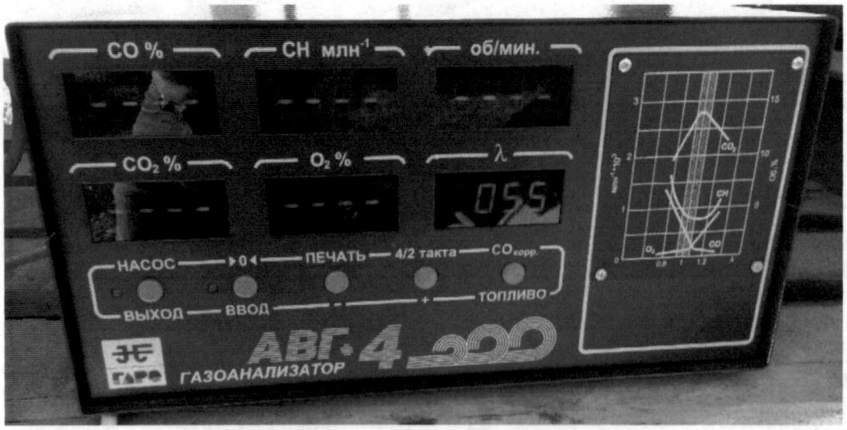

Figure 5. Gas analyzer GARO AVG-4

3. Results and analysis

As a result of an experiment it was succeeded to find out that use of SPH during the long parking of the vehicle allows to reduce emissions of CO_2 on average by 10 times (Table 2). As the engine is in the muted state, and working temperature is maintained by the SPH.

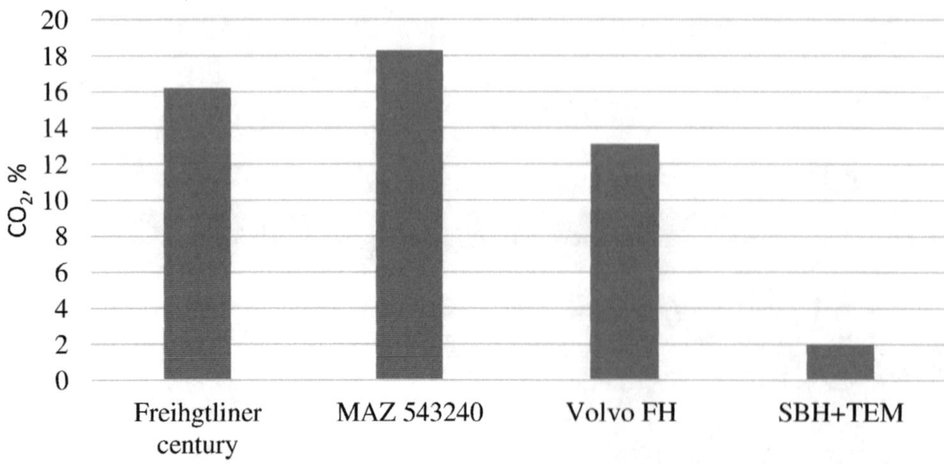

Figure 6. Amount of CO_2 in EG of trucks engines

Apparently from fig. 6 the greatest difference is reached on vehicles with a long service life. Considering that average age of the trucks park in Russia is 10.5 years [16], it is possible to draw a conclusion that in the conditions of the Russian Federation the TEM use will allow to give the maximum effect on decrease in CO_2 emissions. The last researches showed that in winter after start of the cold engine about 90% of all harmful emissions of CO_2 falls on time of the first kilometers of the movement. At start of previously warmed engine the content of harmful gases in an exhaust considerably decreases (by 5 times) therefore the use of the motor heater the in cold season has huge ecological effect. This indicator can still decrease by 1.5 times, at implementation in a design of TEM. The result is achieved because there is no need to start the engine during the long parking to recharge the batteries since their function was undertaken by the TEM.

4. Conclisions

As a result of the TEM installation it becomes possible to operate the SPH for a long time without starting the engine, at long parking under loading/unloading during the winter period. Engine start was necessary for recharge of SB, but with implementation of TEM in design this need disappears. Because of the closed system of power supply from TEM the period of operation is regulated only by a fuel quantity necessary for the heater. Thus, application of TEM in the independent heater will allow to receive the additional electric power and to use it for a power supply of electrical elements of the SPH, thereby reducing harmful emissions from the truck during the parking by 9 times.

References

[1] Alim Nayum Christian A.Klöckner[1]SunitaPrugsamatz[2], Influences of car type class and carbon dioxide emission levels on purchases of new cars: A retrospective analysis of car purchases in Norway // https://doi.org/10.1016/j.tra.2012.10.009

[2] Anisimov, I., Ivanov, A., Chikishev, E., Chainikov, D., Reznik, L. Assessment of gas cylinder vehicles adaptability for operation at low ambient temperature conditions // *WIT Transactions on Ecology and the Environment*, 190 volume **1**, pp. 685-695, 2014, doi: 10.2495/EQ140651.

[3] Bekir SamiYilbas[ab]S.S.Akhtar[a]A.Z.Sahin[ab], Thermal and stress analyses in thermoelectric generator with tapered and rectangular pin configurations // https://doi.org/10.1016/j.energy.2016.07.168

[4] Bielaczyc, P., Szczotka, A., and Woodburn, J. Regulated and Unregulated Exhaust Emissions from CNG Fueled Vehicles in Light of Euro 6 Regulations and the New WLTP/GTR 15 Test Procedure // SAE International Journal of Engines – Volume **8**, Issue 3, pp. 1300-1312, 2015, doi: 10.4271/2015-01-1061.

[5] Bondarenko, E.V., Filippov, A.A. Evaluation of use of certain types of motor fuel for environmental safety criteria // Autogas filling complex + alternative fuel – 4 (**52**), pp. 31-35, 2010.

[6] Baranov E.F., Bugakova N.S., Gelvanovsky M.I., Gohberg L.M. Russian Statistical Yearbook 2016 729 pages // http://www.gks.ru/free_doc/doc_2016/year/year16.pdf (accessed September 4, 2017).

[7] Chainikov D., Chikishev E., Anisimov I., Gavaev A. Influence of ambient temperature on the CO2 emitted with exhaust gases of gasoline vehicles Innovative Technologies in Engineering VII International Scientific Practical Conference. Conference Proceedings. National Research Tomsk Polytechnic University. 2016. C. 12109.

[8] GOST R 41.24-2003 (Regulation № 24) Uniform provisions concerning: I. The approval of compression ignition (C.I.) engines with regard to the emission of visible pollutants; II. The approval of motor vehicles with regard to the installation of C.I. engines of an approved type; III. The approval of motor vehicles equipped with C.I. engines with regard tothe emission of visible pollutants by the engine; IV. The measurement of power of engines – M.: Standartinform, 36 p., 2007.

[9]HulinZhang[a]YuhangXie[a]XiaomeiLi[a]ZhenlongHuang[a]ShangjieZhang[a]YuanjieSu[a]BoWu[a]LongHe[a]WeiqingYang[b]YuanLin[a], Flexible pyroelectric generators for scavenging ambient thermal energy and as self-powered thermosensors // https://doi.org/10.1016/j.energy.2016.02.002

[10] Ling-YunHE[abc]Lu-YiQIU[d1], Transport demand, harmful emissions, environment and health co-benefits in China // https://doi.org/10.1016/j.enpol.2016.07.037

[11] Statistics of harmful emissions CO2 time series 1990-2014 per region/country// http://edgar.jrc.ec.europa.eu/overview.php?v=CO2ts1990-2014&sort=des9 (accessed September 4, 2017).

[12] Transport Strategy of the Russian Federation for the period up to 2030, https://www.mintrans.ru/documents/detail.php?ELEMENT_ID=22371&list=41307,39859,32563,22371,20611,19188,13008,13026,13016,13028 (accessed September 4, 2017).

[13] Karnaukhov N.N., Reznik L.G., The differentiated correction of charge for air pollution // *Advances in transport*, Volume **16**, pp. 701-705, 2004, doi: 10.2495/UT040681.

[14] Volkov N.A., Chaynikov D.A. Transformation of the engine's thermal losses into electric energy

for the automobile // International Scientific and Practical Conference of Young Researchers named after Yu. DI. Mendeleev conference materials. Tyumen Industrial University, Institute of Industrial Technology and Engineering. 2016. P. 444-447.

[15] Volkov N.A, Chainikov D.A. Conversion of heat loss engine electricity for car // In the collection: International Scientific and Practical Conference of Young Researchers. DI. Mendeleev conference materials. Tyumen Industrial University, Institute of Industrial Technology and Engineering. 2016. P. 444-447.

[16] Volkov N.A. Improvement of the start process of the engine in the conditions of low temperatures and development of the technology of application of thermoelectric modules for increasing the autonomousity of the pre-heater worker // In the collection: Transport and transport-technological systems materials of the international scientific and technical conference. 2017. P. 102-106.

IOP Conf. Series: Earth and Environmental Science **115** (2018) 012023 doi:10.1088/1755-1315/115/1/012023

Research of Acoustic Properties of Materials with the Purpose of Their Use at Design of Special Noise Protective Clothes for Oil and Gas Industry

E A Drofa, E Yu Lipilina

Technological Institute of Service (branch) Don State Technical University, Stavropol, Russian Federation

E-mail: lipilina07@mail.ru

Abstract. The article is devoted to the substantiation of the choice of a rational package of materials, which has the greatest noise-protective properties when designing special clothes with reference to the oil and gas industry. Studies were conducted to assess the factors that have the most significant effect on the noise-protective properties of clothing. Conclusions are made about the possibility of using the developed technique for studying the noise-protective properties of materials in selecting rational packages of materials for the production of special clothes with high noise-protective properties.

1. Introduction

Modern technology develops in the direction of increasing power, increasing the speed regime and, naturally, this leads to the appearance of large dynamic loads, which, in turn, determines the noise intensity. In the fight against such negative phenomena, special attention is required, since the weakening of industrial noise greatly improves working conditions, increases efficiency, and positively affects people's health. That is why scientific research in this field has become very active lately.

The tick of the basic directions of the solution of this problem concerns not only the creation of techniques that work with less noise; isolation of equipment creating increased noise, but also the development of tools that protect people from noise [1,2]. For each of the above tasks, in large volumes and more actively. We will focus on the third direction. At the same time, as an example, the research will be carried out with reference to the oil and gas industry, where the noise factor, as unfavorable, is very weighty. The working technological equipment (pumps, compressors, electric motors) is a source of increased noise and vibration, which adversely affects the person. To combat dangerous and harmful production factors special clothing is called for.

2. Results and Discussion

The choice of material is one of the important parameters of the process of designing special clothes. When studying and analyzing the quality of tissues, the evaluated properties are grouped according to certain characteristics, depending on the goals and tasks set [3,4].

Research in the field of noise protection is based on the definition of the most rational package of materials, which has the greatest noise-protective properties [5,6,7,8].

As an optimization criterion, the final sound power in *dB* was chosen for the materials that make up the package, face up – Y.

The factors that have the most significant effect on the noise-protective properties of clothing:

- X_1– Surface density;
- X_2– Thickness of the material package;
- X_3– Frequency of sound [9,10].

These factors are compatible and independent, which makes it possible to use them for setting up a full-factor experiment.

Based on the technological possibilities of the experiment, we varied the factors at two levels: upper and lower. The number of experiments for setting up such a full-factor experiment is determined from the expression:

$$N = 2^k , \quad (1)$$

where N is the number of experiments in the plan, k is the number of factors.

In a particular case, $N = 23$ experiments should be performed for one package of materials. In each of the experiments the measurements were carried out repeatedly, which makes it possible to increase the accuracy of the study. The determination of the necessary number of repetitions of each experiment was found on the basis of the results of a preliminary series of experiments (Table 1). In this case, the values of the confidence probability (0.95) and the required deviation of the average measurement result from the true value of the measured quantity $\varepsilon(\overline{y})=0.1\,\overline{y}$.

Table 1- Results of a preliminary series of experiments.

№ experience	The measured value Y, dB
1	75.3
2	76.5
3	76.5
4	77.2

After mathematical processing of the results of the preliminary series of experiments. All the results of the experiments were statistically processed.

The calculated required number of repetitions of each experiment to obtain a reliable result of the experiment with a significance level of 0.95 is equal to 7. Thus, a minimum of 7 packets of materials should be investigated.

Based on the analysis of the study of the acoustic properties of tissues [9,10,11,12,13] and materials, the following packages were identified:

1 package:Greta + batting + flannel
2 package: Greta + cotton wool clothing + artificial lining
3 package: Leader comfort 250 + woolstone + flannel
4 package:"Coverstat-RW"+ woolstone + flannel
5 package: Titanreqular 190+ woolstone + flannel
6 package: "Coverstat-RW"253+ antistatic impregnation + woolstone + flannel
7package: "Coverstat-RW"190 + woolstone + flannel

For each of these packets, a full factorial experiment was carried out, including 7 experiments on 7 measurements in each.

Levels of factor variation in the implementation of a full-factor experiment for all packets of tissues and materials are presented in Table 2.

Table 2 - Factors and levels of variation.

Name factor	Unit measurements	Variation levels			Variation intervals
		lower	null	upper	
		1 package:Greta + batting + flannel			
Surface density X_1	g /m^2	214.0	232.0	250.0	18.0
Packing thickness X_2 of materials	m	0.00747	0.007845	0.00822	0.000375
Frequency of sound X_3	Hc	500.0	1750.0	3000.0	1250.0

IOP Conf. Series: Earth and Environmental Science **115** (2018) 012023 doi:10.1088/1755-1315/115/1/012023

2 package: Greta + cotton wool clothing + artificial lining					
Surface density X_1	g /m ^{2}	214.0	232.0	250.0	18.0
Packing thickness X_2 of materials	m	0.00827	0.008575	0.00888	0.000305
Frequency of sound X_3	Hc	500.0	1750.0	3000.0	1250.0
3 package: Leader comfort 250 + woolstone + flannel					
Surface density X_1	g /m ^{2}	214.0	232.0	250.0	18.0
Packing thickness X_2 of materials	m	0.00876	0.008785	0.00881	0.000025
Frequency of sound X_3	Hc	500.0	1750.0	3000.0	1250.0
4 package:"Coverstat-RW"+ woolstone + flannel					
Surface density X_1	g /m ^{2}	190.0	221.5	253.0	31.5
Packing thickness X_2 of materials	m	0.00869	0.00874	0.00879	0.00005
Frequency of sound X_3	Hc	500.0	1750.0	3000.0	1250.0
5 package: Titanreqular 190+ woolstone + flannel					
Surface density X_1	g /m ^{2}	190.0	200.0	210.0	10.0
Packing thickness X_2 of materials	m	0.00867	0.00871	0.00874	0.00003
Frequency of sound X_3	Hc	500.0	1750.0	3000.0	1250.0
6 package: "Coverstat-RW"253+ antistatic impregnation + woolstone + flannel					
Surface density X_1	g /m ^{2}	253.0	256.5	260.0	3.5
Packing thickness X_2 of materials	m	0.0087	0.00883	0.00895	0.00013
Frequency of sound X_3	Hc	500.0	1750.0	3000.0	1250.0
7package: "Coverstat-RW"190 + woolstone + flannel					
Surface density X_1	g /m ^{2}	190.0	200.0	210.0	10.0
Packing thickness X_2 of materials	m	0.000869	0.00479	0.00871	0.00392
Frequency of sound X_3	Hc	500.0	1750.0	3000.0	1250.0

The possibility of applying a linear regression equation for the processing of experiments is confirmed by the fact that for the range of amplitudes of sound waves that do not lead to the destruction of the protective device, its characteristics (density, thickness, etc.) can be considered independent of the sound level. Thus, the noise transmittance of the protective device will not depend on the amplitude of the sound wave. Mathematical models of the investigated packages of fabrics and materials, obtained on the basis of formulas for calculating the coefficients of the regression equation, are presented in Table 3.

Table 3 - Regression equations that characterize the influence of the factors studied on the noise-protective properties of clothing.

Package number	The equation
1	$\tilde{Y}= 67.25 - 0.675X_1 - 0.1X_2 - 7.966X_3 + 0.475X_1X_2 - 0.025X_1X_3 - 0.05X_2X_3 + 0.125X_1X_2X_3$
2	$\tilde{Y}= 70.0375 + 0.7375X_1 - 0.1285X_2 - 7.3125X_3 - 0.3125X_1X_2 + 0.9375X_1X_3 - 0.1625X_2X_3 - 0.0125X_1X_2X_3$
3	$\tilde{Y}= 67.75 + 0.05X_2 - 9.49X_3 - 0.05X_2X_3$
4	$\tilde{Y}= 66.9625 - 0.7625X_1 - 0.0285X_2 - 9.7125X_3 - 0.1375X_1X_2 - 0.4875X_1X_3 + 0.0125X_2X_3 - 0.0625X_1X_2X_3$
5	$\tilde{Y}= 65.9875 - 0.0125X_1 + 0.0215X_2 - 10.0875X_3 + 0.0875X_1X_2 - 0.0375X_1X_3 + 0.0625X_2X_3 - 0.0375X_1X_2X_3$
6	$\tilde{Y}= 59.45 - 0.35X_1 - 0.1X_2 - 3.45X3 + 0.25X1X_3 - 0.25X_2X_3 + 0.25X_1X_2X_3$
7	$\tilde{Y}= 66.9125 - 0.7625X_1 - 0.0125X_2 - 9.7625X_3 - 0.0875X_1X_2 - 0.4875X_1X3 + 0.0125X_2X_3 - 0.0125X_1X2X_3$

To assess the significance of the coefficients of the regression equations, a confidence interval was calculated, which made it possible to exclude from the model non-significant coefficients. Coefficients are significant if their absolute values correspond to and satisfy inequality

$$|b_i| > t\sqrt{S^2_{(\beta)}}, \qquad (2)$$

where t - is the Student's test; $S^2_{(\beta)}$ – variance associated with errors in the determination of regression coefficients, the value of which is determined by the formula:

$$S^2_{(\beta)} = \frac{S^2_u}{Nm}, \qquad (3)$$

where S^2_u – the experimentally average variance of reproducibility.

Dispersion of reproducibility S^2_u is calculated as:

$$S^2_u = \frac{\sum_{i=1}^{n}\sum_{j=1}^{m}(Z_{ij} - Y_i)^2}{N(m-1)}, \qquad (4)$$

where N is the total number of experiments, m is the number of observations in one experiment, Y_i– arithmetic mean of m observations at one point, Z_{ij}– result of a separate j observations in i experience.

Criterion Student t is determined from the degree of freedom of distribution $n = N(m-1) = 48$ and the level of significance with confidence 0.95: t=2.021.

Taking into account the number of repetitions of the experiment in each experiment, m = 7 and the results of the study, the significance of each individual coefficient is determined. Table 4 shows the inequalities, the fulfillment of which guarantees the significance of the coefficients being checked.

Table 4 - Results of the study of the significance of the coefficients of the regression equation.

Package number	Condition
1	$\lvert b_i \rvert > 0.4892$
2	$\lvert b_i \rvert > 0.2669$
3	$\lvert b_i \rvert > 0.0929$
4	$\lvert b_i \rvert > 0.0748$
5	$\lvert b_i \rvert > 0.0793$
6	$\lvert b_i \rvert > 0.0705$
7	$\lvert b_i \rvert > 0.0699$

Thus, the regression equations acquire the form presented in Table 5.

Table 5 - Mathematical models of influence of investigated factors on the noise-protective properties of clothing without taking into account insignificant coefficients.

Package number	The equation
1	$\tilde{Y} = 67.25 - 0.675X_1 - 7.966X3$
2	$\tilde{Y} = 70.0375 + 0.7375X_1 - 7.3125X_3 - 0.3125X_1X_2 + 0.9375X_1X_3$
3	$\tilde{Y} = 67.75 - 9.49X3$
4	$\tilde{Y} = 66.9625 - 0.7625X1 - 9.7125X3 - 0.1375X1X2 - 0.4875X1X3$
5	$\tilde{Y} = 65.9875 - 10.0875X3 + 0.0875X1X2$
6	$\tilde{Y} = 59.45 - 0.35X1 - 0.1X2 - 3.45X3 + 0.25X1X3 - 0.25X2X3 + 0.25X1X2X3$
7	$\tilde{Y} = 66.9125 - 0.7625X1 - 9.7625X3 - 0.0875X1X2 - 0.4875X1X3$

The adequacy of the obtained models was checked using the F-test of Fisher. The results allow us to conclude that the models obtained are adequate. Consequently, the regression equations from Table 5 correspond to real processes. Graphical interpretation of the dependence on the regression equation for the three variables of the recommended package of materials No. 6 is shown in Figure 1.

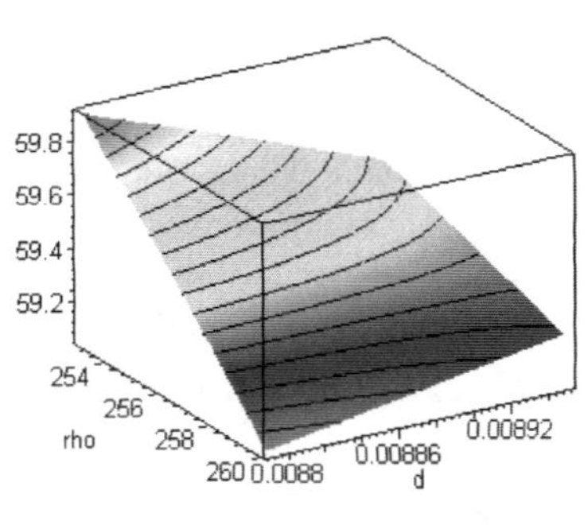

Figure 1. Graph of the dependence on the regression equation for a package of materials № 6.

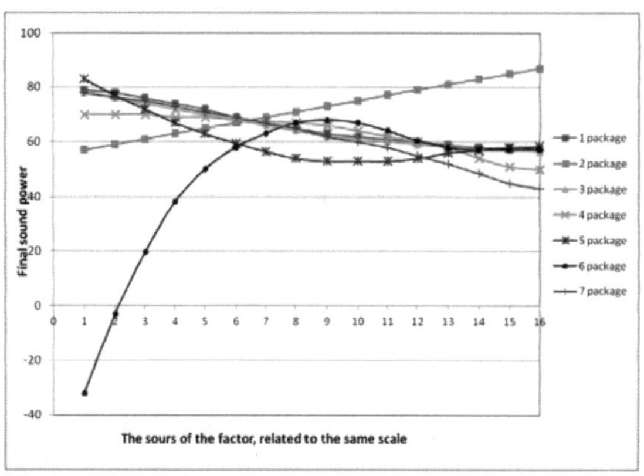

Figure 2.Graphs of regression equations related on a single scale.

Analysis of regression equations shows that the X_2 фактор frequency of sound has the greatest influence on the noise-protective properties of clothing, with the increase in which the optimization criterion Y decreases. A slightly smaller influence on the criterion of optimization is due to the factor X_1, the surface density of the tissue.

The factor X_2 is the thickness of the packet, to a lesser extent affects the noise-protective properties of clothing. However, the joint contribution of factors X_1, X_2 and X_1, X_3 to the investigated process significantly decreases the optimization criterion Y.

The received mathematical models can be used for selection of materials in a package possessing the greatest noise-protective properties. To do this, graphically, the values taken by the functions obtained in one interval, where factor X_1 varies from 190 to 260 with a step of variation of 5, factor X_2 from 0.006 to 0.009 in increments of 0.0002, and factor X_3 from 500 to 3500 in increments of 200.

3. Conclusions

As follows from the graph in Figure 2, the smallest final sound power transmitted through a packet of materials is observed in experiments with package No.6: "Coverstat-RW" 253 + antistatic impregnation + woolstone + flannel, and therefore it can be recommended as the main one Manufacture of noise-protective special clothes for operation in enterprises with increased noise load. The developed technique for studying the noise-protective properties of materials makes it possible to develop rational packages for the manufacture of special clothes with high noise-protective properties.

References

[1] Haase J 2013 14 Flame resistant clothing standards and regulations *In Woodhead Publishing Series in Textiles* ed by F. Selcen Kilinc(Woodhead Publishing). pp 364-414

[2] Dolez P I and Izquierdo V 2018 11 Specific testing of protective clothing *In The Textile Institute Book Series* ed by Patricia Dolez, Olivier Vermeersch and Valério Izquierdo (Woodhead Publishing) pp 301-349

[3] Bragança S, Fontes L, Arezes P, Edelman E R and Carvalho M 2015 The Impact of Work Clothing Design on Workers' Comfort *In Procedia Manufacturing* **V3** pp 5889-5896

[4] Langenhove L 2013 1 Smart textiles for protection: an overview In Woodhead Publishing Series in Textiles ed by R.A. Chapman (Woodhead Publishing) pp 3-33

[5] Roy Choudhury, Majumdar P K and Datta C 2011 1 Factors affecting comfort: human physiology and the role of clothing *In Woodhead Publishing Series in Textiles* ed by Guowen Song (Woodhead Publishing) pp 3-60

[6] Mao N 2014 3 High performance textiles for protective clothing In *Woodhead Publishing Series in Textiles* ed by Carl A. Lawrence (Woodhead Publishing) pp 91-143

[7] Kim Bo-Seung and Park Junhong 2017 Double resonant porous structure backed by air cavity for low frequency sound absorption improvement *In Composite Structures*

[8] Tang Xiaoning and Yan Xiong 2017 Acoustic energy absorption properties of fibrous materials: A review *In Composites Part A: Applied Science and Manufacturing* **V101** pp 360-380

[9] Crouch I G, Arnold L, Pierlot A and Billon H 2017 6 Fibres, textiles and protective apparel *In Woodhead Publishing in Materials* ed by Ian G. Crouch (Woodhead Publishing) pp 269-330

[10] Potluri P and Needham P 2005 6 Technical textiles for protection *In Woodhead Publishing Series in Textiles* ed by Richard A. Scott (Woodhead Publishing) pp 151-175

[11] El Mogahzy Y E 2009 2 Textile engineering principles and concepts In Woodhead Publishing Series in Textiles (Woodhead Publishing) pp 18-38

[12] Dolez P I and Vermeersch O 2018 1 Introduction to advanced characterization and testing of textiles *In The Textile Institute Book Series* ed by Dolez P, Vermeersch O and Izquierdo V (Woodhead Publishing) pp 3-21

[13] Holmes D A and Horrocks A R 2016 10 Technical textiles for survival *In Handbook of Technical Textiles (Second Edition)* (Woodhead Publishing) pp 287-323

Chitosan Dilutable and Dilactin Forte: Assessment of Their Efficiency for Safety and Quality of Foodstuff

N Iurchikova, O Khlebosolova
Sergo Ordzhonikidze Russian State University for Geological Survey, Miklouho-Maklay St, 23a, Moscow, Russia

E-mail: nikolina.yurchukova@gmail.com, o.hlebosolova@mail.ru

Abstract. The modern natural food preservatives used to process and store foodstuff allow to ensure its safety and high quality. Chitosan and dilactin-forte are among such medicines. These preservatives are not only safe, but also are beneficial to a human body in virtue of their effects onto human digestive system. The article describes the results of the research conducted to identify the impact of these natural preservatives on safety of carrot (*Daucus carota* subsp. *sativus*)

1. Introduction

Traditionally, food products which are to soon appear on the shelves have to meet safety requirements and to avoid any risk of using products which could harm and (or) cause damage to life or health of consumers, any deterioration of environment. At the same time, food industry tends to produce and put into realization safe and qualitative foodstuff. Qualitative products are not only safe, but also correspond to all indicators the Russian Federation Technical regulations of the Customs Union (TR CU – 21–2011) states.

Foodstuff production uses natural preservatives such as salt, vinegar, vegetable oil, honey, pepper, garlic to preserve food from spoiling, poisonous smells and harmful effects. New technologies allow the industry to apply medicines received from natural raw materials, for example, chitosan and dilactin-forte. Their safe and highly effective usage is proved by numerous lab experiments [2–7]. Besides, studying new medicines indicates their clear advantages over traditional supplies due to special characteristics allowing to consider them both as preservatives and the additives improving human digestion.

The research of new natural preservatives' effects on various food products including conditions of their storage and processing is essential for both expanding their application and improving quality of food products. The presented research attempts to estimate efficiency of chitosan dilutable and dilactin-forte for ensuring safety of carrot (*Daucus carota* subsp. *Sativus*).

2. Material and Methods

Experimental research uses two natural preservatives: chitosan dilutable and dilactin-forte, and implies verification of their impact on safety of two types of carrot – whole and cut. Here is a brief description of the preservatives' characteristics and the experiment's features.

2.1. Preservatives description.

As mentioned above, two natural preservatives were chosen. However, the focus was set on studying the influence of chitosan.

2.1.1. Dilutable Chitosan. Chitosan is extracted from Crustacea's shells, insects' cuticles and mushroom cellular walls [8-10]. Chitosan dilutable is produced from chitosan by adding CH2COOH (chloracetic acid) to NH2 or by adding OH in the alkaline environment (this carboxymethyl product fully dissolves in water). Chitosan is widely used, including food industry, medicine, agriculture, cosmetology and bioengineering.

In food industry the preservative is used to thicken dietary foodstuff in production of sauces, pastes, juice, wines. It is referred to food fibers and creates solution of high viscosity in a stomach, shows properties of a sorbent and promotes to removal of radionuclides from an organism, possesses immunostimulating, anti-sclerous and anti-arthrosis abilities, reduces cholesterol and sugar levels [11, 12]. Chitosan creates thin, flexible protective peels on surface of fruits, vegetables, fish, which are antimicrobic filters and have selective permeability for gases and water, which allows to prolong storage life of fresh and frozen products.

2.1.2. Dilactin-forte. This is a nutritional supplement comprised of lactic acid, acetic acid, propionic acid and their salts. Due to its antimicrobic, buffer, antioxidant, stabilizing functions as well as safe for human health.

Dilactin-forte is used to produce and store meat, fish, confectionery, bakery, in sauces, seasonings, mayonnaise.

2.2. Experiment features

The research was conducted to show the influence Chitosan and Dilactin-Forte have on carrot microbiological pollution.

2.2.1. Research preparation. Before the experiment started, substratum, 0.6 cm paper disks were prepared and sterilized, 10 kg of carrots, Chitosan dilutable (in a dry form) and Dilactin-Forte (in the liquid state) were prepared. The packing material (bopp-film) was chosen and 15x20 cm packages were welded at a speed of 2 m/s at the temperature of 140 °C by means of the CBS-900 constant heating welder.

2.2.2. Experiment course. The first part of the experiment was made to compare the natural preservatives (Chitosan dilutable and Dilactin-Forte) influence on E.coli.

E.coli was seeded into agar-agar substratum in the sterilized Petri dishes. Each dish got 1ml3 E.coli and 15-20 ml layer of the substratum was added above. A disco-diffusive method was applied to assess the influence: paper disks containing preservatives were located on a surface of the stiffened agar containing E.coli. Diffusion of the preservative into agar led to formation of a suppression zone around the disks, showing the preservatives efficiency.

The preservatives' efficiency was assessed by mass fractions of 1-5%. After a 24-hour incubation of dishes in the thermostat at a temperature of 35-37 °C the data was taken. The disco - diffusive method showed that most effective on E.coli was chitosan in mass fractions in 1%, 2% and 3%. The results determined these concentrations for the second part of the experiment. The second part of the experiment was aimed to state the effect of Chitosan in controlling microbiological pollution of food products.

According to the results of the first part of the experiment Dilactin-Forte was excluded as a less effective preservative, which allowed to choose carrot *Daucus carota* subsp. *savitus* as a sample of food product.

Carrots were used as whole samples or cut in straws of 2-3 cm wide and 5-6 cm length. Samples of 100 g were processed by pre-prepared preservative, then were packed and kept in the refrigerator at 4 °C. Besides preparation of samples microbiological analysis of product samples was held on the start of the experiment to state the amount of yeast and mold, coliforms, QMAFAnM, Salmonella and L.monocytogenes. Also, the same analyses were carried out for Chitosan by a mass fraction of 1%.

Comparative analyses of samples to identify Salmonella bacteria and Listeria monocytogenes were made by the standard methods of the analysis and the polymerase chain reaction (PCR). PCR research was conducted by extracting the DNA from the microorganisms by selection of specific primers of the microorganisms, temporary and temperature parameters of the analysis. The discriminatory analyses on detection of pathogenic microorganisms of the Salmonella and Listeria monocytogenes group showed their absence in the samples studied.

QMAFAnM, coliforms and yeast/mold analysis were carried out by using method of limit cultivations and deep seeds into substratum.

Coliforms' analysis incubated for 24 hours in the thermostat at epy temperature of 37 °C.

QMAFAnM analysis were incubated in the thermostat at the temperature of 37 °C for 3 days.

Yeast/mold analysis were incubated in the thermostat at the temperature of 25 °C for 5 days.

The analysis of microbiological pollution of foodstuff was carried out by selecting 1 gr of a sample and performing 5 tenfold cultivations. First and second cultivations were used for yeast/mold analysis; 2-5 were used for yeast/mold analysis. (The results are specified in Table 1).

All analyses were made three times.

Table 1

0 day analysis results (average numbers), colony-forming units (CFU)

Coliforms	QMAFAnM		Yeast/mold		*Salmonella*	*Listeria monocytogenes*	Chitosan dilutable 1%
Negative	4c	5c	1c	2c	None	None	1
	9	1	80	8			

During the experiment Coliforms, QMAFAnM and yeast/mold analysis were carried out on first, third, fifth, seventh and tenth days of storage. (The results are specified in Table 2).

Table 2

Analysis results, CFU

		1 day											
		Coliforms			QMAFAnM						Yeast/mold		
		1s	2s	3s	1s		2s		3s		1s	2s	3s
					2c	4c	2c	4c	2c	4c	2c	2c	2c
1 %	Whole	Negative			75	0	180	2	110	1	10	2	7
	Cut				227	3	100	1	150	2	32	5	15
2 %	Whole				15	0	4	0	10	0	1	0	0
	Cut				93	3	550	2	240	3	30	10	15
3 %	Whole				16	0	10	0	11	0	3	0	1
	Cut				27	0	8	0	15	0	2	3	2
		3 day											
		Coliforms			QMAFAnM						Yeast/mold		
		1s	2s	3s	1s		2s		3s		1s	2s	3s
					3c	5c	3c	5c	3c	5c	2c	2c	2c
1 %	Whole	Positive			100	0	0	0	110	1	5	1	6
	Cut				90	0	119	1	150	1	6	5	8
2 %	Whole				70	1	0	0	85	0	30	25	30
	Cut				100	0	100	0	120	1	400	500	450
3 %	Whole				1	0	13	0	5	0	0	0	0
	Cut				3	0	2	0	2	0	2	4	3
		5 day											
		Coliforms			QMAFAnM						Yeast/mold		
		1s	2s	3s	1s		2s		3s		1s	2s	3s
					3c	5c	3c	5c	3c	5c	2c	2c	2c
1 %	Whole	Positive			210	2	2	0	115	1	3	4	3
	Cut				200	1	85	1	150	2	0	12	8
2 %	Whole				110	1	250	1	165	1	35	35	40
	Cut				10	1	400	15	150	2	25	500	30
3 %	Whole				20	0	15	0	11	0	0	7	5
	Cut				23	0	35	0	25	0	5	7	6

		7 day											
		Coliforms			QMAFAnM						Yeast/mold		
		1s	2s	3s	1s		2s		3s		1s	2s	3s
					4c	5c	4c	5c	4c	5c	2c	2c	2c
1	Whole	Positive			150	0	90	1	110	10	4	5	7
%	Cut				25	1	10	0	15	1	6	7	9
2	Whole				35	2	10	1	10	1	50	80	65
%	Cut				0	0	28	3	35	3	300	440	400
3	Whole				10	0	5	0	11	0	5	3	4
%	Cut				10	0	7	0	8	0	7	10	5
		10 day											
		Coliforms			QMAFAnM						Yeast/mold		
		1s	2s	3s	1s		2s		3s		1s	2s	3s
					4c	5c	4c	5c	4c	5c	2c	2c	2c
1	Whole	Positive			270	3	110	1	110	1	5	7	8
%	Cut				30	2	35	3	32	2	10	12	15
2	Whole				35	4	10	3	10	1	125	150	115
%	Cut				15	1	35	3	27	3	500	450	500
3	Whole				15	1	10	1	11	1	4	5	5
%	Cut				10	1	7	1	15	1	9	10	8

For results analysis average numbers were counted. (The results are shown in Table 3).

Table 3

Average numbers, CFU

QMAFAnM		Whole				
	1%	12167	70000	100000	116667	166667
	2%	967	51667	100000	183333	266667
	3%	1233	6333	15333	86667	100000
		1 day	3 day	5 day	7 day	10 day
		Cut				
	1%	15900	66667	133333	166667	233333
	2%	29433	106667	600000	210000	233333
	3%	1667	2333	27667	83333	100000
		1 day	3 day	5 day	7 day	10 day
Yeast/mold		Whole				
	1%	633	400	400	467	667
	2%	33	2833	3667	6500	13000
	3%	133	0	400	400	467
		1 day	3 day	5 day	7 day	10 day

	Cut					
	1%	1733	633	667	733	1233
	2%	1833	45000	18500	38000	48333
	3%	233	300	600	733	900
		1 day	3 day	5 day	7 day	10 day

During the experiment cultivations used for analysis could change due to the results of the previous ones.

3. Results and Discussion

The results of the conducted research are presented in Figures 1 – 4. They demonstrate that the effect of chitosan dilutable on microbiological pollution of carrot has the following characteristics:

Comparison of chitosan dilutable and dilactin-forte influence on E.coli revealed higher efficiency of the first in suppression of microorganisms' activity.

The greatest efficiency during the second part of an experiment was shown by 3% solution of chitosan dilutable. Its use allowed to increase expiration date of samples by four days for QMAFAnM indicator comparing to 1% and 2% solutions. 3% solution of Chitosan also had the greatest efficiency on yeast and mold bacteria and allowed to increase the product's expiration date by 10 days.

It also should be noted that Chitosan is more efficient to samples of the whole carrots in comparison with cut carrots.

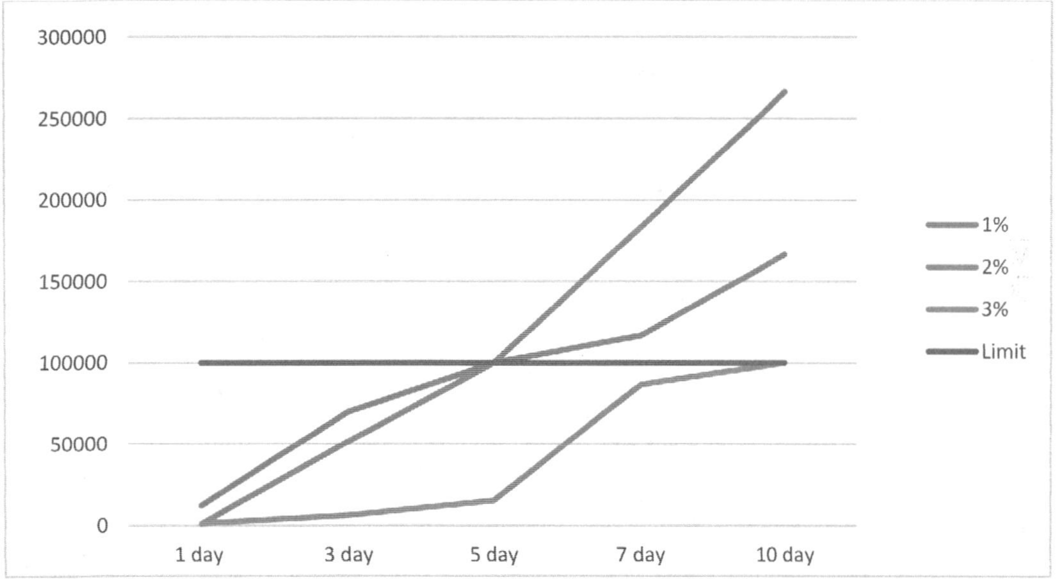

Figure 1. QMAFAnM (average numbers) whole carrot, CFU

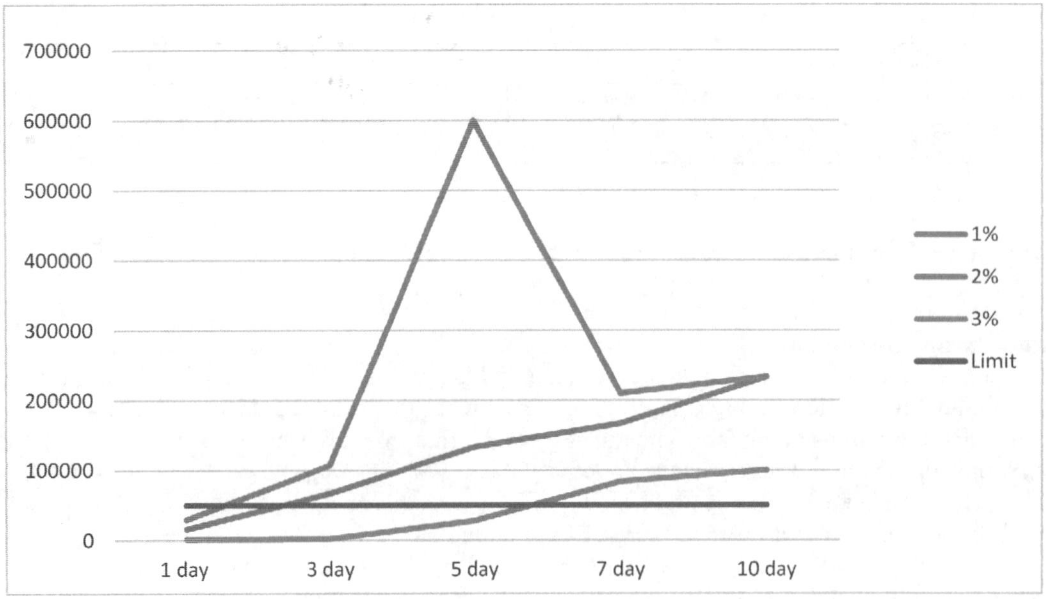

Figure 2. QMAFAnM (average numbers) cut carrot, CFU

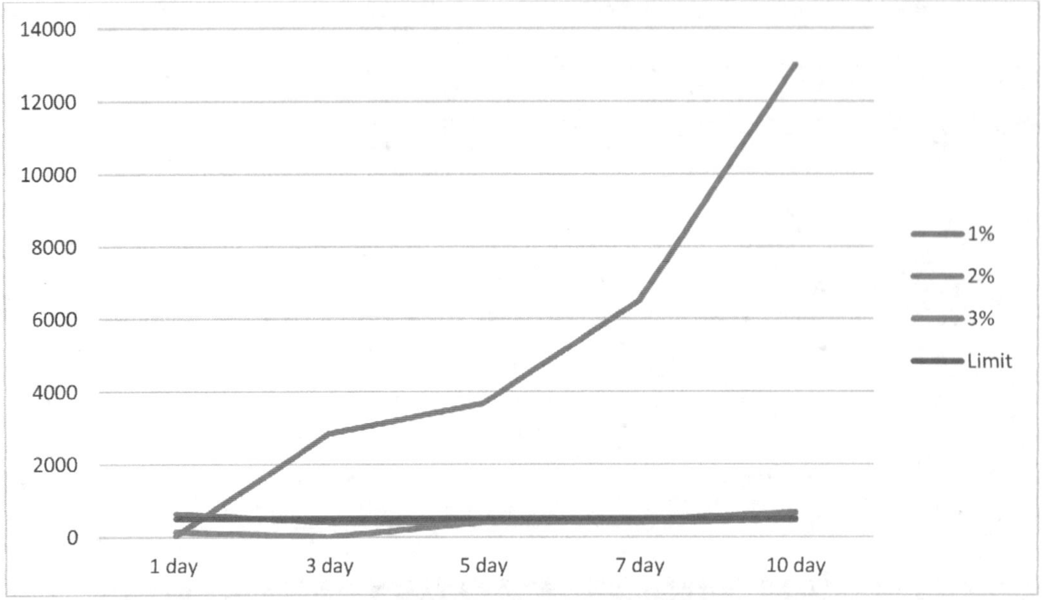

Figure 3. Yeast/mold (average numbers) whole carrot, CFU

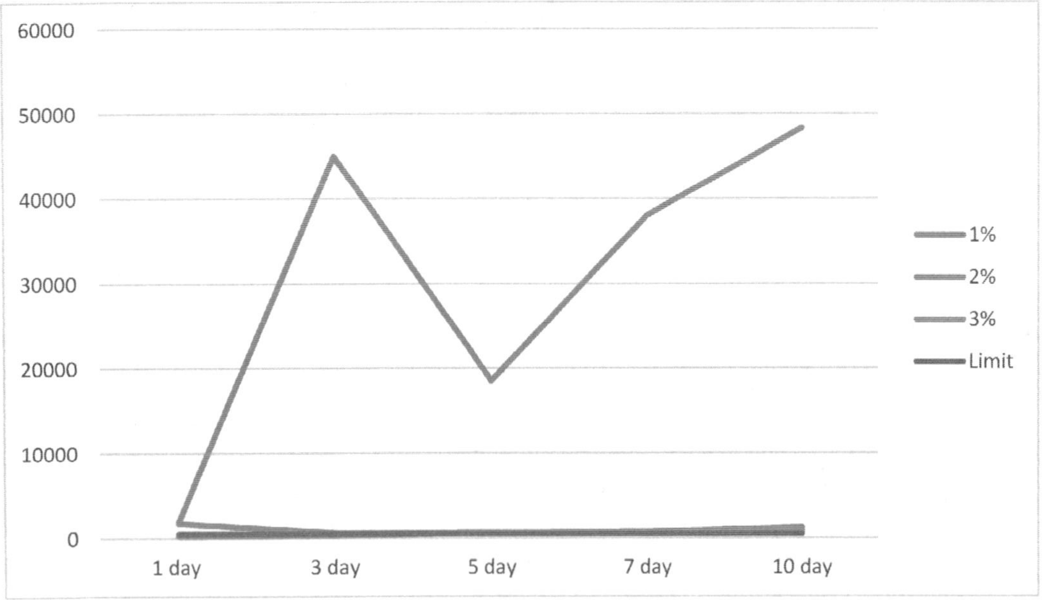

Figure 4. Yeast/mold (average numbers) cut carrot, CFU

4. Conclusion

The analysis of the experiment results allows to draw the following general conclusions:

1) Results of world-wide research and practical use of chitosan show that in comparison with other preservatives (including natural ones) this medicine has considerable advantages: it is safe for human body and improves food quality due to its unique properties.

2) During the carrot experiment it was revealed that the chitosan's influence on E.coli is much higher than that of dilactin-forte.

Chitosan solution in 3% mass fraction turned out to be the most efficient one, its use allowed to extend the QMAFAnM indicated samples expiration date by 4 days in comparison to solutions in 1% and 2% mass fraction.

The use of this medicine compared to contents indicator in yeast and mold samples showed that chitosan of 3% mass fraction is also most effective, its use increased the product's expiration date by 10 days).

The experiment proved natural preservatives are most efficient for whole samples of carrot (in comparison with cut ones).

3) It should be noted that the effect of chitosan in different mass solutions on various foodstuffs needs further research and studying. It's important to outline the most effective concentration of solution for each product type. Besides, research focused on developing new technologies to extract chitosan from different types of raw materials should also be conducted as it could allow to decrease its prime cost.

5. Acknowledgement

The authors are grateful to All-Russia Scientific and Research Institute of Conservation Technologies for help in experimental part of the research. Special thanks goes to A. Gracheva for valuable advice and support.

References

[1] Rabea E. Chitosan as antimicrobial agent: applications and mode of action /E. Rabea, M. Badawy, C. Stevens, G. Smagghe [et al.] //Biomacromolecules. 2003. Vol. 4, 1. P. 14–57.

[2] Kyung, W. K. Antimicrobial activity of native chitosan, degraded chitosan-carboxymethylated

chitosan / W. K. Kyung, R. I. Thomas, L. Chan // Journal of Food Protection. 2003. 66. P. 1495–1498.

[3] Badawy M. A biopolymer chitosan and its derivatives as promising antimicrobial agents against plant pathogens and their applications in crop protection./Badawy M., Rabea E.//International Journal of Carbohydrate Chemistry. 2011. P. 1-29.

[4] Shahidi F. Food applications of chitin and chitosans. Trends in food sciens/Shahidi F., Kamil J., Arachi A. [et al.] //Tecnology. 1999. № 10. P. 37-51.

[5] Badawy M. Potential of the biopolymer chitosan with different molecular weights to control postharvest gray mold of tomato fruit /Badawy M., Rabea E. //Postharvest Biology and Technology. –2009. vol. 51, no. 1. P. 110–117.

[6] Ojagh S. M. Effect of chitosan coatings enriched with cinnamon oil on the quality of refrigerated rainbow trout /S. M. Ojagh, M. Rezaei, S. H. Razavi, S. M. H. Hosseini //Food Chemistry. 2010. vol. 120, No. 1. P. 193–198.

[7] Varun TK. Extraction of chitosan and its oligomers from shrimp shell waste, their characterization and antimicrobial effect /Varun TK., Senani S., Jayapal N., Chikkerur J., Roy S., Tekulapally VB., Gautam M., Kumar N. //Vet World. 2017 P. 170-175.

[8] Sarbon NM. Chitosan extracted from mud crab (Scylla olivicea) shells: physicochemical and antioxidant properties. /Sarbon NM., Sandanamsamy S., Kamaruzaman SF., Ahmad F. //Journal of food science and technology 2015. Vol. 7 P. 4266-4275.

[9] Auerswald L. Simultaneous extraction of chitin and astaxanthin from waste of lobsters Jasus lalandii, and use of astaxanthin as an aquacultural feed additive /L. Auerswald, G. Gäde. //AfricanJournal ofMarineScience. 2008. Vol. 30 №1 P. 35-44.

[10] Hashemi Gahruie H. Antioxidant, antimicrobial, cell viability and enzymatic inhibitory of antioxidant polymers as biological macromolecules /Hashemi Gahruie H., Niakousari M. //International Journal of Biological Macromolecules. 2017. Vol. 104. P. 606-617.

[11] Abdel-Latif M. Immunoprotective Effect of Chitosan Particles on Hymenolepis nana – Infected Mice. /Abdel-Latif M., El-Shahawi G., Aboelhadid SM., Abdel-Tawab H. //Scandinavian Journal of Immunology 2017. Vol 86. P. 83-90.

The Determination of Metals in Welding Fume by X-Ray-Spectrometry

O V Kuznetsova [1], L A Begunova [1], S V Romanenko [2, a], S A Solodsky [2]
[1] Irkutsk National Research Technical University, Irkutsk
[2] Tomsk Polytechnic University, Tomsk

E-mail: [a] svr@tpu.ru

Abstract. Analysis of the current hygienic situation in the welding production showed that the intensification of welding processes involves the deterioration of air quality, which negatively affects the welders health. Respiratory effects seen in full-time welders have included bronchitis, airway irritation, lung function changes, and a possible increase in the incidence of lung cancer. The metal concentration in the air of the working area have been determined using the photometric method of analysis, which involves the stage of decomposition of the sample material before analysis. However, losses of the analyzed elements are possible when the sample is decomposed. The X-ray fluorescence method of analysis has the advantage of being non-destructive. The investigations shown the data of photometric determination of metals in welding aerosols is 1.5÷2 times lower than the results of X-ray fluorescence analysis.

INTRODUCTION

Welding is one of the main processes widely used to join metals in many industries, but this method of processing metals are the most unfavorable in terms of hygiene. Welding is one of the more important sources for fine and ultrafine particles the inhalation of which adversely affects human health. The particles of welding fumes may contain manganese, beryllium, cadmium, chromium, vanadium, antimony, zinc, nickel, molybdenum, mercury, lead, iron and cobalt [1, 2]. The chemical composition of the welding fumes depends upon various factors like type of welding, metal coating, material of the electrode and type of metal being weld. Welding fume pulmonary effects have been associated with bronchitis, metal fume fever, respiratory tract irritation, fibrosis, cancer and functional changes in the lung [1].

The data on the qualitative and quantitative composition of welding aerosols are needed to improve the working conditions of welders. It should be noted that the problem of quality laboratory monitoring of the air condition of the working area is very relevant and its role increases with the wear of technological equipment and the introduction of modern technologies.

A samples of welding fumes are collected on polyvinyl chloride filters or mixed cellulose ester filters. A number of analysis methods have been used to analyse welding fumes collected from workplace air in industrial enterprises. These have included atomic emission spectroscopy, atomic absorption spectroscopy, photometric method of analysis, X-ray fluorescence spectrometry and others. Photometric method (PhM) of analysis is mainly used, but one requires a large amount of time for preliminary sample preparation, while in the course of a long analysis changes in the structure of the aerosol can occur and loss of analyte is observed when the sample material is decomposed.

Non-destructive X-ray fluorescence (XRF) method of analysis is more perspective. The XRF method is the express and allows to study the welding aerosols immediately after its selection. However, the introduction of rapid XRF methods for determining the metal content in welding aerosols is limited to the absence of reference materials (RM) of composition of welding fumes samples collected on the filter. There are only a few copies of reference materials found in the foreign analytic practice, for example, BCR-545 (IRMM, Belgium), HSL MSWF-1 and HSL SSWF-1 (UK). Sample BCR-545 [3] is a glass fiber filter loaded with welding dust; it is certified only for the content of Cr (VI) and total Cr. Samples HSL MSWF-1 [4] and HSL SSWF-1 [5] are welding fumes powder particles collected by welding of mild steel and stainless steel respectively, packed in glass bottles of 1 g each; they are certified for the content of Fe, Mn and Zn.

PROBLEM

Particular complexity in the creation of CO of welding aerosols is represented by wide variations in particle size, as well as the complexity of the phase and chemical composition of TSSA. The solid particles of the welding spray are generally 0.1 to 1 μm in size, regardless of the welding technique [6]. The phase and chemical composition of the TCCA depends on the welding conditions, therefore it is required to create standard samples for each type of welding separately.

The certification of RM as real materials collected on the filter is not possible as it is required to produce a large number of samples copies with the same composition, in order to use a part of them for certification analyses and to use the remaining ones as certified RM. However, the uneven distribution of pollutants in the monitored facility does not allow to obtain a samples group of the same chemical composition. Thus, the solution of the problem is seen only through the creation of synthetic reference materials, which would be adequate to real samples by their physico-chemical characteristics.

The X-ray fluorescence analysis of welding aerosols are graded most often using samples obtained by applying solutions of the analyzed elements to a filter. For example, a filter with a dried aliquot of $K_2Cr_2O_7$ solution is used in the determination of Cr (VI) [7]. The advantage of this method is the simplicity of the preparation, and the main drawback is the inadequacy of such samples to real samples of welding fumes by physico-chemical characteristics.

The samples prepared using a specially designed air samplers or aerosol generators are more closely approximated to real samples of welding fumes loaded onto aspiration filters. In the works [8, 9] a large number of welding aerosols samples were collected in parallel to prepare the samples of welding fumes composition, then some samples were analyzed by different methods and the contents of the components on the filter were determined. The remaining part of samples was used as RM. The authors [8] have found that there are small differences in the mass of the aerosol on the filter and, consequently, the content of the elements in the sample, and therefore, there was an additional analysis of the loaded filters was carried out using an X-ray fluorescence spectrometer with wave dispersion to exclude different specimens. In [10, 11] for the control of the metal content in the air of the working zone when welding steel structures using an generator of fumes, synthetic samples were obtained from artificial fume from metal wire using a robotic system. It should be noted that these techniques make it difficult to prepare a large number of identical samples with different variations of chemical composition typical for samples of welding fumes.

The technology of manufacturing synthetic RM of the atmospheric aerosols composition in the form of a polyvinyl alcohol polymer film containing a finely divided powder carrier of the identified components is proposed in the work [12]. The developed technology makes it possible to prepare the required number of identical RM samples with wide variations in the chemical composition and mass of welding fumes particles. The error in determining the content of elements in RM is 5÷18 % depending on the element, it meets the requirements for the analysis of welding aerosols.

The purpose of this study is to investigate the possibility of using the XRF technique for control of the metal content in the welding fumes and the air of the working area.

EXPERIMENT

The studies were carried out using synthetic samples based on polyvinyl alcohol films [13, 14] containing powder material of welding fumes collected on the process of welding.

The content of metals in the samples was determined using by RFA and PhM techniques. X-ray fluorescence measurements were performed on an energy-dispersive X-ray fluorescence spectrometer "EDX-8000" (Shimadzu, Japan) based on silicon drift detector with thermoelectrical cooling; an X-ray tube with an Rh-anode (work mode: air cooling, voltage 4÷50 kV, current 1÷1000 μA).

Photometric measurements were performed on a spectrophotometer "PromEcoLab PE-5400B" (Shanghai Mapada Instruments, China). Preparation of samples for photometric analysis includes ashing of samples in a muffle furnace at the temperature of 750÷800 °C during 40÷60 min. The fly ash residues were fused with the smelt (Na_2CO_3 and KNO_3 in a ratio of 2:1) the weighting 0.5÷1 g at 800÷850 °C during 25÷30 min. The fusion were leached with 25 ml of 10% H_2SO_4 [15]. The calibration characteristics of the PhM technique were determined from the series of single-element standard solutions. The optical density of the solutions was measured in cuvettes with an absorbing layer thickness of 10 mm at room temperature (20 ± 5) °C.

The table 1 provides a comparison of XRF and PhM resalts of determination of iron content in the samples.

Table 1

Comparison of XRF and PhM results

Sample	Content of Fe, %	
	XRF	PhM
1-1	0.380 ± 0.020	0.320 ± 0.030
1-2	0.330 ± 0.017	0.270 ± 0.027
2-1	0.340 ± 0.017	0.250 ± 0.025
2-2	0.340 ± 0.017	0.240 ± 0.024
3-1	0.71 ± 0.04	0.60 ± 0.06
3-2	0.69 ± 0.03	0.59 ± 0.06
4-1	0.76 ± 0.04	0.60 ± 0.06
4-2	0.74 ± 0.04	0.57 ± 0.06
5-1	1.46 ± 0.07	0.81 ± 0.08
5-2	1.44 ± 0.07	0.79 ± 0.08
6-1	1.26 ± 0.06	0.89 ± 0.09
6-2	1.20 ± 0.06	0.85 ± 0.09

The analysis of the samples showed that the results of the photometric determination of metals in solutions after digestion the sample material according to the methodical instructions [15] are 1.5÷2 times lower than the results of X-ray fluorescence analysis of samples without digestion. Underestimation of the results of photometric measurements may be due to the analyte losses on digestion process or incomplete decomposition of some analyte compounds. Sources of the PhM results errors are discussed in detail in [16].

CONCLUSION

XRF spectrometry is a simple, accurate and reliable method for the evaluation of metals concentration in weld fume samples and require little sample preparation. The energy dispersive XRF method allows to carry out non-destructive rapid definition of a wide range of elements.

REFERENCES

[1] Korczynski R.E. Occupational health concerns in the welding industry // Applied Occupational and Environmental Hygiene. 2000. V. 15, № 12. P. 936-945.
[2] Antonini J.M. Health Effects of Welding // Critical Reviews in Toxicology. 2003. V. 33, № 1. P. 61-103.

[3] Chauhan S.K., Gupta P.K., Shukla A., et.al. Recent developments of certified reference materials for road transportation // Environmental Monitoring Assessment. 2009. V. 156, № 1-4. P. 407-418.
[4] Butler O., Rimmer D., Davies P. Reference material HSL MSWF-1. Elements in mild steel welding fume // United Kingdom: Health and safety laboratory, 2013. 44 p.
[5] Butler O., Musgrove D. Reference material HSL SSWF-01. Elements in stainless steel welding fume // United Kingdom: Health and safety laboratory, 2013. 50 p.
[6] Berlinger B., Benker N., Weinbruch S., et.al. Physicochemical characterisation of different welding aerosols // Analytical and Bioanalytical Chemistry. 2010. V. 399. P. 1773-1780.
[7] Oláh Ĺ., Pivoluska J. Determination of chromium /VI/ in welding fumes by radionuclide X-ray fluorescent analysis // Journal of Radioanalytical and Nuclear Chemistry. 1985. V. 96, № 6. P. 575-584.
[8] Stacey P., Butler O. Performance of laboratories analysing welding fume on filter samples: results from the WASP proficiency testing scheme // Annals of Occupation Hygiene. 2008. V. 52, № 4. P. 287-295.
[9] Anglov J.T.B., Holst E., Dyg S., et.al. Iron, manganese, copper and titanium in welding fume dust on filters for internal and external quality assurance // Fresenius' Zeitschrift für Anallytische Chemie. 1993. V. 345, № 2-4. P. 335-339.
[10] Antonini J.M., Afshari A.A., Stone S., et. al. Design, construction, and characterization of a novel robotic welding fume generator and inhalation exposure system for laboratory animals // Journal of Occupational and Environmental Hygiene. 2006. V. 3, № 4. P. 194-203.
[11] Cena L.G., Keane M.J., Chisholm W.P., et. al. A novel method for assessing respiratory deposition of welding fume nanoparticles // Journal of Occupational and Environmental Hygiene. 2014. V. 11, № 12. P. 771-780.
[12] Kuznetsova O.V., Kuznetsova A.N., Begunova L.A. Synthetic reference materials based on polymer films for the control of welding fumes composition // IOP Conf. Series: Materials Science and Engineering. 2017. V. 189. P. 1-4.
[13] Physicochemical Study of Ta_2O_5 and SiO_2-Ta_2O_5 Film Formation from Film-Forming Solutions Gryaznov, R.V., Borilo, L.P., Kozik, V.V., Mal'chik, .G. 2001 Russian Journal of Applied Chemistry 74(1), c. 18-20.
[14] Synthis and Phisical and Chemical; Properties of SiO_2 - B_2O_3 and SiO2 - P_2O_5 Thin Film Systems and Powders Mal'Chik, A.G., Litovkin, S.V., Seregin, V.I., Rodionov, P.V., Kryuchkova, S.O. 2016 IOP Conference Series: Materials Science and Engineering 142 (1), 012066
[15] MI № 4945-88. Methodical instructions for the definition of hazardous substances in welding fumes (solids and gases). – Moscow, Rarog Publ,, 1988. 232 p. [in Russian].
[16] Kuznetsova O.V., Korzhova E.N., Shmeleva E.I., et. al. Accuracy of Mn and Fe determination in air of the working zone using photometric and voltammetric methods of analysis // Zavodskaya Laboratoriya. Diagnostika Materialov. 2017. V. 83, № 4. P. 66-71.

Diagnosis of Electric Submersible Centrifugal Pump

M S Kovalchuk and D A Poddubniy

Saint-Petersburg Mining University, 2, 21 Line Ave., Saint-Petersburg, 199106, Russia

E-mail: cheremushkina-ms@yandex.ru

Abstract. The paper deals with the development of system operational diagnostics of electrical submersible pumps (ESP). At the initial stage of studies have explored current methods of the diagnosis of ESP, examined the existing problems of their diagnosis. Resulting identified a number of main standard ESP faults, mechanical faults such as bearing wear, protective sleeves of the shaft and the hubs of guide vanes, misalignment and imbalance of the shafts, which causes the breakdown of the stator bottom or top bases. All this leads to electromagnetic faults: rotor eccentricity, weakening the pressing of steel packs, wire breakage or a short circuit in the stator winding, etc., leading to changes in the consumption current.

1. Introduction

The installation of electrical submersible pumps (ESP) plays a significant role in the complicated and energy-intensive technical oil and gas facilities and used for extraction production oil wells. Failures in the functioning modes of ESP are the leading cause of unplanned downtime wells stock, leading to significant financial losses.

Currently, the quality and efficiency of accepted diagnostic decisions about the status of ESP, taking into account the great range of the analyzed parameters and analyze huge amounts of information about the modes of operation of submersible electric centrifugal pumps, are not large enough, which increases the probability of error evaluation of its condition and failure to take appropriate operational decisions.

In this regard, the actual task is the development of new effective methods of diagnostics of the ESP oil producing wells.

2. The purposes of research

In this paper were set and solved the following scientific and technical objectives:

• Comparative analysis of existing methods of submersible centrifugal pumps diagnosis.

• Identified the relationship between the presence of typical harmonics and the corresponding defects of submersible centrifugal pumps through computational and physical experiments.

• On the basis of the obtained dependences developed a technique of ESP diagnostics. There are the following maintenance strategies of submersible centrifugal pumps:

- In terms of preventive maintenance or equipment failures, which is economically unjustified and is a major cause of production downtime.
- 'On the state', by applying the recognition method to the actual technical state of equipment on set of diagnostic characteristics to identify existing or developing defects and to determine the optimal timing of repair work.

3. The existing methods of submersible centrifugal pumps diagnosis

Condition monitoring and technical diagnostics of ESP includes assessment of the thermal status of equipment (thermodiagnosis), vibration measurement of equipment (vibration analysis), modeling, etc. [1].

Thermodiagnosis is based on the fact that in the place of occurrence of the defect of the electrical equipment the temperature rises and, as a consequence, increases the intensity of infrared radiation, which is recorded by thermal imaging devices. Simulation, including the stage of development of a

IOP Conf. Series: Earth and Environmental Science **115** (2018) 012026 doi:10.1088/1755-1315/115/1/012026

computer model of the electrical submersible pump connected to the engine by using sensors. Provides an opportunity to measure multiple operating signals to the pump that is used to solve equation of state in time. This method is implemented only with direct access to the motor and the pump, i.e. it does not provide remote diagnostics; has low accuracy; the complexity of the required measurements that illustrates it impossible to use for diagnostics [2].

Vibration diagnostics, during which register and analyze the signals generated by the vibration of the motor. According to the obtained data analyze the form and amplitude of the received signal and compare with threshold values, to assess the possibility of further operation of the unit. However, the need to install additional vibration sensors on the pump housing, and additional communication lines to transmit information about vibration of the pump, which reduces the reliability of the entire diagnostic system and increases its cost, thus limiting the possibility of applying this method [1].

However, such diagnostic methods are not applicable in all situations, and make it impossible to determine the exact cause of origin of a process malfunction. Diagnosis of all the above types of faults can be accurately and timely implement on the basis of spectral analysis of stator current.

Despite the fact that the proposed method of spectral diagnostics of electric parameters is widely used in many industries, in oil production, this issue has received little attention.

To identify patterns, the spectral structure of the motor current and the corresponding types of pump faults, a mathematical model of ESP was developed, including models of the frequency converter, the submersible motor, centrifugal pump and hydraulic system (load). The dynamic variables of the models of the component parts of a submersible centrifugal pump located in relationship between them and provide mutual influence on each other. In addition, ESP has developed mechanical and hydraulic subsystems, which greatly affect the dynamics of the whole complex [3].

The results of calculations of the model are indicators for which determined a critical value. When the deviation of the values for the boundaries of critical values are needed to identify the causes of deviations and develop corrective actions.

4. The influence of parameters deviations of centrifugal pumps

One of the important parameters to determine the performance of the pumping unit is the relative deviation in efficiency from the nameplate. Coefficient characterizes the degree of wear consider, and fluctuation, associated with the use of the bypass. The lowest coefficient values enable to select pumps with the highest wear and/or the greatest influence of the bypass regulation. After defining the checklist, you need to check the source data of the selected aggregates. The value of the accumulated energy and operating time can be taken in different time periods, which can lead to errors. It is necessary to verify the actual power of the unit. The power value of pumps can be obtained from the experts. The optimal solution is plotting the power of pumps, which can be used to compare values.

After checking the source data, checks the presence and the use of the unit using the bypass. The use of bypasses due to the necessity of operation of the pumps with too small volumes of liquid. Units that can be installed by-passes, have a significant mismatch between the passport and the actual performance of the pump with the real output pressure of pumps. Corrective activity may be the exception of the bypass regulation by means of choose of the optimal pump for the pumped fluid volume.

The second factor of the deviation of actual efficiency from the nameplate is wear. If it is revealed that bypases regulation is not used or has little impact, units must be checked according to the method of the optimal time of removal the equipment for repair, then takes a decision on the necessity for an early overhaul of the pump [2].

Relative excess of pressure pumping units above the yearly maximum line pressure – the ratio of the sum of the actual pressure at pump inlet and passport for the actual volume to the maximum line pressure for the 12 months increased by 35 m (normative losses from the pump to the line). Coefficient characterizes the redundancy of the pump head in relation to the maximum for the object. The highest coefficient values enable to define the pump units with excessive pressure. These pumps should be checked for the possibility of reducing the number of working wheels.

Corrective activity can be the replacement of all or part of the pumps with the pumps with lower nominal pressure. After replacing the operating pump, at maximum should be used the units with lower pressure. In this case, if replaced only a part of the pump during the increase of pressure in the line can be used reserve pumps, having a higher pressure.

The coefficient of irregularity of pressure – the ratio of the maximum pressure in the line to the middle. Coefficient characterizes the oscillation amplitude of the pressure in the line (elevation of the maximum line pressure on the average linear pressure during the period under review). The highest coefficient values allow to select objects with the most irregular mode of operation. As well as pumps is usually selected to work with the maximum pressure lines, as shown on the diagram, on such installations pumps main time working in a pre-loaded condition. This leads to excess of energy consumption [3].

At such installations it is necessary to develop activities for alignment of the pumping mode. This can be applied to the redistribution of flows between installations, coordinated delay of pumps start, so that they were not carried out simultaneously.

Also, such problems can be solved by running multiple pumps with different pressure or with large and constant fluctuations by installing of variable frequency drives (Figure 1).

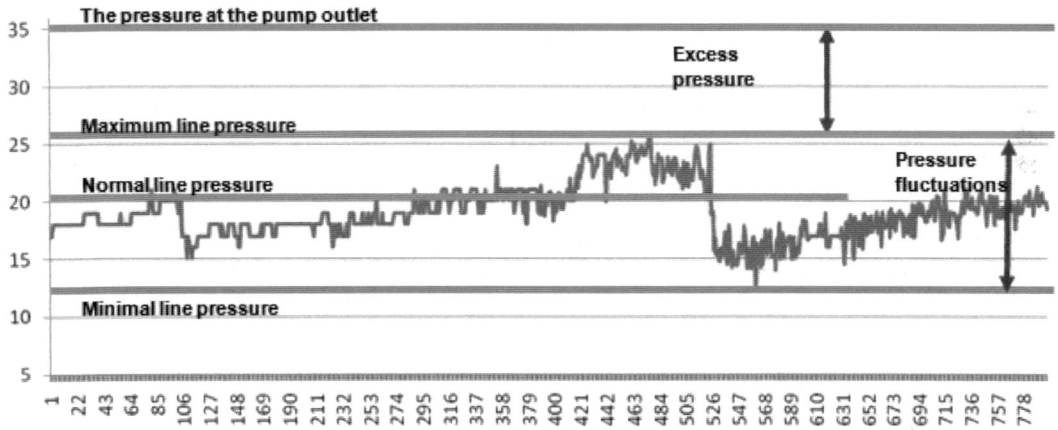

Figure 1. The pressure of electrical submersible pump

Load factor – the ratio of the actual and the nominal capacity of the pumping pump. The coefficient indicates the deviation of the operation of the pump in the left or right area.

With the coefficient deviation it is needed to analyze redistribution of volumes between units for more efficient loading or selection of optimal combinations of equipment.

Efficiency of the pump – the ratio of hydraulic energy, transferred to the line, to energy consumption during the period under review. Coefficient characterizes the overall energy efficiency. The indicator influenced by the following critical factors [4]:

- deviation of the pump in the left/right area;
- excess pressure;
- technical condition.

The indicator allows to select the pump with the greatest power overconsumption. The first step in the analysis it is necessary to determine the reasons for low efficiency, and offer solutions for increasing it. A common solution for a pump with low efficiency is the selection of optimal equipment combinations for the parameters of the current pump. In the list you need to determine the most efficient configuration of equipment available in reserve. A similar check must be done on the parameters of the pumping unit [5].

The boundary values for the criteria are reconsidered at least 1 time per year.

IOP Conf. Series: Earth and Environmental Science **115** (2018) 012026 doi:10.1088/1755-1315/115/1/012026

5. The faults of electric motors of centrifugal pumps

Consider one of the faults of electric motors centrifugal pumps. Rotor eccentricity is quite often occured in practice, the reason of increased vibration of electric machines arising from the wear of rolling bearings.

In the presence of rotor eccentricity in the distribution of the electromagnetic field in the gap of the engine there is a number of features. The density of the electromagnetic field around the circumference of the gap changes with rotation of the rotor, leading to uneven traction engine. When the overlap of the axis of the stator field with the area of increased gap, traction is somewhat reduced, with increasing the slip value of frequency.

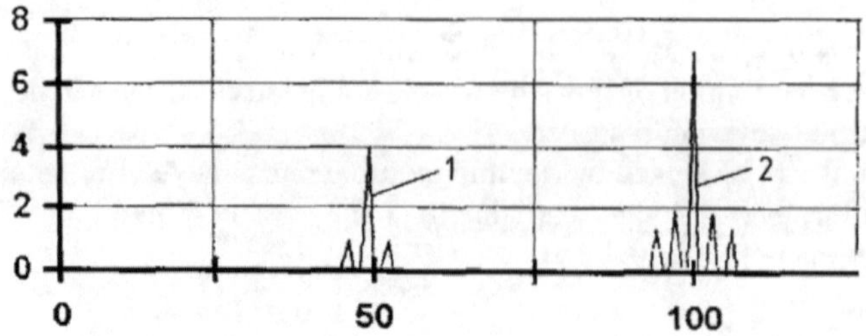

Figure 2. Spectrum of current with rotor eccentricity

During the displacement of the axis fields in the lower zone of the gap, pulling force increases, the slip frequency decreases. When the number of pairs of poles of the stator is greater than one, this process is repeated 'p' times.

Figure 2 shows the range of current with rotor eccentricity.

In the interval of displacement of the rotor from the zone with the increased gap zone to the zone with the reduced gap, the rotor is accelerated in its speed at the small value. In the interval of transition of the rotor back to the area with an increased gap, the rotor is slowed to the same value. This is noticeable in the spectrum [6].

On current spectrum around the fundamental frequency of rotation of the rotor, appear symmetrically located side peaks (harmonics), resembling the teeth of a crown. The symmetry of the peaks relative to the fundamental frequency is a consequence of the accelerations and decelerations of the rotor speed around its mean value.

Similar to the teeth of higher intensity appear around the peak electromagnetic power at a frequency equal to the second harmonic of the mains [7].

The eccentric rotation of the rotor modulates the conductivity of the gap with twice the frequency. When the number of pairs of poles equal to one, the frequency of rotation of the field equal to 50 Hz, twice the line frequency, the frequency of the electromagnetic vibrations to 100 Hz. The eccentricity of the rotor leads to a modulation of the electromagnetic force. By reducing the number of pairs of poles of the frequency of rotation of the field in the gap is reduced to p times. Variable gap of the rotor during one circulation will modulate the electromagnetic force, 2*p times the frequency of its rotation that corresponds to the frequency of the electromagnetic force.

The eccentricity of the rotor is usually appeared both in vertical and transverse projections of the vibration. Sometimes it is possible to detect even in the axial projection. It is possible in the presence of eccentricity of the rotor along its entire length, but only in the region of one edge of the package electrical steel [8].

The eccentricity of the rotor often is transient in nature, when in the spectrum of the engine has a characteristic pattern, and the practical measurement of the gap doesn't confirm the diagnosis. The reason is usually in thermal processes when the rotor is asymmetrically heated, bent, and gives a picture of eccentricity.

After motor shutdown, in the process of disassembly to measure the gap, the temperature quickly equalized and the diagnosis is not confirmed. Often it happens when breakages of the rods or partial "grazing" of the rotor about the stationary elements when the rotor begins to heat up one-sided [9].

During conducting the research was diagnosed the mode of operation of pumping system with open circuit and phase stator fault. In Fig. 3.1 shows the spectrum of the motor current in normal operation and when one phase is open. In Fig. 3.2 shows the range of current when operating in normal mode and the misalignment of one of the phases.

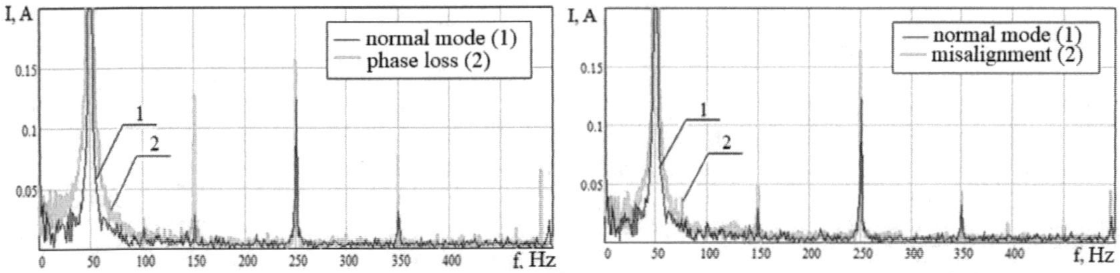

Figure 3.1 and 3.2. The spectrum of the motor current in normal operation and in failure mode

Analysis of the graphs in Fig.3.1 reveals that the open phase of AD leads to an increase of the harmonics number 3 (3 dB) No. 5 (1 dB) No. 7 (1.7 dB), as well as to the emergence of a peak at the area of 490 Hz. A comparison of the spectrograms in Fig.3.2 revealed that when imbalance is present the growth of harmonics number 3 (1.6 dB) No. 5 (1.3 dB), No. 7 (1.6 dB).

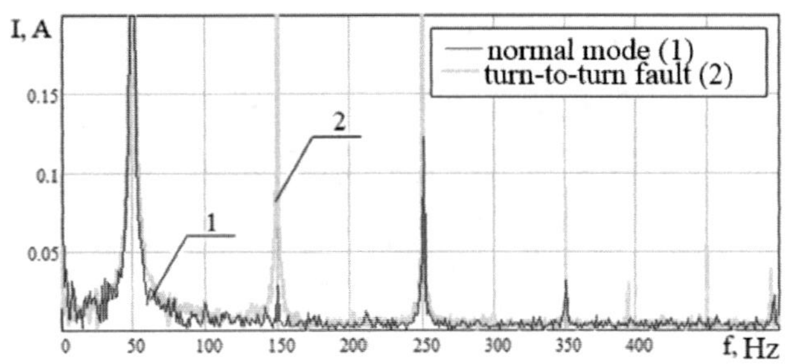

Figure 4. The spectrum of the motor current in normal operation and during turn-to-turn fault

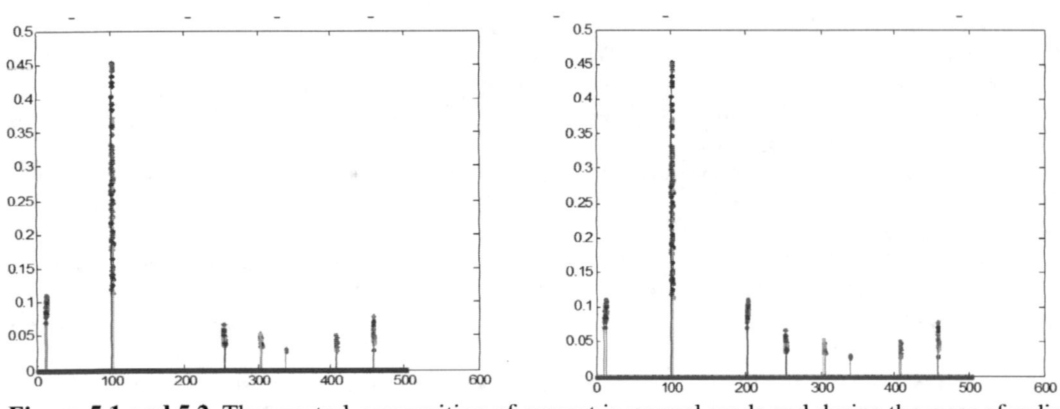

Figure 5.1 and 5.2. The spectral composition of current in normal mode and during the wear of radial bearings

When turn-to-turn fault (Figure 4) harmonica No. 3 increased by more than 4 dB. Also revealed the other odd harmonics: No. 5 (1.5 dB), No. 7 (2 dB) No. 9 (10 dB). Additionally in the spectrum is

IOP Conf. Series: Earth and Environmental Science **115** (2018) 012026 doi:10.1088/1755-1315/115/1/012026

observed 8 (400 Hz) and 10 (490 Hz) harmonics.

In experimental researches were obtained spectrograms of current during normal operation (Figure 5.1), as well as during the wear of the radial bearings (Figure 5.2).

Analysis of the obtained data allowed to conclude that the wear of the radial bearings of the pump leads to the appearance of additional harmonics at the frequency of 200 Hz, which is twice the frequency of rotation of the motor shaft of the pump.

6. Conclusion

The comparative analysis of existing methods of diagnosis ESP system showed that the most reliable diagnostic method is a method based on spectral analysis of stator current [10]. Obtained the relationship between the presence of the characteristic harmonics and certain types of faults ESP. The results of the experiments for the same mode do not differ more than 5%. Analysis of the obtained spectra showed the differences between the modes of operation of motor that allows to diagnose in the initial stages the appearance of a fault in the operation of ESP.

Acknowledgements

The presented results were obtained as a part of scientific researches according to the contract № 13.3746.2017/8.9 within the scope of the State task "The designing on the base of systematic and logic probability evaluations of rational and economically proved structure of centralized, autonomous and combined power supply systems with high reliability and stability level with usage of alternative and renewable power sources for uninterrupted power supply of enterprises with continuous technological cycle".

References

[1] Cheremushkina M S., Poddubniy D A., Reducing The Risk Of Fires In Conveyor Transport / IOP Conference Series: Earth and Environmental Science, № 50, T 50, 2017. C 1 - 6.

[2] Baburin S V, Ustinov D A 2016 *International Journal of Applied Engineering Research* **11(9)** 6402-6406

[3] Cheremushkina M. S., Baburin S. V., Modelling and control algorithms of the cross conveyors line with multiengine variable speed drives / IOP Conf. Series: Materials Science and Engineering, № 177, T 177, 2017. C 1 - 7.

[4] Abramovich B N, Sychev Yu A, Ustinov D A and Shkljarskiy A Ya 2014 *Neftyanoe khozyaystvo - Oil Industry* **8**, 110-112

[5] Liu Q, Peng L, Kang Y, Tang S, Wu D and Qi Y 2014 *IEEE Transactions on Industrial Electronics* **61(8)** 4000-4010

[6] Bardanov A I and Skamin A N 2015 *International Conference on Mechanical Engineering, Automation and Control Systems (MEACS)* **1** 1-3

[7] Araujo-Vargas I, Salas-Duarte S, Ramirez-Hernandez J, Del-Muro-Cuellar B and Rivera M 2015 *IEEE International Symposium on Predictive Control of Electrical Drives and Power Electronics (PRECEDE)* **1** 79-84

[8] Chen Z, Wang Z and Li M 2014 *The 40th Annual Conference of the IEEE Industrial Electronics Society (IECON)* **1** 1041 - 1047

[9] Jimenez F R, Salamanca J M and Cardenas P F 2014 *IEEE The 5th Colombian Workshop Circuits and Systems (CWCAS)* **1** 1-10

[10] Kovalchuk M. S., Skamyin A. N., Developing the System of Monitoring and Diagnostics to Increase the Availability of Equipment / IOP Conference Series: Earth and Environmental Science, № 66, T 66, 2017. C 2 - 8.

The Survey of Sorption Ion-Exchange Properties of Paleozoic Zeolites Before and After Regeneration

I R Valieva[a], T V Germanova[b], V A Nefedov
Tyumen industrial University, 625000, Ural Federal District, Tyumenskaya oblast', Tyumen, St. Volodarsky, 38

E-mail: [a] irina.valieva@list.ru, [b] ecogtv@mail.ru

Abstract. The presented results of experimental studies on the removal of iron ions, manganese, ammonium nitrogen, strontium and cadmium from aqueous solutions by the paleozoic zeolites of the Urals in dynamic conditions. Experiments have been carried out on the regeneration of mineral zeolite filters in order to restore the sorption properties of the explored minerals. The results of the researches show that ion-exchange minerals can be used as highly effective sorbents, with a large ion-exchange capacity and a multifaceted ability to restore the original properties.

1. Introduction

Investigation of ion-exchange properties of Paleozoic zeolites is necessary for the possible use of natural minerals in the field of environmental protection, with the aim of increasing the time of protective action.

Surface and underground waters of the West Siberian artesian basin are characterized by increased contamination with substances of natural and technogenic origin: ammonium nitrogen, manganese, iron, strontium, cadmium, etc. Removal of pollutants from natural and waste water is provided by a complex of technological purification processes, one of which is filtration . When filtering, granular materials of natural or artificial origin are used. In the world practice, research is being conducted on the use of sorbents possessing ion-exchange properties of zeolites, zeolite-montmorillonite rocks, etc. [1,2]. Paleozoic zeolites of the Urals for the purification of polluted waters are highly promising. Zeolite ores contain 90-95% ion-exchange minerals, do not contain toxic and heavy impurities [3].

2. Equipment and devices used in studies

2.1 Sorption and ion-exchange properties of natural minerals

The purpose of this work was to study the sorption-ion exchange properties of Paleozoic zeolites in removing impurities of various nature from solutions and methods for regenerating the filter material. Studies have noted that ion-exchange properties depend on the nature of the sorbent and the composition of the impurities [8-10]. Sorption properties have been studied under various physicochemical conditions (size of fraction, temperature, impurity concentration in solution). In the study of Paleozoic zeolites, a technique developed by the Institute of Mineralogy, Geochemistry and Crystallochemistry of Rare Elements was applied [3,5].
Work performed under the subject study of the chemical and physical properties of zeolite rocks of the Paleozoic of the Urals. The experiments were carried out in the laboratories of the Tyumen Industrial University, on the basis of the Tyumen Central Laboratory [1,2].

3. Results and discussion

3.1. Discussion of research processes

In the study of ion-exchange and sorption properties of zeolites, the ability of regeneration processes, the time the protective actions were carried out experiments on sorption in dynamic conditions before regeneration of the ions of iron, manganese, strontium, cadmium, ammonia-nitrogen purified by passing

solutions through the sorbent layer with a height of 50 cm. Solutions were prepared on the basis of tap water of the city of Tyumen with the following content of the main components (mg/ dm^3): K^+ =3.7; Mg^{2+} = 11.8; Ca^{2+}=31.0; NH_4^+= 0.4 (averaged composition). The water was injected salt of the investigated pollutants found in the West Siberian artesian basin according to other authors. [1-5]
The experiments were carried out by filtration of solutions through a layer of mineral sorbent fraction 0.3-0.6 mm, placed in a burette with a diameter of 1.2 cm. The flow rate of the solution was regulated so that the filtration rate was 2.5 m/hour. Samples for chemical analyses were taken every 0.1 dm^3 filtered liquid.
Research of sorption ability of filters to remove from the solution the ions NH_4^+ were carried out at the introduction into water of 10 mg/ dm^3 ammonium nitrogen. The first traces of ammonium nitrogen in filtered water was noted in 23 dm^3. 26.8 dm^3 filtered liquid experience was discontinued. And the regeneration of 5% NaCl solution. There were four of regeneration, which showed that the mineral filter is increased exchange capacity. The results are shown on the chart. (*Figure 1*) After fourth regeneration of exchange capacity increased 1.8 times compared to the first experience.

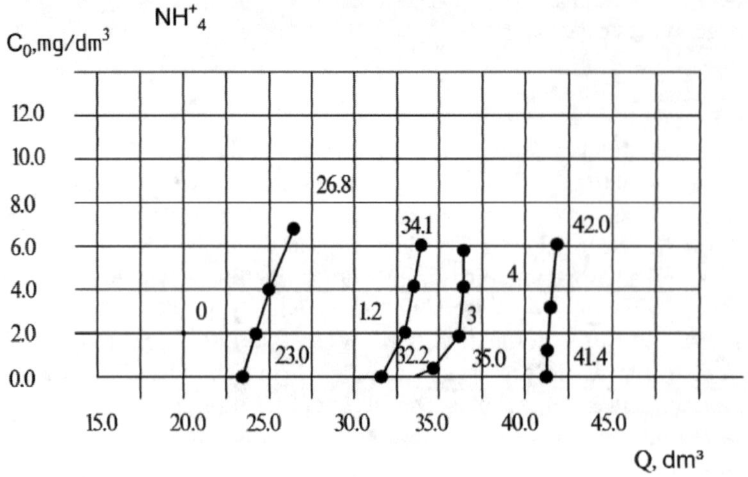

Figure 1 Dependency of sorption ability zeolite-montmorillonite breeds on regenerations` quantity over ammonium nitrogen. The maintenance of $NH_4^+$ in filtered water 10.0 mg/ $dm^3$; the filtration rate 4.0 ml/min, the granular filter fraction 0.3-0.6 mm, the sorbent`s weight 50.0 g, the height of the filterable layer 50 cm.

Notion: 0-4 – curves of sorption: 0 – regeneration; 1 – after the first regeneration; 2 – second; 3 – third, 4 – fourth regeneration.

The sorption capacity of mineral filters in dynamic conditions for extracting ions of iron and manganese from solution was investigated similarly to the experiments on the sorption of ammonium nitrogen. Filtration of a solution containing 20.0 mg/ dm^3 of iron, using zeolite sorbent showed it to a very high ion exchange capacity. (Figure 2)

IOP Conf. Series: Earth and Environmental Science **115** (2018) 012027 doi:10.1088/1755-1315/115/1/012027

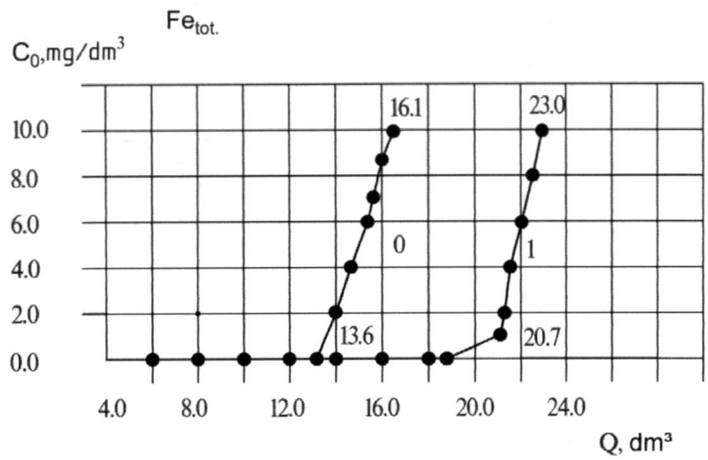

Figure 2 Dependency of sorption ability zeolite-montmorillonite breeds on regenerations` quantity over the iron. The maintenance of iron in filtered water 10.0 mg/dm$^3$; the filtration rate 4.0 ml/min, the granular filter fraction 0.3-0.6 mm, the sorbent`s weight 50.0 g, the height of the filterable layer 50 cm.

Notion: 0, 1 - curves of sorption: 0 - before regeneration, 1 - after the first regeneration.

13.6 dm^3 filtered water the iron content was not recorded. 16.2 dm^3 in filtered water detected a sharp increase in the content of Fe^{2+} and 10 mg/dm^3. Carried out desorption with 5% solution of NaCl in a volume of 1.5 dm^3 fully regenerated granular filter. After regeneration experiments were conducted repeatedly. The presence of iron in the filtrate were recorded in 20.8 dm^3;the adsorption capacity after regeneration compared to the non regenerated filter is increased 1.5 times. Repeated second regeneration led to a sharp decrease in the exchange capacity, already on 5.0 dm^3 in the filtrate there were traces of Fe^{2+}. Study on the purification of water from manganese were carried out on solutions containing 0.5 mg/ dm^3 of manganese. The first traces of manganese found in 29.9 dm^3 of the filtrate, 36 dm^3 Mn^{2+} content was 0.45 mg/dm^3. A grainy filter has been activated 5% NaCl solution. In 0-6.5 filtrate ion content of manganese ranged from 0.01-0.1 mg/dm^3. In the following amounts 7.5-15.0 dm^3 filtered water fluctuations in the content of manganese was in the range of 0.03-0.3 mg/ dm3. Further filtering of the recorded concentration of manganese 0.28-0.8 mg/ dm^3. (*Figure 3*)

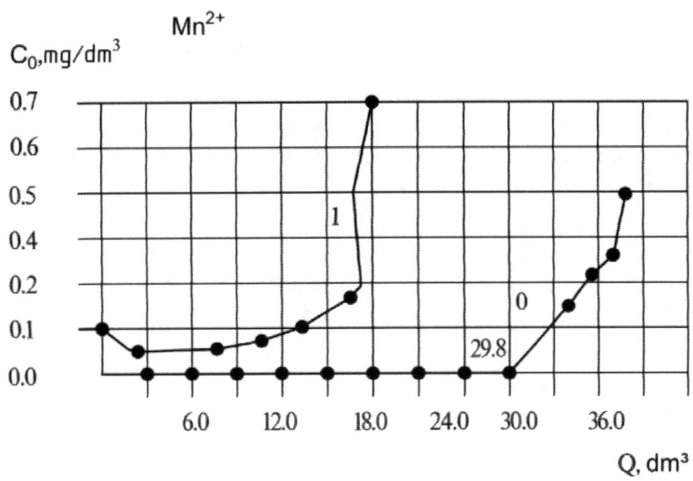

Figure 3 Dependency of sorption ability zeolite-montmorillonite breeds on regenerations` quantity over manganese. The manganese`s maintenance in filtered water 0.5 mg/ dm^3; the filtration rate 4.0 ml/min; the sorbent`s weight 50 g; the height of the filterable layer 50 cm.

Notion: 0,1 – manganese`s curved maintenances in filtered water after: 0-before regeneration; 1-first regeneration

Repeated regeneration has not led to the recovery of the sorption capacity of granular filter.
Studies on the removal of strontium from solutions of spent granular filters by filtration through them

of water enriched with strontium in the amount of 20.0 mg/ dm^3. T V Germanova and I R Valieva was investigated sorption processes for extraction from solutions of strontium and lead in the presence of competing cations in solution are Na^+, K^+, Ca^{2+}, Mg^{2+}, but at lower concentrations of strontium. [1-3] the First traces of strontium recorded in 17.3 dm^3 filtered water. (Figure 4)

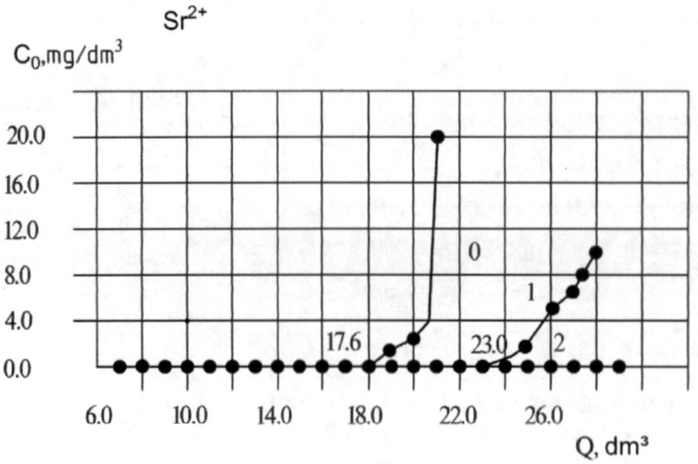

Figure 4 The sorption in dynamic conditions by the zeolite-montmorillonite granular filter of strontium ions from water; the fraction 0.3-0.6 mm; the filtration rate 4.0 ml/min. The original distilled water enriched with ions Sr^{2+} (mg/ dm^3) – 20.

Notion: 0,1,2 - content of strontium cations in filtered water after: 0 – before regeneration; 1 – first regeneration; 2 – second regeneration.

In samples 202-210 contents of strontium in the filtrate was 1.8 mg/ dm^3.In the process of experimental work was conducted 2 regeneration. Studies indicate that after each regeneration of zeolite sorption capacity of the filter increased by 15-20%. These figures make zeolite filters are highly promising for their application purification of highly polluted waters strontium, ammonia, cadmium etc.
The ion-exchange properties of zeolites with cations of cadmium. The experiments had two aims: determination of exchange capacity of the filters cadmium ions and the number of possible regenerations. Filtered solution containing 5.0 mg/dm^3 of cadmium. The first traces of cadmium were recorded in 3.3 dm^3 of filtrate. With further filtering, a gradual increase in the content of cadmium in filtered solution: in 3.65 dm^3-1.12 mg/dm^3, in 4.7-5.0 mg/ dm^3. During the experiments, there were two regeneration. After the first regeneration the cations of cadmium in filtered solution was found in 56 (2.9 dm^3) sample. Further, there was a rapid increase in the solution and 68 (3.4 dm^3) amounted to 6.75 mg/dm^3.The second regeneration. The third filtration have shown that after the second regeneration of sorption properties of granular filter was not recovered. Already in the fourth sample (0.2 dm^3) content of cadmium in leachates exceeded 5 mg/dm^3. (Figure 5).

3.2. Discussion of research results
The analysis of own research and the data of other authors [1-4] on the sorption capacity and the ability to regenerate zeolites show that the ion exchange processes depend on the concentration, temperature of the solutions and the composition of the counterions.
There are two types of regeneration under which the sorption capacity of the filters changes. When cleaning the liquid from ammonium nitrogen, strontium, iron at each subsequent regeneration, the ion exchange capacity increases by 15-40%, and when filtering solutions with cadmium and manganese decreases, which increases or decreases the protective action time of the filters, which should be taken into account when selecting and calculating resources technological equipment. The ion-exchange capacity calculated under specific physico-chemical conditions indicates the high sorption capacity of the zeolites. (Figure 1-5).

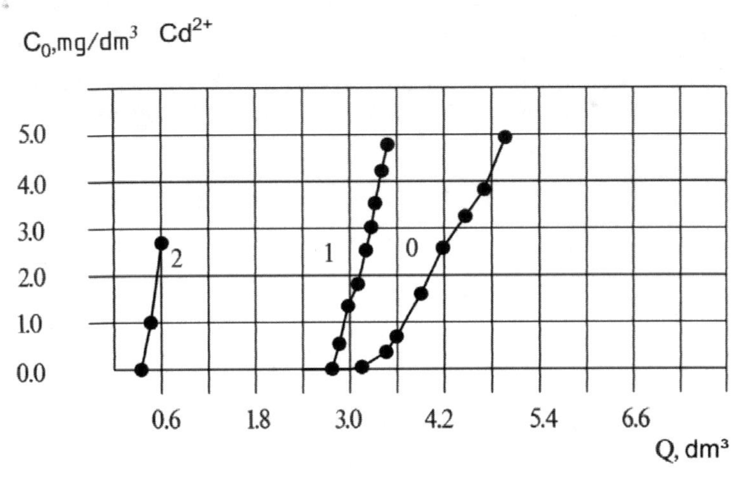

Figure 5 Dependency of sorption ability zeolite-montmorillonite breeds on regenerations` quantity over cadmium. The cadmium`s maintenance in filtered water 5.0 mg/ dm^3; the filtration rate 4.0 ml/min, the granular filter fraction 0.3-0.6 mm, the sorbent`s weight 50.0 g, the height of the filterable layer 50 cm.

Notion: 0,1,2 – cadmium's curved maintenances in filtered water after: 0-before regeneration; 1-first regeneration; 2- second regeneration.

The results of studies of sorption capacity and the ability to regenerate mineral filters shown in *figures 1-5*.

4. Conclusion

The presented results of studies on the sorption ability and the ability to regenerate Paleozoic zeolites during the filtration of solutions containing ammonium nitrogen, iron, manganese, strontium, cadmium indicate their high efficiency. Even with contamination of solutions tens of times higher than normal, the protective action and sorption capacity does not decrease. Low cost, acid resistance, ecological safety of zeolite sorbents allows using them in exchange processes both with regeneration and without regeneration, followed by the use of spent zeolites as cement raw materials or flux.

References

[1] Germanova T V, Valieva I R 2014 Application of zeolites and zeolite-montmorillonitovyh of breeds of Ural Mountains for effective sewage treatment from аммнийного nitrogen and accompanying катионов *News of the Samara centre of science of the Russian Academy of Sciences.* **B 16 № 1-7** 1828-1

[2] Valieva I R, Germanova T V 2015 Ionic an exchange of natural Paleozoic zeolites under various physical and chemical conditions *Natural and engineering science* **№ 6 (84)** 114-119.

[3] Germanova T V, Nefedov V A, Valieva I R 2017 The Survey of Sorption Ion-Exchange Properties of Paleozoic Natural Minerals in the Static Conditions *IOP Conference Series: Earth and Environmental Science»* **66**

[4] Edy Wibowo, Mamat Rokhmata, Sutisnaa, Khairurrijala, Mikrajuddin Abdullah May 1 2017 Reducing the salinity of sea water using natural zeolite (Clinoptilolite): adsorption isotherms, thermodynamics and kinetics *Desalination Volume* **409** 146-156

[5] GU H Y, Wong J W K, Tyagi R D 2017 Bioleaching of heavy metals from sewage sludge for land use. Modern developments in biotechnology and Bioengineering *Solid Waste Management* 241-265

[6] Kun-Yi Andrew Lin, Ah, Zhang-Xuan business woman, Andrew P Jochems April 2017 Adsorption behavior methylimidazolium ionic liquids with Y-type zeolite in water: kinetics, isotherms, thermodynamics and interference *Journal of molecular liquids* **232** 269-276

[7] Tinpan Man Hong Chen Junzhong Lin, Zhang Lin, Junliang Sun Zeolite synthesized from alkaline helped pre-activated halloysite for effective removal of heavy metals in polluted river water and industrial wastewater. *Journal of the science of the environment*

[8] Watts, Zhou H, Liu H, Chay P W, Kuznicki T, Kuznicki S M 15 November 2014 Natural zeolite clinoptilolite-phosphate composite membranes for desalination by pervaporation *Journal of membrane science* **470** 431-438

[9] Ates Ayten, Hardacre Christopher 15 April 2012 The effect of various treatment conditions on natural zeolites: ion exchange, acidic, thermal and steam treatment *Article Journal of Colloid and Interface* Science **372 Issue 1** 130-140

[10] Perego Carlo, Bagatin Roberto, Taliabu Marco, Vignola Rodolfo January 15 2013 Zeolites and related mesoporous materials for multi-purpose environmental solutions *Microporous and mesoporous materials* **166** 37-49

IOP Conf. Series: Earth and Environmental Science **115** (2018) 012028 doi:10.1088/1755-1315/115/1/012028

The complex system of environmental monitoring (CSEM). An analysis of concept

Yury I Kazatsev[a], Igor G Mitchenkov[b]

Department of History and Philosophy, Novosibirsk State University of architecture and Civil Engineering(Sibstrin), 113 Leningradskaya Street, Novosibirsk, 630008, Russia

E-Mail: [a] kazantsev-nsk@rambler.ru, [b] igor.mitchenkov@mail.ru

Abstract. Researches of ecological processes in Russia testify to rather difficult and adverse situation. It is quite obviously that we need a reasonable concept of an exit from the situation. It could become the strategic program for the solution of ecological tasks at the same time. In this regard, it is obviously necessary not just to develop new scientific and technological mechanisms to overcome of critical situations, but to offer a new research platform which will be able to give ideas to predict tendencies of ecological development and to analyses the consequences of their embodiment. Our offer for it is "the composite system of environmental monitoring" (CSEM) of the territory or region. We use a method of the conceptual analysis. Also, we will try to show how definition of the term influences to contents of social practicing, namely, environmental monitoring of the region.

.

1. Introduction

Western Siberia is dynamically developing region, but traditionally there was the difficult ecological situation. In recent years, the regional authorities pay to ecological safety much attention, but nevertheless, measures for the solution of ecological contradictions demand deeper systematic conceptual study. It is apparent that the solution of these problems demands an integration of various areas of researches: naturally scientific, technological, mathematical, political, etc. (About a research of integration approaches to the solution of eco-social problems see: Baumgarten M. I., Mitchenkov I. G., Galanina T. V. Environmental monitoring as the instrument of sustainable development.November 1-2, 2012*Natural and intellectual resources of Siberia – SIBRESURS*, 2012.Materials of XIV International Science. Conf.Kemerovo. Russia.Page 169-171; Baumgarten M. I., Mitchenkov I. G., Galanina T. V. The conceptual framework of environmental monitoring. 2013*Materials for the 9 International Scientific Conference «Новинатазанапредналинаука»*.**Vol. 48**. Екология. (София: БялГРАД-БГ). P. 50-56; Mitchenkov I. G., Mikhaylov V. G., Baumgarten M. I., Taylakov A. A., Sarapulova T. V. Use of WEB technologies for realization of a technique of estimation of environmental problems.2013.*Vestnik of Kuzbass State Technical University*.**Vol 4**. Pp. 136-139; Baumgarten M. I., Galanin T. V. About adequacy of mathematical model to the database of ecological parameters 2014 October 16-17 *Materials of Scientific Conference "Informational and Telecommunication Systems and Technologies"*. Kuzbass State Technical University. Kemerovo. Pp. 353-354). It will provide a possibility of creation of complex system of monitoring of an ecological situation. At the same time formation of the uniform concept capable to submit the complete program for stabilization of a situation and realization of environmental policy, is possible only on condition of its philosophical and theoretical justification. Such approach is especially urgent when forming uniform policy on leveling of ecological contradictions in regional aspect. It is about the complex system of environmental monitoring (CSEM) of the region which is a point of bearing for formation of positive model of the ecological development and environmental relations.

IOP Conf. Series: Earth and Environmental Science **115** (2018) 012028 doi:10.1088/1755-1315/115/1/012028

At the same time, it should be noted that the concept of "monitoring" is not defined conventual, possesses big degree of a theoretical ambiguity and means a philosophical concept, then the settled scientific concept rather.

2.Critical review of concepts and justification of CSEM

Follows from the review of the Russian sources that the concept "monitoring" gains various sense depending on to what object, what phenomenon or process it belongs. (Izrael, Yu. A. *Ecology and monitoring of a condition of the environment.*1984 (Moscow: Gidrometeoizdat); Hotuntsev Yu. L.*Ecology and ecological safety: studies.* 2004 (Moscow: Publishing center "Akademiya").2nd ed.; Bondarik, G. K. *Engineering-geological researches.* 2007(Moscow: KDU)). In the most general view, proceeding from word etymology, monitoring means keeping track of by some objects or the phenomena. It is fair also in relation to the Western sources [1]. Sincethere is an incalculable set of the phenomena, processes and objects, it is necessary to believe existence of the concept "monitoring". Synonyms: observation, tracking, assessment, forecast.

Let's note that monitoring can be carried out at the different space levels, defined by the tasks set for monitoring: global, regional, local. Here a difference only for monitoring, methods of collection of information, volume of the obtained data and ways of their processing. In the same way monitoring, can be carried out at the different temporary levels: the continuous, daily, monthly, annual, periodic (through a time term).

In reference books, it is possible to allocate three types of interpretation of the concept "monitoring". The first definition, the most common, is bound to word etymology. Definition can be short or developed. For example,

- **Monitoring** – process of tracking of a condition of system or the phenomenon (*The Terminological Dictionary of the Librarian on Socio-Economic Issues.* 2011(St. Petersburg: National Library of Russia) (http://window.edu.ru/resource/646/64646)).
- **Monitoring** – the technique and system of overseeing by a condition of an object or process giving the chance to observe them in development, to estimate to quickly reveal results of influence of various external factors [2]. Results of monitoring give the chance to make corrections on management of an object or process (*The Dictionary of Business Terms.* 2017 (http://dic.academic.ru/)). Similar definitions are given "*The Encyclopedia of Sociology*". 2017 (http://onlineslovari.com/entsiklopediya_sotsiologii); D.V. Dmitriyev. *Explanatory Dictionary of Russian Language.* 2017 (http://padabum.com/d.php?id=41428).

The second definition is directly bound to monitoring of the nature (a surrounding medium, the environment, etc.) since for the first time it was offered for the description of system of repeated observations of one and more elements of the environmental environment in space and in time with definite purposes per in advance prepared program. (The similar comprehension can be met, for example, in work Izrael Yu. A.*Ecology and monitoring of a condition of the environment.*1984 (Moscow: Gidrometeoizdat).P. 176). It can also be short and more developed, for example:

- **Monitoring** – (in Latin "monitor" means "supervising") – is the system of long-term overseeing by change of ecosystems (ecological) and biospheres (biosphere monitoring). It is made at express stations (including hydro meteorological) and in biosphere reserves [3] (Very similar definition is given in "*The Ecological Dictionary*". 1983. (http://ecologylib.ru/books/item/f00/s00/z0000003/index.shtml)).
- **Monitoring** – is a complex of the observations and researches defining the changes in a surrounding medium caused by activity of the person (*The Terminological Dictionary of Bank and Financial Terms.* 2011 (http://www.2buh.ru/slovar/financial/32522.html)).
- **Monitoring** – 1) is a complex of the observations and researches defining the changes in a surrounding medium caused by activity of the person; 2) is overseeing by state of environment (the atmospheres, hydrosphere, other geospheres, a soil and vegetable cover, fauna, techno sphere objects) for monitoring and the forecast of its state, and protection. Distinguish global, regional and local levels of monitoring. It is carried out by means of space, air, land and sea

tools; 3) is keeping track of by any object or phenomenon; in the annex to the life environment - keeping track of by its state and the warning of the created critical situations (increase in a gas contamination of air over threshold limit values, etc.), harmful or people hazardous to health, other living beings, communities, natural and anthropogenesis objects (including constructions); 4) is the system of the regular long-lived observations in space and time giving information on state of environment for the purpose of assessment of the past, the present and the forecast in the future of the parameters of a surrounding medium important for the person. At the national and regional levels the organization of monitoring is assigned to the appropriate executive bodies and is regulated by the relevant acts and resolutions. Trial functions of monitoring are quality control of free air, water, the soil, etc. landscape components; definition of the main sources of pollution; prediction of a condition of quality of principal components of a landscape and t of the item. (*The Ecological Encyclopedic Dictionary*. 2009 (http://www.cnshb.ru/AKDiL/0039/base/RM/005606.shtm)).

Similar definitions are given in the following sources: *Ecological Dictionary*. 2017 (http://gufo.me/ekoslov_a); *Dictionary of ecological terms and definitions*. 2017 (http://www.mnr.gov.ru/regulatory/detail.php?ID=143951); *Big Encyclopedic Dictionary*. 2000 (http://www.onlinedics.ru/slovar/bes.html); *Modern encyclopedia*. 2000 (http://7tor.org/viewtopic.php?t=1019264); *Biological encyclopedic dictionary*. 2006. 2nd issue. (http://istudy.su/biologicheskij-enciklopedicheskij-slovar-m-s-gilyarov-1986/); *Encyclopedic dictionary*. 2009 (http://gufo.me/bes_a); *Natural sciences. Encyclopedic Dictionary*. (http://estestvoznanie.slovaronline.com/); *Civil Protection. Conceptual and Terminological Dictionary*. 2001 (http://иванов-ам.рф/manual_slovar/index.html) and others.

The third definition is synthesis of two first:

- **Monitoring** – 1. Observation, assessment and the forecast of state of environment about economic activity of the person; 2. Systematic overseeing by some process with the purpose to fix compliance (or discrepancy) results of this process to tentative assumptions (Krysin L. P. *The Explanatory Dictionary of Foreign Words*. 1998.(http://megabook.ru/search?SearchText=); Efremova T.F. *Explanatory Dictionary*. 2000 (http://www.efremova.info/)).
- **Monitoring** – [English monitoring]. System of constant observations, assessment and the forecast of what change of a state - M. object natural, social, etc. System of monitoring. (*Big Dictionary of Foreign Words*. 2007. (http://www.bankreferatov.ru/db/B/8EFFE5C2A0CD1A79C325783400717D14#.WLkcbIVOK00)).
- **Monitoring** - 1. Constant overseeing by any process to detection of its compliance to desirable result or tentative assumptions. 2. Observation, assessment and prediction of state of environment about economic activity of the person. (*Modern Dictionary of Foreign Words*. 2015;*Modern dictionary of foreign words*. 2000).

Similar definitions are given in: *Big Legal Dictionary*. 2010 (http://www.twirpx.com/file/1134890/); *Beginnings of the modern natural sciences. Thesaurus*. 2016 (http://www.alleng.ru/d/natur/nat016.htm); *Big Explanatory Dictionary of Russian*. 2016 (http://www.gramota.ru/slovari/info/bts); and others.

Presence at the concept "monitoring" of a polysemy and infinity follows from the above-stated definitions. Perhaps, for the first time the mathematicians trying to construct mathematical models of monitoring [4] faced this ambiguity. Especially brightly it can be seen by consideration of concrete types of monitoring [5]. Let's try to make some classification and structuration of a set "monitoring".

At first, it is possible "to divide" monitoring on:

- Monitoring of objects.
- Monitoring of the phenomena and processes.

Then "to divide" monitoring of objects on:

- Monitoring of the nature.
- Monitoring of society.

IOP Conf. Series: Earth and Environmental Science **115** (2018) 012028 doi:10.1088/1755-1315/115/1/012028

Monitoring of the nature is subdivided on:
- Monitoring of the atmosphere.
- Monitoring of the hydrosphere.
- Monitoring of the lithosphere.
- Monitoring of the biosphere.
- Monitoring of the Universe.
- Monitoring of the techno sphere.

Monitoring of society can be subdivided proceeding from monitoring of the main social institutes (table 1):

Table 1. Types of monitoring.

Mainsocial institutes	Institutes of the lowest level (selectively)	Types of monitoring
Economic	Industry, agriculture, trade, property, financial system, market, exchange, marketing, management, banking system, etc.	Informational, a banking system, an exclusive economic zone, profitability, control devices in audit, electronic databases, objects of town-planning activity, etc.
Political	State, party, social movement, army, police, parliamentary, presidency, court (legal profession, prosecutor's office, refereeing), civil society, etc.	Public opinion, law enforcement in the Russian Federation, ethnological, etc.
Cultural	Church (religion, monkhood, confession), (spiritual sphere education (school, the higher school, average professional), science, institutions of art (theaters, museums, clubs, libraries), etc.	Media, social media, a state and use of historical and cultural monuments, in education, etc.

Then monitoring of society is subdivided on:
- Economic monitoring.
- Political monitoring.
- Social monitoring.
- Cultural monitoring.

Monitoring of phenomena and processes can be subdivided on:
- Monitoring of climate.
- Monitoring of accidents.
- Monitoring of pollutions.
- Monitoring of global habitat condition.
- Monitoring of information, and others.

Each of them is in turn subdivided into more detailed types of monitoring. Let's note that in literature, especially help and encyclopedias, more than a half of definitions of the concept "monitoring" is bound to a surrounding medium (the nature, the biosphere). Often monitoring of a surrounding medium is equated to *environmental monitoring*. (For example, similar approach practices in encyclopedia online "*Academy*": http://dic.academic.ru/). "**Monitoring**" (in Latin "monitor" means "reminding", "supervising") – is keeping track of by some objects or the phenomena. Usually mean complex system of observations, assessment and the forecast of changes of state of environment under the influence of the person by monitoring" [6]. (Rather exponential example of a similar definition can be found in Masur I. I., Chumakov A. N. *Global Studies: Encyclopedia*.2003. (Moscow: JSC Publishing House Raduga).

Several other interpretations of the concept "monitoring" is given in the popular Internet encyclopedia "Wikipedia". Mean by "monitoring": – system of collecting/filing, storage and analysis of a small amount key (apparent or indirect) signs/parameters of the description of this object for removal of judgment of a behavior/condition of this object in general. That is for removal of judgment of an object in general based on the analysis of a small amount of the signs characterizing it.

In the Western sources of a concept monitoring most often has the following values:

Monitoring – is process of observation and data recording about any object on the time slices which are indissolubly adjoining to each other during which values of data significantly do not change (in the same place). Distinguish *Monitoring of parameters* and *Monitoring of a condition of an object*. *Monitoring of parameters* – overseeing by any parameters. The result of monitoring of parameters represents set of the measured values of the parameters received on the time slices which are indissolubly adjoining to each other during which values of parameters significantly do not change [7].

Monitoring *of a condition* – is overseeing by a condition of an object for definition and a prediction of the moment of transition to the limiting condition. The result of monitoring of a condition of an object represents set of diagnoses of the subjects making it received on the time slices which are indissolubly adjoining to each other during which the condition of an object significantly does not change. The fundamental difference of monitoring of a state from monitoring of parameters is existence of the interpreter of the measured parameters in terms of a state – expert system of support of a decision making about a condition of an object and further management [8].

Monitoring – is systematic data collection and processing which can be used for decision-making process improvement, and, indirectly, for informing the public or just like the instrument of feedback coupling for exercise of projects, assessment of programs or development of policy. It bears one or more of three organizational functions:

- Reveals a condition of the critical or being in a condition of change phenomena of a surrounding medium concerning which the course of actions on the future will be developed.
- Establishes the relations with the environment, providing a feed-back, concerning the previous achievements and failures of policy or programs.
- Establishes compliances to rules and contractual obligations [9, 10].

At such approach "monitoring of objects" is equivalent to "monitoring of object state", but "monitoring of processes and the phenomena" - to "monitoring of parameters".

From definitions, it is visible that the concept "monitoring" is the composite, matters both as systems and as process, and depends on object of studying[11]. Systemic orientation of the concept "monitoring" is reflected in the following definition:

Monitoring(in Latin "monitor" - is the one who reminds prevents - monitoring) –is the complex system of the regulated periodic observations, assessment and the forecast of changes of a condition of the environment with the purpose of identification of negative changes and developments of recommendations about their elimination or easing. Monitoring as a multi-purpose information system includes: the bioenvironmental monitoring studying the environment from the point of view of its influence on the state of health of people; the geosystemic, or natural and economic monitoring studying changes of geosystems (including natural) of which the surrounding medium (geo-monitoring) consists; the biosphere monitoring providing observation, control and the forecast of possible changes of the environment on a global scale (biomonitoring). Monitoring of the geological environment is called a "litho-monitoring". (This definition is taken in the Kozlowski E. A. and other.*Mountain Encyclopedia*. 1984 – 1991 (Moscow: Soviet Encyclopedia). In 5 volumes. **Vol.3**, 1987. Page 391. It should be noted that for naturally scientific approaches it is typical).

Considering the above, and relying on our characteristic idea of the maintenance of the concept "monitoring" developed based on integrative and comparatists approaches to the above definitions, we will shortly consider key parameters (algorithm) of the complex system of environmental monitoring which is the practical embodiment of substantial aspect of the concept "environmental monitoring".

IOP Conf. Series: Earth and Environmental Science **115** (2018) 012028 doi:10.1088/1755-1315/115/1/012028

So, an object of a research of social-and-ecological monitoring are: eco-social aspects of environmental problems of the region;

- *research objectives*: creation of complex model of regional social-and-ecological monitoring. To designate philosophy of innovative environmental policy and mechanisms of its realization in the Russian Federation;
- *research problems at a starting stage*: to analyses environmental problems of the region; to investigate regional tendencies in the sphere of nature protection and environmental management; to develop mathematical model of regional environmental monitoring;
- *range of application*: results of a research can be used during creation of plans of economic, social and ecological development of the region;
- *competitive advantages*: lack of similar works;
- *principal specifications*: the project assumes creation and use of databases on ecological and economic indexes of development of the Russian Federation, in parameters and indicators of sustainable development of the region; creation of mathematical model of environmental monitoring of the region; creation of model of the complex analysis of environmental monitoring of the region;
- *tools of research*: computer programs for processing of statistical data like SPSS or Statistic; IDEFO Standard as basis for creation of the functional and structural and analytical models for generalization, systematization and descriptions of interrelations of the composite systems; methods of expected researches; methods of optimization of parameters;
- *result of a research* – creation of mathematical model of the regional eco-social indexes.

The explained algorithm of "environmental monitoring", from our point of view, fullest considers the volume and the maintenance of the concept "monitoring" and removes indeterminacy at its use in the sphere of ecology. In too time, this definition sets the program content of activities for prevention and sanitation of eco-negative consequences. This activity is made out as complex system of eco-social monitoring.

3. Conclusion

Thus, the standard concept "monitoring" completely is defined by expression "keeping track of any objects or the phenomena". At the same time both temporary, and space coordinates implicitly are meant. Separate consideration is demanded by a question of objects of monitoring. If on an incipient state (the 70th – the 90th years of the 20th century) the main object of monitoring, it is rather a field of close attention, there was the environment, then it both economy, and pedagogics, and media, and sociology, and other areas of human activity now. In definitions objects or processes, but all of them – "monitoring" is concretized. It means that when using the term, it is necessary to concretize every time an object, the phenomenon of monitoring, the purpose and ways of its realization. Results of this work can promote more complete comprehension of volume of the concept "environmental monitoring" and those processes which constitute the maintenance of the ecological development and environmental relations, and mechanisms of their reforming and regulation. Also, results of a research can be applicable at correction of regional environmental policy and coordination of the measures directed to overcoming the ecological and social centers of danger (based on the offered eco-monitoring algorithm).The presented algorithm of the complex system of environmental monitoring (CSEM) will help with development of the innovative environmental policy capable to influence design the relation of the person to a surrounding medium, including in realities of the old industrial centers.

References

[1] Shanno C.E. (1948) July & October 1948 A Mathematical Theory of Communication. *Bell System Technical Journal***27**pp 379–423 & 623–656.

[2] Guariso G., Hitz M., Werthner H.1996 An integrated simulation and optimization modelling Environment for decision support *Decision Support Systems*pp. 103–117

[3] PielkeR.A.1984 *Mesoscale Meteorological Modeling*(New York: Academic Press)

[4] Graham W.R., Peraire J., Tang K.Y.1999 Optimal control of vortex shedding using low–order

models. Part I: open–loop model development. *International Journal of Numerical Methods in Engineering***44**(7): 945–972.

[5] Agoshkov V.I., Ambrosi D., Pennati V., Quarteroni A., Saleri F.1993 Mathematical and Numerical modelling of shallow water flow *Computational Mechanics***11**(5–6):280 - 299.

[6] Rinaldi S., Sanderson W., Gragnani A.1996 Pollution control policies and natural resource dynamics: a theoretical analysis *Journal of Environmental Management***48**: 357–373.

[7] Arya S.P. 1999 *Air Pollution Meteorology and Dispersion*. (New York: Oxford University Press)

[8] AkbarzadehA., Ghorbani-DashtakiS., Naderi-Khorasgani M., et al. December 2016 Monitoring and assessment of soil erosion at micro-scale and macro-scale in forests affected by fire damage in northern Iran *Environmental Monitoring and Assessment***2016**: 188:699

[9] Soffianian A., Madanian M. August 2015 Monitoring land cover changes in Isfahan Province, Iran using Landsat satellite data. *Environmental Monitoring and Assessment***2015**: 187:543

[10] Zheng J., Feng X., Liu P., Zhong L., Lai S. 2011 Site location optimization of regional air quality monitoring network in China: methodology and case study. J. *Environ. Monit.* **13**: 3185-3195 DOI: 10.1039/C1EM10560D

[11] Insaf S., Babiker A., Mohamed A., Tetsuya H., Keiichi I., Kikou K. October 2012 Investigating Response of Global Vegetation to ENSO Events Between 1987 and 1997. Using NDVI Data. *International Journal of Environmental Monitoring and Protection***vol. 2**, Issue 5. P. 76-83.

IOP Conf. Series: Earth and Environmental Science **115** (2018) 012029 doi:10.1088/1755-1315/115/1/012029

Development of safe routes for children in urban environment

M E Koryagin [a], V I Medvedev [b] and P G Strykov [c]

Siberian Transport University, 191, Dusi Kovalchuk Street, Novosibirsk, Russia

E-mail: [a] markkoryagin@yandex.ru, [b] medvedevvi@stu.ru, [c] centr-bt@yandex.ru

Abstract. The matter of development of safe travel routes for children between school and home is analyzed. The availability of various applications and devices to identify the location of the child and his/her travel routes is noted. The main factors to be taken into account when planning children travel routes are described. The most popular Russian services for route planning, Google, Yandex, and 2GIS, are discussed. These services are shown to have a number of shortcomings which does not allow them to choose really safe routes. A decision on making the route selection by two criteria (the travel time and the probability of an accident) is obtained. As a numerical example, the Pareto area for possible routes is constructed.

Introduction

The lives of children in urban environment are fraught with many challenges. The necessity of travels between school and home requires solving problems on routes choice and travel control.

The GIS and GPS technologies are widely used for solving those problems. Location tracking systems allow the parents to monitor their children's location [1]. Such systems are helpful within the context of security enhancement, peace of mind settling, and the need for reducing parents' uncertainty. That is why parents are encouraged to use surveillance technologies for keeping their children "safe" [2].

In [3], the following parental control tasks were described:

- to give care;
- to be responsible;
- to be protective;
- to be involved, which task includes control and establishing sets of rules and norms;
- to avoid instilling fear;
- to foster self-reliance;
- to trust;
- to foster communication through dialogue;
- to (co-)regulate their son or daughter in order to help him/her to become a conscientious and self-reliant adult.

Although such parental tasks are important ones, the availability of location-based services (LBS) raises a number of problems that should be borne in mind [3]. Location-based services may create deceiving situations for parents. Also, such services may impede developmental processes towards establishing the children's autonomy and self-reliance.

Location-based services identify the location of a user based on the indications of some handheld, wearable, or implanted device used for a particular purpose. L. Perusco and K. Michael [4] have identified potential risks related with LBS within the context of security and privacy. The major dilemma here consists in privacy-security dichotomy.

J. Boesen *et al.* [5] have reported a study in which using LBS by families was analyzed. The possibility of undermining the trust in children, which may become an upbringing problem, was described. Location-tracking technology can generate anxieties and conflicts between parents and children. Tracking technology must be based on a better understanding of the complex interplay between the technology and the inter-human relationships in various social groups [6].

In this connection, the society has to address the following issues [3]:

- the right to privacy;
- the right to know who controls your data;

- the right to know that you are being tracked;
- the right to agency, in order to become a responsible person;
- the right to learn by trial and error and the right to make mistakes;
- the right to autonomous decision-making.

In the field of health care and education, governments can use monitoring systems to identify and "manage" the behavioral deviations from the norm [2].

However, today the devices and applications monitoring the location of children are not so common tools (or they are no commercial success) because [1]:

- some parents have not ever heard of location-tracking technologies;
- many parents are not aware of the fact that such technologies are indeed available;
- only a minority of parents advocate using the location-tracking technologies;
- most of parents need no location-tracking tools;
- the control effected via location tracking was considered to be a threat to the self-direction of children and to the trust in them.

The physical activity of children

Many studies were devoted to the matter of children's physical activity. Monitoring systems offer a unique possibility to track the physical activities of children and adults. Children's current physical activity levels are disturbingly low in comparison with the recommended levels [7].

The time spent by children outdoors with other children is an important source of children's physical activity [8]. Interventions to intensify the physical activity may benefit from fostering friendship groups and from restricting the time during which the children stay alone.

The researchers have identified the following important factors positively influencing the physical activity among children:

the growth of the walking between home and school [9, 10];
the presence of green belts near the house;
the presence of parks located over the travel route [11];
the presence of green school gardens [10];

On the schoolyard, children's physical activity levels are higher than on the average [7].

Travel route choice

Since the beginning of the XX century, the modern cities have been designed around the use of motor vehicles despite the unfavorable interactions between vehicles and pedestrians.

There is an increasing interest in community walkability [12]:

- initiatives on the development of safe routes to school are being advanced;
- concerns regarding the epidemic obesity are being delivered;
- political initiatives designed to convince travelers to switch from auto trips to more environmentally sustainable bicycle and walking trips are being put forward.

The study [13] was performed to examine the extent to which the street network configuration is associated with the path selection by pedestrians. Designing better connected street networks with a reduced distance between home and school may serve a factor supporting the navigation choices and walking behavior.

Nowadays, a multitude of new applications have been developed for different types of navigation systems [14]. There exist significant differences between the nature of route choice problems for pedestrians and those for motor cars as the navigation services designed for vehicles proved to be inappropriate for pedestrians.

That is why M. Sahelgozin *et al.* [14] have proposed a multi-criteria path optimization method that offers a Ubiquitous Pedestrian Way Finding Service. The developed model allows one to construct routes on the basis of a linear convolution of criteria. Thus, the latter procedure yields alternative routes for people. In particular, it was shown that the most important value for children is the prevention of accidents; those for young people, the inadmissibility of teen crimes; and those for retirees, short walking distances.

The travel route choice depends on many factors [15], which involve
- connectivity (Bus/Taxi stops, Commercial area, Business area, Residential area)
- conspicuous signage (indicating names, directions, distances, etc.)
- conviviality (Food and beverages vendors, Public parks, Recreation/Rest centers)
- shelters (Row of Roofs, Row of Trees)
- convenience (Walking time, Walking distance)
- traffic safety (Presence of crossings, Presence of traffic lights)
- perceived security (Proximity to abandon area, Proximity to construction sites, Proximity to public alleys).

The study of road collision risks is important for the improvement of pedestrian safety [16]. The article [16] discusses the formulation of the problem of route choice with different road crossing options (zebra crossing, controlled intersection of roads). The risks assessment problem is a difficult one due to the lack of information about roads intersection (traffic volumes and amounts of accidents).
The route choice during the active transportation of children to school is under study [17]. Children avoid streets with a high rate of accidents. They prefer residential streets yet avoid zebra crossings. The majority of the differences between the real and shortest routes can be attributed to children and their parents' preferences to avoid walking along busy roads on their way to school.

Services for making routes in Russia

The number of cars in cities exhibits a permanent growth. The interaction between people and cars can be expressed in specific figures. For example, according to the State Traffic Safety Inspectorate statistics, a significant number of road accidents with the injured and killed occur as a result of the violations of traffic rules by pedestrians. This fact is illustrated by Fig. 1.

Figure 1. The number of accidents and victims due to the violation of traffic rules by pedestrians in Russia

It is seen from the figure that systematic activities aimed at reducing the injury rates are being conducted. There are changes in the legislation - the measure of responsibility for drivers and pedestrians is growing, together with the amount of penalties imposed for the violation of rules. Road marking and visibility of pedestrian crossings are being improved significantly. Clothing and individual elements

IOP Conf. Series: Earth and Environmental Science **115** (2018) 012029 doi:10.1088/1755-1315/115/1/012029

with reflective properties are put into use. However, we believe that additional opportunities for reducing the number of accidents and injuries are available.

Wearable electronic devices (phones, smartphones, communicators, and smart watches) greatly simplify the life of modern individuals as those devices enable the voice and video communication, the messaging, the access to information, etc. Also, there exist an opportunity to plan car and foot routes using the public land and underground transport. The most widely used software products are Yandex.Maps, Google.Maps, and 2GIS.

As a rule, most of the existing programs are used by motorists to construct fast or short routes taking into account the busy city roads, repair works and road accidents. Pedestrians are given fewer criteria for choosing their routes. As a rule, here just one, rarely, several options are offered. If we are talking about moving from point A to point B without using public transport, then, as a rule, a shortest route, or a lowest-distance route, will be constructed.

In our opinion, modern software products and tools are presently capable of taking into account more criteria for constructing routes. In addition, among a variety of options for movement, one of the routes can be identified as the safest one. We believe that the so-called "safe route" is a pre-planned route with the lowest risk of injury or illness. We would like to highlight some criteria that can affect the level of route security.

1. Traffic lights, signs, and marking. Separation of transport and pedestrian flows in different planes with the use of overground and underground pedestrian crossings will exclude the contact between cars and pedestrians and, in this way, will reduce the probability of auto-pedestrian accidents and injuries. The probability of injuries occurring as a result of auto-pedestrian accidents on regulated pedestrian crossings is much lower than that on non-regulated pedestrian crossings and, especially, when roads are crossed in the wrong places. The presence of sidewalks on the route is an advantage.

2. Time of day. Movement in the night-time is associated with a load on the visual analyzer, a neuro-emotional strain, an increased probability of stumbling or falling, for example, because of works on the improvement of the yard territory, repair works, etc. In the early morning hours, as a rule, the street is not crowded.

3. City lighting. The central streets are much better illuminated than the secondary and courtyard territories. Good lighting increases the safety on the route.

4. Season. Long-term exposure to low or elevated temperatures can adversely affect the health status of people (hypothermia, frostbites, heat and sunstrokes, sunburns). Under such conditions, when compiling the route, it is necessary to take into account the outdoor temperature, the continuous time to be spent in the open air, the possibility of heating on the route, and the possibility of linking with public transport. For example, the temperature is minus 24^{0}C, the bus stop is 40 minutes away, the traffic schedule is one bus per hour, not all the cars (part of them is frozen or cannot be started, etc.) have come on the route. Here, the pedestrian gets in an extremely difficult situation.

5. Crime and accidents. The safe route must be laid taking into account the criminality and accident foci and, also, the probability of an unfavorable public.

6. Weather. Weather phenomena, - strong wind, rain, fog, snowfall, ice, - this is a warning not only for car enthusiasts, but also for pedestrians. When laying the route, the program should prompt the user of unfavorable weather conditions.

7. Age of the user and his/her personal features. Information about the age of the user is stored in a portable device. Using it, you can make routes that take into account the age (children, adults, elderly or disabled people).

8. Parkland areas. In some cases, it would be necessary to avoid constructing routes through such areas in the spring, during the period of mite activity providing that no adequate anti-maltreatment measures were taken. In all other cases, when moving at daytime, parkland areas located on the route will be an advantage (nature, fresh air, etc.).

9. A route in the group. It is necessary to take into account the number of people (children) on the route. For organized groups of schoolchildren, it is necessary to draw up special routes.

10. Police. Police strongholds and stationary posts are to be indicated in the map as places where you can apply for help if necessary.
11. Medicine. Hospitals, polyclinics, private clinics. Places where you can get medical help.
12. Toilets. Public paid or free toilets for comfortable route planning for long distances or in case of urgent need.

Table 1. Comparison of software products

	1	2	3	4	5	6	7	8	9	10	11	12
Google.Maps	-	-	-	-	-	-	-	+	-	+	+	-
2GIS	-	-	-	-	-	-	-	+	-	+	+	+
Yandex.Maps	+	-	-	-	-	-	-	+	-	+	+	+

It should be noted that Google.Maps offer a schematic representation of urban environment which can be used for identifying the location with high accuracy. However, in compiling the route the user is limited in information; for instance, when viewing the route the user is devoid of information about the presence of traffic lights, pedestrian crossings, paths and courtyard aisles. Everything is fairly approximate and conventional; in fact, here the user deals just with the direction of the movement. On the other hand, Google.Maps can construct an alternative (second) route with time and distance indication.
2GIS offers more detailed information about the terrain and pedestrian paths in parkland areas, but, here, no traffic lights or pedestrian crossings are indicated. Routes are laid without taking into account traffic signs. When following Google.Maps or 2GIS, you should consider with care the choice of highways intersections.
Today, you can get a most complete idea of moving along a route by using the Yandex.Maps service. Here, all kinds of pedestrian crossings, ground, above-ground and underground ones, are indicated. Pedestrian crossings equipped with traffic signaling are shown. An extensive road-and-track network and intra-courtyard passages are represented. Routes constructed by Yandex.Maps use pedestrian crossings when crossing a highway. All maps bear police stations and medical facilities, but only in 2GIS these are shown permanently. In the two other programs, the police stations are displayed on demand. All the above programs lack the function of constructing a safe route.

Example of development of a movement route

Investigating a particular situation, the daily trips of a child from home to school, you can be sure that there are many options to overcome this distance. Some of the most commonly used routes are highlighted on the map below.

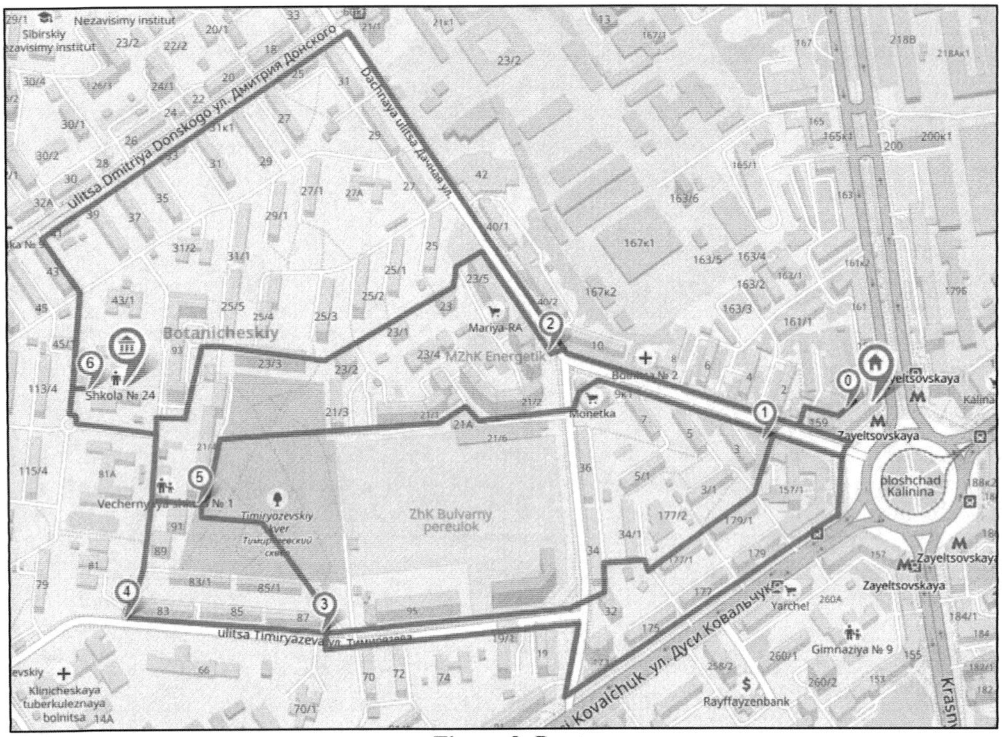

Figure 2. Route map

The school is located at a distance of 950 meters from the house along a straight line. To get to the place of study, you need to cross the road at least once. On the way, there are regulated and non-regulated pedestrian crossings, a parkland area, and yard territories. Schematically, potential routes can be represented as a graph.

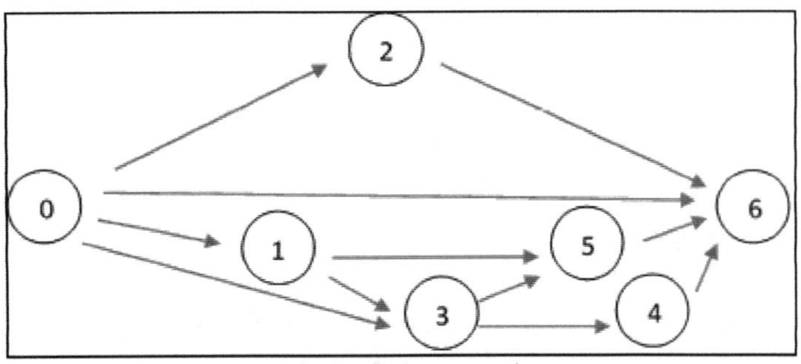

Figure 3. State graph

On each of the route sections, the distance and the time required for covering the distance were measured.

Table2. The results of the measurements

Section	Type	Time, min	Route length, m
0 – 1	2	4	310
0 – 2	2	6	460
0 – 3	1	12	1100
0 – 6	1	19	1700
1 – 3	2	8	720
1 – 5	1	10	890
3 – 4	0	3	280
3 – 5	0	3	270
2 – 6	0	11	950
4 – 6	0	6	450
5 – 6	0	5	410

In addition, for route sections having an intersection with a road through a non-regulated pedestrian crossing, it is a common practice to denote this type of intersection with digit 2, whereas the regulated pedestrian crossings and the places where intersections with motor roads are absent are traditionally denoted respectively with digits 1 and 0. We have performed calculations of the occurrence probability of negative events on each of the routes with the adoption for the probability of a road accident at regulated and non-regulated intersections the values of 0.000001 and 0.00001, respectively.

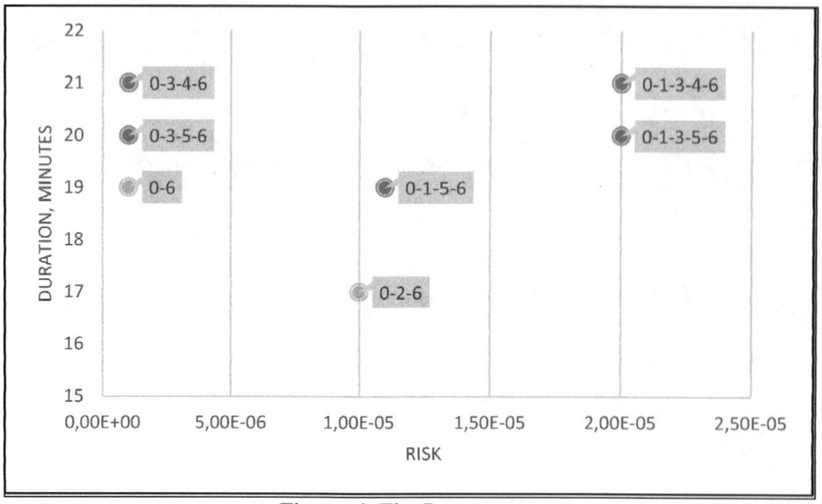

Figure 4. The Pareto area

In Fig. 4, the Pareto area is highlighted in green, so the choice should be made between the two routes, the route 0-6 being less risky yet slower, and the route 0-2-6, more risky yet faster. The final decision on the choice of the final route remains for the person (pedestrian).

Conclusion

The article discusses the issue of choosing secure routes for children. It is shown that there exist many technical solutions allowing the design and control of the routes.

The proposed measures enable an improvement of the existing services in the development of children's walking routes. The proposed approach allows one to find optimal travel routes for children in urban environment with taking into account safety considerations. In the future, we are going to increase the number of used criteria that describe the routes. Also, various transportation purposes, environment conditions, and child conditions will be included in consideration. This approach will offer routes meeting the various needs of the child. In the future, the proposed approach will be incorporated in a navigation system intended for monitoring children's location. Also, the proposed approach will allow solution of the important task of urban planning in transport infrastructure to ensure the safe movement of children.

References

[1] Vasalou, A.; Oostveen, A.-M. & Joinson, A. N. A case study of non-adoption: The values of location tracking in the family. *Proceedings of the ACM 2012 Conference on Computer Supported Cooperative Work, ACM,* **2012**, 779-788

[2] Marx, G. & Steeves, V. From the beginning: Children as subjects and agents of surveillance. *Surveillance & Society,* **2010**, *7*, 192-230

[3] Gabriels, K. 'I keep a close watch on this child of mine': A moral critique of other-tracking apps. *Ethics and Information Technology,* **2016**, *18*, 175-184

[4] Perusco, L. & Michael, K. Control, trust, privacy, and security: Evaluating location-based services. *Technology and Society Magazine, IEEE,* **2007**, *26*, 4 – 16

[5] Boesen, J.; Rode, J. A. & Mancini, C. The domestic panopticon: Location tracking in families. *Proceedings of the 12th ACM International Conference on Ubiquitous Computing, ACM,* **2010**, 65-74

[6] Mancini, C.; Rogers, Y.; Thomas, K.; Joinson, A. N.; Price, B. A.; Bandara, A. K.; Jedrzejczyk, L. & Nuseibeh, B. In the best families: Tracking and relationships. *Proceedings of the SIGCHI Conference on Human Factors in Computing Systems, ACM,* **2011**, 2419-2428

[7] Dessing, D.; Pierik, F. H.; Sterkenburg, R. P.; van Dommelen, P.; Maas, J. & de Vries, S. I. Schoolyard physical activity of 6—11 year old children assessed by GPS and accelerometry. *International Journal of Behavioral Nutrition and Physical Activity,* **2013**, *10*, 97

[8] Pearce, M.; Page, A. S.; Griffin, T. P. & Cooper, A. R. Who children spend time with after school: associations with objectively recorded indoor and outdoor physical activity. *International Journal of Behavioral Nutrition and Physical Activity,* **2014**, *11*, 45

[9] Dessing, D.; de Vries, S. I.; Graham, J. M. & Pierik, F. H. Active transport between home and school assessed with GPS: a cross-sectional study among Dutch elementary school children. *BMC Public Health,* **2014**, *14*, 227

[10] Burgoine, T.; Jones, A. P.; Namenek Brouwer, R. J. & Benjamin Neelon, S. E. Associations between BMI and home, school and route environmental exposures estimated using GPS and GIS: do we see evidence of selective daily mobility bias in children? *International Journal of Health Geographics,* **2015**, *14*, 8

[11] Almanza, E.; Jerrett, M.; Dunton, G.; Seto, E. & Pentz, M. A. A study of community design, greenness, and physical activity in children using satellite, GPS and accelerometer data *Health & Place,* **2012**, *18*, 46 - 54

[12] Agrawal, A.; Schlossberg, M. & Irvin, K. How Far, by Which Route and Why? A Spatial Analysis of Pedestrian Preference. *Journal of Urban Design,* **2008**, *13*, 81-98

[13] Ozbil, A.; Argin, G. & Yesiltepe, D. Pedestrian route choice by elementary school students: The role of street network configuration and pedestrian quality attributes in walking to school. *International Journal of Design Creativity and Innovation,* **2016**, *4*, 1-18

[14] Sahelgozin, M.; Sadeghi-Niaraki, A. & Dareshiri, S. Proposing a multi-criteria path optimization method in order to provide a ubiquitous pedestrian wayfinding service. *ISPRS - International Archives of the Photogrammetry, Remote Sensing and Spatial Information Sciences,* **2015**, *XL-1-W5*, 639-644

[15] Naharudin, N.; S. Ahamad, M. S. & F. M. Sadullah, A. Optimizing pedestrian-friendly walking path for the first and last mile transit journey by using an Analytical Network Process (ANP)

decision model and GIS network analysis. *ISPRS - International Archives of the Photogrammetry, Remote Sensing and Spatial Information Sciences,* **2017**, *XLII-4/W5*, 137-144

[16] Yao, S.; Loo, B. P. & Lam, W. W. Measures of activity-based pedestrian exposure to the risk of vehicle-pedestrian collisions: Space-time path vs. potential path tree methods *Accident Analysis & Prevention,* **2015**, *75*, 320 – 332

[17] Dessing, D.; De Vries, S.; Hegeman, G.; Verhagen, E.; van Mechelen, W. & H. Pierik, F. Children's route choice during active transportation to school: Difference between shortest and actual route. *International Journal of Behavioral Nutrition and Physical Activity,* **2016**, *13*

Reducing the risk of the collapse of the soil by macro system modeling the slopes stability of the quarries

E V Klimova [a], A Yu Semeykin

Belgorod State Technological University named after V.G. Shukhov,
46, Kostyukova str., Belgorod, 308012, Russia

E-mail: [a] lena_1109@mail.ru

Abstract. The urgent task of modern production is to reduce the risks of man-made disasters and, as a consequence, preserve the life and health of workers, material properties and natural environment. In the mining industry, one of the reasons for the high level of injuries and accidents is the collapse of the soil. Macro system modelling of slopes stability of the quarries is based on the compliance with the conditions of physical and mathematical correctness of the application of the model of a continuous medium. This type of modelling allows to choose the safe parameters of the slopes of the quarries and to reduce the risk of collapse of the soil.

1. Introduction

Risk is the probability of harm to the life or health of citizens, property of individuals or legal entities, state or municipal property, environment, life or health of animals and plants taking into account the severity of the harm. Risk is identified with the probability of occurrence of negative events for a certain period of time (usually a year) and the size of the damage in kind (the number of deaths, the area exposed to the dangerous effects, etc.) or monetary terms [1].

Industrial disaster is the major industrial accident, which caused human victims, damage to human health or damage and destruction of objects, material assets in the considerable sizes, and also serious environmental damage [2]. The presence of serious consequences, such as death of people or destruction of objects is a distinctive feature of any disaster.

In accordance with these definitions, the collapse of the soil in the quarry can be attributed to the industrial disaster. Since the collapse of the soil contributes to a variety of factors, to exclude that it is often impossible, the risk of this adverse event is high.

The risk of an adverse event is defined as a measure of risk of adverse events combining the probability of occurrence of the adverse event and its consequences [3].

From a mathematical point of view, the risk R is a function of two variables – frequency F and the consequences U of undesirable event:

$$R = f(F, U). \quad (1)$$

The consequences of undesirable events can be: human casualties, property damage and damage to the environment.

There are more than 7000 quarries for the extraction of ore and non-ferrous metallurgy and gold mining industry and quarries for the extraction of widespread minerals under the supervision in the Russian Federation.

The results of the analysis of the general state of accidents and injuries in the mining industry indicate a high traumatic risk of the industry and a lack of a steady trend towards its decline (figure 1) [4, 5, 6].

The causes of injuries and accidents in the mining industry are of a diverse nature, the collapse of the soil among them occupies about 20% (figure 2).

Adverse factors that affect the level of safety of work in quarries can be divided into three groups: technological, personal and natural factors of the environment.

Let's imagine the technological factors in the form of the following groups:

1. Technological solutions (design and construction);
2. Reliability (optimality) of the engineering structure;

3. Operational technological design of the engineering structure;
4. Controllability of the functioning mode of the engineering structure.

Also here can be attributed mining technical factors (geometric parameters of slopes, the way of loosening of soils, etc.), which affect the stability of soils in the slopes of quarries.

The second group (personal factors) implies compliance with the criteria of reliability (optimality), legislative and regulatory framework for labor protection, industrial safety rules, violation of established rules, erroneous action or inaction of employees involved in this work [5, 6]. These factors are a consequence of the reasons laid down in the individual characteristics of people, namely professional level, psychophysical condition, discipline, personality traits.

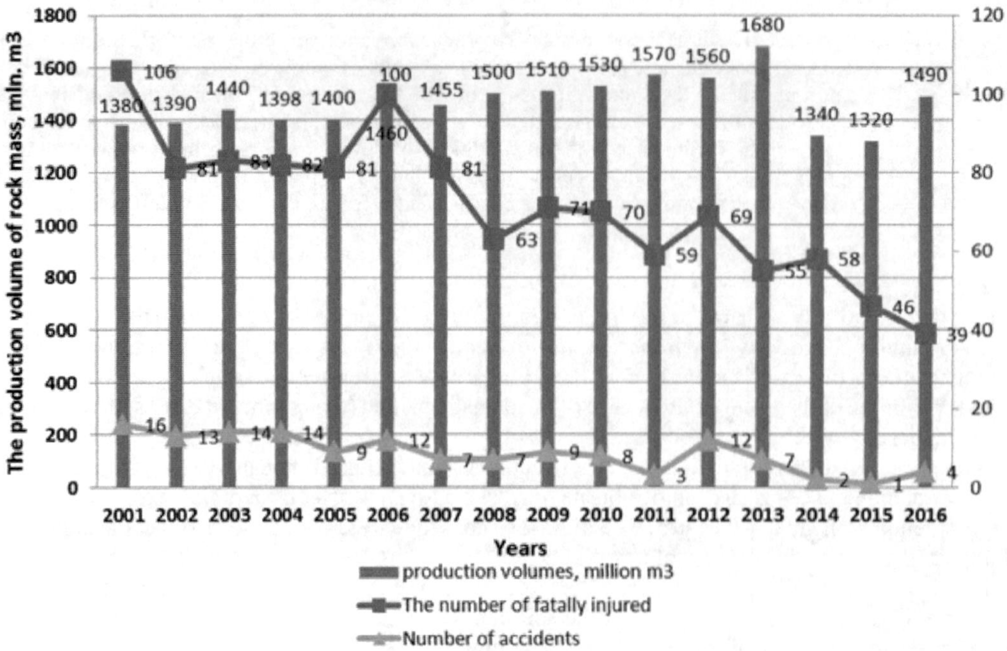

Figure 1. The dynamics of injuries and accidents in the mining and non-metallic industry

Environmental factors imply natural factors that can cause not only caving or landslides, but also human errors. This group of factors includes the following: climatic – air temperature, rainfall, wind regime; geological – composition, structure and strength properties of rocks, hydrogeological – the presence of surface water and aquifers, water cut, etc [7].

The prolonged standing of the slopes in the stress-strain state increases the risk of the occurrence of a technogenic catastrophe. One of the main reasons for this situation is unjustifiably chosen parameters of the slope, as well as inadequate actions of maintenance personnel. Therefore, in order to prevent dangerous situations, it is necessary to develop engineering and technological solutions, which should include issues related to the use of sound methods for selecting safe slope parameters with a specified safety factor and ensuring safety and labor protection in the development of soils in quarries [8].

IOP Conf. Series: Earth and Environmental Science **115** (2018) 012030 doi:10.1088/1755-1315/115/1/012030

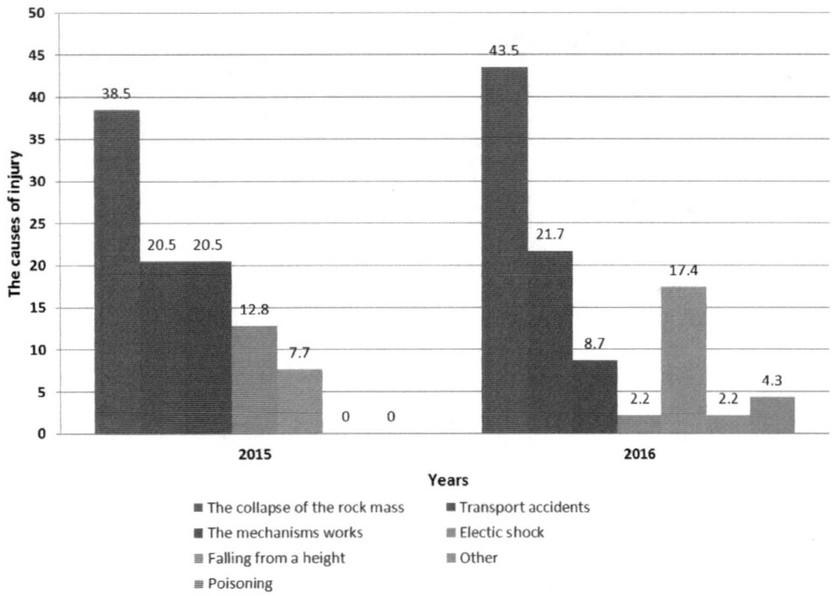

Figure 2. Percentage distribution of accidents by types of causes incidents in the mining and non-metallic industry

2. Results and discussion

Calculation of the maximum permissible parameters of the slopes is based on the choice of the geomechanical model of the soil array. The development of a geomechanical model of a soil massif is based on the study of the physical and mechanical properties of soils as constituents of a geological structure model [9, 10, 11].

Arrays of rocks of small quarries correspond to models of continuous medium, which completely abstract from the discrete structure of matter.

The desire for such a schematization of rock arrays is conditioned by the desire to use effectively the well-developed mathematical apparatus of differential and integral calculus. From the physical point of view, the use of the model of a continuous medium is justified if the smallest of the considered volumes of material preserves the «representativeness» of its mechanical properties as the integral effect of many micro particles, that is, inequality must be fulfilled [12]:

$$\sqrt[3]{\Delta V} >> d_m, \tag{2}$$

where V is the volume of the material; d_m is the characteristic size of the microstructure elements.

The study of the mechanical behavior of arrays using the model of a continuous medium provides for an operation of integration over the volume of the array in question. From this point of view, the volume of the material ΔV is geometrically regarded as an infinitesimal object. In practice, this should be reduced to the fulfillment of inequality:

$$\sqrt[3]{\Delta V} << L, \tag{3}$$

where L is the characteristic size of the array under study.

Inequality (2) can be treated as a physical condition, and inequality (3) is the mathematical correctness of the application of the continuous medium model when choosing safe slope parameters.

The particle sizes are commensurately small in comparison with the size of the studied sections, which reach tens and hundreds of meters, which causes inequalities (2) and (3) to be satisfied.

As a result of the conducted studies, we found that for homogeneous arrays it is recommended to use a circular cylindrical slip surface, which is the most dangerous.

In the development of soils for calculating the slope stability factor of slopes, the position of the potential slip surface in a homogeneous slope is determined by the laws of the static nature of the granular medium [13, 14, 15].

The scheme for calculating the slope stability factor is shown in figure 3.

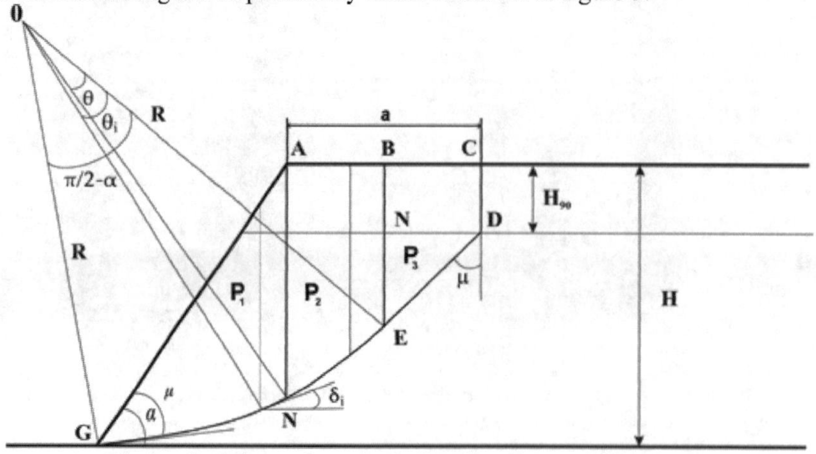

Figure 3. Scheme for calculating the slope stability factor

The stability factor of a homogeneous slope, taking into account the expansion of the forces due to the action of the self-weight of the soil, on the restraining and shearing, is determined by the following analytic expression:

$$n = \frac{tg\left[\sum P_i \cos\delta_i + P_1 \sin\left(\mu + \frac{\theta}{2}\right) + P_2 \sin\mu\right] + c(l_1 + l_2 + l_3)}{\sum P_i \sin\theta_i + P_1 \cos\left(\mu + \frac{\theta}{2}\right) + P_2 \cos\mu}, \quad (4)$$

where φ is the angle of internal friction of the soil;

P_i is the weight of the elementary column of soil at the i-th point of the curvilinear portion of the slip line;

δ_i is the angle of inclination of the tangent at the i point of the curvilinear portion of the slip line;

P_1, P_2 – weight of prism of reference pressure ABEN and weight of prism of active pressure BCDE accordingly;

$$\mu = 45^0 - \frac{\varphi}{2}, \quad (5)$$

θ is the angle contracted by the chord NE;

c – cohesion of the soil that forms the escarpment;

l_1 – length of the section of the circular cylindrical sliding surface GN;

l_2, l_3 – the lengths of two straight sections NE and ED, respectively.

Based on the scheme, which is shown in Figure 3, the radius of the sliding surface of the prism of a possible collapse is:

$$R = \frac{(H - H_{90})\cdot tg\mu + H \cdot ctg\alpha}{\cos\mu - \sin(\alpha - \mu) + tg[\cos(\alpha - \mu) - \sin\mu]}, \quad (6)$$

$$H_{90} = \frac{2c \cdot ctg\mu}{\gamma}; \quad (7)$$

where H is the height of the slope of the ledge;

α is the angle of the slope of the ledge;

H_{90} – height of the elastic layer of soil;

γ is the density of the soil.

The width of the prism of the possible collapse of the AC at the upper platform is determined from the following expression:

$$a = 2\{R[\cos\mu - \sin(\alpha - \mu)] - Hctg\alpha\}. \quad (8)$$

After a number of mathematical transformations, we obtained analytical expressions for a more accurate determination of the slope stability factor of the slope. The developed mathematical apparatus for calculating the coefficient of stability of a homogeneous slope, allowed to create a program for calculating the safe parameters of slopes "Slope1".

This program allows you to calculate the safe parameters of slopes within a few seconds for a given stability factor. These calculations were the basis for plotting the graphs of the dependence of the nominal height of the slope H' on the angle of slope of the slope α and the conditional width of the prism of possible collapse.

Figures 4 – 7 show the graphs plotted taking into account the following characteristics c = 1 t/m^2, γ = 1 t/m^3, φ = 200.

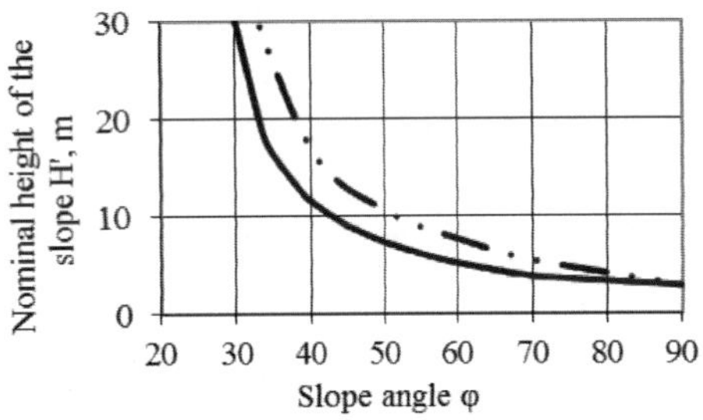

Figure 4. Graph of the dependence of the nominal height of the slope H 'on the slope angle of the slope at φ = 200 and the safety factor n = 1.0 (1 – graph plotted with the help of the Slope program, 1, 2 – graph constructed using the VNIMI method)

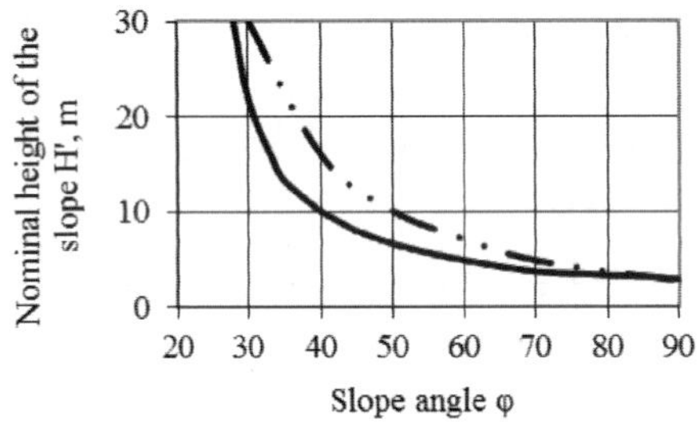

Figure 5. Graph of the dependence of the nominal height of the slope H' on the slope angle at φ = 200 and the safety factor n = 1.15. (1 – a graph constructed with the help of the Slope program, 1, 2 – a graph constructed using the VNIMI method)

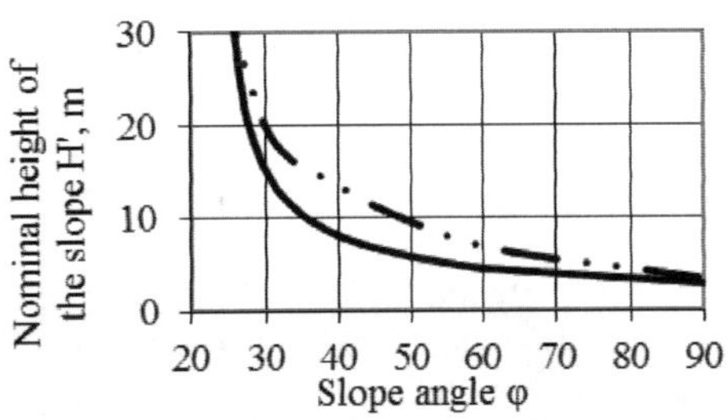

Figure 6. The graph of the dependence of the nominal height of the slope H 'on the slope angle at φ = 200 and the safety factor n = 1,2. (1 – a graph constructed with the help of the Slope program, 1, 2 – a graph constructed using the VNIMI method)

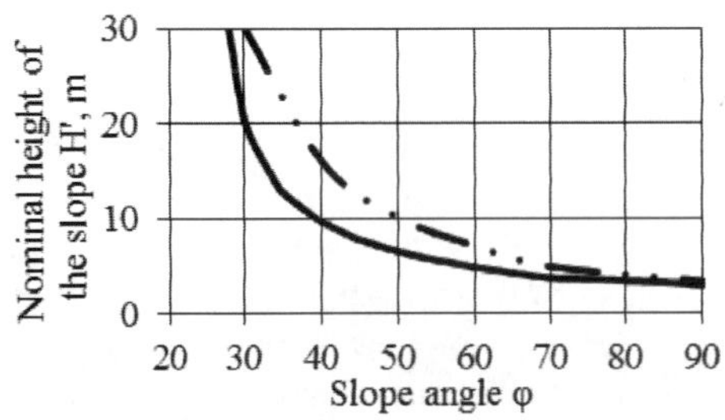

Figure 7. The graph of the dependence of the nominal height of the slope H' on the slope angle at φ = 200 and the safety factor n = 1.3 (1 – a graph constructed with the help of the program Slope, 1, 2 – a graph constructed using the VNIMI method)

Analysis of the results of the comparison shows that the parameters of the slopes, calculated using the graphs developed by VNIMI, are overestimated from 10 to 50%, which is inadmissible, since in this case the coefficient of stability is in fact lower than the normative by 15%.

Dependences of the nominal height of the slope H on the angle of slope of the slope α and the conditional width of the prism of the possible collapse, obtained with the help of the developed technique, make it possible to increase the reliability of the calculation of the safety factor by 20% in comparison with the graphical method developed by VNIMI [15, 16] risk of slope collapse by 20%.

3. Conclusion

The conducted analysis of industrial injuries and accidents in the mining industry showed a high level of traumatic danger of the industry and the absence of a stable tendency to reduce it. It has been established that about 20% of accidents in the mining industry are caused by caving and collapsing of quarry pit slopes, and the main reasons for the occurrence of caving in or collapse are unjustifiably chosen slope parameters, as well as inadequate actions of maintenance personnel.

It was revealed that the calculation of the maximum permissible parameters of the slopes is based on the choice of the geomechanical model of the soil array, the development of the geomechanical model of the soil massif is based on the study of the physical and mechanical properties of soils as constituent parts of the geological structure model.

As a result of the conducted studies, we found that for homogeneous massifs it is recommended to use a circular cylindrical slip surface, which is the most dangerous.

Analytical expressions are obtained for a more accurate determination of the slope stability factor of the slope. A program has been developed for calculating the safe parameters of slopes «Slope1».

The graphs of the dependence of the nominal height of the slope H 'on the slope angle α and the conditional width of the prism of possible collapse have been constructed, it has been established that the slope parameters calculated using the graphs developed by VNIMI are overestimated from 10 to 50%, which is inadmissible, since in this case, the safety factor will actually be lower than the normative by 15%.

Dependences of the nominal height of the slope H on the angle of inclination of the slope α and the conditional width of the prism of possible collapse, obtained with the help of the developed technique, make it possible to increase the reliability of the calculation of the safety factor by 20% in comparison with the graphical method developed by VNIMI and reduce the risk of collapse of slopes by 20% .

Acknowledgments

The work is realized in the framework of the Program of flagship university development on the base of the Belgorod State Technological University named after V.G. Shukhov, using equipment of High Technology Center at BSTU named after V.G. Shukhov.

References

[1] Federal law *On technical regulation*, approved by The Council of Deputies of Belgorod from 15.12.2002 №184 (in Russian)

[2] GOST 22.0.05-97 *Safety in emergencies. Technogenic emergencies. Terms and definitions.* Moscow 1996 (in Russian)

[3] Faleev M I 2016 Risk management of man-made disasters and natural disasters (a handbook for leaders of organizations) (Moscow: All-Russian research Institute on problems of civil defense and emergency situations of EMERCOM of Russia) 270 p (in Russian)

[4] Lubenskaya O A, Klimova E V, Khramtsov B A, Rostovtseva A A 2013 Assessement of emergency and accidents at quarrying way *Bulletin of BSTU named after V.G. Shukhov* **1** 140–4 (in Russian)

[5] Semeykin A Yu, Khomchenko Yu V 2014 Development of monitoring system conditions and occupational health for assessing occupational risks at the enterprises of the Belgorod region *Youth and scientific and technical progress: Proceedings of VII International Conference vol 1* (Staryi Oskol: Assistant Plus) 379–82 (in Russian)

[6] Yastrebinskaya A V, Edamenko A S, Divichenko I V, Matveeva L Y 2017 On the issue of occupational traumatism occurrence rate in the mining industry on the example of the Belgorod region. Bulletin of civil engineers **3 (62)** 273–9 (in Russian)

[7] Latkin M A, Nesterova N V, Shaptala V G, Radoutski V Yu 2017 The choice of measures to respond to technological risks of the enterprise *Bulletin of BSTU named after V.G. Shukhov* **4** 145–9 (in Russian)

[8] Klimova E V, Khramtsov B A, Rybka O A 2006 Security works in the formation of pit walls *News of the Higher Institutions. Mining Journal* **4** 63–5.
[9] Chen C, Xia Y, Bowa V M 2017 Slope stability analysis by polar slice method in rotational failure mechanism *Computers and geotechnics* **81** 188-94
[10] Deng D, Zhao L, Li L 2016 Limit equilibrium method for slope stability based on assumed stress on slip surface *J. Cent. South Univ.* **23** 2972-83
[11] Fu X, Sheng Q, Zhang Y, Chen J, Zhang S, Zhang Z 2017 Computation of the safety factor for slope stability using discontinuous deformation analysis and the vector sum method *Computers and geotechnics* **92** 68-76
[12] Fisenko G L 1965 *Stability of pit walls and dumps* (Moscow: Nedra) 378 p (in Russian)
[13] Sokolovsky V V 1960 *Statics of granular medium* (Moscow: Phismathgis) 244 p (in Russian)
[14] Tsytovich N 1983 *Soil mechanic*. (Moscow: Higher school) 288 p (in Russian)
[15] Methodical instructions 1972 *Definition of the angles of inclination of the sides of slopes of ledges and dumps built and operated quarries* (Leningrad: VNIMI) 164 p. (in Russian)
[16] Sergeev E M 1986 *Theoretical foundations of engineering geology. Mechanics and mathematical foundations* (Moscow: Nedra) 254 p (in Russian)

Increasing the efficiency of the process of snow mass utilization from the urban infrastructure and reducing the negative impact on the environment through the introduction of an installation for the utilization of snow mass

V O Dovbysh [a], L N Burakova [b] and A D Burakova

Tyumen Industrial University, str. Volodarskogo 38, Tyumen, 625000 Russia

E-mail: [a] Vadim932@mail.ru, [b] burakova.1992@mail.ru

Abstract. The article raises the problem of the maintenance of the urban area in winter, namely, the problem of disposal of snow masses from the urban area. The article describes the main disadvantages of the existing snow-melting systems for the utilization of city snow mass and are encouraged to develop an installation for melting of snow, functioning at the expense of power consumption. The developed installation allows to reduce the noise level during its operation, to exclude the influence of exhaust gases on the environment, therefore, to improve the environmental safety of the urban infrastructure.

1. Introduction

The characteristic climatic conditions of the Russian Federation raise a number of problems which associate with the development of the urban areas in the winter in most regions of the country [1]. There are often problems associated with the environmental safety of urban infrastructure. In most works, it is proposed to solve it by reducing emissions of harmful substances with exhaust gases of vehicles [2,3,4,5,6,7,8]. However, studies that are aimed at the utilization of snow mass without the use of exhaust gases are not enough [9,10,11].

In every city there are regulations about the rules of urban areas. These standards marked the period when it carries out snow removal, permissible time of storage and highlights the environmental danger of snow masses and requirements for their utilization. The problem is the complexity of compliance with these standards. This causes a reduction in traffic, an increase in the number of accidents and violation of environmental safety [12].

Currently, the trend of urban growth is increasing every day, which, in turn, causes an increase in the area of landfills for snow. This is an expensive process due to the technical and environmental requirements that are imposed on the design of the urban environment [13, 14, 15]. The main requirement is the need to take into account the influence of a large volume of melt water in the spring on urban infrastructure facilities. Therefore, the continuous system of utilization of an urban snow-melting plant is actively developing and being introduced now.

On the territory of the Russian Federation mobile snow-melting systems that operate at the expense of energy which is obtained from the combustion of liquid fuels is widespread, relative to similar systems that operate at the expense of power consumption. First of all, this is connected with the design of plants for the utilization of snow mass, the structural feature of which is the simplicity of the technological scheme. The process of melting is carried out when exposed to an open flame on the snow mass.

Typically, the process of melting city snow mass flows in a metal container, which is called a melting chamber. The combustion products of diesel fuel or natural gas heat this chamber. Further, the cooled combustion products are discharged into the atmosphere.

A typical snow melting system is shown in Figure 1 [16,17]. It includes a receiving chamber (melting chamber) 1 is a metal structure with an inlet 3, a heat generator 2 that contains a burner and a mixing chamber, a channel for gaseous fuel and a cooled coolant outlet. The inner walls and the lower part of the inclined chamber and the outer walls are located by the exhaust jacket 4. The exhaust jacket closes the heating of the hot gas of the bottom wall of the receiver. The snow mass melts in the funnel during the heating of the chamber wall. The liquid that is formed by melting on inclined walls flows through the outlet pipe 5. The heated gas from the mixing chamber is supplied to the chamber of the exhaust jacket by melting.

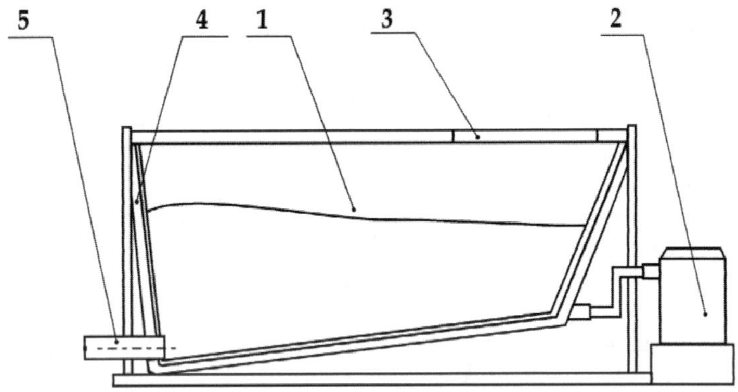

Figure 1. Snow melting system.

Main disadvantages of such structures are:

- small distance between the outer surface of heat exchangers;
- the need for frequent stops melting process because of the cleaning system from debris;
- high complexity of the recycling process, the availability of fuel reserves, the use of which, the pollution of environment with combustion products.

In the modern world, the problem of preserving a clean environment, which casts doubt on the future prospects of the spread of snowmelt systems, is acute and contributes to increasing emissions of pollutants into the atmosphere [18].

2. Results and Discussion

To date, the latest technical developments have serious standards of environmental safety. The conducted studies revealed a significant level of snow pollution in the city [12]. The average values of pollutants are different and are presented in Table 1.

Table 1. Average values of typical contaminants in urban snow mass.

Substances	Dimension	Value
Suspended substance	mg/l	121 – 732
BOD_5	mg/l	2.48-7.61
Ammonium nitrogen	mg/l	0.33-9.6
Chlorides	mg/l	16.11-338.2
Sodium	mg/l	13.76-442.4
Potassium	mg/l	22.3-110.9
Iron	mg/l	0.68-1.45
Manganese	mg/l	0.11-0.67
Zinc	mg/l	0.013-0.09
Lead	mg/l	15.23-37.23
Oil	mg/l	1.4-50.2
Synthetic surfactants	mg/l	0.23-1.012

Their magnitude depends on the urban area. In the snow samples, suspended substances, organic compounds that are not susceptible to oxidation, and solid salt for flour were observed. The chloride content exceeded the permissible concentration by 7-15 times. The amount of sulfates exceeded 3.5 times. The maximum allowable concentration of ions of toxic metals (manganese, iron, copper, cobalt, lithium, zinc, molybdenum, cadmium) exceeded the maximum allowable concentration up to 46 times. The content of phenols and petroleum products exceeded the maximum allowable concentration, respectively, from 1 to 4 and from 25 to 150.

The analysis identified the following pollutants: suspended substance, chlorides, oil products. The averaged concentrations of suspended, which matter equal to 426.5 mg/l, is presented in 27% of cases. The concentration of oil, equal to 25.6 mg/l, is presented in 33% of cases. The concentration of chlorides, equal 177.2 mg/l, is presented in 28% of cases.

The developed installation for the utilization of urban snow mass as a primary energy source uses electricity. This makes it possible to eliminate exhaust fumes during plant operation, which is an important advantage in comparison with similar installations that operate on hydrocarbon fuels. Thus, the use of this unit allows to significantly reduce the negative impact of the operation of the snow melting unit on the environment.

As a result of the study of the problem, a new approach emerged to address the negative impact of the snow utilization facilities on the environment and the Grodsk infrastructure. Thus, the work suggests the development of an installation for the utilization of urban snow. The main purpose of the plant being developed is to intensify the process of melting city snow, provide low energy costs and increase environmental safety in the disposal of snow.

The installation solves the problem of recycling urban snow masses from the urban area. This problem leads to an increase in the traffic flow, a reduction in the number of accidents and pollution of the environment. The development of an installation for urban snow reclamation is a specialized project designed for the active process of melting urban snow mass due to the use of electricity. The electrical energy used in the installation is converted into microwave electromagnetic radiation. This allows during the snow melting to clean and drain the resulting water melt. The scheme of the developed installation of snow melting is shown in Figure 2.

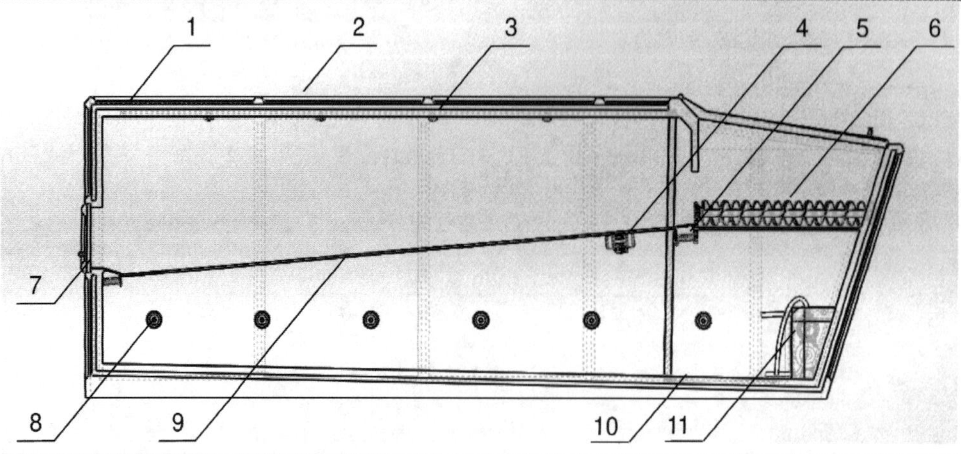

Figure 2. The design of the installation for the utilization of snow: 1 – heat and soundproofing layer, 2 – hull installation, 3 – atomizer, 4 – inertial vibrator, 5 – restraining spring, 6 – screw conveyor, 7 – bunker bulky waste, 8 – electromagnetic emitter, 9 – sifting (protective) grate, 10 – temperature sensor 11 – water distribution pump.

The principle of the installation is based on the progressive process of cleaning and heating the snow mass. It is loaded into the area for receiving and processing snow, which includes a conveyor belt and a storage compartment, where it is subjected to the first stage - loosening and transport of snow mass. The snow mass is weakened due to the rotational-translational motion of the screw 6. Then, in the section for receiving and treating the snow mass, it undergoes preliminary cleaning of the snow cover, which does not contact the transport screw 6. It then interacts with the rigid panel, which provides access to the preliminary cleaning only the snow weakened. In the pre-treatment chamber, the snow mass decreases with the waste.

Further, the mass of snow passes the screening process through the protective grate 9. The sieving process is intensified by using a protective grating 9, which is installed under a slight inclination (150 degrees), and an inertial vibrator. It provides vibration of the protective grid 9. Hydration of the snow mass takes place with the sprayer 3 simultaneously with the sieving process. The moistened and sieved mass of snow enters the melting chamber 9, and the bulk waste that remains on the protective grid enters the waste compartment.

The mixed and purified snow mass is melted in the melting chamber by means of an electromagnetic radiator 8. The water melt enters the water distribution pump 11. Small debris is removed through a special compartment located at the bottom of the unit. 3-D model of installation for the utilization of snow mass is shown in Figure 3.

Figure 3. 3-D model of installation for the utilization of snow mass.

The purpose of creating the 3-D model of the installation is not only the development of modern design, but also the calculation of the minimum costs for its production and the theoretical evaluation of the possibility of implementing the melting process of the snow mass. Technical characteristics of the installation for the utilization of snow mass are shown in Table 2.

Table 2. Technical characteristics of the installation for the utilization of snow mass.

Parameter	Value
Performance in the snow, m^3/hour (t/h)	15 (1.05-1.55)
Electricity consumption, kW·h (MJ)	30 (108)
Dimensions, m	6.2·2.3·2
The volume of the chamber melting, m^3	5.4
Volume load snow, m^3	4.1
Water discharge, m^3/hour	4.65-6.95

Indicators of "performance in the snow m3/h» and «discharge of water m3/h" have variable values due to differences in density of the snow mass in the melting process of snow.

The main novelty is the use of a water distribution system (irrigation and drainage) and melting based on electromagnetic effects as the primary source of electricity, which solves the main problem from the point of view of the energy consumption of the melting process of snow. This is to ensure the phase transition of the solid phase into the liquid. The efficiency of melting is due to the nature of the propagation of electromagnetic radiation. In the proposed design of the installation, it acts directly on the entire volume of snow mass and interacts with water molecules, which cause a resonance effect. This increases the frequency of vibrations of polar molecules and causes the appearance of thermodynamic reactions that provide an intense thermal environment.

The urban snow mass is an adsorbent and contains oil products, heavy metals, household waste, de-icing materials and so on. The change in heat capacity and density of pure snow mass with the addition of pollutants causes a change in its melting temperature. Therefore, the work also included studies of interfacial heat transfer processes in the melting of urban snow mass. During this stage of the experiment, we found that the use of heat exchangers in a local thermal system increases the melting time of contaminated snow compared to a pure analogue. The use of electromagnetic radiation reduces the melting time with the content of pollutants in the snow mass.

Thus, the developed installation is an effective way of recycling urban snow mass, which contains pollutants.

3. Conclusion

This article describes the scheme and principle of operation of existing installations for the utilization of snow masses. In the course of analyzing their technical characteristics and features, we found that the melting of the snow mass is carried out with the help of gases that are formed during the combustion of hydrocarbon fuels and has a negative impact on the environment. Therefore, we have developed an installation for the disposal of urban snow mass with the use of electromagnetic radiation, which has made it possible to improve the environmental safety of urban infrastructure.

Operation of the developed installation allows:

- to reduce the noise level during the operation of the plant due to the presence of a soundproof layer;
- to remove exhaust fumes during operation, since electricity is the source of energy;
- to use it directly in the urban environment, as the technical characteristics of its operation are consistent with the norms of acceptable noise pollution and environmental safety;
- to increase the plant productivity through the use of screening systems, electromagnetic effects and a humidification system and the melting process;
- improve the quality of the process of cleaning urban snow from debris because of the presence of a two-level system.

References

[1] Anisimov I, Magaril E, Magaril R, Chikishev E, Chainikov D, Gavaev A, ErtmanS, Ertman Yu and Ivanov A 2016 Improving vehicle adaptability to the operating conditions of "smart" cities in the northern regions *E3S Web of Conf. Rev.* **6** 02003

[2] Chikishev E, Chainikov D and Anisimov I 2017 Increasing the use of natural gas on motor transport by an efficient location of the NGV RS *IOP Conf. Ser.: Earth Environ. Sci. Rev.* **50** 012004

[3] Chikishev E, Chikisheva A, Anisimov I and Chainikov D 2017 Natural gas use on mimbuses, engaged in the carriage of passengers and baggage on the regular routes, as a measure for decrease I harmful environment effects *IOP Conf. Ser.: Earth Environ. Sci. Rev.* **50** 012008

[4] Chainikov D, Chikishev E, Anisimov I and Gavaev A 2016 Influence temperature on the CO_2 emitted with exhaust gases of gasoline vehicles *IOP Conf. Ser.: Mater. Sci. Eng. Rev.* **142** 012109

[5] Chikishev E, Ivanov A, Anisimov I and Chainikov D 2016 Prospects of and problems in using natural gas for motor transport in russia *IOP Conf. Ser.: Mater. Sci. Eng. Rev.* **142** 012110

[6] Ertman J, Ertman S, Anisimov I, Chainikov D and Chikishev Y 2015 Complex characteristic estimation of vehicle adaptability for low temperature operation conditions *Research Journal of Pharmaceutical, Biological and Chemical Sciences (Electronic Materials* vol 6*)* (India:

RJPBCS) pp 1761-70

[7] Anisimov I, Ivanov A, Chekishev E, Chainikov D and Reznik L 2014 Assesment of gas cylinder vehicles adaptabilty for operation at low ambient temperature conditions *WIT Transactions on Ecology and the Enviroment* vol 190 (UK: WIT Press) pp 685-695

[8] Ertman S, Ertman J and Zaharov D 2016 Raising of operating a motor vehicle effects on enviroment in winter *Innovative Technologies in Engineering Rev.* **142** 12119

[9] Yasuhiro H, Hideki K, Makoto N, Kazuhiro K, Yoshiaki H 2010 Experiments and evaluationof a mobile high-density snow storage system *Energy and Buildings* vol 42 (UK: Elsevier) pp 178-182

[10] Yasuhiro H, Hideki K, Makoto N 2007 Field measurements and analyses for a hybrid system for snow storage/melting and air conditioning by using renewable energy *Applied Energy* vol 84 (UK: Elsevier) pp 117-134

[11] Yasuhiro H, Tsutomu N, Hideki K, Takayuki O, Yoshiaki H 2012 Study on a snow storage system in a renovated space *Renewable Energy* vol 41 (UK: Elsevier) pp 401-406

[12] Dovbysh V O, Sharukha A V, Evtin P V and Vershinina S V 2015 City snow's physicochemical property affects snow disposal *IOP Conf. Ser.: Materials Science and Engineering, Rev.* **93** 012066

[13] Jacyna M and Merkisz J 2015 Estimations in real urban traffic conditions s a deteminant of shaping sustainable urban development *WIT Transactions on The Built Environment* vol 146 (UK: WIT Press) pp 219-231

[14] Galderisi A 2014 Adapting cities for a changing climate: an integrated approach for sustainable urban development *WIT Transactions on Ecology and the Environment* vol 191 (UK: WIT Press) pp 549-560

[15] Ntzeremes P, Kirytopoulos K and Tatsiopoulos I 2016 Management of infrastructure's safety under the influence of normative provisions *WIT Transactions on The Built Environment* vol 164 (UK: WIT Press) pp 51-59

[16] Voitkovskii K 1977 *Mechanical properties of snow* (Moscow: Nauka)

[17] Pavlov N 1999 *On the development of housing and communal services and its reform* (Moscow: Prima-Press)

[18] Kazopoulo M, El-Fadel M and Kaysi I 2004 Emissions Standards Development for an I/M program *WIT Transactions on Ecology and the Environment* vol 74 (UK: WIT Press) p 10

Geo-ecology of surface atmosphere of Tomsk and methodology for the ecological risk calculation

E V Ivanova, M V Anisimov, U N Kuznetsova, N V Taldonova and A V Petrova

Tomsk State University of Architecture and Building, 2, Solyanaya square, Tomsk, 634003, Russia.

E-mail: teploproekt@mail2000.ru

Abstract
The present study presents new methodological approach of environmental assessment of surface atmosphere layer based on principles of non equilibrium dynamics. The role of natural and technogenic factors in forming areas of dust and airborne pollution is determined. The results of the study of ecological risk from atmosphere chemical pollution of the town are presented.

1. Introduction

The Earth's atmospheric mantle suffers from increasing effects of anthropogenic and technogenic factors in global, regional and local scale. The situation is particularly critical in natural and technogenic geosystems and urban areas of Western Siberia. The limits of atmosphere self-purification and self-regeneration from a wide range of toxicants are exceeded several dozen of times, and according to macro-components many times, which is accompanied by the ecological risk increase.

The international community, concerned about the state of the environment, is making efforts to improve the current ecological situation. Most states issued the concept of sustainable development, which is implemented in their long-term plans. It provides for the prevention of further destruction of the natural environment by prioritizing environmental and social requirements over economic ones.

In the sustainable development policy of Tomsk the following sustainable development criteria have been developed: 1) the level of air pollution; 2) drinking water quality; 3) the level of waste management; 4) morbidity of respiratory system; 5) morbidity of digestive system; 6) infectious and parasitic diseases.

One of the most important hygiene problems in Tomsk is atmospheric pollution. According to the data from the Center for Hydrometeorology and Monitoring of Tomsk Region, the index of atmospheric pollution (IAP) in Tomsk is the highest in the region.

The territory of Tomsk is confined to the joint of two large geological and, correspondingly, hydrogeological complexes. The lower one is represented by Paleozoic metamorphic stales, which are folded and broken by deep tectonic faults into horst-keystone blocks. The upper (Mesozoic-Cenozoic) complex is represented by sedimentary rocks of the platform cover. After considering the results of drilling, geophysical measurements and satellite image interpretation, the cover rocks on the northwestern slope overlap the submerged surface of the basement rocks and have a slightly inclined monoclinal folding with angle up to 5°. At the same time, a large number of disjunctive dislocations have been revealed in the cover rocks with the absence of characteristics of plicative tectonics. They are recorded by geological and geophysical data, and become clearly obvious in the landscape, and by the sharp difference in hypsometric marks of stand-alone formations. The system of large tectonic fractures of the east-north-east extension prevails, which include diabase dikes the right tributary valley of the river Tom, the paleochannels and geoactive linear zones adjacent to the foundation block boundaries and deep faults. Another leading characteristic of disjunctive tectonics is the embedded nature of valleys and river terraces.

As a result, a peculiar landscape with several hills and lowland areas was formed within the limits of the town. There are showings of numerous springs and even lakes on both types of landscapes within the platform cover. And lacustrine-boggy sediments predominate in the lowland areas. So the *natural*

IOP Conf. Series: Earth and Environmental Science **115** (2018) 012032 doi:10.1088/1755-1315/115/1/012032

landscape of the territory was formed in the geological history. Eventually, the *landscape and the river network* played a leading role in the development of the *town's structure*. They were the main limiting factors in the development of urban systems, including residential areas, industrial sites, traffic arteries and recreational areas.

The complex of factors responsible for the specific features of atmospheric pollution within the town includes natural *weather and climatic conditions* as well as features of *architectural solutions* and the *spatially functional structure* of the town.

Tomsk has a wide range of industries (with nested and mostly impregnated structure in residential areas), in contrast to the predominantly mono-profile neighbour towns (Kemerovo, Novokuznetsk, etc.). There are several thousand stationary sources of atmospheric pollution, which belong to industrial enterprises, in the town. Some areas of Tomsk possess a long-outdated street and road network and a low traffic capacity. The study of natural and technogenic factors, mechanisms of direct communications and feedback between them for ecologically significant sources of pollution allows us to reveal general patterns of the atmospheric air state and to work out measures for environment and human protection. Thus, the density of uneven distribution of atmospheric pollution largely depends on the landscape, the location of industrial facilities and their hazard category, the density and number of stores in buildings and structures, microclimatic parameters, and so on.

The surface atmosphere of Tomsk differs by exceeding the maximum permissible concentration (MPC) of dust in 2 times, carbon monoxide in 1.4 times and nitrogen dioxide by 1.5 times. The exceedings in the deposit media (in snow blanket) in relation to the background values are the following: dust from 5 to 100 times, toxic metals from 20 to 10000 times. In recent years, the environmental characteristics are significantly worsen by increasing pollution with formaldehyde, phenol, ammonia, methanol. The number of motor transport and the volume of pollutants from mobile sources is rapidly growing (more than 5% per year), their proportion in Tomsk exceeds 78%. Concentration factor of lead along transport lines and their intercrossing areas is 5-6 MPC.

The implementation of the concept of sustainable development (SUSTAINABLE DEVELOPMENT), approved at Heads-of-government conference in December 1992 in Rio de Janeiro, stimulated searching for new approaches in methodology and interpretation of environmental data. The statutory environmental requirements are established in legislative and regulatory documents and are aimed at ensuring environment conservancy, environmental rehabilitation and human health protection.

Russian standards for environmental requirements and environmental certification conform to international standards of ISO14000 series according to environmental management systems and are called GOST-R ISO 14000-98. Various criteria for the integral assessment of the environment within urban areas were developed taking into account these requirements [1, 2].

Numerous mathematical models and computer technologies for accounting aerosol transfer and the surface atmosphere quality assessment were created. As a rule, they are used under different conditions and oriented towards assessment of the damage caused or prevented. The ranking methods considering the degradation factors of ecological systems and combinatorial methods got widespread use [3, 4].

A special breakthrough of atmosphere ecology research is associated with the implementation of nonlinear dynamics principles, especially those devoted to the study of chaos in determinate systems. The promising direction described in the present article suggests a comprehensive approach with the participation of meteorologists and geologists-geochemists. It allows to obtain the most complete and timely information about the state of the surface atmosphere layer, taking into account climate dynamics, meteorological conditions and sources of pollution in warm and cold seasons, as well as geomorphological and town-planning features of the area. Under comprehensive approach all changes in the surface atmosphere quality in all time scales must be considered. Whereas even a single emission of pollutants may cause serious negative effects on people and the environment.

2. Results and its discussion

We performed the postcasting of the surface atmosphere quality for the last 25 years on the stationary hydrometeorological facilities of the natural and technogenic geosystem of Tomsk. It made it possible to us establish a complex, periodically (from diurnal and circadian to perennial) recurring spatiotemporal dynamics of the internal distribution structure of macro- and microimpurities of the surface atmosphere. Time-series trends of the surface atmosphere quality being studied, the long-term trend indicating a change trend of the surface atmosphere quality and the frequency of these changes, is calculated . We chose a linear polynomial as a function approximating a long-term trend. It allows us to optimize the process of searching for new empirical facts of changing the operating modes of the system generating the time series.

The variability of the surface atmosphere quality and the risk of pollution is determined by the combined effects of synoptic, weather and climate conditions and the emission regime, with the emission regime having a stronger effect. The repeatability of pollutant concentrations above $MPC_{c.c.}$ was calculated for various types of synoptic processes: anticyclone, cyclone, intermediate field, low-gradient field with increased pressure, low-gradient field with reduced pressure, contrast zone. We found that the maximum number of pollutions occur in cold seasons with anticyclones, but with their rapid displacement the content of impurities does not rise significantly.

The ratio of multi-temporal cycles depends on the location of posts, landscape and geomorphological conditions of different zones of the natural and technogenic geosystem. Practically, diurnal and intra-monthly cycles have the greatest variability for all impurities and at all survey points. However, their contribution depends heavily on the type of pollutant and geomorphological conditions. Mean values and dispersion of pollutants are calculated. The ratio of the maximum concentrations to the mean values varies according to the type of impurity and the natural and technogenic geosystem zone in wide range from 6.2 to 548.1. We found that the content of dust and main aerosol-gaseous impurities ranges from zero to values exceeding the $MPC_{c.c.}$ in 7-8 times.

The proposed methodology also includes the calculation of ecological risk, including a carcinogenic risk of the area within the geoecology of the natural and technogenic geosystem. The risk assessments, in particular, along with the system of integrated environmental monitoring, are currently the main content of the problem of ensuring environmental safety.

The WHO references define risk as "the expected frequency of adverse effects arising from a specified pollutant effects". According to the US Environmental Protection Agency's (US EPA) Glossary, risk is "the probability of injury, disease or death in certain circumstances. Quantitatively, the risk is expressed by values from zero (no harm will be done) to one (reflecting the certainty that the harm will be done). "

The methodology of risk assessment is the universally recognized and most important tool for characterizing the effects of environmental factors on the population health and making management decisions [9].There is a large number of studies on the health risk assessment with chronic (long-term) exposures of atmosphere pollutions, as well as those belonging to the authors of the present study [10-15].

As for research and practical application, the main task of risk assessment is to obtain and summarize the information on the possible effect of human habitat factors (in the present study - atmosphere pollution) on his health. The risk assessment is a key element of the risk analysis procedure. The general scheme of risk assessment includes several stages:

1. hazard identification;
2. assessment of dose-response relationship;
3. exposure assessment;
4. risk profile.

3. Hazard identification

Hazard identification is the first stage of health risk assessment.

The following blocks are important for hazard identification:

- finding the sources of hazardous substances emissions into the environment;
- defining the hazardous emission;
- identification of all potentially hazardous factors;
- listing of priority (the most dangerous) factors.

IOP Conf. Series: Earth and Environmental Science **115** (2018) 012032 doi:10.1088/1755-1315/115/1/012032

One of the important activities at the stage of hazard identification is the determination of the chemicals toxicity. This indicator is the key in further study of the peculiarities of the substance distribution in the environment and risk assessment of its health effect.

4. Assessment of dose-response relationships

The dose-response relationship assessment is the relation between the exposing dose (the amount of chemical effecting the body), the order, the duration of the exposure and the intensity, the prevalence of the studied adverse effect in the exposed population.

Characteristics defining the dose-response relationship:

- Carcinogens
  - carcinogenic potential factor (slope factor), SF (mg / kg-day) -1;
  - a single risk, UR, (mkg / m^3)$^{-1}$, (mkg / l) $^{-1}$.
- Non-carcinogens
  - reference dose (RfD), mg / kg;
  - reference concentration (RfC), mg / m^3

Carcinogenic potential (carcinogenic potential factor, SF) is a measure of additional individual carcinogenic risk or the degree of carcinogenesis increase under carcinogen exposure.

The reference dose is the daily lifetime exposure of a chemical, which is defined by taking into account all available up-to-date research data and probably does not result in an unacceptable health risk for sensitive population groups.

5. Exposure assessment

The exposure (effect) is the contact of a body (receptor) with chemical, physical or biological agents.

The exposure can be expressed as the total amount of substance in the environment (in units of mass, for example, mg); as the exposure value, that is the mass of the substance, referred to a unit of time (for example, mg / day); the exposure value standardized with account of the body weight (mg / kg-day). Quantitative exposure assessment can be obtained on the basis of the exposure concentrations assessment during the exposure period, as well as the calculation of the intake.

In assessing the risk of inhalation exposure the pollutant exposure concentrations are used, compared with the limit concentration for continuous exposure (inhalation concentration, units of measurement are mg of pollutant per m^3 of air).

It provides controlled concentration for inhalation exposures, the units of measurement are mg of pollutant per m^3 of air.

The average daily lifetime dose with inhalation exposure should be calculated by the formula:

$$LADD = \{C \times CR \times ED \times EF | BW \times AT \times 365\}, \quad (1)$$

where $LADD$ is a daily lifetime dose, mg/(kg×day);
C is a substance concentration in air, mg/m^3;
CR is air intake rate 20 m^3 / day;
ED is an exposure duration, 70 years ;
EF is an exposure frequency – 365 days per year;
BW is a body weight, 70 kg;
AT is an averaging period of exposure – 70 years (average human life expectancy);
365 is a number of days per year.

6. Risk profile

The risk profile is the final part of the risk assessment and the initial stage of risk management.

The method of comparison of the received dose / concentration with the reference dose (safe dose / concentration) is used to characterize the non-carcinogenic risk, that is the calculation of the hazard quotient (HQ);

Hazard quotient (HQ) is the ratio of the exposure dose (or concentration) of a chemical to its safe dose (reference level of exposure). The calculation of the hazard quotient is carried out according to the following formula:

$$HQ = \frac{AD}{RfD}, \quad (2)$$

where HQ is a hazard quotient; AD is an average dose (mg/kg-day); RfD is a reference (safe) dose (mg/kg-day).

The carcinogenic risk profile is based on the assessment of an additional number of cancerous diseases throughout the lifetime resulted from carcinogen exposure. The amount of the individual carcinogenic risk is determined by the formula:

$$CR = LADD \times SF, \quad (3)$$

where CR is the individual carcinogenic risk; $LADD$ is a lifetime average daily dose, mg/(kg x day); SF is the carcinogenic potential factor, reference data are used, (mg/(kg x day))$^{-1}$.

The laboratory data of air environment in Tomsk regularly conducted by the hydrometeorological service at stationary sites were analyzed during the study. This allows us to obtain reliable information about the actual conditions of atmospheric air pollution throughout the town. Based on these data, the daily, annual average and maximum concentrations of the main air pollutants were calculated for the years 1993-2014.

To assess the ecological risk in the territory of Tomsk conditioned by chemical pollution of the surface atmosphere layer, the normative documents of the Sanitary and Epidemiological Supervision Center of the Ministry of Health of the Russian Federation "Guidelines for the population health risk assessment from pollutant exposure" [5] , the software EPA US [6], the software system "RISK ASSISTANT" [7] were used. In the present study, we consider the ecological risk of harmful substances intake through the respiratory tract (inhalation exposure). A group of people exposed was examined (an "average inhabitant" weighing 70 kg, being 70 years old and with the 30 years exposure experience was selected). It is accepted that Tomsk is his permanent residence, except for 2 weeks a year, and the exposure lasts 3 hours a week.

Carcinogenic risk calculation is carried out according to the formula:

$$CR = 1 - \exp(-SF \times LADD), \quad (4)$$

where CR is an individual carcinogenic risk;
$LADD$ is a lifetime average daily dose throughout lifetime, mg/(kg × day);
SF is the cancerogenic potential factor (mg/(kgxday))$^{-1}$, (the software uses the database of substance cancerogenic properties http://www.epa.gov/iris/subst/

These assessments represent a theoretical over-background carcinogenic risk. For example, if the calculated risk is one per 1.000.000, it means that a person has a chance of a million to get cancer due to this chemical exposure, in addition to his chances of getting cancer for other reasons.

The ecological risk assessment was carried out for all ingredients controlled at Tomsk stationary sites: dust, sulfur oxide, nitric oxide, carbon monoxide, hydrogen sulphide, phenol, formaldehyde, ammonia, hydrogen chloride. The individual carcinogenic risks of chemicals, with the exception of formaldehyde, were less than $1 \cdot 10^{-6}$. According to the classification of risk levels given in [5-7], the individual lifetime risk, equal to or less than $1 \cdot 10^{-6}$, corresponds to one additional case of serious disease or death per 1 million exhibited persons; it characterizes such levels of risk that are perceived as negligible. Such risks do not require additional measures to reduce them, but their levels are subject to periodic monitoring.

However, atmospheric air in Tomsk is polluted with formaldehyde, its content may exceed $MPC_{c.c.}$ in 4 to 6 times. It is crucial to note that formaldehyde is also the only controlled chemical pollutant inducing individual carcinogenic risk. In this context, a hygienic assessment of the additional carcinogenic risk of atmosphere pollution in the residential areas of Tomsk was performed. The term "additional carcinogenic risk" means an addition to the risk that some person has already possessed, getting cancer caused by other reasons.

According to the performed study, the greatest contribution to levels of carcinogenic risk in this area is made by formaldehyde after atmosphere pollution. On average, across the city the individual carcinogenic risk caused by formaldehyde atmosphere pollution is $2 \cdot 10^{-5}$ and it annually adds 1 case of malignancy to the general morbidity in the town.

The ranking of individual carcinogenic risk levels for 2014 is shown in Table 1. The minimum level of risk is observed at Stationary site-11 (crossing of st. Proletarskaya and sst. Baranchukovsky), Stationary site-5 (st. Herzen, 68), Stationary site-13 (st. Vershinin) and Stationary site-14 (st. Lazo), due to the data for 2002. According to the risk classification, such risk is characterized as low. The individual risk throughout the lifetime is in the range of less than $1 \cdot 10^{-4}$ – more $1 \cdot 10^{-6}$ and corresponds to the area of conditionally acceptable (permissible) risk; the majority of foreign and sanitary-hygienic standards recommended by international organizations for the population in general are established at the same level.

The maximum risk level is observed at Stationary site-2 (Lenin Square) and at Stationary site-12 (sett. Svetliy). According to the current regulations, it is characterized as average, and is acceptable for occupational areas pollution and exceeding for residential areas; the administrative and technological and sanitary-hygienic measures should be taken to reduce such risk and develop curative measures.

The average level of risk in the town is $7.7 \cdot 10^{-5}$. It means that within seven decades, seven additional cases of cancer among population of 100.000 people exposed to formaldehyde inhalation exposure may occur. A low level of risk with the values of $8 \cdot 10^{-5}$ and $9 \cdot 10^{-5}$ is noted in the districts of st. Herzen and st. Lazo, respectively, and it is characterized as minimal, due to the lack of significant stationary technogenic sources of atmosphere pollution.

Table 1. The individual carcinogenic risk levels ranking for 2014.

stationary site #	Address	Individual lifetime carcinogenic risk	Risk level
Stationary site - 2	Lenin square	2e-004	Average
Stationary site - 5	st. 68, Herzen	8e-005	Low
Stationary site - 11	crossing of st. Proletarskaya and sst. Baranchukovskiy	1e-004	Low
Stationary site - 12	settl. Svetliy	2e-004	Average
Stationary site - 13	st. Vershinin	1e-004	Low
Stationary site - 14	st. Lazo	9e-005	Low

According to the research results the individual risk value $2 \cdot 10^{-4}$ makes possible two additional cases of cancer among population of 10.000 people within 70 years. Such risk is observed in the center of the town on Lenin Square, as well as in the Svetliy. settlement. In accordance with the classification of risk levels, such individual lifetime carcinogenic risk has an average level and is unacceptable to the population. Similar studies were carried out in St. Petersburg [8]. Their results showed that the values of carcinogenic risk in 20 districts of the city are in the range from $4.9 \cdot 10^{-5}$ to $11.2 \cdot 10^{-5}$. This results approximate to our data for Tomsk, except for the areas with the average risk level. We may presume that the level of individual carcinogenic risk from formaldehyde defined in the present study is typical and peculiar for urban areas with a high technogenic load.

7. Conclusion

Thus, owing to the development of the methodical base following the principles of nonlinear dynamics, including the assessment of individual ecological risks, we could zone the town according to the carcinogenic risk level, predict the possible health consequences for the population by comparing quantitative risk levels over a long period of observations, under different scenarios for the development of industrial production, motor transport and economic activities.

At the same time, it is clear that a direct extension of foreign methods prevents us from giving a reliable assessment. Therefore, it is necessary to find a hygienic and epidemiological justification of the risk assessment techniques application in a specific region, taking into account climatic, physical, geographical and social conditions.

References

[1] Adam A M 2000 Ecological problems of the Regions of Russia The Tomsk Region *Information Issue* 135–147.

[2] Adam A M 2002 Natural Resources and Environmental Safety of Western Siberia *Nature* 115–124.

[3] Kozlovskaya O V 2003 Sustainable development indicators of Tomsk oblast *STT* 15–21.

[4] Mikhalchuk A A 2006 Statistic analisys of ecological-geochemical data *Publishing house TPU* 193 – 210.

[5] Guidelines for the population health risk assessment from pollutant exposure Guidelines 2004 R. 2.1.10.1920-04 *Federal Sanitary and Epidemiological Supervision Center of the Ministry of Health of the Russian Federation* 273 p.

[6] U.S.Environmental Protection Agency 1998 Guidelines for ecological assessment. EPA/630/R-95/002F. Washington, DC.

[7] Tarasova N P, Malkov A V, Krapchatov V P 1996 Risk assistant Guidelines. Russian version *RCTU after D.I. Mendeleev., Alexandria: The Hampshire Research Institute* 270 p.

[8] Scherbo A P 2002 Environment and health: risk assessment approaches *MAPO* 370 p.

[9] Onischenko G G, Novikov S M, Rakhmanin J A, Avaliani S L, Bushtueva K A 2002 Population health risk assessment basement on pollutants exposure *NII EC and GOS* 408 p.

[10] Ivanova E V, Osipova N A 2007 Health risk assessment on atmosphere exposure for population of Tomsk The problems of geology and resources development *The works of XI international symposium of students and young scientists named after academician M.A. Usov* 579–581.

[11] Ivanova E V, Osipova N A 2011 Health risk assessment as a result of high pollution level of the atmosphere The problems of geology and resources development *The works of XV international symposium of students and young scientists named after academician M.A. Usov* 389–393.

[12] Ivanova E V, Bykova V V, Osipova N A 2006 Assessment of human health risk caused by chemical pollutants of the atmosphere *Atmospheric and Oceanic Optics* 867–869.

[13] Menshikov V V, Shvyriaev I A 2004 The problems of risk assessment for population and environment under atmosphere pollution *Publishing house of MGU* 202 p.

[14] Shvyriayev I A, Menshikov V V 2006 The algorithm of comparative assessment of risk from industrial enterprises emissions *RUDN bulletin, Ecology and health protection* 5–13.

[15] Novikov S M, Shashina T A, Abalkina I L, Skvortsova N S 2003 The risk of chemical pollurion exposure on population health: from assessment to practice *Adamant* 84 p.

[16] Anisimov M V, Babuta M N, Kuznetsova U N, Safonova E V and Minaeva O M 2016 The mathematical simulation of the temperature fields of building envelopes under permanent frozen soil conditions *IOP Conf. Series: Materials Science and Engineering* 124 (2016)

Microorganisms in the deposits of cold carbon mineral waters of the Russian Far East and their habitats

E G Kalitina[1, a], N A Kharitonova[2], T V Kuzmina[1], G A Chelnokov[1]

[1]Far Eastern Geological Institute, FEB RAS, Vladivostok, 690022, Russia
[2]Lomonosov Moscow State University, Moscow, 119991, Russia

E-mail: [a] microbiol@mail.ru

Abstract. Study of the chemical composition of carbon mineral waters has shown the prevalence of calcium, magnesium and sodium among the cations, sulfate, nitrate and chloride ions among the anions, and ferric iron, strontium and manganese in the microelement composition. Results of the microbiological studies have revealed that carbon mineral waters contain various microorganisms that can transform the physical and chemical composition of mineral waters by interfering with geochemical cycles. The sanitary and microbiological properties of carbon mineral waters have been evaluated thus proving that the waters of Medvezhii (Shmakovskoe deposit) are microbiologically clean.

1. Introduction

The geologically complex territory of the Russian Far East is home to extensive resources of mineral ground waters. With respect to chemical and gas composition, and their temperature range, the region's mineral waters fall into the groups: carbon, nitric or nitric-methane by their chemical composition, and thermal or cold temperature-wise [1]. In recent decades the chemical composition of mineral waters as well as their genesis has been sufficiently studied including the works by O.V. Chudaev, G.A. Chelnokov, and N.A. Kharitonova [2-5]; however, the formation mechanisms of many water types remain undiscovered. Recent years have brought many works that show show that microbiological processes actively change the hydrogeochemical system thus partially transforming the physical and chemical properties of ground waters and porous space of water-bearing rocks [6-8].

Microorganisms are a huge group of life forms; however, bacteria are most relevant for ground waters. They can be suspended in the ground waters ranging from 0.2 to 1.0 μm in size. In certain physical and chemical conditions bacteria form large agglomerates and become part of the porous medium. Bacteria often become lodged with the surface of water-bearing rocks thus staying in place and not travelling together with the water medium. It is adherent bacteria that cause the major transformations of the chemical composition of water-bearing rocks [9-10]. The most important biochemical processes in the ground waters that involve microorganisms include the production of methane, sulfate reduction, biogenic generation of CO_2, nitrification and denitrification [11-15]. Mineral waters can also contain contaminant pathogenic and potentially pathogenic microorganisms that can survive for long periods of time and cause a serious threat to human health [16]. Since mineral waters are actively used in balneology, they have to be monitored both for their chemical and microbiological composition in order to reveal the natural and disease-causing organisms. Study of the microorganisms residing in ground waters is required for predicting possible biochemical processes in the water-rock system that involve microorganisms as well as for assessing the sanitary and microbiological quality of ground drinking waters.

Since there is no data on the qualitative composition of microorganisms in the ground mineral carbon waters of the Russian Far East, the present research is aimed at studying the distribution, structure and quantity of various physiological groups of bacteria and indicative microorganisms, as well as at defining their habitat conditions.

2. Material and methods

2.1. Sampling

Carbon mineral waters of four major deposits – Lastochka, Shmakovskoe (Medvezhii and Vostochno-Ussurskii locations), Pokrovskoe and Nizhnie Luzhki have been selected as objects of research (fig.1). The deposits are located in various hydrogeological conditions and have different types of water-bearing rocks and geochemical types of waters. Water samples were collected in conditions of sterility in July 2014 into 1500 ml sterile glass vials, in triplicate.

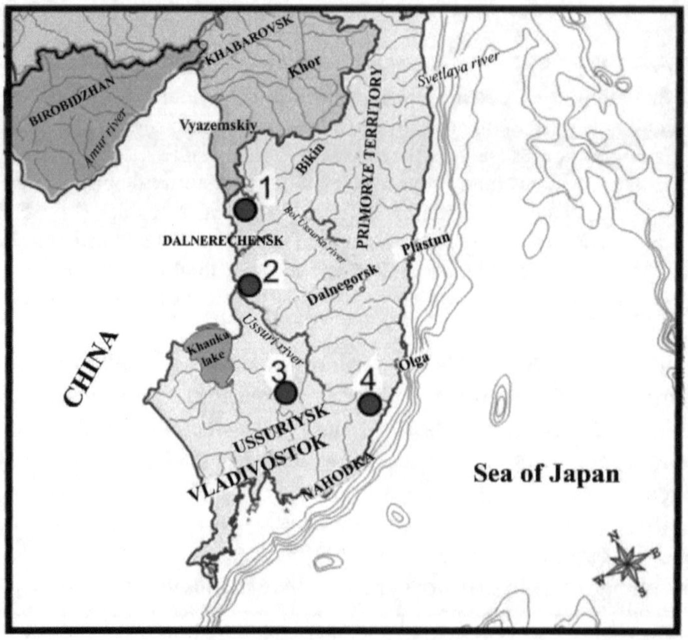

Figure 1. Location of the surveyed deposits of carbon mineral waters in the Russian Far East (1 – Lastochka; 2- Shmakovskoe; 3 - Pokrovskoe; 4 – Nizhnie Luzhki).

2.2. Sample preparation and analysis

Water samples for analysis of cations and sulfates were filtered through membrane filters (cellulose nitrate, 0.45 μm, Sartorius) at the site of sampling for removing the suspended matter and were subsequently acidified with nitric acid. Water samples for analysis of anions were also filtered through membrane filters (cellulose nitrate, 0.45 μm, Sartorius) and then placed into plastic vials (without acidification). The cations and anions were separated via the method of liquid ion chromatography (HPLC-10AVp, SHIMADZU); the microelements were analyzed via plasma optical emission spectrometry (ICP-AES, Plasmaquant-110) and inductively-coupled plasma mass spectrometry (ICP-MS, Agilent 7500c).

The quantity of bacteria of different physiological groups was calculated via the method of limiting dilution in different selection media [17]. Mac-Credi tables were used to identify the most probable number of bacteria cells. The Koch's technique was used to count the colony-forming types of indicative microorganisms [18]. Saprophytic bacteria (QMA&OAMO – quantity of mesophyll aerobic and optional-anaerobic microorganisms) were seeded on meat-peptone agar with incubation at 30°C for 48-72 hours. Total coliforms and thermotolerant coliforms were identified in accordance with methodological guidelines № 96/225-1997 via the method of membrane filtration [19]. The bacteria *Pseudomonas aeruginosa* were separated in the Endo selective and differential medium with their prior

IOP Conf. Series: Earth and Environmental Science **115** (2018) 012033 doi:10.1088/1755-1315/115/1/012033

accumulation in the lactose peptone broth [20]. Bergey's manual of bacteria classification was used to identify the bacteria.

3. Results and discussion

3.1. Chemical composition of carbon mineral waters

In the Lastochka mineral waters, sodium and calcium are the prevailing cations; nitrate and sulfate ions dominate among the anions, and ferric iron and strontium are the major microelements (table 1). Carbon waters of the Shmakovskoe deposit show prevalence of calcium and magnesium among the cations, sulfate, chloride and nitrate ions among the anions, and ferric iron, strontium and manganese among the microelements (table 1). In the Pokrovskoe and Nizhnie Luzhki deposits the prevailing cations include calcium and sodium, the dominating anion is a nitrate ion; the carbon waters are significantly enriched in strontium, ferric iron and manganese (table 1).

Table 1. Content of cations, anions and microelements in the deposits of cold carbon mineral waters of the Russian Far East

Components	Deposits of carbon mineral waters in the Russian Far East				
	Lastochka	Shmakovskoe		Pokrovskoe	Nizhnie Luzhki
		Medvezhii	Vostochno-Ussurskii		
Cations, anions, mg/l					
F^-	0.80	0.61	0.64	0.20	0.35
Cl^-	1.71	1.09	1.18	0.68	1.32
NO_3^-	4.32	0.14	1.72	10.2	16.33
SO_4^-	2.50	3.09	2.40	0.78	0.64
Li^+	1.18	0.09	0.12	0.23	0.60
Na^+	264.0	21.9	32.7	53.1	67.2
NH_4^+	0	0.28	0.26	0.52	1.10
K^+	29.1	3.40	2.52	3.80	2.96
Ca^{2+}	74.7	139	127.0	70.9	110.0
Mg^{2+}	60.8	54.6	48.6	6.13	10.3
Microelements, µg/l					
Be	1.52	2.79	6.83	6.37	10.3
Al	6.0	157.0	100.0	93.0	19.3
Cr	1.12	0.495	0.154	1.16	7.16
Fe	7544.0	25030.0	2232.0	1073.0	2137.0
Mn	214.0	427.0	732.0	539.0	1640.0
Cu	1.45	3.6	6.18	0.271	1.55
Sr	5332.0	487.0	620.0	1249.0	3408.0

3.2. Counts of bacteria functional groups in carbon waters

Each source of mineral waters has a specific set of prevailing physiological groups of microorganisms that can be attributed to the difference in the chemical composition of ground mineral waters as well as to the different composition of water-bearing rocks.

Physiological groups of microorganisms in the carbon waters of Lastochka deposit are not numerous on average ranging from 0 (denitrifiers) to 2.5×10^3 c/ml; as this takes place, the microbiocenosis includes a prevailing number of chemolithoautotrophic thionic bacteria *Thiobacillus denitrificans* that can anaerobically oxidize reduced sulfur compounds to elemental sulfur and subsequently to sulfuric acid. Earlier researchers showed the ability of *Thiobacillus denitrificans* to cause anaerobic oxidation of pyrite nanoparticles by using nitrates as electron acceptors [21]. The thionic bacteria activity may cause generation of sulfates in the Lastochka carbon waters by the following reaction equation:

$$5Na_2S_2O_3 + 4O_2 + H_2O \longrightarrow 5\,Na_2SO_4 + H_2SO_4 + 4S. \quad (1)$$

High numbers of thionic bacteria identified only in the Lastochka carbon waters may be caused by the active processing of sulfide minerals (pyrite) that are found in sufficient quantity in the water-bearing rocks of the deposit [22]. The numbers of other ecologo-trophic groups of bacteria in the Lastochka reservoir are quite low, ranging from 0.1×10^2 to 0.4×10^2 c/ml (table 2).

The carbon waters of Medvezhii (Shmakovskoe deposit) are characterized by low total numbers of all physiological groups of microorganisms that comprise 0.6×10^2 c/ml on average. Chemoorganotrophic heterotrophic nitrifiers are more widespread here; they oxidize nitrogen compounds to nitrites, nitrates and byproducts (table 2). The carbon waters of Vostochno-Ussurskii location (Shmakovskoe deposit) contain high numbers of colorless sulfur bacteria (4.5×10^3 c/ml), heterotrophic nitrifiers (1.3×10^3 c/ml) and an insignificant amount of manganese-oxidizing microorganisms (1.4×10^2 c/ml) (table 2). High numbers of sulfur bacteria can probably be attributed to the favorable conditions for propagation of microorganisms (presence of organic substances, oxygen, metallic sulfides and hydrogen sulfide), as well as to the specific hydrogeological structure of ground waters. High amount of heterotrophic nitrifiers found in the Vostochno-Ussurskii spring is indicative of the ammonia and organic nitrogen compounds oxidation processes. Growth of manganese-oxidizing bacteria only in the carbon waters of Vostochno-Ussurskii is probably caused by a higher oxygenation of ground waters (6.15 mg/l), higher concentrations of manganese (732 μg/l) combined with an average amount of organic carbon (3.0 mg/l) that altogether create favorable conditions for bacteria propagation. The identified manganese-oxidizing bacteria were morphologically similar to *Metallogenium*.

The bacterial count of microorganism physiological groups in the Pokrovskoe deposit ranges from 0 to 4.5×10^4 c/ml with iron-oxidizing chemoautotrophs prevailing (table 2). High content of iron-oxidizing microorganisms in the mineral waters and lower concentrations of ferric iron in the ground waters may be indicative of oxidation processes of Fe^{2+} to Fe^{3+} and its subsequent sedimentation as goethite or lepidocrocite:

$$4Fe^{2+} + O_2 + 4H^+ = 4Fe^{3+} + 2H_2O. \quad (2)$$

The isolated iron-oxidizing microorganisms were identified as *Thiobacillus ferooxidans, Leptospirillum, and Siderococcus sp.*

The carbon mineral waters of Nizhnie Luzhki contain prevailing numbers of iron-oxidizing autotrophs (4.5×10^3c/ml), sulfate-reducing microorganisms (2.5×10^3c/ml) and heterotrophic nitrifiers (2.7×10^2c/ml) (table 2). Iron-oxidizing bacteria are morphologically similar to *Siderocapsa sp.* Anaerobic sulfate-reducing bacteria that reduce sulfates to hydrogen sulfide are found in high numbers only in the carbon waters of Nizhnie Luzhki which is probably related to the more favorable conditions for propagation of anaerobic bacteria (lower concentrations of dissolved oxygen (3.7 mg/l), presence of organic substances and sulfates (0.8 mg/l and 0.64 mg/l respectively), as well as to the specific hydrogeological structure of ground waters. The prevailing amount of ammonia-nitrogen (1.10 mg/l) in the ground waters of Nizhnie Luzhki triggers the growth of heterotrophic nitrifiers; however, the low amount of organic carbon (0.8 mg/l) does not let the bacterial count reach high numbers (table 2).

Table 2. Average amounts of different ecologo-trophic groups of bacteria in cold carbon mineral waters of the Russian Far East.

Physiological groups of microorganisms: c/ml	Deposits of carbon mineral waters:				
	Lastochka	Shmakovskoe		Pokrovskoe	Nizhnie Luzhki
		Medvezhii	Vostochno-Ussurskii		
Saprophytes-copiotrophs	0.5×10^2	0	0.9×10^2	0.8×10^2	0.3×10^2
Saprophytes-oligotrophs	0.7×10^2	0.4×10^1	0.3×10^1	0	0
Ammonia-oxidizing bacteria	0.2×10^2	0	0	0	0
Nirtite-oxidizing bacteria	0.4×10^2	0	0	0	0
Denitrifiers	0	0	0.2×10^2	0	0
Nitrogen fixers	0.1×10^2	0	0.2×10^2	0	0.5×10^2
Heterotrophic nitrifiers	0.2×10^2	0.7×10^3	1.3×10^3	0.8×10^2	2.7×10^2
Ammonifiers	0	0	0	0	0
Thionic bacteria	2.5×10^3	0.4×10^2	0	0	0
Sulfate-reducing bacteria	0.2×10^2	0.2×10^2	0	0	2.5×10^2
Colorless sulfur bacteria	0.4×10^2	0.4×10^2	4.5×10^3	0	0
Iron-oxidizing autotrophs	0.4×10^2	0.2×10^2	0	4.5×10^4	4.5×10^3
Iron-oxidizing heterotrophs	0	0	0	0	0
Heterotrophic manganese-oxidizing bacteria	0	0	1.4×10^2	0	0

3.3. *Sanitary-microbiological analysis of carbon mineral waters*

The total count of mesophyll aerobic and optional-anaerobic microorganisms in the carbon mineral waters is not high ranging from 0 CFU/ml (Medvezhii spring, Shmakovskoe deposit) up to 0.9×10^2 CFU/ml/ (Vostochno-Ussurskii spring, Shmakovskoe deposit) with the quantity of bacteria remaining below standard values (table 3). The count of total coliforms varied between $0.2\times10^2\pm0.63$ CFU/100 ml (Lastochka) and $2.1\times10^2\pm5.7$ CFU/100 ml (Nizhnie Luzhki) exceeding the maximum permissible values by 20-210 times that shows an anthropogenic pollution of the ground carbon waters (table 3). Total coliforms were not found in the carbon mineral waters of the Medvezhii spring (Shmakovskoe deposit) that shows the purity of ground waters. Thermotolerant coliform bacteria that are indicative of recent fecal pollution and also show the presence of bacteria *Escherichia coli* were not found in the carbon mineral waters (table 3). Pathogenic bacteria *Pseudomonas aeruginosa* were also absent in the samples of cold carbon ground waters thus proving their purity.

Table 3. Quantity of sanitary-indicative microorganisms in the ground cold carbon waters in the Russian Far East.

Sanitary-indicative microorganisms:	Deposits of carbon mineral waters					Standard[1]
	Lastochka	Shmakovskoe		Pokrovskoe	Nizhnie Luzhki	
		Medvezhii	Vostochno-Ussurskii			
QMA&OAMO*, CFU/ml	$0.5\times10^2\pm$ 1.62	0	$0.9\times10^2\pm2.59$	$0.8\times10^2\pm0.63$	$0.3\times10^2\pm$ 1.45	Not > 1×10^2 CFU/ml
Total coliform bacteria, CFU/100ml	0.2×10^2 ±0.63	0	$0.3\times10^2\pm1.51$	$1.5\times10^2\pm4.2$	2.1×10^2 ±5.7	None in 100 ml
Thermotolerant coliform bacteria, CFU/100ml	0	0	0	0	0	None in 100 ml
Pseudomonas aeruginosa CFU/1000ml	0	0	0	0	0	None in 1000 ml

*QMA&OAMO – quantity of mesophyll aerobic and optional-anaerobic microorganisms.
[1]Standardization in accordance with [19].

4. Conclusions

- The waters that have been studied are characterized by the prevalence of calcium, magnesium and sodium among the cations, sulfate, nitrate and chloride ions among the anions, and ferric iron, strontium and manganese in the microelement composition.
- The carbon mineral waters contain various physiological groups of microorganisms that can transform the physical and chemical composition of mineral waters by interfering with geochemical cycles. However, each source of mineral waters has its specific prevalent ecologo-trophic groups of microorganisms related to the differences in chemical composition of carbon waters and water-bearing rocks as well as to the specific hydrogeological structures of ground waters. The carbon waters of Lastochka, Medvezhii, Vostochno-Ussurskii (Shmakovskoe) showed the prevalence of microorganisms of sulfur and nitrogen cycles; iron-oxidizing chemolitotrophic bacteria were predominant in Pokrovskoe and Nizhnie Luzhki.
- Microbiological indicators of the carbon mineral waters have been evaluated through the analysis of indicator microorganisms. Total coliforms have been found in the ground waters of Lastochka, Vostochno-Ussurskii (Shmakovskoe), Pokrovskoe, Nizhnie Luzhki thus showing the evidence of their anthropogenic pollution. Neither thermotolerant coliform bacteria nor the pathogenic bacteria *Pseudomonas aeroginosa* have been found in the ground waters.

References

[1] Chudaev O V, Kharitonova N A, Chelnokov G A, Bragin I V 2016 Water and mineral resources of Primorsky Krai *Bulletin FEB RAS.* **5** 11-20
[2] Chelnokov G A, Chepkaya N A, Karabtsov A A, *et al* 2006 Geochemistry of carbon mineral waters and water-enclosing rocks of the Lastochka deposit *Russian journal of Pacific geology* **25(3)** 89-97
[3] Kharitonova N A, Chelnokov G A, Vach E A, Goryachev V A 2011 Geochemistry of Nizhniye Luzhki carbonaceous mineral deposit (Primorye) *Russian journal of Pacific geology* **30(1)** 108-118
[4] Chelnokov G A, Kharitonova N A, Bragin I V, Vasileva M 2013 Deuterium, oxygen-18 and tritium in precipitation, surface and groundwater in the Far East of Russia *Procedia earth and planetary science* **7** 151-154

[5] Kharitonova N A, Chelnokov G A, Chudaev O V, *et al* 2017 The fate of rare earth elements and yttrium in groundwater and bedrocks from Sikhote-Alin Ridge, Primorye *Procedia earth and planetary science* **17** 412-415
[6] Flynn T M, Sanford R A, Bethke C M, *et al* 2013 Functional microbial diversity explains groundwater chemistry in a pristine aquifer *BMC microbiology* **13** 146-161
[7] Sirisena K A, Daughney C J, Moreau-Fournier, *et al* 2013 National survey of molecular bacterial diversity of New Zeland groundwater: relationships between biodiversity, groundwater chemistry and aquifer characteristics *FEMS microbiology ecology* **86** 490-504
[8] Chelnokov G A, Kalitina E G, Bragin I V, Kharitonova N A 2014 Hydrochemistry and genesis of thermal waters of the Goryachii Klyuch spring in Primorskii Krai (Far East of Russia) *Russian journal of Pacific geology* **8(6)** 475-488
[9] Lehmar R M, Roberto E F, Earley D, *et al* 2001 Attached and unattached bacterial communities in a 120-meter corehole in an acidic, crystalline rock aquifer *Applied and environmental microbiology* **67** 2095-2106
[10] Griebler C, Mindi B, Slezak D, Geiger-Kaiser M 2002 Distribution patterns of attached and suspended bacteria in pristine and contaminated shallow aquifers studied with in situ sediment exposure microcosm *Aquatic microbial ecology* **28** 117-129
[11] Grundger F, Jimenez N, Thielemann T, *et al* 2015 Microbial methan formation in deep aquifers of a coal-bearing sedimentary basin, Germany *Frontiers in microbiology* **6** 200
[12] Miao Z, Brusseau M L, Carroll K C, *et al* 2012 Sulfate reduction in groundwater: characterization and applications for remediation *Environ. Geochem. Health.* **34(4)** 539-550
[13] Moore T A, Xing Y, Lazenby B, et al 2011 Prevalence of anaerobic ammonium-oxidizing bacteria in contaminated groundwater *Environ. Sci. Technol.* **45(17)** 7217-7225
[14] Soares M I 2000 Biological denitrification of groundwater *Water, air and soil pollution* **123(1-4)** 183-193
[15] Vodyanitskii Y N 2016 Biochemical processes in soil and groundwater contaminated by leachates from municipal landfills (mini review) *Annals of agrarian science* **14(3)** 249-256
[16] Aditi F Y, Rahman S S, Hossain M M 2017 A study on the microbiological status of mineral drinking water *Open microbiology journal* **11** 31-44
[17] Kuznetsov S I, Dubinina G A 1989 Methods of studying aquatic microorganisms Moscow: Nauka 285 p.
[18] Egorov N S 1995 Guide to practical exercises on microbiology Moscow: Moscow State University 224 p.
[19] Methodical manual No. 96/225 1997 Control of the quality and safety of mineral waters by chemical and microbiological indicators Moscow Ministry of Health of Russia 18 p
[20] Methodical instructions 1986 Detection and identification of Pseudomonas aeruginosa in environmental objects (food, water, sewage) Moscow publisher of standards 22 p
[21] Bosch J, Lee K Y, Jordan G, *et al* 2012 Anaerobic, nitrate-dependent oxidation of pyrite nanoparticles by Thiobacillus denitrificans *Environmental science and technology* **46** 2095-2101
[22] Kharitonova N A, Chelnokov G A, Karabtsov A A, Kiselev V I 2007 Geochemistry of Na-HCO_3 groundwater and sedimentary bedrocks from the central part of Sikhote-Aline - 44 - mountain region (Far East of Russia) *Applied geochemistry* **22** 1764-1776

IOP Conf. Series: Earth and Environmental Science **115** (2018) 012034 doi:10.1088/1755-1315/115/1/012034

Assessment of the dynamics of urbanized areas by remote sensing

S A Yeprintsev[a], M A Klevtsova[b], L A Lepeshkina[c], S V Shekoyan[d], A A Voronin[e]

Voronezh State University,
Universitetskaya pl. 1, Voronezh, Russia, 394018

E-mail: [a]esa81@mail.ru, [b]klevtcova@geogr.vsu., [c]rulilez1980@mail.ru, [d]shekoyan.syuzanna@mail.ru, [e]voronin@bio.vsu.ru

Abstract. This research looks at the results of a study of spatial ecological zoning of urban territories using the NDVI-analysis of actual multi-channel satellite images from Landsat-7 and Landsat-8 in the Voronezh region for the period 2001 to 2016. The results obtained in the course of interpretation of space images and processing of statistical information compiled in the GIS environment "Ecology of cities Voronezh region" on the basis of which carried out a comprehensive ecological zoning of the studied urbanized areas. The obtained data on the spatial classification of urban and suburban areas, the peculiarities of the dynamics of weakly and strongly anthropogenically territories, hydrological features and vegetation.

1. Introduction

The increasing pace of urbanization and industrialization in addition to positive effects lead to an increase in anthropogenic impact on the environment. Therefore, the magnitude of environmental risk, and reduced social comfort to the population of large cities. In Russia and most other developed countries this problem is particularly acute since the mid-twentieth century with a high rate of development of industry, oil and petrochemical industries, due to the increased power of thermal power plants, vehicles, etc.

The increasing geochemical contamination of the environment in the population of many large industrial cities of the manifest of environment-related diseases. The attention of environmentalists focused on the study of mechanisms of formation of zones of technogenic pollution and the search of effective ways of improvement of the urban environment [1-3].

A natural frame around the urban settlements is a key determinant of environmental comfort in urbanized areas. Operational spatial assessment of the ecological framework and anthropogenically territories is conducted according to remote sensing [4-6]. This method allows to assess the dynamics of urbanized areas and environmental quality.

The application of satellite imagery data greatly enhances the study of the environmental factors. The development of remote sensing techniques has greatly simplified the process of mapping land and water resources, soil, forests, agricultural crops and urban infrastructure, assessment of crop and much more.

In Russia and other countries of methods of remote sensing are highly relevant and are widely used for assessing spatial dynamics of urban territories. Using geographic information systems allow us to efficiently develop ecological basis for the organization of urbanized areas [7-10].

2. Equipment and devices used in studies

Classification of the studied spatial objects performed by means of NDVI (Normalized Difference Vegetation Index) of relevant multichannel space images Landsat-7 and Landsat-8. Investigated spatial correlations of weakly and strongly anthropogenically territories, hydrologic objects, green space (natural frame). The first slice of data was obtained by satellite images 2001 for the cities and towns of Voronezh, Liski, Rossosh', Pavlovsk, Borisoglebsk (Voronezh region) and their 10-kilometer buffer zones. Another slice of data was obtained for the same areas from satellite imagery of 2016. Thus was obtained the assessment of changes in different areas over a fifteen-year period.

IOP Conf. Series: Earth and Environmental Science **115** (2018) 012034 doi:10.1088/1755-1315/115/1/012034

Spatial classification of objects according to the method of NDVI cities of Voronezh and Liski was conducted using satellite images: LE71760242001222KIS00 of Landsat 7, August 10, 2001, LC881760242016240LGN00 of Landsat-8, August 16, 2016**;** for towns of Rossosh and Pavlovsk was conducted using satellite images: LE71750252001215SGS00 of Landsat-7, August 3, 2001, LC81750252016233LGN00 of Landsat-8 August 20, 2016; for Borisoglebsk town and its suburban area was conducted using satellite images: LE7174024001224KIS00 of Landsat-7, August 12 2001, LC81740242016242LGN00 of Landsat-8, August 29, 2016.

3. Results and Discussion

The results of the spatial dynamics of urban territories on the example of *Voronezh city* are presented in figures 1 and 2. The analysis of the spatial zoning of Voronezh and its suburban ten kilometer areas (a total of 1246 km2) using NDVI, showed that the majority of the study space (40-50 %) applies to weakly anthropogenically area.

Figure 1. The spatial zoning of Voronezh and its suburban ten kilometer zone (NDVI method by satellite image LE71760242001222KIS00 of Landsat-7, August 10, 2001) color coding is according to table 1

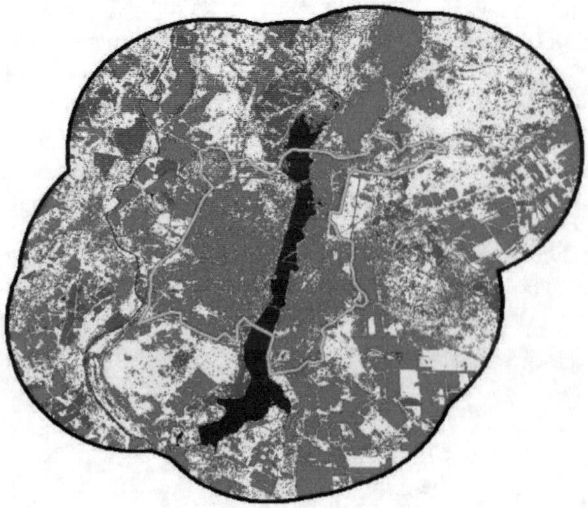

Figure 2. The spatial zoning of Voronezh and its suburban ten kilometer zone (NDVI method by satellite image LC881760242016240LGN00 of Landsat-8, August 16, 2016 color coding is according to table 1

Names and descriptions of the classification units indicated by different colors in figures 1, 2 and are given in table 1.

Table 1. The color symbols in figures the spatial zoning of Voronezh city and its suburbs (NDVI method)

Color	Classification of territory	Description
Blue	Water objects	Rivers, lakes, reservoirs and other hydrological objects
Green	Dense green vegetation	The territory constituting the nature frame – forests, gardens, squares, natural tracts and other lush green vegetation
Yellow	Weakly anthropogenically territories	Open soil, agricultural land, weak green vegetation
Red	Strongly anthropogenically territories	Anthropogenic structures – buildings, roads and other objects. The classification groups of (excluding residential ecological-functional zone) are considered as objects of ecological risk

The main share of weakly anthropogenically territories is formed by the agricultural fields of the Ramonsky, Novousmansky and Semiluksky administrative districts, which adjoin of Voronezh (figures 1, 2). The share of natural frame is 8-10% of total square of the study area. The green areas located mainly in its northern part, which significantly reduces their positive impact on the microclimate of the urban area. It is connected with the fact that movement of air masses over the city happens mainly in the northeast direction [11]. Strongly anthropogenically territories that are considered as objects of ecological risk, primarily located in the central part of the city and in the small urban settlements of the suburban zone (figures 1, 2).

The analysis of the spatial dynamics of the various zones for a 15 year period found little (within the error of the technique) reduction of water objects, increase of 8% strongly anthropogenically territories. Such tendency is caused by active construction of housing estates in city boundaries of Voronezh and adjacent satellite towns – Bobyakovo, Somovo. There was a slight increase in the natural frame (less than 5%) that is associated with the implementation of federal and regional programs for the reconstruction of the forested areas. Weakly antropogenically territory of Voronezh and its suburbs ten-kilometer zone decreased by 10% due to the increase in strongly antropogenically and natural frame.

The analysis of the spatial dynamics of the *town Liski* and its suburban area (a total of 770 km^2) shows a slight reduction in hydrologic objects and an increase of 5% weakly antropogenically territories is caused by active construction of new residential complexes.

Unlike the Voronezh city, the most part of Liski and its suburban zone consists of a strongly anthropogenically territories. It is caused by existence of the large city-forming enterprise – Locomotive depot Liski southeastern railway. During the last 10 years, active construction of new residential and industrial complexes is underway. Over a fifteen-year period, the territory of the natural frame of Liski has increased by more than 10%. Placing green spaces from the windward southeast side of the town contributes to improving the microclimatic conditions of its territory.

The dynamics of urbanized territories according to the data of multi-channel space images in the period 2001-2016 is presented in table 2

Table 2. Dynamics of urbanized areas according to multi-channel space images (2001-2016)

Objects/squares	2001				2016			
	SAT,%	WAT,%	NF,%	HO,%	SAT,%	WAT,%	NF,%	HO,%
Voronezh 1246 км^2	32	55	9	4	40	44	12	4
Liski 770 км^2	41	51	7	1	45	38	16	1
Rossosh' 827 км^2	35	40	24	<1	37	44	19	<1
Pavlovsk 521,15	30	44	25	1	45	35	19	1
Borisoglebsk 639,19	31	54	14	<1	36	34	29	<1

SAT – strongly anthropogenically territories; WAT – weakly anthropogenically territories; NF – natural frame; HO – hydrological objects

Spatial analysis of the territory of the *town Rossosh'* (the center of the chemical industry in the Voronezh region) and its suburban zone (totaling 827 km^2) showed a predominance of weakly anthropogenically territory. It is caused by domination of low floor building and development of homestead economy. In 2001 strongly antropogenically territories were located mainly outside city line. The active construction of multi-storey residential buildings already by 2016 led to an increase in highly anthropogenically territories by 15%.

In 2001, the green structure of Rossosh' was more than 200 km^2, which is almost 25% of total square of town territory and its suburban zone (827 km^2). By 2016 the area of the territory of a natural framework was reduced for 10% that is caused by natural and anthropogenic wildfires during the summer period of 2010. As a result these territories have turned into category a weakly antropogenically. The transfer of air masses north-west direction and features of distribution of green space have a beneficial effect on the microclimate of the town. Spatial dynamics of water objects in the Rossosh town and in its suburban zone are located in the limits of error of the technique.

Spatial analysis of territory of the *town Pavlovsk.* The territory of the settlement from the North to the south is crossed by the highway of federal importance of M-4 Don dividing it into two parts: western and east. In the eastern part of the town there are major industrial and communal enterprises, an array of modern individual buildings and agricultural land. In the western part - blocks of residential and public buildings, forest tracts along the Don River, floodplain areas of Osered' River and agricultural lands.

During the period from 2001 to 2016 strongly antropogenically territory has increased by 1.5 times. Significant changes are observed in urban areas along the M4-Don highway, where roadside infrastructure is actively developing. The weakly anthropogenically territory, allotted in 2001, is transformed into a highly antropogenically territory by 2016. This fact is explained by the increasing role of the automobile route and the increased transport load, the peak of which was observed the day before and during the Winter Olympic Games in Sochi (2014).

Increase in a share of strongly antropogenically territory is observed also in a suburban area of Pavlovsk where mining objects are located. Among them OJSC «Pavlovskgranit» especially actively develops. From 2001 to 2016, the growth of highly anthropogenically territory is due to the reduction of the slightly anthropogenically territory.

It is revealed that a natural framework of Pavlovsk and its suburban zone were reduced by 35 km^2 that makes about 8% of the explored territory. This fact is caused by the strong fires in the summer of 2010 which have led to loss a valuable forests. Territories of Rossosh' and Pavlovsk belong to a steppe zone with an insufficient amount of precipitation, therefore natural restoration of forests happens here extremely slowly and demands development of special reforestation programs. Almost equal ratio of winds of east and western direction and also location of green plantings to the west from the town in a flood plain of the Don River in general positively influence on the climatic features of the settlement.

Spatial analysis of the territory of *Borisoglebsk.* The Borisoglebsk town district is located in a southeast part of Oka-Don plain on confluence of the Vorona River and Khopyor River. The Borisoglebsk and 24 rural settlements make uniform municipal unit – the town Borisoglebsk's district. Almost a quarter of the territory of the district (26 thousand hectares) is occupied by forests, including oak forests. Through the town are the highways Voronezh-Saratov, Moscow-Volgograd and the railway (line Moscow-Volgograd).

Over a fifteen-year period, the area of strongly anthropogenically territories within Borisoglebsk and its suburban zone increased by 5% of the total area, which is a regularity for the urban territories of the developed countries of the world

The square of weakly anthropogenically territory of Borisoglebsk and its suburban zone for a fifteen-year period decreased by 125 km^2 due to an increase in the area of green plantations. The spatial dynamics of water bodies in the study area is within the error of the methodology.

Among the cities and town of Voronezh region Borisoglebsk and its suburban area has the largest a natural frame. Spatial dynamics of different areas over a fifteen-year period shows a significant increase (almost 2 times) of square of green space. It is connected with introduction of the ecological program which includes actions for restoration of the forest lands in the southern part of the city, restriction of access of motor transport to the woods, strict control of stay in forests of the population and other. Despite loss of a part of forest fund as a result of the fires in the summer of 2010, the total area of forests continues to increase.

Spatial location of the natural frame around Borisoglebsk has a beneficial effect on microclimatic indicators and the ecological situation in the explored territory.

4. Conclusion

Spatial zoning of urban areas of Voronezh region by NDVI method shows differentiation of the studied territories by the sizes of a natural frame – from 7 to 30%, weakly anthropogenically –34-55% and strongly anthropogenically –30-45% of territories.

The smallest value of the square a natural frame is noted in the regional center – the Voronezh city that is caused by the social and economic reasons. The price of land and its profitability in Voronezh and its suburban zones is significantly higher than in other studied urban sites. Local business elites actively develop suburban areas, which leads to reduction of forests and natural areas. General town plan (2006) provides for the removal of objects of environmental risk in suburban areas. On the one hand it has positive environmental effect, and from another also leads to reduction of the territory of a natural framework.

The greatest area of green plantings of urbanized territories of Voronezh region is in Borisoglebsk. The arrangement of a natural framework concerning the city is important. Priority is the location of the natural framework from the windward side of the town. In the territory of the Voronezh region, the best way for this indicator is the green zones around Liski and Borisoglebsk.

IOP Conf. Series: Earth and Environmental Science **115** (2018) 012034 doi:10.1088/1755-1315/115/1/012034

For all the studied urban and suburban areas showed a stable growth trend strongly anthropogenically territories, which is a feature of rapidly developing cities of Russia.

The studies were carried out with the financial support of the RFFI grant (project 16-45-360284)

References

[1] Zubareva O N Skripal'shchikova L N Greshilova N V Kharuk V I 2003 Zoning of landscapes exposed to technologenic emisssions from the norilsk mining and smelting works (Russian Journal of Ecology vol 34 № 6) pp 375-380

[2] Ashfaq A Sharma P 2013 Environmental effects of air pollution and application of engineered methods to combat the problem (Journal of Industrial Pollution Control 29(1)) pp 25-28

[3] Qiu H Ho K F Pun VC (...) Yu I T S Tian L 2015 Air pollution and mortality: Effect modification by personal characteristics and specific cause of death in a case-only study (Environmental Pollution 199) pp 192-197

[4] Stossel Z Kissinger M Meir A 2015 Assessing the state of environmental quality in cities - A multi-component urban performance (EMCUP) index (Environmental Pollution 206) pp 679-687

[5] Krishnakumar M Ganesh Kumar B 2012 Landfill requisite and optimal allocation for municipal solid waste disposal in madurai city - An assessment of urban environmental issues using gis modeling for sustainable development (Pollution Research 31(3)) pp 481-486

[6] Javed W Murtaza G Ahmad H R 2015 A GIS based study on air quality assessment along the roadside environment in Faisalabad, Pakistan (Pollution Research 34(1)) pp 1-7

[7] Gualtier G Crisci A Toscano P Gioli B Tartaglia M 2015 A statistical model to assess air quality levels at urban sites (Water Air and Soil Pollution 226(12))

[8] Polezhaev A N 2009 Vegetation of northern Russian Far East in geographic information systems (Russian Journal of Ecology vol 40 №3) pp 166-171

[9] Kravtsova V I Loshkareva A R 2013 Dynamics of vegetation in the tundra-taiga ecotone one the Kola peninsula depending of climate fluctuations (Russian Journal of Ecology vol 44 № 4) pp 303-311

[10] Li W Saphores J D M Gillespie T W 2015 A comparison of the economic benefits of urban green spaces estimated with NDVI and with high-resolution land cover data (Landscape and Urban Planning 133) pp105-117.

[11] Yeprintsev S A Kurolap S A Komov I V Minnikov I V 2013 Monitoring of factors of ecological safety of urbanized territories' population (by example of settlements of Voronezh Region) (Life Science Journal Volume 10 - Special Issue 12) pp 846-848.

Study of ants as bioindicators of industrial pollution in Kemerovo Region, Russia

S V Blinova[1] and T I Dobrydina[2]

[1] Kemerovo State University, Department of Ecology and Nature Management, 650000, 6 Krasnaya Street, Kemerovo, Russian Federation
[2] Kemerovo State University, Department of Foreign Languages in Professional Communication, 650000, 6 Krasnaya Street, Kemerovo, Russian Federation

E-mail: sv_blinova@mail.ru

Abstract. The myrmecocomplexes of five industrial cities in the territory of Kuzbass, Russia, were studied. The general trends in the reaction of ants to the pollution by gaseous (chemical industry) and solid (coal and cement dust, lead and zinc waste and metallurgical industry) emissions were revealed. It was found out that the proportion of species of subfamilies, species richness and density of the settlement can be used as bioindicators. The predominance of L. niger and M. rubra nests in myrmecocomplexes immediately indicates a high degree of man-made impact.

1. Introduction

Ants are one of the essential components of most terrestrial ecosystems. Settled in urbanized areas, they are exposed to many factors of anthropogenic and recreational nature. There is an opinion that urban conditions are more severe for the life of ants: only some species settle and survive [1]. Therefore, the myrmecocomplexes of cities differ from natural biocenoses by poor species diversity [2-3] and the presence of 1-2 mass widespread species [4]. The abundance of relatively rare species often increases at a time when the number of nests of background species decreases [5].

To adapt the urban conditions of existence, ants are noticed to go into the soil [6], where they often create multi-species communities, and even on anthropogenically disturbed sites [7].

Emissions of industrial enterprises are one of the main types of human impact on the environment. According to the observations [8-9], industrial pollution leads to limiting the number of species, reducing the density of nests and their richness. However, the reaction of ants to man-caused impacts can be different. Thus, the diversity of dominant species and their quantity increase as they approach the source of emissions of sulfur production [10]. The domination of ants among predators – geobionts is detected in the area of the sulfur basin [11]. In the immediate vicinity of the nitrogen complex, no ants are recorded [11].

In the areas of industrial pollution, the spatial structure of ants' populations changes, the proportion of larvae and pupae in families increases, the mass decreases [9], the chemical composition of the body (including the accumulation of various elements in the body of ants) changes [12-13].

Close to the Finnish metallurgical industry, a decrease in the relative abundance of two of the five identified species Formica s.str was recorded when compared with the areas not exposed to the metallurgical industry. The volume of nesting hills in the contaminated area is 34% lower than in relatively clean areas [14]. In this case, the morphological features of Formica s.str. do not change [14].

In the Czech Republic, on the dumps formed after the extraction of brown coal, the same kinds of ants as in natural cenoses were found. However, a number of species were recorded only on the dumps and not found in natural cenoses [15]. In Italy, the representation of the individual genera Formicidae and the number of forest cold-hearted species vary on the coal dumps, which were formed in different times [16].

According to Puszkar T. [17] and Majer J.D. [18], Formicidae appear to be an informative component of land ecosystems and can serve as indicators of habitat conditions thanks to their high abundance, species diversity, complex structure of communities and stability in habitat.

It is shown that the ants' communities can be used as of the environmental consequences of land use in mangrove forest systems [19] and post-fire regeneration [20] in Brazil. Yamaguchi T. [3] suggests using Messor aciculatus as bioindicators of the changes in soil in the cities of Japan.

The example of Italy shows [21] that indicators of species richness and abundance of ants can act as bioindicators of the rehabilitation / disturbance of various biotopes. The other authors suggest using a species composition to control the development of agrocenoses. At the same time, it is pointed out that for each locality there should exist its own set of indicator species [22].

Obviously, the bioindicator for each locality comes from the specific environmental conditions and available biodiversity. Therefore, the purpose of this work is to identify possible indicators in the characteristics of myrmecocomplexes of industrial cities.

2. Methods and materials

The studies were conducted in the territory of Kuznetsk-Salair mountain region of Altai-Sayan mountain country in the administrative boundaries of Kemerovo Region (Kuzbass), Russia. The region is characterized by a high degree of the environment transformation as a result of mining and processing of minerals, for example, coal, lead-zinc rock, etc. The materials were collected in relatively clean areas, as control data, and in 5 cities with well-developed industry (Table 1).

Table 1. Some characteristics of the cities studied.

city	dominant industry	dominant emissions
Kemerovo	chemical	gaseous emissions
Prokopyevsk	mining coal	coal dust
Topki	cement production	lime (cement) dust
Novokuznetsk	metallurgical	integrated (solid and gaseous)
Salair	mining	solid

In each studied city, there were identified model areas, located along a transect passing through opposite the wind rose along the gradient of pollution reduction. The model area (1), located in close proximity to the sources of emissions of the main pollutants, is taken for the zero mark. As the distribution of solid and gaseous pollutants occurs at different distances, the model area (2) with medium impact is located for cities with gaseous emissions at a distance of 5.1-10 km from the emission sources, for solid ones – 1.31-6 km; the model area (3) with a weak degree of impact, respectively, 10.1 and further, 6.1 and further. Similar biotopes, located 30-45 km from the city limits opposite the wind rose, were served as the control. At each area, 6-15 test sites 5x5 m were laid. All the selected areas are grass-cereals meadows.

All the materials were conducted in 1996–2015 by standard myrmecological methods. 10-15 working individuals were taken from the nest for faunistic studies. Quantitative counting of the nests was carried out on the routes and test sites measuring 5x5 m. The sites were inspected and, if necessary, the soil was dug out. The nest density, calculated for one test site, was used as an indicator of abundance. All received materials were processed in Microsoft Excel 2016 and Statistica v.6.0. For numerical data, the mean values for the study time ± standard deviation are given.

3.Results and discussion

The conducted studies made it possible to reveal the changes in the taxonomic and ecological structure of myrmecocomplexes in industrial contaminated areas in comparison with relatively clean zones. The general tendencies on all studied areas were revealed. Thus, a decrease in species richness, which is associated with a change in the taxonomic structure of the myrmecocomplexes, was recorded: most of the myrmecocomplexes included Formicinae and Myrmicinae. In the most contaminated areas, a decrease in the richness of Myrmicinae subfamily species, compared to natural cenoses, was observed. The contribution of subfamilies to the faunistic structure dramatically changes when exposed to solid and complex pollutants. For example, in the areas, affected by complex industrial emissions, there are 16% Myrmicinae species and 84% Formicinae species in the most polluted zone; and, respectively, 31% and 69% in natural ecosystems. In the areas, affected by gaseous emissions, the contribution of subfamilies to faunal structure is less expressed, but in any case the richness of Myrmicinae in the affected area is always lower than in the control: there are 29.2% Myrmicinae subfamily species and 70.8% Formicinae; and respectively, 34.2% and 65.8% in natural ecosystems.

However, the proportion of the Formicidae subfamilies' species in myrmecocomplexes differs not only in general comparing the territory of the city and the control. The ratio of the species of the ants' subfamilies to the model areas located at different distances from the sources of industrial emissions is changing. Thus, in areas strongly effected by solid and complex industrial emissions, the minimum proportion of ants of the subfamily Myrmicinae (zero) is recorded. At a distance from the sources of pollution, the richness of Myrmicinae species increases, reaching a maximum in the relatively clean zone of the city (on average 40%). In the control, the richness of the species somewhat decreases again (an average of 31%). According to our observations, the absence of nests of representatives of the subfamily Myrmicinae in model areas of meadows in the conditions of Kuznetsk-Salair mountain region indicates a strong degree of man-made impact on them.

In addition, an inverse correlation between the gross industrial emissions and the number of species of the Myrmicinae subfamily ($r = -0.82$ at $p < 0.05$) was detected. In years with a relatively high level of environmental pollution, the nests of this subfamily ants were not recorded near the sources of industrial emissions (Figure. 1). With a decrease in industrial emissions due to a fall in the production capacity of enterprises, in 2–5 years, individual nests of Myrmicinae were found in the model areas in the zone of strong impact. With prevailing solid emissions, Tetramorium caespitum L. is first recorded, with gaseous – Myrmica rubra L. With relatively low stable volumes of industrial emissions, during seven years, one to three Myrmicinae species appear in the model areas.

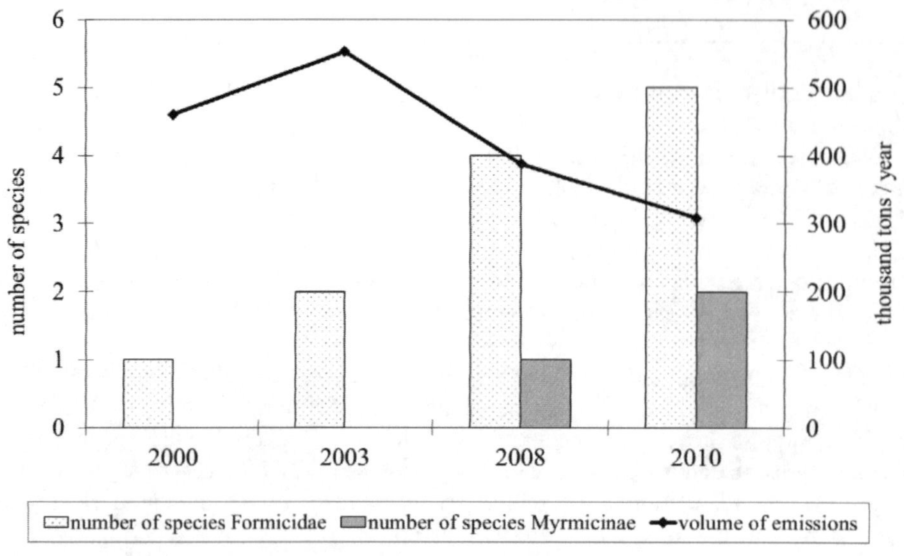

Figure 1. The number of the Formicidae near the source of emissions and the volume of complex industrial emissions in different years; designation of areas is in the text.

The species richness of the myrmecocomplexes of the studied areas can also act as an indicator of pollution of meadows. So, a decrease in the species richness of ants is observed at any type and level of pollution of model areas. Near the emission sources, 1–3 species of Formicidae were recorded. With a decrease in the degree of impact, the species richness of the myrmecocomplexes increases: in the zone of weak impact, there are on average 5.3 ± 2.1 species; in the control their number reaches 8.1 ± 1.2. Significant differences (at $p < 0.01$) of the species richness of Formicidae in the zone of strong man-caused impact and the control zone were revealed.

The change in the density of ants' settlements on grass-cereals meadows located along the gradient of pollution reduction can be used as another indicator. Two main trends in the density of species were identified. The first is quasi-adaptive: the species reach a maximum density at some distance from the emission sources; the density again decreases somewhat to the control (Figure 2). Such a change in density is characteristic of areas exposed to gaseous emissions and coal dust emissions. In the second case (cement dust, lead-zinc production waste, and complex emissions), the density changes according to the non-adaptive type: the minimum is recorded near the emission sources, the maximum is recorded in the control (Figure 3). However, in any case, the density of the nests in the zone of strong impact of industrial emissions is always significantly lower (at $p < 0.01$) than in the zone of weak impact.

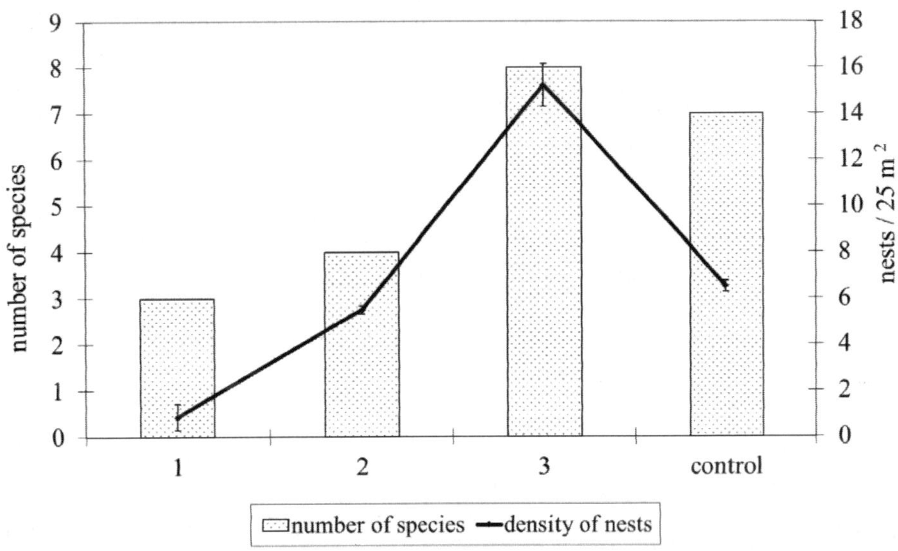

Figure 2. The density of nests and the number of ants' species on the areas located at different distances from coal dust sources, on average, without taking into account Formica s.str nests; designation of areas is in the text.

Representatives of different genera react differently to man-caused impacts. The Lasius genus is the most responsive: a direct correlation was found between the degree of pollution and the quantity of Lasius nests in the myrmecocomplex ($r = 0.78$ at $p < 0.05$). Lasius niger L. is urbophilic and the most flexible species, which is noted by the russian researchers [5-7]. So, in a zone of strong impact on the model areas, a mono-species settlement of L. niger was recorded. With increasing distance from the pollution sources or industrial emissions, the other species of the genus Lasius (L. flavus Fabr., L. alienus Först., L. platythorax Seifert) appear. However, in general, with a decrease in the degree of influence, the proportion of Lasius genus nests decreases from 100% in the area of strong pollution to 25% in the control.

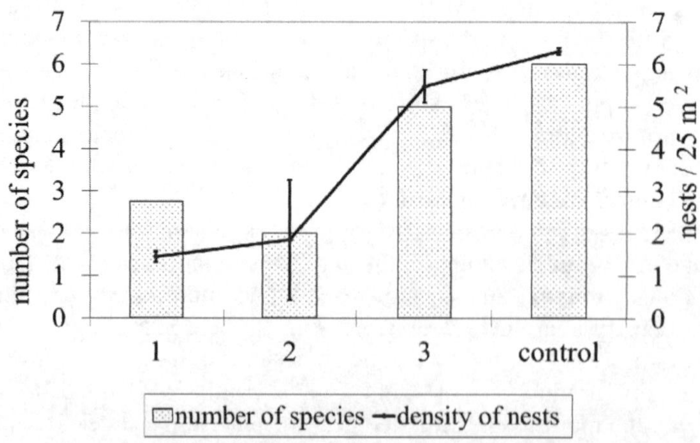

Figure 3. The density of nests and the number of ants' species on areas located at different distances from the sources of complex emissions, on average; designation of areas is in the text.

The genus Myrmica ants exhibit a non-adaptive response to solid emissions and quasi-adaptive – to gaseous. In the most polluted areas, the total density of the settlement of this genus does not exceed 0.27 nests / 25 m^2. With a reduction in industrial emissions, the density of the Myrmica nests increases. For example, in the period from 2004 to 2010, in the areas exposed to complex industrial emissions, the reduction of emissions from 457 to 308.3 thousand tons was noted; the density of Myrmica settlements increased from 0 to 0.1 nests / 25 m^2.

According to our observations, the ants of M. rubra tend to the areas exposed to anthropogenic influence. In the most polluted area (near the sources of man-made pollution), the other species of this genus were not found. In natural cenoses, the density of M. rubra nests is minimal. It is interesting, that on the dumps formed as a result of the extraction of brown coal in the Czech Republic, the nests of Myrmica rubra were detected only in natural cenoses [16]. Probably, such differences are due to the difference in climatic conditions and competition between species.

For the Formica genus, an inverse correlation ($r = -0.68$ at $p < 0.05$) was found between the degree of pollution and species richness. However, from the genus Formica, the species of the subgenus Serviformica are adapted to live in the meadows in the area of industrial emissions impact (F. fusca L. and F. cunicularia Latr. settle in the zone of medium influence). Formica ants. str. (F. rufa L., F.pratensis Retz., F. aquilonia Yarr., F. polyctena Först.) were recorded only in the area of weak impact and in the control.

4. Conclusion

Thus, the conducted studies of myrmecocomplexes under the impact of solid (coal and cement dust, lead and zinc waste and metallurgical industry) and gaseous (chemical industry) emissions have revealed indicators that can be used for bioindicative evaluation of the degree of anthropogenic pollution. Such parameters are:

1) The ratio of the number of species (proportion) of the Myrmicinae and Formicinae subfamilies: an increase in the proportion of the former indicates a decrease in the degree of impact.

2) Species richness and density of settlement: with an increase in the degree of pollution, a decrease in these parameters is observed until complete absence. At the same time, a quasi-adaptive reaction is observed for gaseous emissions, while for solid emissions it is not adaptive.

3) Reactions of individual genera: with increasing distance from the sources of man-made pollution, species of the genus Lasius appear first, and then – the representatives of the genera Myrmica, Formica (subgenus Serviformica) and Tetramorium.

4) Ants' nests of L. niger and M. rubra: a decrease in the proportion of these species in the myrmecocomplex indicates a decrease in the degree of man-made impact.

References

[1] Vepsalainen K and Wourenrinne H 1978 *Mem. zool.* **29** 191–202
[2] Majer J D and Brown K R 1987 *J. R. Soc. West. Aust.* **69** 13–17
[3] Yamaguchi T 2005 *Entomological Science* **8** 1 17–25
[4] Pisarski B, Czechowski W 1978 *Mem. zool.* **29** 109–128
[5] Antonov I A 2013 *Russian Journal of Ecology* **44** 6 523–526
[6] Malozemova L A and Malozemov Y A 1999 *Russian Journal of Ecology* **30** 4 283–286
[7] Blinova S V 2008 *Russian Journal of Ecology*. **39** 2 148–150
[8] Petal J M 1978 *Mem. zool.* **29** 99–108
[9] Petal J M 1980 *Pol. Ecol. Stud. (PRL)* **6** 4 665–672
[10] Puszkar T 1979 *Bull. Acad. Pol. Sci. Ser. sci. biol.* **27** 6 481–485
[11] Puszkar T 1982 *Ann. UM*CS. 1982 **37** 9 105–116
[12] MacKay W P, Mena R, Gardea J and Pingatore N 1997 *J. Kansas Entomol. Soc.* **70** (4) 329–334
[13] Belskaya E, Gilev A and Belskii E 2017 *Environmental Science and Pollution Research* **24** 11 10768–10777
[14] Eeva T, Sorvari J and Koivunen V 2004 *Environmental Pollution* **132** 3 533–539
[15] Holec M and Frouz J 2005 *Pedobiologia* **49** 4 345–357
[16] Ottonetti L, Tucci L and Santini G 2006 *Restoration Ecology* **14** 1 60–66
[17] Puszkar T 1978 *Mem. Zool.* **29** 129–142
[18] Majer J D 1983 *Env. Man.* **7** 4 375–383
[19] Delabie J H, Paim V R, Do Nascimento I C, Campiolo S and Mariano C S 2006 *Neotrop. Entomol.* **35** 5 602–615
[20] Teixeira M C, Schcimento J H, Nascimento J T, Louzada J N C 2005 *Sociobiology*. **45** 3 631–641
[21] Ottonetti L, Tucci L and Santini G 2004 *Redia* **87** 199–202
[22] Castracani C, Piotti A and Grasso D A 2007 *Redia* **90** 67–70

The story of the creation and monitoring of the Russian Desman (Desmana moschata L.) population reintroduced of in the Kerzhenets river floodplain in the Nizhny Novgorod region

S V Bakka[1, a], **N Yu Kiseleva**[2], **I I Pankratov**[2], **I A Tarasov**[2] **and P M Shukov**[2]

[1]State Nature Reserve "Nurgush", Kirov, Lenina str, 129 a, Kirov region, 610000,Russia
[2]Minin Nizhny Novgorod State Pedagogical University, Nizhny Novgorod, ul.Uljanova, 1, 603950, Russia

E-mail: [a] sopr@dront.ru

Abstract. The article summarizes the results of creating the reintroduced population of the Russian desman (Desmana moschata L.) in the Nizhny Novgorod region in the floodplain of the Kerzhenets river (the left tributary of the Volga) and monitoring of its status in 2005-2017. In 2001-2002, a total of 51 individuals were released in the Kerzhenskiy State Nature Reserve. In subsequent years, the desman iinhabited the floodplain of the Kerzhenets river 20 km upstream and 60 km downstream of the river. The number of reintroduced population was 35-40 individuals in 2012, 17 individuals – in 2013. Probably from 30 to 50% of the reintroduced population of desmans inhabit the territory of the Kerzhenskiy State Nature Reserve. Population numbers of desmans in the Reserve varied from 25 individuals in 2005 to 3 in 2015. The positive population trend was recored in 2016-17. Also the paper discusses the limiting factors, the relationship between desmans and muskrats. Now the number of reintroduced population is at a critically low level. However, it is essential for conservation of this endangered species. Recommendations for continued monitoring of the desman status in the valley of the Kerzhenets river are presented.

1. Introduction

The Russian desman is a relic of the tertiary period and endemic to Eastern Europe. The species is listed in The IUCN Red List of Threatened Species (Category Endangered) [1], in the Red Data Book of Russia (category 2 – species decreasing in numbers), in the Red Data Book of the Nizhny Novgorod region (category B – endangered species with rapidly declining numbers). A great reduction in the species number in different parts of the home range in the early of 21st century was noted by many specialists [2-6]. In the XIX century in the Nizhny Novgorod region, as well as throughout European Russia, the desman number was great; it was a common game species. At the turn of the millennia in the region, the species has remained mainly in the floodplain of the Oka river, where its numbers have decreased rapidly. In the floodplain of the Kerzhenets river (left tributary of the Volga), the species has extinct by the early 1990s [7]. The State Nature Reserve "Kerzhensky" was established in the Nizhny Novgorod region in 1993, the Russian desman has become one of the model species,which was selected in the Kerzhenskiy State Nature Reserve for the implementation of the programme of recovery of extinct species and study the problems associated with this process. This article summarizes the results of reintroducing the desman population in the floodplain Kerzhenets and monitoring of its status in 2005-2017.

According to the "Chronicles of nature" of the Kerzhenskiy State Nature Reserve we present the information about the catching and release of animals in wild to reintroduce the desman population [8]. Monitoring of its number and distribution in the Kerzhenets flood plain was carried out from 2005 to

2017.Census of desman was conducted in autumn according to the standard technique [9, 10]. Terms and the volume of surveys are presented in Table 1.

Table 1. Terms and the volume of surveys.

Year	Surveyed waterbodies		Length of routes, km	
	in the Kerzhenets Reserve	In adjacent territories	in the Kerzhenets Reserve	In adjacent territories
2005	14	-	22.82	-
2006	41	-	52.58	-
2012	32	35	51.52	51.31
2013	59	59	66.54	105.87
2015	57	-	66.85	-
2016	19	-	34.46	-
2017	20	-	36.84	-

The first assessment of success rates of the desman recovery we carried out on the territory of the KerzhenskiyReservein 2005-2006. The Kerzhenets river floodland within the boundaries of the reserve was recognised as a model territory where the surveys were conducted regularly. According to the results of these surveys, we estimate the desman population trend. In 2012-13, we surveyed the whole valley of the lower reaches of the Kerzhenetsriverto determine the borders of the reintroducing population of desman.

2. Results and Discussion

2.1. Reintroduction of the Russian Desman

Activities aimed at the re-acclimatization of the Russian desman were carried out in the Kerzhenskiy State Nature Reserve in 2001-02. A total of 51 animals was released in nine floodplain lakes of the Reserve (Table 2). Animals were caught in the floodplain of the Oka river in Vacha, Vyksa and Pavlovo districts of the Nizhny Novgorod region. Locations of catching and release of animals have shown in Figure 1.

Table 2. The number and structure of the group of desmans, which was released in the Kerzhenets river valley

Year of release	Number of released individuals			
	total	♂	♀	juv
2001	11	3	3	5
2002	40	16	12	12
Total:	**51**	**19**	**15**	**17**

In 2001, 11 animals were released into the Krasnyy Yar lake. Some of them successfully overwintered. Desmans were recorded in spring and summer 2002, both at the site of release and in the neighbouring lakes, located 1-2 km upstream and downstream of the Kerzhenets river. Furthermore, desmans were noted in the lake, located 12 km upstream of the Kerzhenets river from the place of release.

In 2001, all 11 animals were released into the lake. Krasnyy Yar. Some of them successfully overwinter. Meeting desman were recorded in spring and summer 2002, both at the site of release and in the neighbouring lakes, located 1-2 km upstream and downstream of the river Kerzhenets. Furthermore, was the presence of muskrat in the lake, located 12 km upstream (as the crow flies) downstream of the river Kerzhenets from the place of issue.

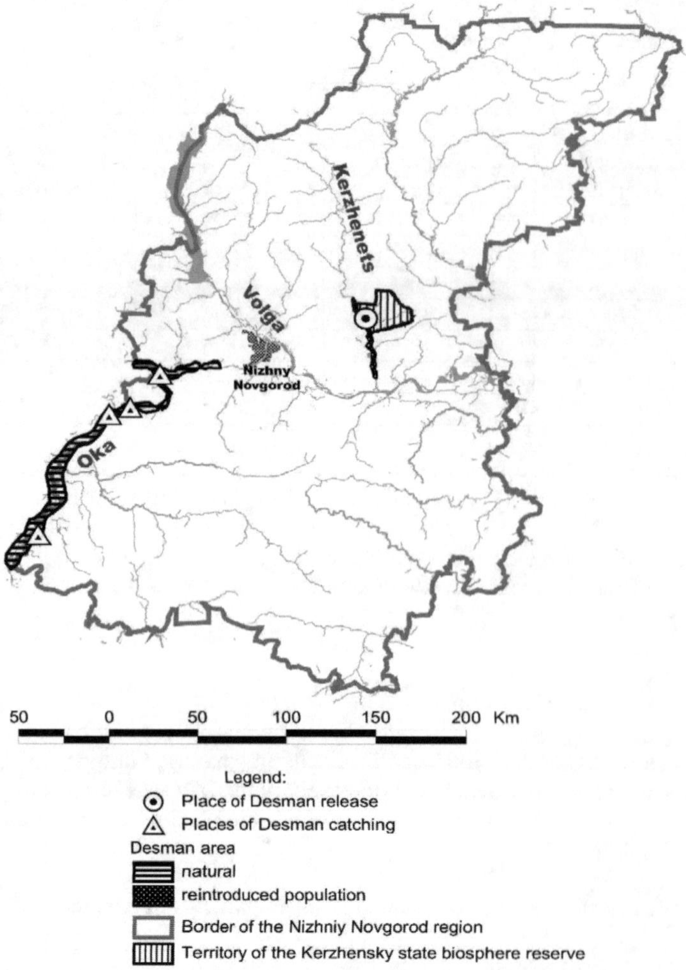

Figure 1. Locations of catching and release of desmans in the Nizhny Novgorod region

To track individuals released in 2002 we used transmitters SOM 2190 (weight about 6 g, a radius of transmitting about 1 km, the battery duration on average 270 days) and receivers TRX-48 with three-element feeder antenna. Transmitters were fit to the base of the tail. A total of 10 individuals was tagged and released in 6 lakes. Movements of tagged animals were recorded daily. It was found that 5 individuals left the release site in the early days. They moved at the long distances up to 3.9 km from the release site, reaching other floodplain lake or channel of the river Kerzhenets. Of the 10 tagged animals, 5 were died and 4 individuals with transmitters during the first weeks left the zone the reception and in spite of searches, were not found. One individual until December 2002 had been reliable alive. After the waterbodies were covered with thick ice the signal of its transmitter had been lost. In addition to the death of 5 tagged animals, 4 desmans moved outside the Reserve and died in fishing nets in 2002. Thus, in 2001-2002, at least 9 individuals were reliably died (17.6% of the total number of released animals).

2.2. Monitoring of the desman reintroduced population

Monitoring, which had been conducted for 15 years after releasing, has showed the succes of reintroduction activities. Desmans stable inhabit the Kerzhenets river valley. They survive on the site of the floodplain in the border of the Kerzhenskiy State Nature Reserve and have settled along the river valley at the length of at least 50 km (about 90 km along the river). Points of animal registrations in 2012-13 are located 20 km upstream and 60 km downstream of the river Kerzhenets from the place of release. The desman population trend surveyed by us in the floodplain in the Kerzhenskiy State Nature Reserve is presented in Table 3.

Table 3. The desman population trend in the Kerzhenskiy State Nature Reserve in 2005-2017.

Year of surveys	Density, ind./km	Population number (ind.) in the Kerzhenskiy State Nature Reserve
2002	No data	No more than 20
2005	0.74	25
2006	0.47	16
2012	0.14	11
2013	0.05	9
2015	0.04	3
2016	0.19	5-8
2017	0.22	6-10

Considering the number of individuals, which died before the breeding season, the reintroduced population has become to develop from 40 animals. Because in the first months after releasing the desmans actively settled throughout the area, probably no more than half of the survived individuals has remained in the Reserve. In 2005, the population was estimated at no less than 25 individuals. This indicates successful reproduction. The number of animals has probably increeased in the sites of release. The next decade, there was a steady decline in density and numbers. For the last two years the population trend has been towards stabilization and slow growth of the species. Assessing the poplation status in general, it is necessary to consider not only the model area in the release site within the Reserve, but changes in the home range and the total number. This survey encompassed all over the valley of the lower reaches of the Kerzhenets river in 2012-13. In 2012, signes of 18 animals were found in the surveyed waterbodies. The total population number is estimated at 35-40 individuals (including 11 in the Reserve). According the survey results we projected favorable future for stability of the reintroduced population of the species.

However the results of surveys in 2013 had shown that the forecast was unduly optimistic. Despite the extensive surveys, taking into account the extrapolation the total number of desman in the Kerzhenets river valley, was 17 individuals (including 9 species in the Reserve). Probably from 30 to 50% of the reintroduced desman population inhabit the territory of the Reserve.

Decreasing the population number in twice cannot be the result of errors of counts. The desman had disappeared in several waterbodies, where was registered in 2012. Decrease in the species number was recorded in the Reserve territory in lesser degree than neighboring areas. The desman population numbers decreased in the surveyed area apparently due to the impact of limiting factors. Considering the fact that the Kerzhenets river mouth and the Volga floodplain have been flooded with waters of the Cheboksary reservoir the impact of spreading and redistribution of animals seem to be not significant,. Large reservoirs are not a habitat for the desman.

Surveys conducted in the Reserve in 2015 showed a continued negative population trend. However, in 2015-16, there was some increase in density and desman population number in the Reserve. During summer 2017 (before surveys), according to reports of local people, 5 individuals died in the illegal fishing nets in the vicinity of the site of release. However, the species number have remained extremely low. The threat of death of the reintroduced population have existed.

Low success of the desman population recovery may be caused by several reasons. An important limiting factor is the fish poaching with nets. The animal deaths for this reason in the Kerzhenets river valley was validly registered. Indirect evidence of this is the fact that the desman inhabits the lakes of the Reserve at the site of release constantly, but appears and disappears in some waterbodies in adjacent areas in different years. At the same time, despite on the species protection in the Reserve, its number in the protected area of the Kerzhenets river valley, remains extremely low. In the mid-twentieth century it was here that the density of the species reached extremely high values. In 1940, in separate lakes, there were 10-20 burrows per 100 m of the river bank [11]. Apparently, one of the reasons that the density of reintroduced desmans in the lakes of the Kerzhenets river floodplain is unable to reach former values is due to changes to its habitat, particularly the reduction of biomass and productivity of zoobenthos. According to our observations, even over the last 15 years the signs of aging waterbodies have been noted: a reduction in their width and depth. A number of shallow lakes, which were with open water in 2005, have overgrown with sedges and willows, and young black alders have appeared in some places. In 2005, a significant number of gastropods (Lymnaea stagnalis L., Planorbarius corneus L., Viviparus viviparus) was noted in the majority of waterbodies. Near almost all the desman burrows visited by us their empty shells were registered. In 2015-17, on particular water bodies both living and eaten gastropods were recorded only a few times. Perhaps reintroduced desmans in the first years of their life in the new place had reduced the prey species numbers. The question of capacity of the desman habitats, and the rate of aging waterbodies in the Kerzhenets valley needs further study.

The muskrat and beaver had been absent in the Kerzhenets river valley until 1940. Reacclimatized desmans have lived together with these semi-aquatic rodents. According to our observations, the desman in the site of release and in the remained habitats in the Oka floodplain, actively uses the burrows of beavers and muskrats. In separate waterbodies the desman practically not digs its own burrows, but uses ones of other semi-aquatic rodents, where makes holes at a distance from the entrance from 10 cm to several meters. The ratio between surveyed burrows that were digged by desmans themselves and made by beaver and muskrats and occupied by desmans in the Kerzhenets floodplain has shown in Table 4.

Table 4. The desman records and its burrows encountered in the Kerzhenets floodplain in 2005-17.

Year	Encountered burrows				Encountered animals
	by desmans	by muskrats	by beavers	total	
2005	10	0	19	29	4
2006	11	8	9	28	4
2012	18	2	9	29	2
2013	11	2	9	22	0
2015	4	0	0	4	0
2016	6	0	5	11	1
2017	4	0	5	9	0

In dry years cohabitation with beavers has a significant positive value to desman. Beavers make waterbodies deeper, building dams, paving the channels to a depth of 0.5-1.0 m across shallow water to the burrows and foraging places, creating conditions for the desman survival.

It is widely believed [12] that muskrats displace desmans from suitable habitats. According to our long-term observations (since 1987) in the Oka river floodplain the muskrat could not be recognized as a competitor or enemy for the desman. These species have been living together in a single waterbody in shared burrows for decades.

The presence of burrows by beaver and muskrat conceals the desman existence and significantly complicates its census. At high abundance of gastropods the burrows of these rodents occupied by desmans are easily identified by the presence of empty shells at the entrance. Under conditions of the almost complete absence of gastropods the desmans living in burrows by beaver and muskrat may be significantly underestimated.

3. Conclusion

Thus, a reintroduced population of desmans have been created and have existed in the Kerzhenets river valley for 15 years. The population is at a critically low level. Recovery of the species density to values that occurred in the mid-twentieth century is the apparently impossible as a result of habitat changes and limited prey numbers. However, even the limited additional population is of great importance to conserve this endangered species.

It is necessary to continue annual monitoring of the desman population in the model area within the Kerzhenskiy State Nature Reserve. It is advisable to carry out surveys aimed at the study of the species distribution and numbers along all the Kerzhenets river valley in 2018-19, covering the sampling of waterbodies in the lower and upper reaches of the river. It is also necessary to study the limiting factors and to calculate the habitat capacity for the species in the Kerzhenets river floodplain, to assess the feasibility of additional release of animals caught in the Oka floodplain.

References

[1] Kennerley, R. & Turvey, S.T. 2016. *Desmana moschata*. The IUCN Red List of Threatened Species 2016: e.T6506A22321477. http://dx.doi.org/10.2305/IUCN.UK.2016-2.RLTS.T6506A22321477.en. Downloaded on **28 October 2017**

[2] Smirnov N.G., Ponomarev D.V. 2007. News about the Past Distribution of the Desman (*Desmana moschata* L.) *Doklady Biological Sciences* 414: 219–220

[3] Okulova N.M., Onufrenya A.S., Onufrenya M.V. 2008. Analysis of monitoring data on the Russian desman (*Desmana moschata*) in the Oka state biosphere reserve in relation to the problem of species' population decline. *Russian Journal of Ecology* 39 (7): 510–515. https://doi.org/10.1134/S1067413608070072

[4] Oparina O.S., Filinova E.I., Sonina E.E., Oparin M.L. 2013 Current status of the Russian desman habitats in small rivers of the Don River Basin in Saratov oblast and the abundance of this species. *Biology Bulletin* 40(10): 854–861.

[5] Onufrenya A.S., Onufrenya M.V. 2016. Russian desman in Oka basin. *Proceedings of the Oka State Natural Biosphere Reserve* 37: 1–204. [In Russian]

[6] Rutovskaya M.V., Onufrenya M.V., Onufrenya A.S. 2017. Russian desman (*Desmana moschata*: Talpidae) at the edge of disappearance // Nature Conservation Research. Заповедная наука. Т. 2(Suppl. 1). С. DOI: 10.24189/ncr.2017.020

[7] Bakka S.V., Kiseleva N.Yu., Denisov D.A., Pankratov I.I. Current status of russian desman (Desmana moschata) in Nizhny Novgorod region /The International Scientific-Practical Conference on the Humanities and the Natural Science ISPC 2013. The International Scientific-Practical Conference on the Humanities and the Natural Science Held by SCIEURO in London, 26. December. 2013. P. 310-319

[8] The Chronicle of Nature of the Kerzhensky State Nature Reserve. 2002, Vol. 9. N. Novgorod, 2003. – 211 p. (Manuscript, State Nature Reserve "Kerzhensky")

[9] Kudryashov, V.S. 1976. *Metodicheskie ukazaniya po uchetu vykhukholi i ondatry v poimennykh ugod'yakh* (Methodological Guidelines for Taking Census of Desmans and Muskrats in Floodplain Areas), Moscow, 1976.Google Scholar

[10] Marchenko N.F. 2011. Methodical Peculiarities of Russian Desman Counts. Povolzhskiy Journal of Ecology, 1, 97-102.

[11] Report of the expedition of Professor S. S. Turov and Assoc. A. P. Protopopov on surveys and selection of territory for the establishment of the Reserve near the Kerzhenets river, Gorky region. 1940. manuscript, 30 p. (archive of the Kerzhenskiy State Nature Reserve).

[12] Borodin L.P. 1963. *Russian desman*. Saransk: Mordovia Book Publishing House. 301 p. [In Russian]

IOP Conf. Series: Earth and Environmental Science **115** (2018) 012037 doi:10.1088/1755-1315/115/1/012037

Visualization of electromagnetic exposure near LTE antennae

M Yu Zvezdina, Yu A Shokova, O Yu Nazarova, H T A Al-Ali and G H A Al-Farhan

Radioelectronic Department, Don State Technical University, 1 Gagarin Square, Rostov-on-Don, 344000, Russian Federation

E-mail: zvezdina_m@mail.ru

Abstract. Technical progress in wireless data transfer has given an opportunity to apply information and communication technologies in various areas of economics. Digital economy is linked to the 4th and 5th generation mobile network deployment. The peculiarities of the abovementioned standards decrease BTS antenna range three times in dense developed areas and worsen electromagnetic background in big cities. In the paper the comparative assessment results for rooftop electromagnetic exposure near BTS LTE and BTS GSM antennae are given. It is shown, that at the same level of transmitter power, energy flux density for LTE standard is three times less than the one for GSM. Moreover, the conclusion is made that the rooftop could be considered safe for people for indefinite time if antenna is placed more than 5 meters above the rooftop. The value of antenna height is taken to be on the safe side, as it is required by an application of "preventive principle".

1. Introduction

Wireless data transfer upgrade allowed an increasing data transfer rate up to 100 Mbps for highly mobile users (users taking a train or riding in a car) and up to 1000 Mbps for moderately mobile users (pedestrians or fixed users) within the standards of LTE-Advanced and WirelessMAN-Advanced (WiMAX). It has opened up a business implementation of mobile technologies [1-3]. As a result, after a year of the 4th generation mobile network deployment the index of mobile technologies implementation business in has scored 57% in the developed countries in 2011 [3]. Five years later, in 2016, Great Britain's information and communication technologies adoption in GDP equaled to 12.4%, while the country has gained the leading positions in digital economy contribution to GDP. In the same time in Russia the same index equaled to 2.8%. In July of 2017 the Government of the Russian Federation has established a program "Digital economy of the Russian Federation" to overcome the development delay.

An implementation of the program is linked, among other things, to expansion of mobile technologies application, thus, is calling for substantial quantity increase for Base Transceiver Station (BTS) to provide broadband Internet access. It is caused by a number of reasons; first of all, by a direct increase of LTE users. For example, in 2020 it is scheduled to provide Internet access to 97% of households and 100% of health care, educational and business organizations. Secondly, LTE standard features the usage of higher frequencies as compared to GSM standard, which leads to distance range decrease. Thus, in dense developed areas, where the population density is the highest, distance range lowers almost three times from 0.7 km to 0.25 km [5]. Negative result of GSM mobile communications usage lies within the complete cover with BTS nonionizing radiation, as it was mentioned in [6, 7]. This radiation has a constant twenty-four-hour impact on health of the area residents, including children, elderly and unhealthy [8]. Locating BTS atop of inhabitant buildings or atop of annexed stores (at the level of 2-3 floor of inhabitant building) worsens electromagnetic environment in residential areas under the conditions of altering building height.

It is impossible to eliminate the antenna harmful factor - radiated electromagnetic field. Thus, regulatory documents [9-12] are adapted; they state that energy flux density for habitable territories should not bypass maximum permissible level of power equal to 10 $\mu W/cm^2$. At the same time a socially-oriented electromagnetic monitoring [13] is carried out for these areas. This kind of monitoring encircles large areas and promotes timely and situational provision of information to the public concerning electromagnetic pollution level; it limits sanitary protection zones (SPZ) measured at height of 2 m above ground or above top floor of building if antenna is located at the rooftop. Sanitary protection zone limits correspond to maximum permissible level of 10 $\mu W/cm^2$. Population comprehensibility and capability of continuous adjustment to new electromagnetic environment problems are core requirements to socially-oriented monitoring results.

In works [6, 14, 15] the results of electromagnetic exposure near GSM antennae were presented. However, LTE channels differ from GSM in technical characteristics. Thus, the paper is aimed to estimate the electromagnetic exposure near BTS LTE with the visualization of the results obtained with calculation forecast method.

2. Equipment and devices used in studies

Electromagnetic exposure near BTS LTE antenna was estimated basing on regulatory documents [9-12]. The technique is described in [16]. Its peculiarity is an application of "preventive principle", which is recommended by the World Health Organization [8]. This principle allows for preventive actions to reduce possible harmful effect of electromagnetic radiation on human beings even if the effect itself is not proven, but is assumed with contemporary level of scientific knowledge. As a result, sanitary protection zone limits would exceed the real ones.

3. Results and Discussion

3.1. Electric parameters of BTS antennae for mobile communications

In the process of electromagnetic exposure research near BTS LTE antennae and results comparison to BTS GSM data the operational schemes differences of the abovementioned standards were considered. LTE operation features closed-loop MIMO system [5]. It presumes multielement antenna usage on both ends of a channel - transmitting and receiving ones - to form the effective beam for a given radio path channel. Thus, the number of required transmitters is increased to two, and total. It increases the number of transmitters to 2, and total power output equals to 46 dBm or 40 W [5]. The beam of the obtained directivity patterns is narrower than the one of the GSM antennae.

For an example, table 1 shows electrical characteristics for one of this kind of antennae - ASTRA 3G/4G MIMO panel antenna [17]. A picture of BTS LTE antenna configuration is given on Figure 1.

Table 1. Electrical characteristics for ASTRA 3G/4G MIMO antenna [17]

Characteristics	
Telecommunication standard	LTE 800, LTE 900, LTE 1800, LTE 2600
Frequency range, MHz	790-960 / 1700-2700
Gain, dB	10-15
Half power beamwidth in horizontal plane, degrees	36
Half power beamwidth in vertical plane, degrees	36
Recommended distance to mobile tower, km	<10

Figure 1 – BTS LTE antenna configuration

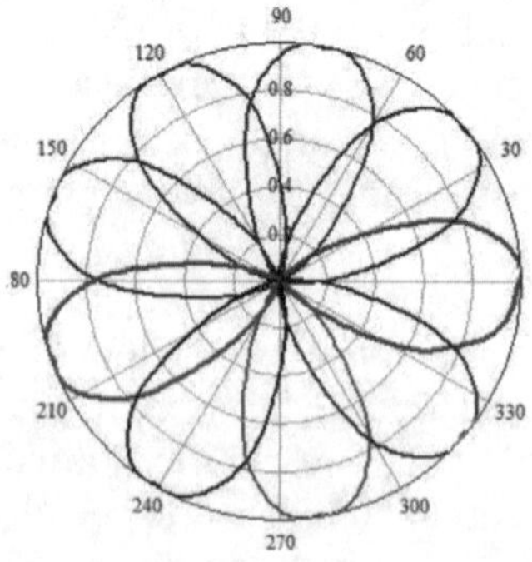

Figure 2 – Directivity pattern of BTS antenna set

3.2. Electromagnetic exposure estimation by forecast calculation

Electromagnetic exposure level near rooftop LTE antennae was estimated with the program set developed by the paper authors. The coordinates of roof angles were set equal to $A(0,0,30)$, $B(-7,4,30)$, $C(-13.5,-4,30)$, $D(-2.5,-8.5,30)$, the same way it was done for GSM antennae research [6, 14]. The research on electromagnetic field radiated by antennae was carried out in horizontal plane 2 m above the rooftop.

The simulation presumed that BTS LTE was a set of nine antennae of the same type with half power beamwidth equaled to 36 degrees (ASTRA), and BTS GSM was a set of three antennae with half power beamwidth equaled to 120 degrees. Directivity pattern for the considered BTS LTE is shown on Figure 2. Antenna gain for both BTS systems was considered equal to 13 dB.

The considered frequency range was limited to 860-900 MHz, because sanitary protection zone area should be maximal for these frequencies [14].

Figures 3 and 4 show simulation results for energy flux density distribution for different transmitter power values; antenna phase center was located 3 m above the rooftop. Figure 3 shows energy flux density distribution for transmitter power equal to 20 W for BTS LTE antennae (Figure 3,a) and BTS GSM antennae (Figure 3,b).

Figure 4 shows energy flux density distributions near BTS LTE antenna for transmitter power equal to 40 W (Figure 4,a) and 10 W (Figure 4,b). Table 2 contains maximum values of energy flux density for BTS LTE and BTS GSM in horizontal plane 2 m above the rooftop.

Table 2. Maximum radiation level for BTS GSM and BTS LTE antenna depending on total power of transmitters

Total power, dB	Maximum energy flux density value, μW/cm^2		Antenna height for no SPZ at the rooftop, m	
	GSM	LTE	GSM	LTE
40	419	159	5	4.5
20	210	80	4.5	4
10	105	40	4.3	3.8

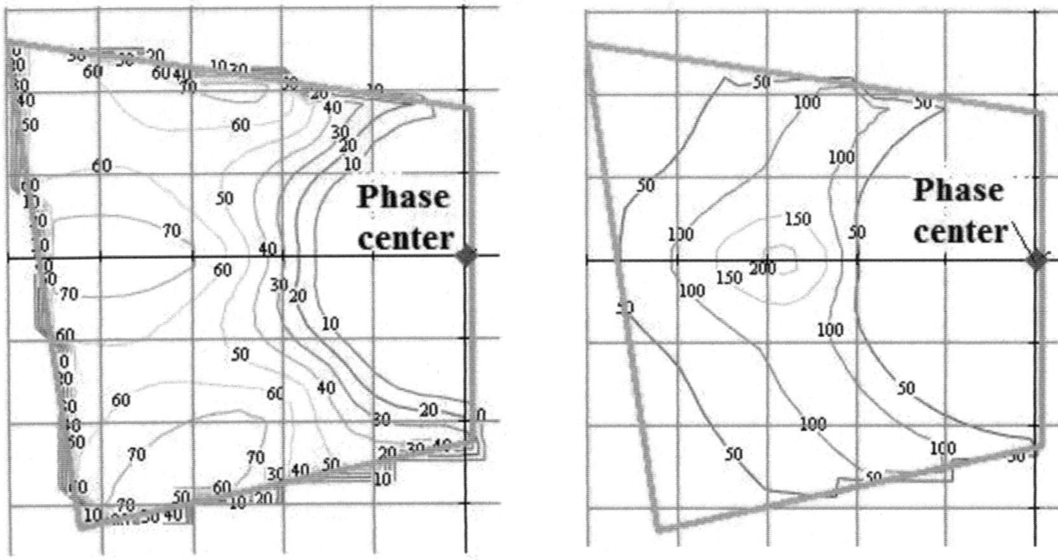

a b

Figure 3 – Directivity pattern for GSM and LTE antennae, transmitter power equals 20 W:

a – nine-element LTE antenna; b – three-element GSM antenna

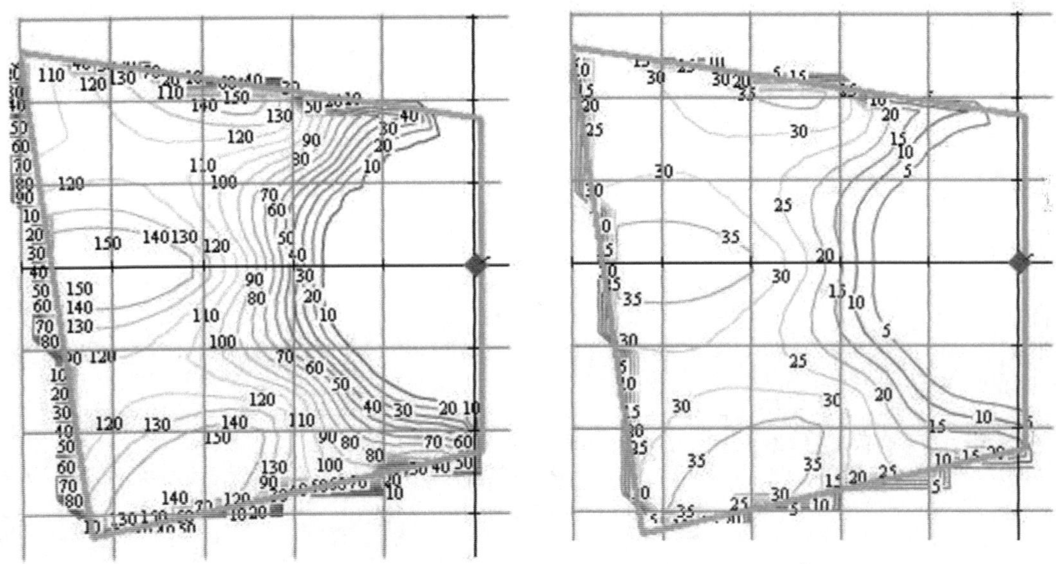

a b

Figure 4 – Directivity pattern of nine-element LTE antenna for different transmitter power :

a – maximum power equals 40 W; b –transmitter power equals 10 W

The obtained results analysis show that it is required to declare practically all the rooftop as sanitary protection zone for BTS of both systems. Safe zone is located directly under antenna location; however, the area is rather small. As it is shown in table 2, electromagnetic exposure near BTS LTE antennae in terms of energy flux density is almost 3 times lower than the one near BTS GSM ones. However, even if the transmitter power is lowered four times to the value of 10 W, it is still required to declare practically all the rooftop as sanitary protection zone. It is possible to make the roof safe for people for indefinite time if antenna is placed higher. Nonetheless, BTS GSM antenna have to be located 0.5 m higher than LTE ones for any considered value of transmitter power.

Thus, BTS LTE antennae are considered safe for people on the rooftop for indefinite time if they are placed five or more meters above the roof surface. The value of antenna height is taken to be on the safe side, as it is required by an application of "preventive principle".

Conclusion

Wide spread of information and communication technologies in all fields of human activity has led to the necessity of development of digital economy. The adaptation of the Government of the Russian Federation strategies in digital economy is directly linked to wireless communication usage to provide broadband Internet access, and thus, requires substantial increase in BTS LTE numbers. Its negative ecological result lies within worsening of electromagnetic background caused by BTS density increase; it is due to LTE standard features, that in dense developed areas antenna distance range lowers almost three times if compared to BTS GSM.

The analysis of rooftop electromagnetic exposure near BTS LTE antennae and the comparison to the same kind of results for BTS GSM antennae has shown that the rooftop can be considered safe for people for indefinite time if antennae are placed five or more meters above the roof surface. This value is taken to be on the safe side, as it is required by an application of "preventive principle".

The comparison of energy flux density levels for BTS LTE and BTS GSM antennae at the same level of transmitter power has shown that maximum value of energy flux density for LTE standard is almost three times less than the one for GSM standard.

References

[1] ITU (International Telecommunications Union) 2010 ITU-R IMT-Advanced 4G standards to usher new era of mobile broadband communications. Web-site: http://www.itu.int/net/pressoffice/press_releases/2010/40.aspx#.Vkpb87-ns5 (accessed October 24, 2017)

[2] Ali S, Khusro S 2016 *Indian J. of Sci. and Techn.* **9** (19) doi: 10.17485/ijst/2016/v9i19/53088

[3] The New Digital Economy 2011 *Oxford Economics*. Web-site: http://www.pwc.com/mt/en/publications/assets/the-new-digital-economy.pdf (accessed October 24, 2017)

[4] Government of the Russian Federation 2017 Program Digital Economy of the Russian Federation (Moscow) Web-site: http://static.government.ru/media/files/9gFM4FHj4PsB79I5v7yLVuPgu4bvR7M0.pdf (accessed October 24, 2017)

[5] Sesia S, Toufik I, Baker M. LTE – the UMTS long term evolution: from theory to practice 2011 2nd edition. NY: John Wiley & Sons Ltd ISBN 9780470660256

[6] Zvezdina M Yu, Shokova Yu A, Kundrukova N I, Kutukova V D, Pozdnyakova A V 2017 *IOP Conf. Series: Earth and Environmental Science* **50** 012029 doi: 10.1088/1755-1315/50/1/012029

[7] Zvezdina M Yu, Shokova Yu A, Cherckesova L V, Golovko T M, Cherskaya A A 2017 *IOP Conf. Series: Earth and Environmental Science* **66** 012005 doi: 10.1088/1755-1315/66/1/012005

[8] WHO Library Cataloguing-in-Publication Data 2004 Establishing a dialogue on risk from electromagnetic fields. Web-site: http://www.who.int/peh-emf/publications/en/russian_risk.pdf (accessed October 14, 2017)

[9] Ministry of Health of the Russian Federation 2003 *SanPin 2.1.8/2.2.4.1190-03 Hygienic requirements on land-based mobile communication location and operation* (Moscow)

[10] Ministry of Health of the Russian Federation 2002 *MG 4.3-1167-02 Methodological guidelines for energy flux density estimation for electromagnetic field for location of radio facilities with operating band of 300 MHz – 300 GHz* (Moscow)

[11] ICNIRP (International Commission on Non-Ionising Radiation Protection) 1998 Guidelines for limiting exposure to time-varing electric, magnetic and electromagnetic fields (up 300 GHz) *Health Phys.* **74** 494-522
[12] IEEE (Institute of Electrical and Electronic Engineers) 2005 C95.1-2005 IEEE Standart for Safety Levels with respect to Human Exposure to Radio Frequency Electromagnetic Fields, 3 kHz to 300 GHz (New York IEEE) Web-site: https://standards.ieee.org/findstds/standard/C95.1-2005.html (accessed September 24, 2017)
[13] Maslov M Yu, Spodobaev M Yu, Spodobaev Yu M 2014 *Telecommunications* (10) 39-42
[14] Zvezdina M Yu, Shokova Yu A, Shokov A V, Dymchenko A A, Parhomenko P A 2015 *Int. J. Theor. & Appl. Sci.* **03** (23) 10-7 doi: 10.15863/TAS.2015.03.23.3
[15] Zvezdina M Y, Shokova Yu A, Shokov A V 2015 *Proc. Int. Conf. on Application of Information and Communication (Rostov-on-Don)* 399-403 doi: 10.1109/ICAICT.2015.7338588
[16] Spodobayev Yu M, Kubanov D P 2000 Electromagnetic ecology basics. Moscow: Radio i svyaz'.
[17] Antenna ASTRA 3G/4G Web-site: https://gsm-repiteri.ru/prodazha/antenny-yota-lte/antenna-astra-3g-4g-mimo-panelnaja-2x15db (accessed September 24, 2017)

IOP Conf. Series: Earth and Environmental Science **115** (2018) 012038 doi:10.1088/1755-1315/115/1/012038

Development of Automated Monitoring and Management System of Municipal Solid Waste Landfill Based on the Industrial OMRON Controller

S N Kostarev[1,2,3,a], **T G Sereda**[1]

[1]Perm State Agricultural Academy named after academician D.N. Prianishnikov, 23 Petropavlovskaja St., Perm, Russia, 614000
[2]The Perm Military Institute of the National Guard's Forces of the Russian Federation, 1, Gremjachij Log St., Perm, Russia, 614112
[3]The Perm Institute of the FPS of Russia, 125, Karpinskogo St., Perm, Russia, 614012

E-mail: [a] iums@dom.raid.ru

Abstract. The application of the programmable logic integrated circuits (PLC) for creating the software and hardware complexes of the medium complexity is an economically sound solution. The application of the OMRON controller to solve the monitoring and management tasks of safety of the municipal solid waste (MSW) landfill with the use of technology of the filtrate recirculation and the landfill maps irrigation is shown in the article. The article contains the technical solution connected with the implementation of the 2162059RU invention patent for the municipal solid waste landfill management in the Kurgan region of Russia. The calculation of maps and ponds was made with consideration of the limited sanitary and protection zone. The GRUNDFOS dosing and reactor equipment was proposed to use in the project.

1. Introduction

The municipal solid waste landfills (MSW) are the source of the long-term negative impact on the environment [1]. Despite the fact that the municipal solid waste landfills have a limited operating period (an average of 30 years) after its closure and reclamation the complete life cycle of this natural and technical system lasts for thousands of years during which the emission products hazardous to the environment will be released [2]. Qualitative determination of energy potential and methane generation from municipal solid waste (MSW) was considered in the article [3]. Therefore it is relevant to formalize the generalized models of the process management at the municipal solid waste landfills with consideration of the analysis of the material-energy and information flows and to develop the algorithms for the structural-parametric synthesis and the automatic control for setting up an automated monitoring and management system of the municipal solid waste landfills. Now the operational management of the waste recycling process at the landfills in Russia is almost not used which justifies expediency of the development of the control automation system of the municipal solid waste landfill [4].

The monitoring and analysis of the environmental status of the municipal solid waste landfill incorporated in the software and technical solutions of the management provide the operational control over the processes occurring at the given facility. The complex automated system of the landfill management will significantly improve the monitoring quality. The modular construction of the system will provide the flexibility of the structural implementation with the possibility of the maximum consideration of the individual characteristics of the region where the municipal solid waste landfill is located. The application of this software product will provide the informational support throughout the life cycle of the municipal solid waste landfill and increase the management efficiency and its operation safety.

2. Material and methods of the research

When justifying the methods and algorithms of the management of waste depositing objects, the methods of system analysis, the general theory of systems, the theory of automatic control, the methods of mathematical and simulation modeling were used.Considered as objects of the modeling and management, the material-energy and information flows occurring at the waste recycling facilities were the object of research.

3. The results of the study and their discussion

According to the functional feature, the projected system is logically divided into five modules representing design, technological and management solutions in the life cycle of the landfill. The design module is designed to calculate the main parameters of the landfill, which also includes the dynamics of emission flows of the filtrate and biogas. The automated information processing module is an automated workplace of the polygon controller. The ASUTP module is designed for automated control of the technological process at the landfillMSW.

3.1. Schematic diagram and algorithm for polygon management

The purpose of the work was the development of technical solutions for the use of patent 2162059RU, aimed at managing local treatment facilities of the landfill. As a hardware, a programmable Omron controller was chosen, which is a compact product for solving automation problems of low and medium complexity. The use of Omron controller software also allowed the development of a human-machine interface for managing local wastewater treatment facilities for the MSW landfill [5].

The automated system of operational control included the following components: control of the sections of the landfill by recirculation of the leachate; pH adjustment stage, also purification of the leachate and surface waste on hydrobiological sites is also proposed [6, 7]. The scheme of automation of local structures of the solid waste landfill is shown in Figure 1.

The landfill consisted of two maps (sections). Local treatment facilities for neutralizing the leachate are represented by a storage pond-averageizer and a settling pond. To stabilize the moisture content of the solid waste landfill, an algorithm has been developed that implements the management of processes at the MSW landfill with feedback. To accelerate the stabilization of the pH environment, the necessary amount of lime solution added to the recycled filtrate is calculated, depending on the pH of the solid waste massif and the volume of the recirculatedleachate (Figure 1). From the maps of the landfill, the leachate enters the accumulation pond-averager, consisting of three sections connected by locks. Drivers for and against municipal leachate recycling was considered in the article [8]. According to this scheme, the leachate formed in section 1 of the waste disposal site of the landfill (and later in section 2 of the of the landfill will be fed to section 1 of the accumulation pond-averager. When the 1 section is filled, the leachate will flow over into the 2 section of the accumulation pond-averager, and then to the 3 section of the accumulation pond-averager.

By the pump (H1) the leachate from the collecting pond sections is fed to the neutralization station consisting of three vessels: a reaction chamber (TU) where also a prepared lime solution from the tank will be supplied. The scoop is driven by a screw with an electric drive to a dilution with a water pump (H3). Then the prepared solution of $Ca(OH)_2$ will be sent to the vessel (ND). Then the solution of Ca $(OH)_2$ is fed to the reactor-mixer. All tanks are equipped with mixers. In the reactor-mixer, a constant pH value of 9 is maintained using an automatic control system, which will allow automatic pH-regulation of the effluents at the outlet of the mixer. After processing the effluent, they enter a sedimentation pond. In the settling pond, the pH will be controlled by the pH meter (pH4), the pH of the filtrate coming from the I–II sections of the landfill waste, pH meters, pH1 and pH6 respectively. When the pH of the leachate arrives from the landfill sites of the values typical for 10–15 years of operation of the landfill site, the leachatewill be fed directly from the pond-averager by the H1 pump to the settling pond (in Figure 1 as a dashed line).

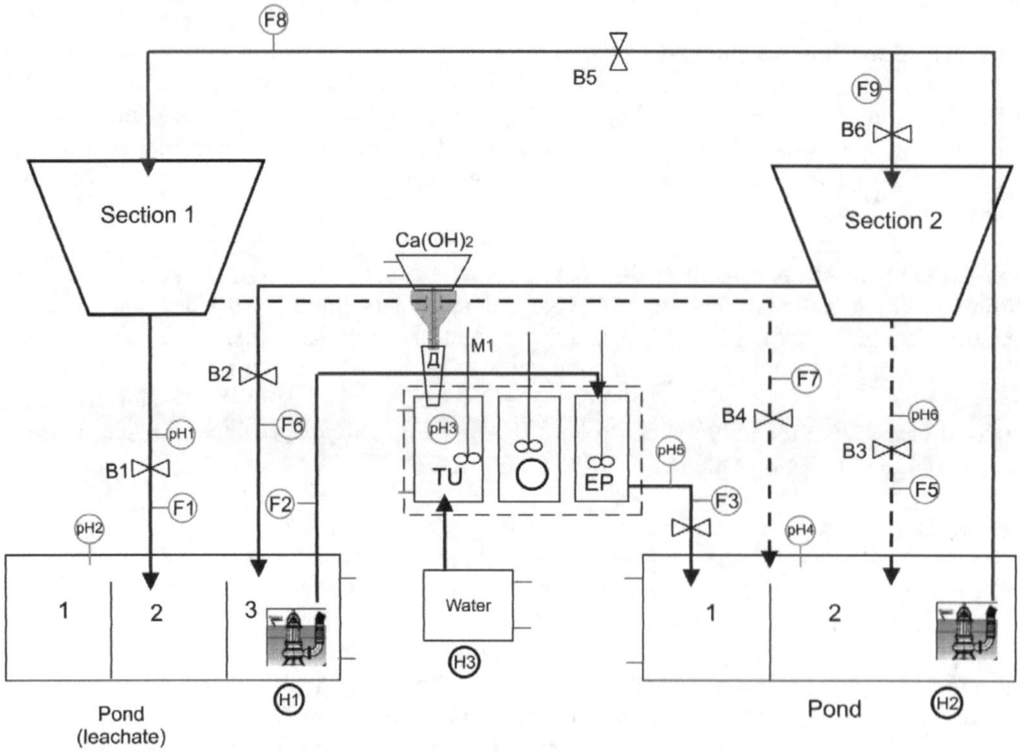

Figure 1. Scheme of automation of local structures of solid domestic waste:
F1–F9 – consumption; B1–B6 – the adjustable valve.

Irrigation of the polygon maps is made by filtrating from the settler pond at a pressure of 1–2 atmosphere through perforated pipelines with a diameter of 1.5–2" based on the required flow rate. Software control of the technological mode is to maintain and reduce the deviation from the specified process. Physical and chemical quantities are monitored. The algorithm for controlling the technological process was previously described in [8].

When choosing a reactor station, the metering equipment of GRUNDFOS was considered. The dosing equipment is connected to the general control system of local facilities of the landfill. With the help of the pump H3, the reagent preparation tank (TU) is filled with water, the required amount of dry reagent is filled, after which the mixer M1 is turned on until the powder is completely dissolved. Then the prepared solution is pumped to the dosing container (ND). In the dosing tank, the agitator works continuously to maintain undissolved particles in a suspended state. The filling of the reactor (EP) occurs in an automatic mode. The system tracks the upper and lower levels of the liquid in the tanks. When the reactor (EP) is filled, the microcontroller monitors the pH in real time and deviates from the set one, gives a command to the pump dispenser, which brings the pH level to the preset level. When the required pH is established, the reactor is emptied and the treated effluents are fed to a settling pond.

3.2. Development of technological scheme, implementing patent for invention No. 2162059 RU
Programming of the algorithm of the technological scheme is carried out with the help of the PLC Omron.

The station of the operator controller of the landfill site includes the following modules:
- Sensor monitor of the operator of control of local structures;
- Programmable Omron microprocessor.

The dosing station is a separate standard industrial module and has its own control panel. The control program for the Omron controller is written in the language of Ladder diagram (LD) (standard IEC

61131-3). Language LD is a software implementation of electrical circuits based on electromagnetic relays. The program in the language of LD has a clear and intuitivegraphical interface, representing logical operations as an electrical circuit with closed and open contacts.The CX-Programmer Support Software is equipped with an SFC editor, and can be used to create, monitor, and debug SFC programs for CS/CJ-series and CP-series PLCs. The CX-Programmer supports SFC language in addition to conventional ladder and structured text (ST) programming languages. The main blocks of the developed program and a brief explanation of the blocks are shown in Figure 2 [9].

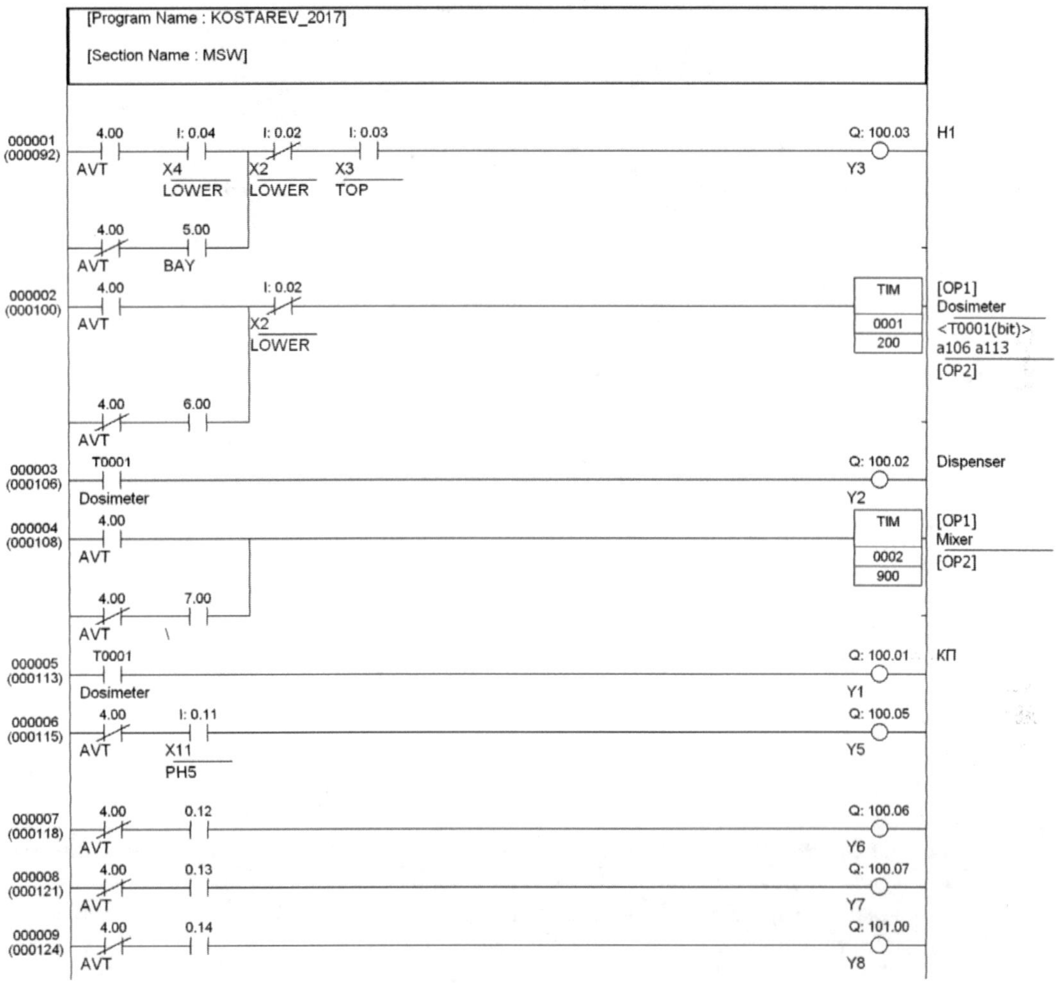

Figure 2. Ladder diagram.

The filling of the lime milk preparation tank is described in the first section (see Figure 2). Automatic activation of the pump H3 when emptying the tank (TU), checking the conditions for the presence of clean water, the bay to the upper level of the capacity of the EP; manual control from the operator panel – contact 4.00 (see Figure 2). The process of controlling the dosage of the $Ca(CO)_2$ in the second section. The control of the mixer shows the third section [10]. The operator panel (touch screen) is designed using the CX-Designer (Omron) program. The dialog box for the operator of the local leachate decontamination facilities contains a duplex control screen (Figure 3) and monitoring of physicochemical parameters.

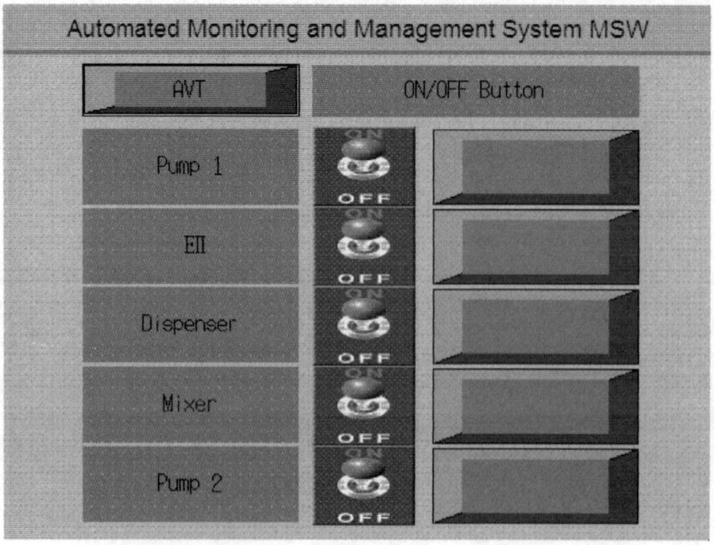

Figure 3. Touch screen monitor recirculation system control leachate.

Conclusions

The application of the industrial Omron controller and the CX-Programmer and CX-Designer software complex showed the feasibility and flexibility of the implementation with the use of technology of the leachate recirculation and the MSW landfill irrigation.

References

[1] Akinade OO and Oyedele L O 2017 Design for Deconstruction (DfD): Critical success factors for diverting end-of-life waste from landfills *Waste Management* **60** 3–13

[2] Pan H, Lei H, Liu X, Wei H and Liu S 2017 Assessment on the leakage hazard of landfill leachate using three-dimensional excitation-emission fluorescence and parallel factor analysis method *Waste Management* **67** 214-221

[3] Kostarev S N, Sereda T G and Elancheva E N 2014 Study safety environmental protection landfills using Models anaerobic digesters *Applied Mechanics and Materials* **682** 339–345.

[4] Yang F, Li G, Li H, Zhao D 2006 Development of PLC-based monitor and control system for the water supply on passenger trains *Journal of Huazhong University of Science and Technology (Natural Science Edition)* **34** 29–31

[5] Nawi I, Aisham B M Z, Pauline O, Jaffery N 2016 Review on system operation of auto-feeder for door panel production using programmable logic controller *ARPN Journal of Engineering and Applied Sciences* **11** 5326–30

[6] Arev I A, Marchuk A S and Bogolyubov N V 1983 Monitoring of the quality of natural water by a 'Nayada' automatic station *Soviet Journal of Water Chemistry and Technology* **5** 107–110

[7] Kunz N C, Fischer M, Ingold K and Hering J G 2016 Drivers for and against municipal wastewater recycling: A review *Water Science and Technology* **73** 251–259

[8] Mboowa D, Quereshi S, Bhattacharjee C, Tonny K and Dutta S 2017 Qualitative determination of energy potential and methane generation from municipal solid waste (MSW) in Dhanbad (India) *Energy* **123** 386–391

[9] Kostarev S N and Sereda T G 2013 Automated process control of sanitary municipal solid waste landfill *World Applied Sciences Journal* **22** 64–69

[10] Kostarev S N, Sereda T G and Elancheva E N 2014 Application Models of Continuum Mechanics Control Landfills *Applied Mechanics and Materials* **682** 251–255

The electrical conductivity of the flame front, as a characteristic of the rate of heat release and composition of gas fuel in SI engines

N M Smolenskaya
Togliatti State University, Belorusskaya st. 14, Togliatti, Russian Federation, 445020

E-mail: nata_smolenskaya@mail.ru

Abstract: The paper considers the possibility of using the electrical conductivity of the flame front as a characteristic of the rate of heat release and composition of gas fuel in a SI engines. Based on the analysis of the experimental data, the dependences of the parameters of the electrical conductivity of the flame front on the rate of heat release are obtained with the variation of the chemical activity of the gas fuel in a SI engines. The influence of the composition of the mixture and the effect of the amount of added hydrogen on the increase in the rate of heat release and, consequently, on the increase in the electrical conductivity of the flame. The obtained dependences will allow to increase the efficiency and reduce the toxicity of the SI engines operation during the regulation of the working process by ionization sensors.

1. Introduction

The phenomenon of the electrical conductivity of the flame front is characterized by the presence in the combustion process of a large number of ionized molecules. So in the combustion zone the following are the main ions: CH_3^+, CHO^+, OH^+, H^+, H_3O^+ [1 – 5]. Therefore, the flame front is capable of conducting an electric current. In this case, the characteristic of electrical conductivity is determined, first of all, by the density of ions in the zone of the ionization sensor. On the basis of this, many researchers suggest that the characteristics of the electrical conductivity of the flame front allow us to more accurately estimate the parameters of the flow of the working process. Therefore, it is possible to implement a system for controlling the operation of the engine by ionization sensors [6]. This control system will allow to more accurately maintain efficient engine operating modes while reducing toxicity. This is necessary with further tightening of toxicity standards and active introduction of gas fuel. As shown by numerous studies and projects that have been practically realized, natural gas is called environmentally friendly fuel with the addition of a chemical activator of combustion in the form of hydrogen.

For this purpose, this work is an extension of knowledge about the possibility of estimating the rate of heat release from the characteristics of the electrical conductivity of the flame front near the spark plug in the initial phase of combustion and remotely from it in the main combustion phase.

2. Experimental technique

Experimental studies were carried out on a single-cylinder UIT-85 unit. Geometric parameters of the engine: the cylinder capacity is 0.652 liters, the cylinder diameter is 85 mm, the piston stroke is 115 mm, the compression ratio is 7. The electric motor keeps the revolutions constant, uniformity of the fuel-air mixture is provided by heating the intake manifold.

Registration of the movement of the flame front inside the UIT-85 cylinder (Figure 1) was carried out by ionization sensors. The ionization sensor 2 was located near the spark plug 1 (Figure 2a), and the ionization sensor 3 was installed in an adapter with a pressure sensor 4 (Figure 2b). This arrangement of the ionization sensors makes it possible to evaluate the propagation parameters of the flame front in the initial and final phases of combustion [7, 8].

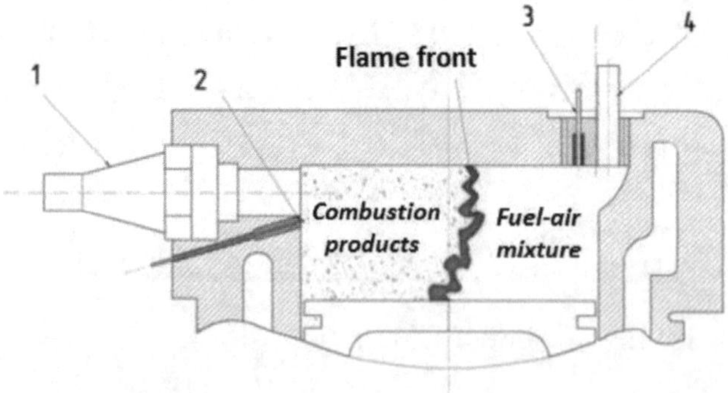

Figure 1 – The diagram of the combustion chamber of the UIT-85 unit, where 1 – spark plug; 2 – ionization sensor; 3 – ionization sensor; 4 – pressure sensor

(a) (b)

Figure 2 – The applied ionization sensors: (a) ionization sensor 1, installed next to spark plugs; (b) ionization sensor 2, installed next to the pressure sensor

The installation was performed at a fixed ignition timing angle of 13 BTDC and a rotation speed of 909 min^{-1}. The experiment was based on parallel recording of signals by a multichannel analog digital converter L-783M L-CARD. Signals were recorded from the following sensors: ionization, spark ignition, the position of the crankshaft, the pressure in the engine cylinder, the mass flow of air. The toxicity and the coefficient of excess air (α) were determined from the gas analyzer and the lambda sensor LSU 4.9. Natural gas and hydrogen were fed separately into the intake manifold for the carburetor, the mass flow rate of gas was determined from the calibrated duty ratio of the injectors.

3. Results and Discussion

To determine the possibility of estimating the rate of heat release from the electrical conductivity of the flame, experimental data were processed. The characteristics of the change in work and internal energy in the elementary section of the indicator diagram were found to be 0.27 degrees of the angle of rotation. The change in the work and internal energy at the elementary section characterizes the rate of active heat release:

$$\Delta Q_{act} = \Delta A + \Delta U = p\Delta V + \frac{C_V}{Z_r(p,T)\cdot R_\mu}\cdot V\Delta p \quad (1)$$

where ΔA – the elementary work; ΔU – the change in internal energy in an elementary area; p – the average pressure on the elementary section; ΔV – the volume change in the elementary section; C_V – the average molar heat capacity at a constant volume on an elementary section; $Z_r(p,T)$ – the real gas compressibility factor; R_μ – the universal gas constant; V – the average volume on an elementary site; Δp – the change in pressure on an elementary section.

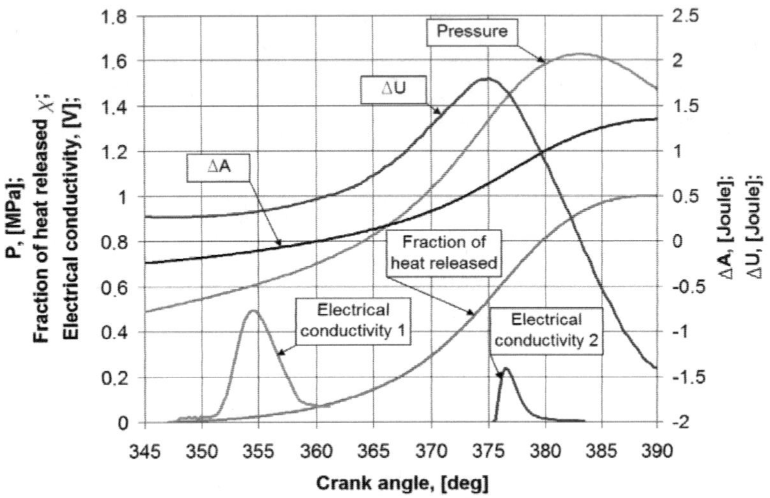

Figure 3 – The change in the basic thermodynamic parameters of the cycle characterizing the rate of heat release and the electrical conductivity of the flame for UIT-85 when working on CNG = 85%, H_2 = 15%

As a result of the thermodynamic analysis of the indicator diagrams [9 – 12], the characteristics of the amount of actively evolved heat in the combustion process. Where Q_{act} this is the sum of the change in work and internal energy from the start of the combustion process to its completion. It characterizes the amount of actively released energy from the moment of ignition during the entire combustion process. The results of the influence of the composition of the mixture, with a variation in the fraction of hydrogen in the fuel, on the amount of heat released are shown in Figures 4a, 5a, 6a, 7a. In these figures, we see that an increase in the proportion of hydrogen increases the rate of heat release, and, consequently, the efficiency of supply of heat. The conductivity characteristics corresponding to these regimes are given in the Figures 4b, 5b, 6b, 7b. In these figures, we see that with an increase in the proportion of hydrogen, the signal intensity and its stability increase. This is observed both on the sensor at the spark plug and on the ionization sensor 2 located in the part of the combustion chamber remote from the spark plug.

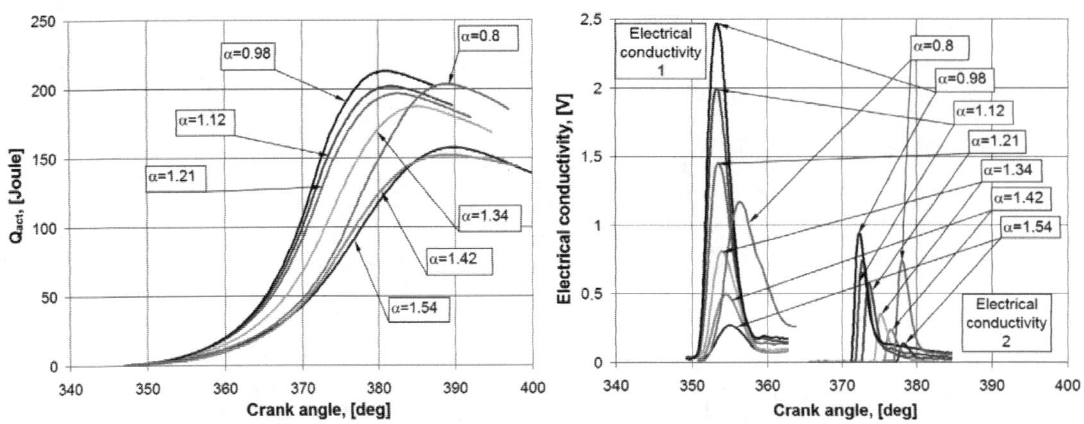

(a) (b)

Figure 4 – The effect of the excess air factor for the composition of the mixture CNG = 85%, H_2 = 15%:(a) on the amount of actively released heat in the combustion process; (b) on the electrical conductivity of the flame

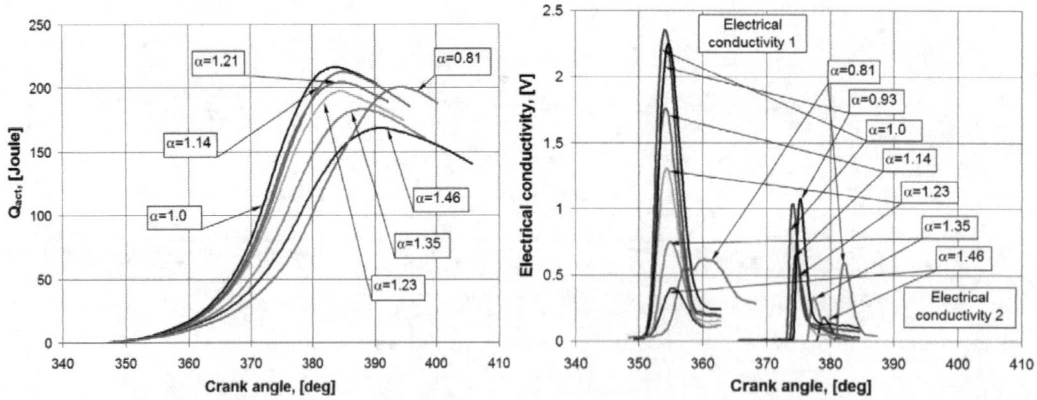

(a) (b)

Figure 5 – The effect of the excess air factor for the composition of the mixture CNG = 90%, H_2 = 10%: (a) on the amount of actively released heat in the combustion process; (b) on the electrical conductivity of the flame

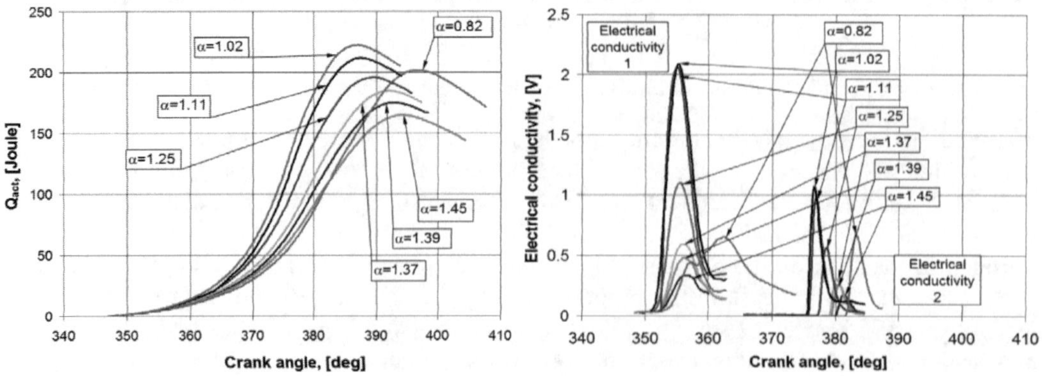

(a) (b)

Figure 6 – The effect of the excess air factor for the composition of the mixture CNG = 95%, H_2 = 5%: (a) on the amount of actively released heat in the combustion process; (b) on the electrical conductivity of the flame

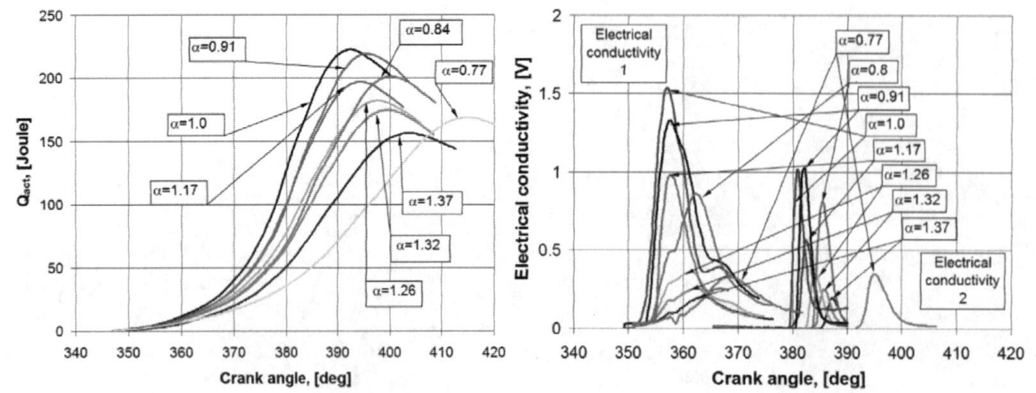

(a) (b)

Figure 7 – The effect of the excess air factor for the composition of the mixture CNG = 100%: (a) on the amount of actively released heat in the combustion process; (b) on the electrical conductivity of the flame

Carrying out the analysis of the change in the amount of actively released heat in the combustion process from the excess air factor (Figure 8a), one can see that on the regimes where there is an active combustion, the values approximately on one curve. The exception is the depletion zone when working on CNG (α = 1.26 – 1.37) where there is a sluggish burning, which affects the reduction of heat dissipation efficiency.

The dependence of the amplitude of the ion current of the electrical conductivity of the flame front on the composition of the mixture is shown in Figure 8b. Where it can be seen that, for the ionisation probe 2 remote from the spark plug, all the points lie on one curve, regardless of the amount of hydrogen added. For the ionization sensor 1 located near the spark plug, the amplitude of the signal when operating on the CNG is slightly lower than the corresponding values with the addition of hydrogen.

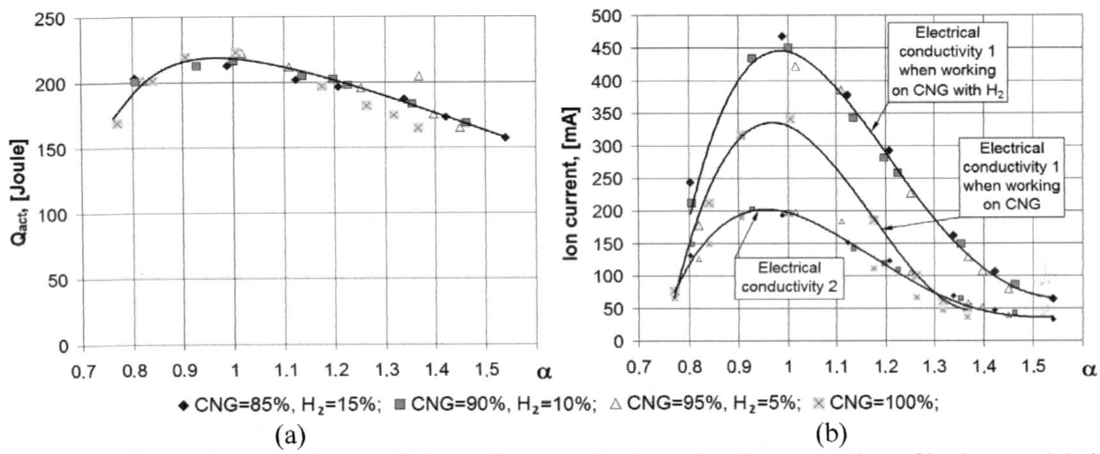

Figure 8 – Dependence on the coefficient of excess air at varying the proportion of hydrogen: (a) the amount of actively released heat in the combustion process; (b) the amplitude of the ion current of the electrical conductivity of the flame front

Estimating the possibility of monitoring the progress of the combustion process by ionization sensors, an analysis of the electrical conductivity. The following regularities are revealed, which relate them to the heat release parameters. On the Figure 9 shows the dependencies of the amount of actively evolved heat at the moment the signal appears on the sensor from the amplitude of the ion current signal when the fraction of hydrogen.

For the ionization sensor 1, some decrease in the amount of released heat from the moment of ignition until the appearance of a signal on the ionization sensor. This is due to the fact that the combustion process, which develops more actively, quickly forms a stable flame front and begins to spread through the combustion chamber. For this reason, a stratification is observed between the values in Figure 9a when working on CNG and CNG with hydrogen additives. Considering the second ionization sensor, we note the presence of a characteristic dependence. Which shows that for large amplitudes of the ion current signal, an increase in the amount of released energy is characteristic at the time of the appearance of the signal at the ionization sensor. This is due to the fact that the increase in the amplitude of the ion current signal is associated with high combustion rates. Consequently, at the time the signal appears on the sensor, the volume of the fresh mixture is less. The absence of a clear correlation between the amount of released energy at the moment of the appearance of the ion current on the first sensor is explained by the close arrangement of the electrode to the spark plug (distance between them 7 mm) and the high stochastic ignition and combustion process. The obtained data show that for evaluation of heat release in the cylinder of the engine, the sensor located in the more remote part of the combustion chamber is more informative. Since it summarizes the effect of hydrogen on the combustion process in the first and second phases.

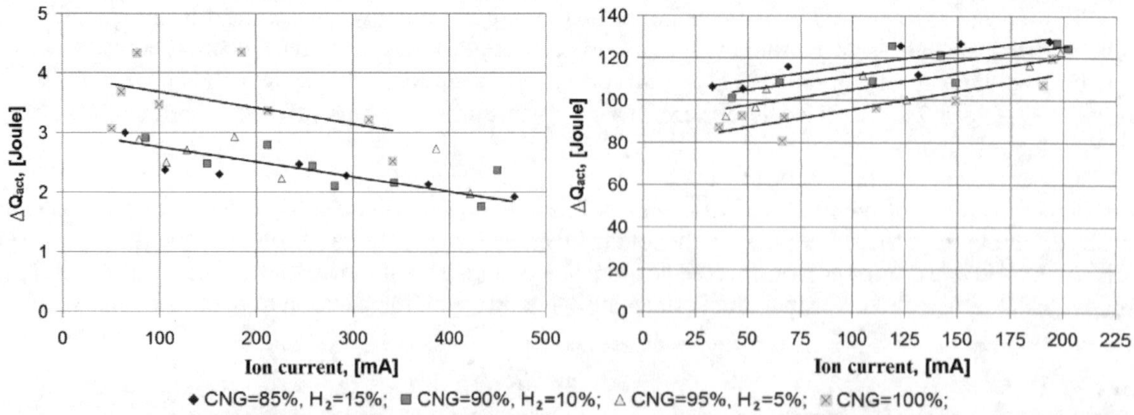

(a) (b)

Figure 9 – Dependence of the amount of actively released heat at the moment of the appearance of the ion current on the amplitude of the ion current signal with a variation in the fraction of hydrogen: (a) on the ionization sensor 1; (b) on the ionization sensor 2

The estimating the relationship of finer parameters of electrical conductivity to the rate of heat release, the notion of "current density" on the sensor, that is this is the average value of the signal. Also, the average rate of heat generation in the signal zone at the ionization sensor is shown, indicating the average amount of released heat on the 0.27 degree steps of the rotation angle. The dependence of these parameters is shown in Figure 10, where Figure 10a is the dependence for the first ionization sensor, and Figure 10b for the second ionization sensor.

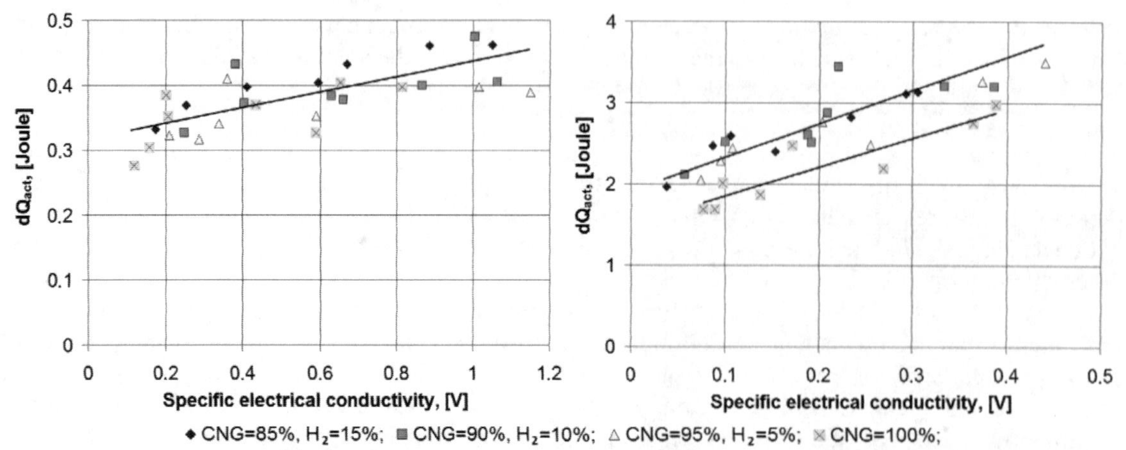

(a) (b)

Figure 10 – The dependence of the specific heat released during the time of the ion current signal on the average voltage density of this signal: (a) on the ionization sensor 1; (b) on the ionization sensor 2

It can be seen from the graphs that the increase in the average value of the ion current signal reflects the increase in the average rate of heat release in the zone of the ionization sensor. The principal effect of the change in the share of hydrogen in gas fuel on the increase in the average specific heat release rate in the ion current detection zone on the first ionization sensor was not detected. At the same time, for the second ionization sensor, there is a difference between the given dependence when working on CNG and CNG with hydrogen additives.

It is more correct to evaluate the characteristics of the entire flame conduction signal for the intensity of chemical oxidation reactions in the flame front, but for engine control systems, the parameter of the amplitude of the ion current signal. Therefore, the dependence of the magnitude of the amplitude of the ion current signal on the average voltage density of this signal was shown, as shown in Figure 11.

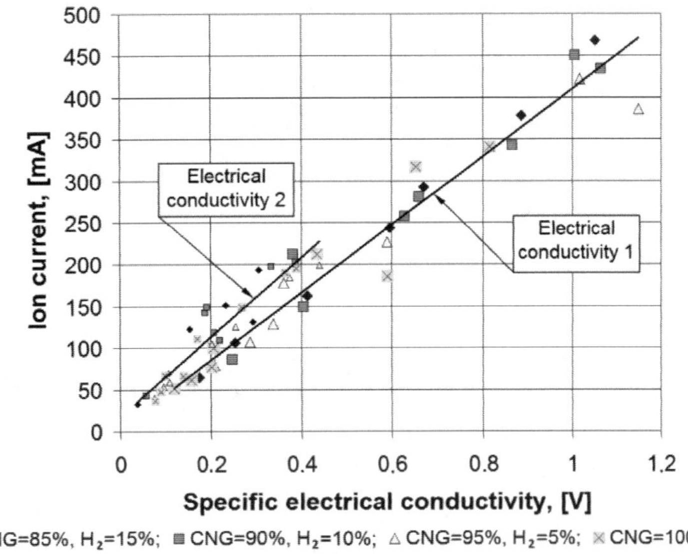

Figure 11 – The dependence of the amplitude of the ion current signal on the average voltage density of this signal

The dependencies for the ionization sensor 1 in the spark plug and ionisation probe 2 observed in Figure 11 have a similar picture, and in rough approximation they can be put on one straight line. Thus, the most easily determined and easily correlated parameter of the electrical conductivity of the flame remains the amplitude of the ion current signal. As a result, the most informative is the sensor remote from the spark plug, where the flame front has already formed and the signal has an easily readable shape.

4. Conclusion

1. The presented results of experimental studies show a significant effect of hydrogen on the rate of heat release. And also, as the analysis of the electrical conductivity shows, the effect on the rate of chemical oxidation reactions in the flame front, which is reflected by an increase in the amplitude of the signal and an increase in its uniformity.

2. The possibility of determining the parameters of the heat release rate from the electrical conductivity. As the specific, that is, from the start of the combustion process to the appearance of the ion current, and locally, in the detection zone of the flame signal at the sensor electrode. It is shown that the stability of the signal removed from the spark plug of the ionization sensor is higher and it is easier to interpret it in the characteristics of the heat release rate.

3. The possibility of estimating the efficiency of the combustion process by signals from ionization sensors is shown, which makes it possible not only to estimate the fuel surplus, the composition of the gas fuel, but also the rate of heat release and, consequently, the efficiency of the combustion process, which shows the wide possibilities for regulating the working process.

5. Acknowledgments

This work was financially supported by the Russian Foundation for Basic Research (project no. 16-38-00331).

References

[1] Upadhay D., Rizzoni G. AFR Control on a Single Cylinder Engine Uzing the Ionizaition Current. SAE 980203
[2] Hellring M., Munter T., Rögnvaldsson T., Wikström N., Carlsson C., Larsson M., Nytomt J. Robust AFR Estimation Using Ion Current and Neural Networks. SAE 1999-01-1161
[3] Saitzkoff A., Reinmann R., Mauss F., Glavmo M. In-Cylinder Pressure Measurements Using the Sparg Plug as an Ionization Sensor. SAE 970857
[4] VanDyne E.A., Burcmyer C.L., Wahl A.M., Funaioli A.E. Misfire Detection from Ionization Feedback Utilizing the Smartfire Plazma Ignition Tecnology. SAE 2000-01-1377
[5] Auzins J., Johansson H., Nytomt J. Ion-Gap Sense in Misfire Detection, Knock, and Engine Control. SAE 950004
[6] Smolenskaya N.M., Korneev N.V. Modelling of the combustion velocity in UIT-85 on sustainable alternative gas fuel, IOP Conf. Series: Earth and Environmental Science 66 (2017) 012016 doi:10.1088/1755-1315/66/1/012016.
[7] Shaikin A.P., Galiev I.R. On the effect of temperature and the width of the turbulent combustion zone on the ionization detector readings, Technical Physics, 61(8) (2016) 1206-1208. DOI: 10.1134/S1063784216080247.
[8] Shaikin A.P., Galiev I.R. Relationship of flame propagation speed for methane-hydrogen fuel of a Internal Combustion Engine with parameters of ion current and hydrogen concentration, Russian Aeronautics (iz VUZ). 2016. V.59(2). P.249-253 DOI: 10.3103/S106879981602015X
[9] Smolenskaya N.M., Smolenskii V.V., Bobrovskij I. Research of polytropic exponent changing for influence evaluation of actual mixture composition on hydrocarbons concentration decreasing on deep throttling operation, IOP Conf. Series: Earth and Environmental Science 50 (2017) 012016 doi:10.1088/1755-1315/50/1/012016.
[10] Heywood J.B. Internal Combustion Engine Fundamentals, McGraw-Hill, – New York : McGraw-Hill, 1988. – 931 p.
[11] Li Y., Jia M., Chang Y., Kokjohn S.L., Reitz R.D. Thermodynamic energy and exergy analysis of three different engine combustion regimes, Applied Energy, 180 (2016) 849-858. DOI: 10.1016/j.apenergy.2016.08.038.
[12] Zhao Z., Wang S., Zhang S., Zhang F. Thermodynamic and energy saving benefits of hydraulic free-piston engines, Energy, 102 (2016) 650-659. DOI: 10.1016/j.energy.2016.02.018.

The metrological assurance of quality petroleum products. Methods for determining octane and sulfur numbers in the petroleum products

A V Shalkov[a], Y A Fadeev

Branch KuzSTU in the city of Prokopyevsk, 19a, Nogradskaya street, Kemerovo region, Russia Federation, 653039

E-mail: [a] prk-s@yandex.ru

Abstract. At present, solving environmental problems in industrially developed regions with a large concentration of mining and machine building enterprises is one of the main socially important tasks. Taking into account the increase in the volume of mining, there is an increase in the environmental burden, which affects the internal migration of the population. This is particularly evident in the examples of single-industry towns, in which a gradual decrease in the young workable population occurs. The article presents an analysis of the sources of maximum pollution of the environment by coal mining enterprises. Modern methods of controlling automobile fuel were analyzed. The analysis of fuel quality and the environmental assessment of combustion products was carried out. The equipment used in the article makes it possible to exclude substandard fuel and to reduce harmful emissions of vehicles to the atmosphere.

Introduction

A research aim is development of the metrological assurance of account of amount and quality of petroleum products in depositories, namely in reservoirs.

Because measuring of two parameters is in-process conducted: qualities and amounts of petroleum products, then it is necessary clearly to put the aims of researches, to obtain successful results. In hired next tasks are put.

For determining the amount of petroleum products:

1) to investigate the existent methods of implementation of gaging and conduct a comparative analysis;

2) to analyze the modern metrological assurance of petroleum storage depot, both in Republic of Kazakhstan and in other countries.

For measuring of quality of petroleum products:

1) to investigate the methods of determination of quality of petroleum products in oil storage tanks;

2) to offer the method of determination of quality of petroleum products, suitable for petroleum storage depot.

Analyzing and comparing methods, it is necessary to choose most exact and going near the terms of measuring in static reservoirs method.

Results and Discussion

On petroleum storage depots at the use of reservoirs for measuring of mass of petroleum products in the commercial account of measuring error caused by the construction of reservoir, exploitation and setting of measuring equipment, and also by methods and procedures of measuring of level, closeness and temperature. The considerable errors, conditioned by measuring instruments, are usually small. Most errors at measuring are stipulated by a human factor at the hand method of measuring, especially if the chosen method is used without the proper supervision.

Presently most enterprises use the hand method of implementation of measuring. In this case the best error protection is complete or partial replacement of method of implementation of measuring automatic.

IOP Conf. Series: Earth and Environmental Science **115** (2018) 012040 doi:10.1088/1755-1315/115/1/012040

CAS of control and account of mass of petroleum products in stationary reservoirs is used for determining the amount of petroleum products in every controlled reservoir on storage of fuel. Software and devices of account of amount of petroleum products are usually included in the system. The device used for development of methodology of account of amount of petroleum products in a reservoir in dissertation is radar liquid level indicator.

Thus, the use of CASS considerably simplifies the account of petroleum products and allows realizing the any, even most difficult methods of implementation of measuring of mass in him.

What amount is important any consumer (in this case to the motor-car enthusiast) and what quality a fuel he inundates in the tank of the car. Therefore for a consumer the not lab test of fuel is needed, and methods of express analysis directly at the NGV-refueling station. Various devices are presently developed for this purpose, including portable octane meters. Principle of work of Octane meters is simple. On the size of measureable inductivity of fuel determine an antiknock value.

Together with the scientific laboratory of motor transport of the Kuzbass State Technical University the problem of determination of quality of commodity brands of petrol was set. Studies of quality of trade petrol were undertaken by the method of determination of antiknock value.

As to trace the way of fuel from OR (Oil Refinery) it is practically impossible to the NGV-refueling station, then petrol must be subject to the obligatory certification and certain party of petrol must have a passport of quality.

Petrol is composition more than 250 meaningful components of hydrocarbon connections and not all of them are useful. According to ecological requirements in the fuel of sulphur must not exceed 0.05 %, and benzol (strongest poison of blood) no more than 5%.

Metals, such as lead, iron, manganese in automotive gasoline determined by optical methods on atomic-absorbing spectrometers.Analyses show that forbidden octane-increasing of metal-containing additions is not present.

A question about quality of fuel every day becomes all more actual, because the expense of fuel, resource of thruster-on, ecological safety, depends on quality. A question about the amount of fuel also plays a not insignificant role the trade and economic aspect of production and consumption of oil product. It would be desirable to mark that determinations of quality and amount in petroleum industry are very associate.

The quality of petroleum products depends on the amount of impurities contained in its composition. The main impurities determination in the oil products are:

- determination of sulfur mass fraction;
- determination of benzene volumetric fraction;
- determination of iron concentration;
- determination of manganese concentration;
- determination of lead concentration;
- determination of oxygen mass fraction and oxygenate volumetric fraction;
- determination of hydrocarbon volumetric fraction (aromatic, olefin);
- determination of octane number.

In this article we estimate the petrol quality on such characteristics as the sulfur mass fraction and octane number. The organic compounds of sulfur appear to be the natural constituent of crude oil [8]. During thermal influence in the process of crude oil refining, the sulfur and its compounds fall into oil products in different concentrations. The basic forms of sulfur compounds presence in oil products are:

- hydrogen sulfide H2S produced by high-temperature deterioration of sulfur-containing compounds;
- elemental sulfur, sulfide oxidation product;
- mercaptans R-SH;
- sulfides or thioether or R-S-R;
- disulfides and polythioesters R-S....S-R;
- thiophen C_4H_4S and its derivatives;
- bicyclical and polycyclic sulfur-containing compounds;

• acid and middle esters of sulfuric acid and, sulfuric acids produced in the process of the petroleum distillates refining [1].

The presence of these compounds is undesirable as they impart an unpleasant odor oils products cause corrosion of equipment and pollute the atmosphere when burned. Sulfur compounds poison expensive catalysts of oil refining and releasing sulfur oxides into the atmosphere during combustion pose environmental issues [1].

Methods of determination of sulphur in an oil product it is possible conditionally to divide into 3 groups:

1. Methods based on oxidization of sulphur and subsequent determination of oxides: ASTM D129, D1266, D1551, D1552, D3120, D5453, D6920 and their analogues.
2. Methods based on renewal of sulphur to H_2S: ASTMD4045, UOP 357, STATE 13380.
3. Spectralmethods:

a) based on the x-rayed radiation: ASTMD2622, D4294, D6334, D6443, D6445, D7039, D7212, D7220 and their analogues;

b) atomic-emission: ASTMD4951, D5185 [1].

The conducted laboratory researches are in the Kuzbass State Technical University; on the energy dispersible analyzer of «Spectroskan S» (St. Petersburg) showed that in a fuel producible maintenance of sulphur is considerably exceeded in Russian Federation. According to the standard of Evro-5 in petrol maintenance of sulphur must not exceed 0.0001 %. . The analyses repeatedly conducted by us on maintenance sulphur in a fuel showed that maintenance of sulphur in petrols of AI-92 and AI-95 made the 3 mg/ kg ± of 1 mg/ kg, which corresponds to the brand of Evro-5. We will mark that on this description of AI-92 near to the fuel of AI-95.

An antiknock value is determined by the selection of mixture of standard hydrocarbons - isooctane, at that an antiknock value is equal to 100 and N-Heptane (normal heptane) at that an antiknock value is equal to 0. At the equal terms of test detonation firmness is equal to detonation firmness of the tested petrol [7]. A percentage of isooctane in the got mixture just and is the antiknock value of petrol.

A research method is determined on the single-cylinder setting with the variable degree of compression, called UIT-65 or UIT-85, at frequency of rotation of crankshaft 600 turn/min, to the temperature of the sucked in air 52° with and coal of passing of lighting 13 hails. It shows, as petrol behaves in the modes of the small and middle loading.

A motor method is determined similarly on the single-cylinder setting, at frequency of rotation of crankshaft 900 turn/min, to the temperature of the sucked in mixture 149° with and variable coal of passing of lighting. MON has more subzero values, than RON.

MON characterizes behavior of petrol on the modes of the large loading. Has influence on high-rate and detonation at a partial choke acceleration and thruster-on on-loading, motion uphill and etc.

A tent-bed test is a single-cylinder combustion engine with a carburetor. Start him on the investigated petrol, and the level of detonation is fixed by the special sensors. After mixture of standard fuel sneaks up - isooctane and N-Heptane on that an engine works how and on the investigated fuel. Got percentage of isooctane in neat standard mixture and is description of detonation firmness of petrol. Id est if in mixture of 95% isooctane, an antiknock value will be 95.At test method modes and parameters of the motor installation can detect explosive properties of gasoline in the operation of a vehicle in an urban environment (movement with variable speed) [6]. The research method is less rigorous test mode that allows you to explore the process of combustion of gasoline cars in operation at constant engine operation modes. In this way, the octanes number the research method 5-10 units higher than the motor.

In the scientific laboratory of the road's transport of KSTU (Kuzbass State Technical University) laboratory tests of motor fuel were carried out different brands for compliance to the octane number. Measuring octane petrol was conducted using Octane meter «Octane IM».

The method is based on the measurement of dielectric constant fluids. Below are graphs of the dielectric constant (X-axis) of the octane number (Y-axis) (Figure 1).

This dependence has become possible to obtain due to the fact that now the main stream of petrol is produced by blending (compounding) the large number of components.

The measurements showed that for petrol AI-92 octane the research method of 92.4 units, and by the motor method 84.3 units. For petrol of AI-95 octane the research method of 89.5 units and 82.9 for motor method. So it can be concluded that the fuel AI-95 in the selected part corresponds to the studies STATE.

The methods of metrology control of the oil storage depot, generals of measuring of mass of oil, account of oil products on modern petroleum bases, are in-process studied and investigational [9]. The analysis of existent types of reservoirs, conducted in-process, is applied for the study of quality of product at storage on the oil storage depot. Also on maintenance of quality of oil product the review of the special events was conducted.

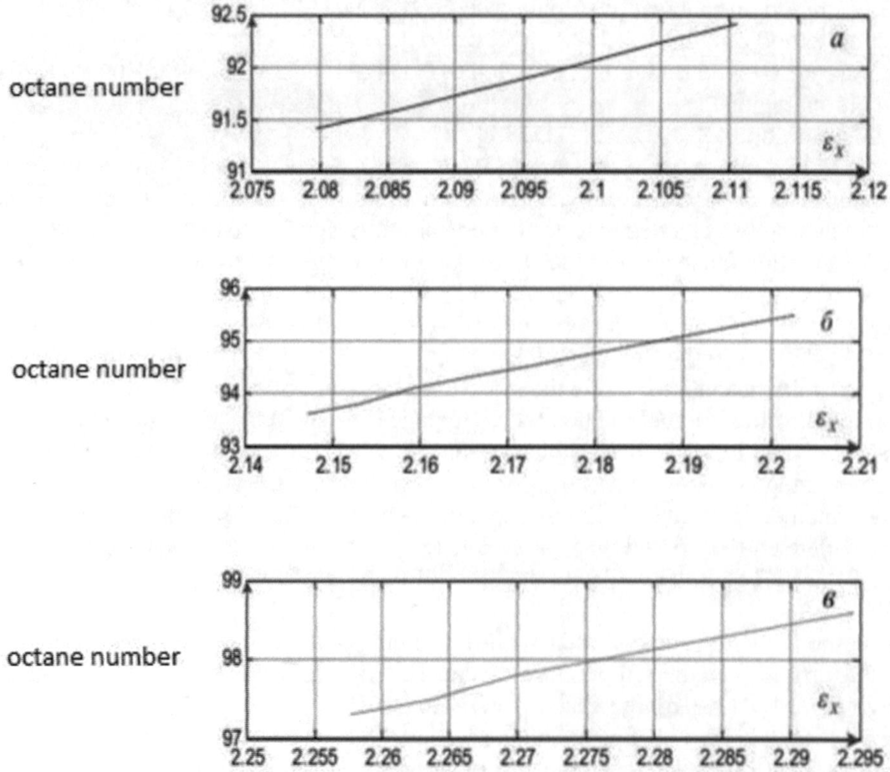

Figure 1. Graphs of the dependence of the dielectric constant on the octane number.

Work on the review of methods of account of quality and amount of oil product and facilities of their measuring is conducted. Investigating problems and methods of measuring of amount of oil products, was analyzed and chosen methodology of account of amount of oil product [10]. On undertaken a study of technological process of account of quality and amount of oil product it is possible to draw conclusion in oil storage tanks, that in course of time the automatic devices of account will master the great number of enterprises of petroleum bases, because it will allow quickly and exactly to define the level of oil product real-time.

Conclusion

Factors influencing on inaccuracy of gaging in a reservoir and way of their decision were studied. Errors, conditioned by measuring instruments, influences of environment, human factor, it is possible to minimize, using the automatic systems of account.

Work is executed on research of technological process, principle of development of algorithms of account of quality and quantitative descriptions of oil products.

On petroleum bases can working facilities of measuring be: octanometer - for determination of antiknock value, areometer with a built-in thermometer - for determination of closeness and temperature, radarlevel gage - for the automatic account of level of oil product in a reservoir.

The pre-production model of portable device of «Octane-IM» is investigational for the expressanalysis of antiknock value. This analyzer allows measuring the concentration of antiknock value in an oil product in a range from 67 to 98 units.

The method of determination of closeness of oil product is also investigational by means of areometer of «ANT-1». This device is used as means of measuring of closeness in ranges 650-710, 710-770, 770-830, 830-890, 890-950, 950-1010, 1010-1070 kg/m3.

Radar level gage of BM 70a allows a noncontact method to define the level of oil product in a reservoir. The radar signal released from aerial is reflected from a liquid and acts on aerial. Difference of frequencies between radiated and reflected by signals proportional to distance to the surface of liquid.

On the basis of research of devices methodology of implementation of measuring of antiknock value, closeness and level of oil product is worked out.

The put tasks and aims are in-process attained on the metrology providing of the oil storage depot. Offer methodologies are applicable in the conditions of real OR (Oil Refinery) and NGV-refueling station.

References

[1] E.A. Novikov «Determination of sulphur is in oil products. Review of analytical methods», magazine «World of oil products.Announcer of petroleum companies» №1, 3, 4, 5, 2008.
[2] V.E. Emelyanov«All about a fuel. Motor-car petrol»,M.:LTDpublishing house «Astrel», 2003, 79p.
[3] Y.A.Fadeev, A.G. Prikin «Electronic-vibration structure of molecules and crystals», Tomsk: TSU, 1999, 174 p.
[4] V.N. Korolev, A.V. Margulin, V.B. Cagedravsky «Method of determination of detonation descriptions of oil products on the basis of regressive analysis of spectrums of absorption in a near infra-red range», magazine of the applied physics, 2000, T.70, 9, 90-95 p.
[5] Y.A.Fadeev «Influence of intramolecular cooperations on intensity of IR-lines of identical connections in the conditions of near to resonance», magazine of the applied spectroscopy, 1993. T.59. 1-2, 67-71 p.
[6] O. S. Vitkovskii and N. R. Nikolaeva, "Analyzers of the pressure of saturated vapors of petroleum products," Prib. Sreds. Avtomatiz., No. 8, 45–49 (2010).
[7] Nadkarni, R. A., Ed., Elemental Analysis of Fuels and Lubricants: Recent Advances and Future Prospects, ASTM STP 1468, ASTM, 2004.
[8] Rand, S. J., Ed., Manual on Significance of Tests for Petroleum Products, 7th ed., ASTM, West Conshohocken, PA, MNL 1, 2003.
[9] Speight, J.G., and Ozum, B. 2002. Petroleum Refining Processes. Marcel Dekker, New York.
[10] Riazi, M.R.; Roomi, Y.A. Preprint. Pap. presented at the Division of Petroleum Chemistry, ACS National Meeting, Chicago, Illinois, August 28-31, 2001.

IOP Conf. Series: Earth and Environmental Science **115** (2018) 012041 doi:10.1088/1755-1315/115/1/012041

Effect of Inverter Power Source Characteristics on Welding Stability and Heat Affected Zone Dimensions

D P Il'yaschenko[1, a], **D A Chinakhov**[1], **R A Mamadaliev**[2]
[1]Yurga Institute of Technology, National Research Tomsk Polytechnic University, Leningradskaya 26, Yurga, Russia, 652055
[2]Tyumen Industrial University, Volodarskogo str., 38, Tyumen, Russia

E-mail: [a] mita8@rambler.ru

Abstract. The paper presents results the research in the effect of power sources dynamic characteristics on stability of melting and electrode metal transfer to the weld pool shielded metal arc welding. It is proved that when applying inverter-type welding power sources, heat and mass transfer characteristics change, arc gap short-circuit time and drop generation time are reduced. This leads to reduction of weld pool heat content and contraction of the heat-affected zone by 36% in comparison the same parameters obtained using a diode rectifier.

Introduction

At present, manufacturers of welding equipment (including Fronius (Austria), Lincoln Electric (USA), ESAB (Sweden), UKP (Tekhnotron, Russia), VKZ (ITS, Russia), etc.) produce a wide range of shielded metal arc welding power sources. Almost all the world leaders in the field of welding production are focused primarily on designing and producing inverter welding power sources. However, production plants still use welding diode rectifiers are because they lack knowledge of the dynamic characteristics of inverter power sources.

There is a number of research works [1-7] devoted to improving stability of shielded metal arc welding. It is also important to increase efficiency of new power sources for shielded metal arc welding and to assess their influence on heat and mass transfer and on performance properties of permanent joints.

The aim of the work is to assess how dynamic properties of inverter power sources affect stability of melting and electrode metal transfer, the later being the main heat and mass transfer parameters affecting the area of a welded joint heat affected zone.

Methodology

To study the characteristics of heat and mass transfer depending on dynamic properties of a power source we used an ARC-250 inverter rectifier which implements a high-frequency energy conversion method, a VD-306E diode rectifier based on the traditional energy conversion method and a VDE-160single-phase welding rectifier. We also used the most common electrodes coated with LB-52U (KOBELCO) and UONI-13/55 (ESAB). Table 1 shows surfacing conditions for test samples.

IOP Conf. Series: Earth and Environmental Science **115** (2018) 012041 doi:10.1088/1755-1315/115/1/012041

Table 1. Surfacing conditions

Rectifier type	Electrode classification	Average parameters values	Number of short circuits during surfacing
Diode	LB-52U	Current 89 ± 2.7 A Voltage 20.8 ± 0.6 V Predicted welding rate 0.25 m/min	17
Invertor			22
Single-phase			21
Diode	UONI-13/55	Current 88 ± 2.7 A Voltage 21.5 ± 0.6 V Predicted welding rate 0.29 m/min	17
Inventor			22
Single-phase			27

Results and Discussion

Stability of the welding process was assessed using static processing of welding current and voltage oscillograms of inverter, diode and single-phase rectifiers (Fig. 1) using the method developed earlier [7]. The results are shown as diagrams in Figures 2 and 3. The oscillograms were obtained using AKIP-4122 / 1V digital storage oscilloscope, Pintek Electronics DP-50 differential probe; PR 1030current probe remover; OWON_Oscilloscope_2.0.8.26 software.

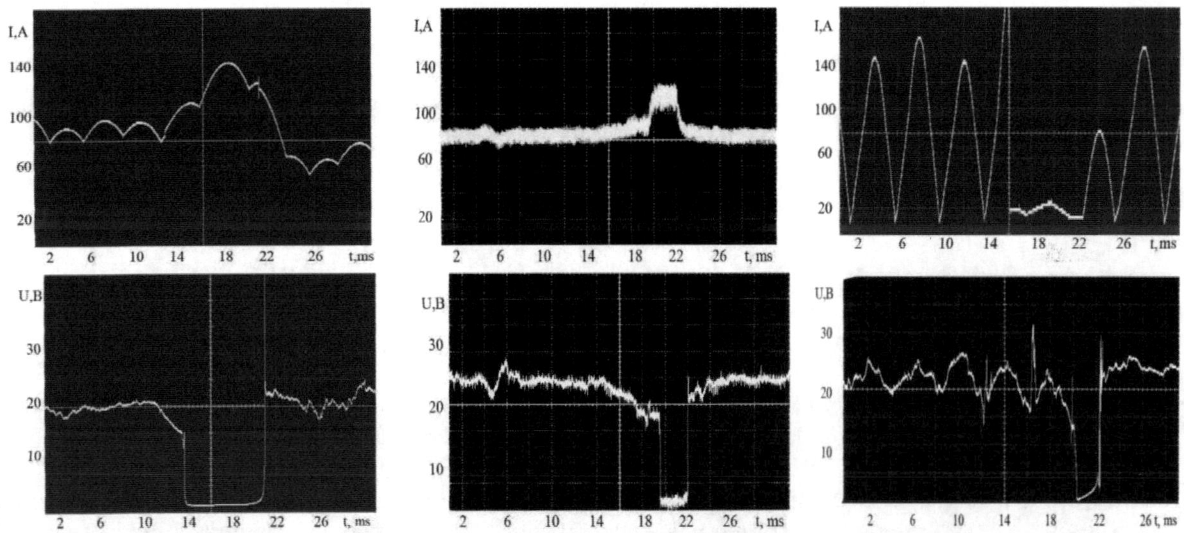

Diode rectifier Inventor rectifier Single-phase rectifier

Figure 1. Oscillograms of welding circuit current and voltage between the electrode and the product, (3.2 mm in diameter LB-52U electrodes)

The experiment data analysis (Table 1 and Figure 1) shows that under identical surfacing conditions, frequency of short circuits in the arc gap increases and short circuit time shortens. It shows the advantages of the inverter rectifier before the diode and single-phase ones and may prove that the electrode metal is transferred in smaller drops.

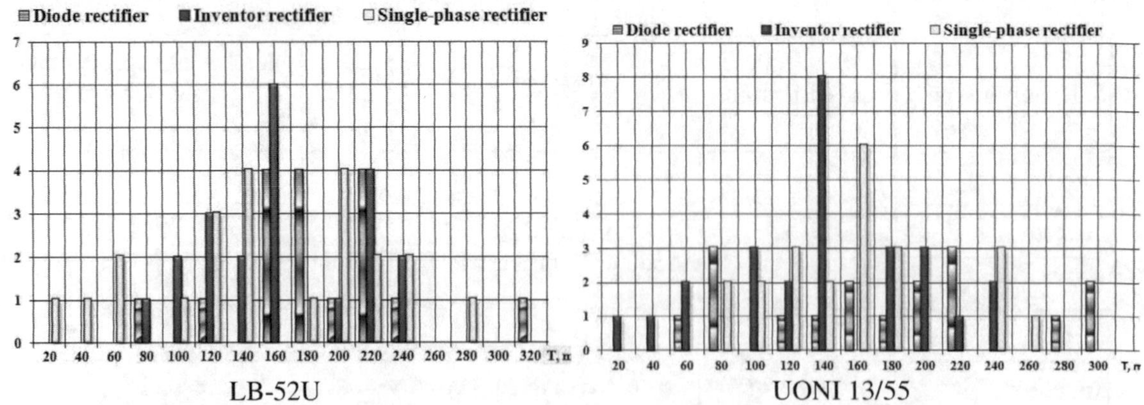

Figure 2. Diagram of time between short circuits for coated electrodes

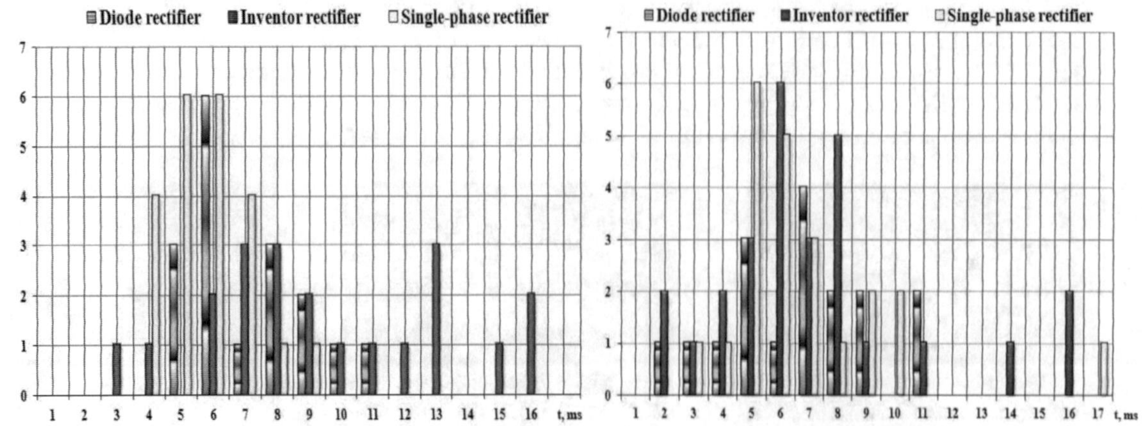

Figure 3. Diagrams of short circuits time for coated electrode

The diagrams analysis (Figures 2 and 3) shows a greater stability of the electrode metal transfer to the welding pool when welding with an inverter power source. It means that the drop transfer time shortens and the number of short circuits increases, i.e. a finer drops of electrode metal drop are transferred [8, 9] which can be explained by a high short-circuit current increase rate with the inverter rectifier in contrast to the diode (three-phase) one in which the current increases slower and in a random way.

Reducing the arc burning time when welding with an inverter power source changes the heat input to the welded product which aligns with the results of [10, 11]. We have made a series of welds using different types of equipment: inverter, diode and single-phase rectifiers. A pipe of low-alloyed steel containing 0.09% C, less than 2% Mn, less than 1% Si with dimensions 159 × 6 mm was welded (S17 joint) with the following electrodes: bottom run – LB-52U (d = 2.6 mm), welding current I = 50...60 A; filling – LB-52U (d = 3.2 mm), welding current I = 80...90 A.

The results of macrosection treatment (Fig. 4, a) using the procedure described in [12, 13] implemented as shown in Fig. 4, b, show that when using the inverter rectifier, the heat affected zone area decreases by 18% and its width is reduced by 25% in comparison with the diode rectifier.

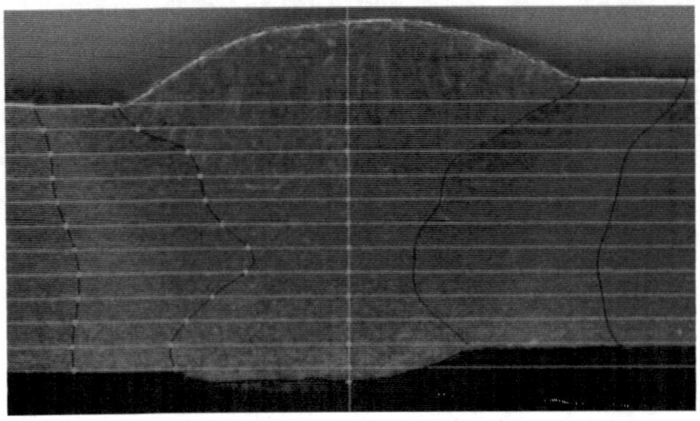

a

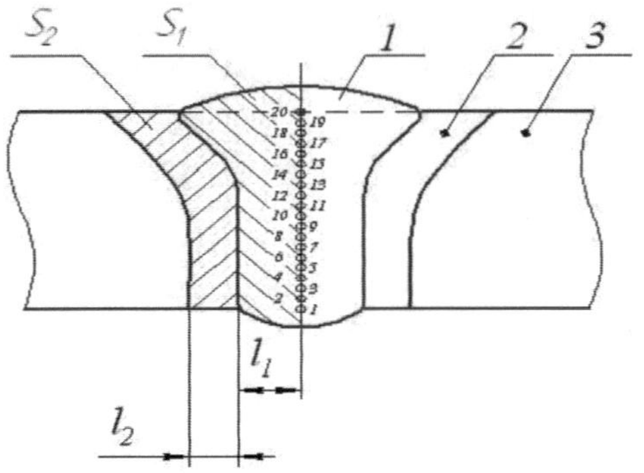

b

Figure 4. Macrosection processing scheme obtained by Compass 3D software:
a – steel sample; b – measuring scheme (point spacing is 0.5 mm):
1 – welded seam; 2 – HAZ; 3 – base metal; l_1 – distance from the weld axis to the weld-fusion line;
l_2 – HAZ width; S_1 – welded seam area; S_2 – HAZ area

Table 2. Experimental data of macrosections processing

Steel brand	Welded joint type according to the State Standard	Rectifier type	S_1, mm^2	l_1, mm	S_2, mm^2	l_2, mm
45	C17 thickness10 mm	Diode	133±5.1	4.06±0.04	95.5±2.9	5.17±0.04
		Invertor	129±2.6	4±0.04	80.6±2.8	3.34±0.03
		Single-phase	132±3.1	4±0.04	84±2.8	4.1±0.03
09G2S	C17 thickness 6 mm	Diode	23.6±0.4	3.33±2.6	51.8±0.53	2.08±0.06
		Inventor	21.4±0.3	2.89±0.94	42.6±0.28	1.58±0.04
		Single-phase	22.1±0.3	2.93±1.1	44.7±0.3	1.7±0.05

Analysis of the data in Table 2 shows that the HAZ area decreased by 15% and the average HAZ width lowered by 36% when using an inverter rectifier in comparison with the diode one. As proved in [14-17], destruction of welded structures most often occurs within HAZ, its structure, area and properties depending on the thermal cycle conditions. Thus the decrease in the HAZ area increases operational strength of the welded joint.

Conclusion

It is proved that the use of an inverter rectifier in comparison with a conventional diode one reduces an average time of arc burning at the periods of electrode melting and electrode metal drops generation by 36% on average, which helps to reduce overheating of the welded product. It also cuts down the HAZ area by 15% and the HAZ width by 36%, which contributes to improving welded joint performance.

The research is carried out at Tomsk Polytechnic University within the framework of Tomsk Polytechnic University Competitiveness Enhancement Program grant.

References

[1] Milyutin V.S., Polukhin A.V. Evaluation of the stability of the process when testing power sources for manual arc welding. Welding and diagnostics. 2012. No. 6. P. 30-37.
[2] Yushin A.A., Gladkov E.A. Criteria development for rapid assessment of welding properties of modern welding equipment. Welding and diagnostics. 2010. 4. P. 52-55.
[3] Bardin V.M., Borisov D.A., Zemskov A.V. Dynamics of transients in welding inverter rectifiers. Practical power electronics. 2012. 3. P. 52-55.
[4] Bardin V.M., Zemskov A.V. Load characteristic of a welding inverter rectifier. Practical power electronics. 2013. 2. P. 52-55.
[5] Iljyashchenko D.P., Chinakhov D. A. Investigating the Influence of the Power Supply Type upon the Weld Joints Properties and Health Characteristics of the Manual Arc welding. Materials Science Forum. 2012. Vol. 704. Pp. 608-611.
[6] Slazak B., Slania J., Węgrzyn T., Silva A.P. Process Stability Evaluation of Manual Metal Arc Welding Using Digital Signals. Materials Science Forum. 2013. Vol. 730–732. P. 847–852.
[7] Saraev Y. N., Chinakhov D. A., Iljyashchenko D. P., Kiselev A. S., Gordynets A. S. Investigation of the stability of melting and electrode metal transfer in consumable electrode arc welding using power sources with different dynamic characteristics. Welding International. 2017. Vol. 31. 10. p. 784-790.
[8] Zaruba I.I. New welding power sources. Kiev: Publishing House "Visha school", 1992. 144 p.
[9] Dedyukh R.I. Physical and technological properties of the welding arc. Tomsk: Publisher TPU, 2002. 92 p.
[10] Kirichenko K.Yu., Kazarin O.A., Gridasov A.V., Kosyanov D.Yu., Karabtsov A.A., Golokhvast K.S. The Influence of Electrode Coating Type on Key Parameters of PM10 Fraction of the Welding Aerosol. AIP Conf. Proc., 2017. Vol. 1874. P. 040017-1-040017-4.
[11] Kirichenko K.Yu., Drozd V.A., Chaika V. V., Gridasov A.V., Kholodov A.S., Golokhvast K.S. Nano- and microparticles in welding aerosol: granulometric analysis. Physics Procedia, 2017. Vol. 86. P. 50–53.
[12] Krampit A.G., Krampit N.Yu. A method for the determination of the geometrical dimensions and area of the welded joint. Welding International. 2013. 27. P. 834-836.
[13] Gryaznov R.V., Borilo L.P., Kozik V.V., Mal'chik A.G. Physicochemical Study of Ta_2O_5 and SiO_2-Ta_2O_5 Film Formation from Film-Forming Solutions. 2001 Russian Journal of Applied Chemistry 74(1), p. 18-20.
[14] Krivonosova E.A. Multifractal analysis of heat affected zone structural composition of steels with carbonitride hardening / O.A. Rudakova, G.V. Vstovsky // Factory laboratory. Diagnostics of materials. 2010. T. 76, No. 6. P. 26-30.
[15] Ilyaschenko D.P., Chinakhov D.A., Danilov V.I., Schlyakhova G.V., Gotovschik Y.M. Increasing Strength and Operational Reliability of Fixed Joints of Tubes by MMA Welding. IOP Conf. Series: Materials Science and Engineering. 91 (2015) 012007.
[16] Krivonosova E.A. Predicting the properties of the metal of welded joints on the basis of the results of quantitative parametrisation of the structure. Welding International. 2016. Vol. 30. pp. 459–462.
[17] Korotkova E. et al., Investigation of Structure and Magnetic Properties of 36NiCrTiAl Alloy and 12Cr18Ni10Ti Steel Welded Joints. Key Engineering Materials. 2017. Vol. 743, pp. 31-36.

Application of Non-Destructive Testing Methods for Analysis of Parameters of the Surface Layer State after Mechanical Processing

Maxim V Pimonov [a], Andrey A Krechetov, Nikolay V Ababkov

T.F. Gorbachev Kuzbass State Technical University, 28 str. Vesennyaya, Kemerovo, Russia

E-mail: [a] makc130685@gmail.com

Abstract. The paper presents a study aimed at describing the effect of machining parameters on the structural state of the surface layer on the basis of the mechanics of technological inheritance. The samples were controlled by the methods of nondestructive testing and the study of the values of microhardness. Dependencies between the results of the control and the method and treatment regimes are given.

Introduction

At present, there are many approaches to ensuring the ex-operational properties of machine parts. A number of studies have been carried out to determine the influence of the parameters of machining on the quality of the surface layer, as well as the quality of the surface layer on the operational properties [1, 2]. A more integrated approach to the development of technological processes, it is possible to develop on the basis of the apparatus of the mechanics of technological inheritance [3, 4]. By now in most works the description of technological inheritance boils down to difficultly applicable mathematical formulas that do not allow one to unequivocally determine the effect of inheritance on the parameters of the surface layer.

The goal is to develop the mechanics of technological inheritance in the direction of describing the parameters of the mechanical state in terms of the parameters of the structural-phase state of the material. To achieve this goal, a number of tasks are formulated:

1. To develop a structural-analytical model of the hereditary formation and transformation of the structural-phase state of the metal of the surface layer at the stages of machining.
2. To develop a methodology for experimental studies of the hereditary formation and transformation of the structural-phase state of the metal of the surface layer at the stages of mechanical processing.
3. To develop a complex of mathematical models characterizing the hereditary influence of the regularities of plastic flow on the formation of the structural-phase state of the metal of the surface layer, as well as the dependence of the parameters of the structural-phase state on the accumulated deformation of shear and the depletion of the plasticity reserve.
4. Develop a functional model for the design of technological processes that provide specified parameters of the structure and, as a consequence, certain operational properties.

In this paper, we present a structural-analytical model, the results of using non-destructive testing methods, as well as the results of measuring microhardness.

In the course of solving the set tasks, a structural-analytical model was developed reflecting the concepts of "Formation and transformation of the structural state" (Fig. 1) [5].

IOP Conf. Series: Earth and Environmental Science **115** (2018) 012042 doi:10.1088/1755-1315/115/1/012042

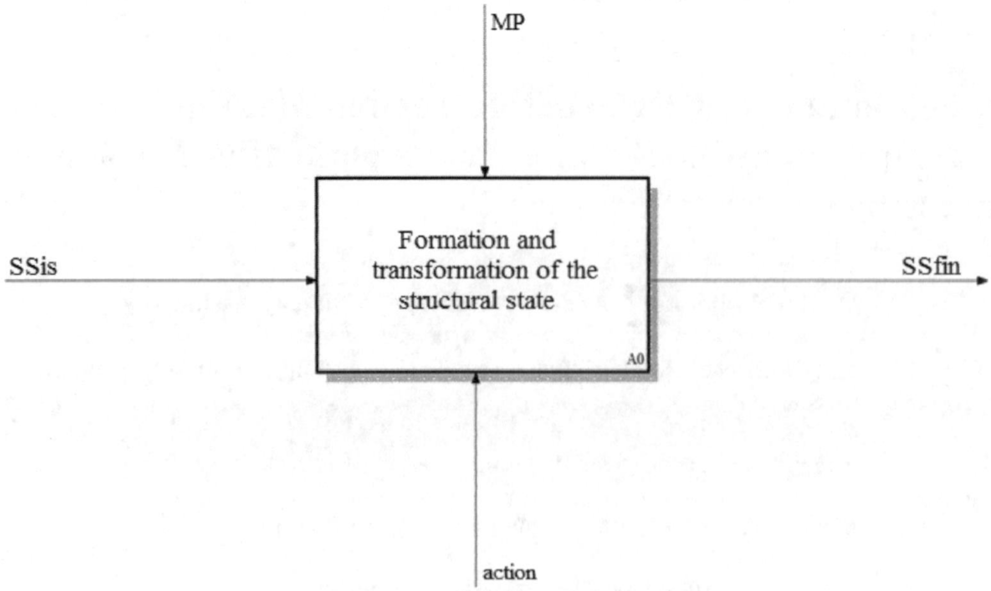

Fig. 1. A0 "Context function of the model":
SSis – the initial structural state of the material, SSfin – the final structural state, MP – is the material properties

In accordance with the developed model, at the stages of the life cycle, the structural state of the surface layer of the article is formed and transformed from the initial state to the final one. These processes occur due to the impacts of both technological and operational, the regularities are determined by the properties of the material.

The most interesting for us is "the formation of a structural state at the stage of machining" (Fig. 2), the block A1.1 at the output forms the parameters of the deformation focus (geometry, qualitative and quantitative picture of the distribution of the parameters of the stress-strain state) in accordance with regimes processing, the parameters of the deformation focus are, in turn, the mechanism for forming the loading program A1.2. Block A1.3 gives the output indicators of the structural state formed depending on the properties of the material and the mechanism of a certain loading program. Exhaustion of the stock of plasticity A1.4 is determined by the properties of the material, the process mechanism is the indicators of the structural state, the output is a structural state at the stages of processing and loading history.

The type of loading program can be considered the determining factor for the formation of a structural state.

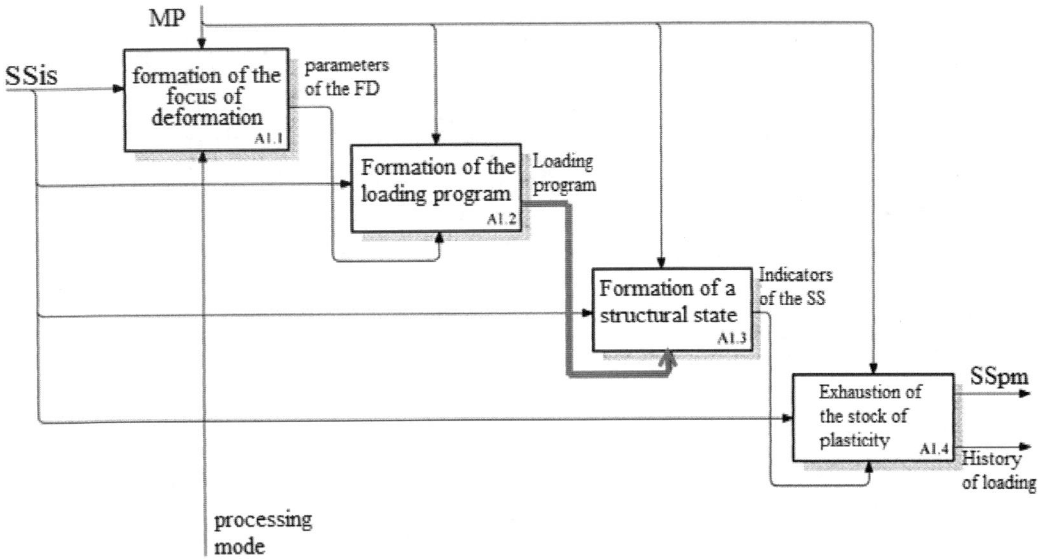

Fig. 2. Decomposition A1 "Formation of a structural state during machining":
SSis – the initial structural state of the material, MP is the material properties, SSpm - structural state of the metal after machining

Material and methods of investigation

To use this model, it is necessary to describe the connections between its elements from the mathematical point of view. Development of mathematical models requires a number of experimental studies, including:

1. Mechanical processing of certain modes, which will allow establishing relationships between the parameters of the stress-strain state of the surface layer after machining and the parameters of the structural-phase state.
2. Study of the parameters of the mechanical and structural-phase state of the processed samples by methods of nondestructive and destructive testing.

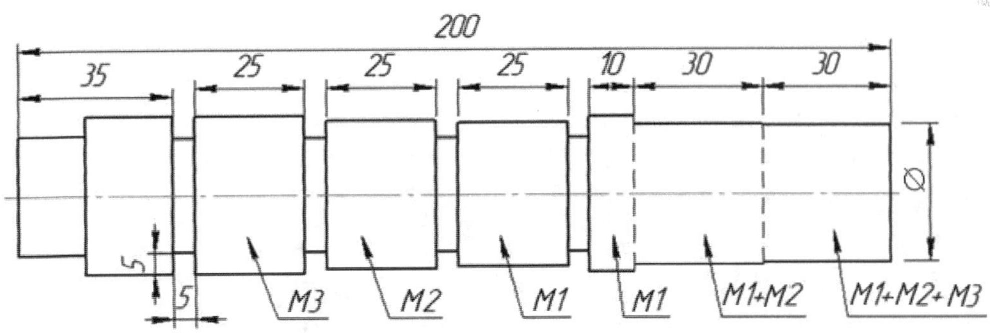

Fig. 3. Appearance of the sample

In Fig. 3 shows the appearance of the sample, the designations M1, M2 and M3 correspond to modes 1, 2 and 3 (Table 1), the processing will be performed both operationally (ie, in certain sections of the sample the cutting will be performed according to the corresponding regime) and sequentially this is successively cut by the modes M1 and M2, M1 and M2 and M3).

The sample material is 40X steel. Before machining by cutting, the workpiece was subjected to heat treatment: recrystallization annealing T = 600°C, time 60 min.

Table 1

Processing Modes

	Mode 1 Rough cutting	Mode 2 Semi-pure cutting	Mode 3 Finishing cutting
Diameter of the workpiece ∅, mm	36	34	35
Depth of cut, t, мм	1	0.5	0.25
Innings, s, мм/об	0.4	0.25	0.1
Number of revolutions, n, об-1	650	900	1200

After the mechanical processing, studies of samples with spectral-acoustic control and magnetic-noise control are planned, these monitoring methods were adopted as well-proven in the diagnosis of accumulated damage to FEC equipment [6, 8] and evaluation of the state of the metal in simple deformation [9]. From destructive testing methods, measurements of the microhardness values [10] of the surface layer have been performed, and optical and transmission electron microscopy is planned.

The control of samples by the spectral-acoustic method was carried out by the automated acoustic system ASTRON. The control of samples by the method of magneto-noise control was carried out by the stress analyzer and the "Introscan" structure. Measurement of the microhardness values of the surface layer was performed using a Durascan 20 microhardnesser [7, 10].

Results and its discussion

The control zones of non-destructive testing methods are shown in Fig. 4, and the results of the control are presented in Table 2 and Table 3, as well as in Fig. 5-7.

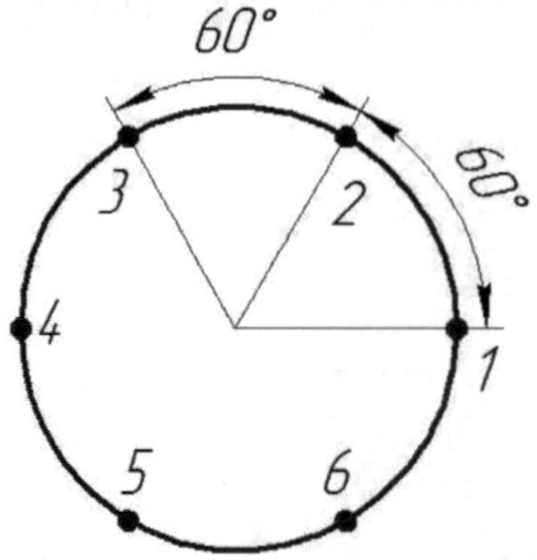

Fig. 4. Zones of research by non-destructive methods of control

As studies have shown by the method of spectral-acoustic control, processing from the initial state according to the regime of "rough" cutting causes the delay time and the average amplitude of the signal to increase, which indicates changes occurring in the surface layer under the action of cutting forces.

In sequential processing by modes corresponding to the "rough – semi-pure" processing, the value of the signal delay time is higher than in the "semi-pure – finished" mode for operational processing. With "finished" processing, the delay time of the signal decreases, both in the case of sequential and operational processing. The values of the average amplitude of the signal for sequential processing are somewhat lower than in the operating room, which may be due to the introduction of high voltages into the surface layer of the metal.

Table 2

The results of sample control by the spectral-acoustic method

№	Processing mode	The delay time of the signal R, ns	The attenuation coefficient K	The average amplitude of signal A
1	**The initial state**	4622	0.42	46
2	**M1**	4655	0.13	181
3	**M2**	4651	0.08	232
4	**M3**	4651	0.08	187
5	**M1+M2**	4652	0.13	176
6	**M1+M2+M3**	4647	0.15	145

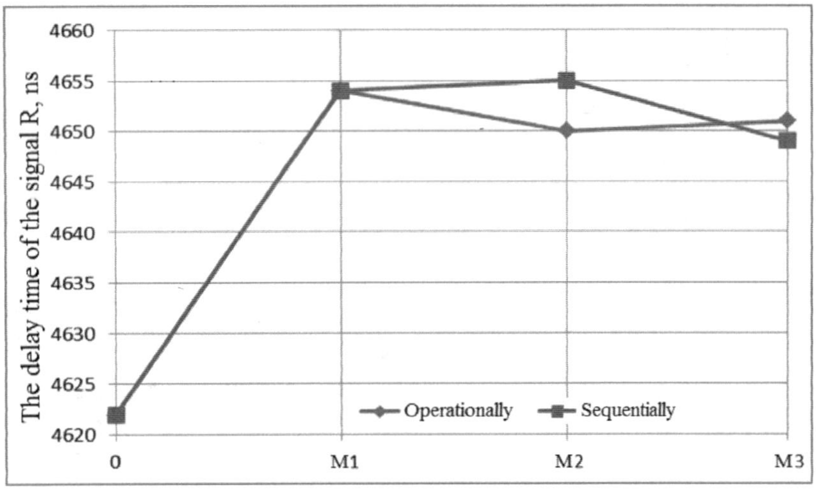

Fig. 5. Dependence of the signal delay time on the mode and the processing sequence

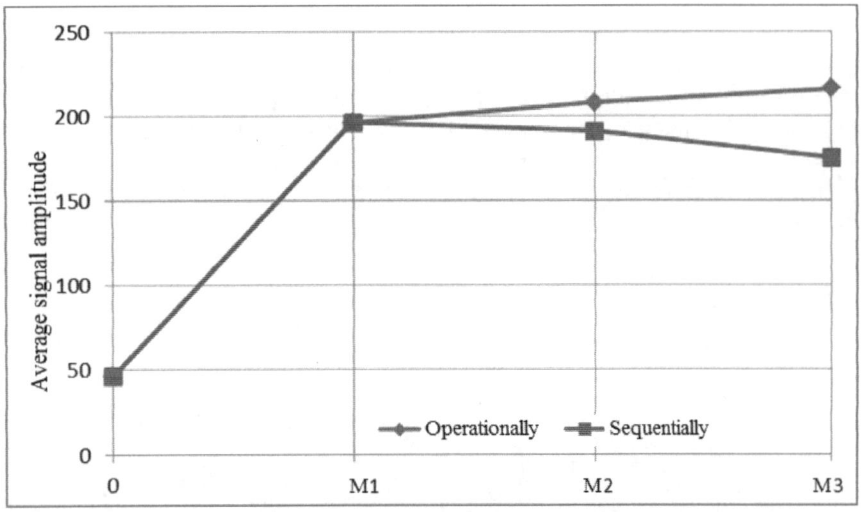

Fig. 6. Dependence of the average signal amplitude on the mode and sequence of processing

IOP Conf. Series: Earth and Environmental Science **115** (2018) 012042 doi:10.1088/1755-1315/115/1/012042

Table 3

The results of the control of samples by the method of magneto-noise control.

№	Processing mode	The average value of the intensity of magnetic noise
1	**The initial state**	109.5
2	**M1**	201.7
3	**M2**	198.2
4	**M3**	197.0
5	**M1+M2**	218.3
6	**M1+M2+ M3**	202.2

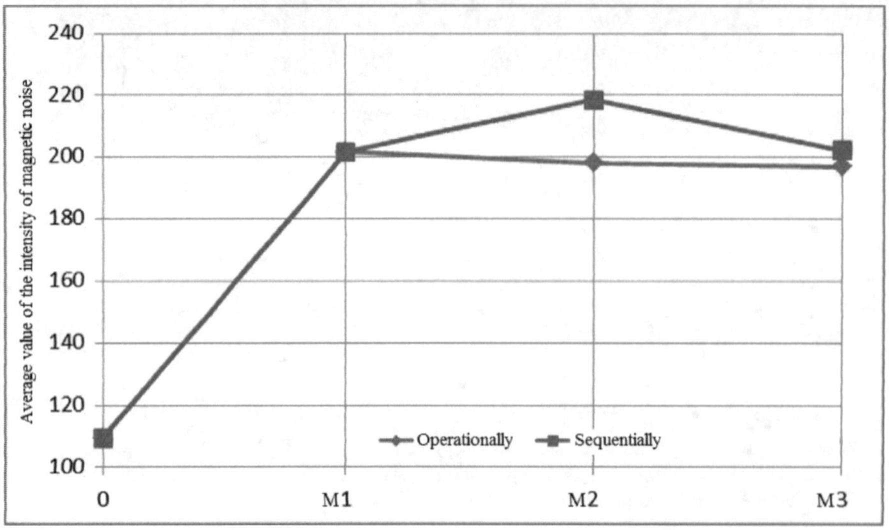

Fig. 7. Dependence of the intensity of magnetic noise on the mode and sequence of processing

For all processing methods, the magnetic noise increases from the initial state to the "rough" mode. In sequential processing by modes corresponding to the "rough – semi-pure" processing, the intensity of magnetic noise is higher than in the "semi-pure – finished" mode for operational processing. With "finished " processing, the intensity of magnetic noise decreases, both in the case of sequential and in operating processing.

The microhardness measurement scheme is shown in Fig. 8 [7], and the results of the control in Fig. 9.

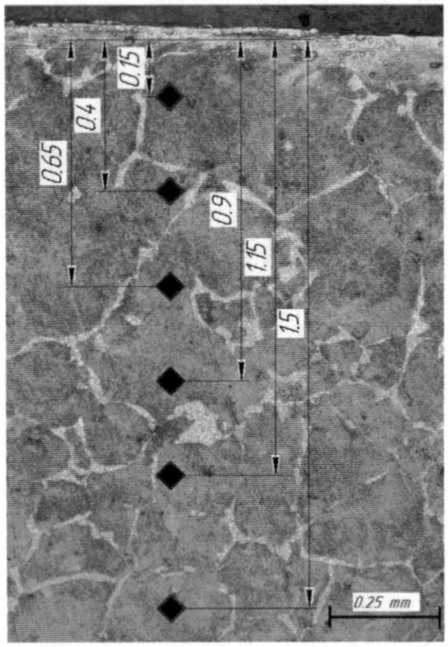

Fig. 8. Scheme of measurements of microhardness

According to the results of the study, it is seen that when processing by "cutting" modes, a decrease in the value of microhardness is observed, which indicates a slight softening of the surface layer. With further sequential processing by the appropriate "rough – semi-pure " processing, an increase in microhardness is observed compared to processing in " semi-pure -finished" mode for operating processing, however, when machining in " finished " mode for all processing methods, the microhardness value is equalized.

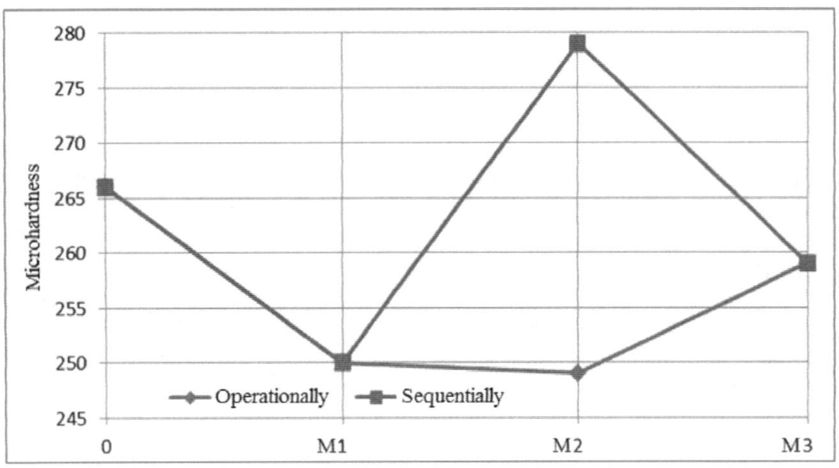

Fig. 9. The dependence of the average microhardness of the surface on the regime and the processing sequence

IOP Conf. Series: Earth and Environmental Science **115** (2018) 012042 doi:10.1088/1755-1315/115/1/012042

Conclusion

1. A structural-analytical model has been developed that reflects the concept of "Formation and transformation of the structural state" in accordance with which the main factor determining the parameters of the structural-phase state after machining is the loading program.

2. Studies were carried out using spectral-acoustic and magnetic-noise control. It is established that, depending on the method and treatment regime, similar dependencies are observed.

3. Investigations of the distribution of the values of microhardness are performed. It was found that the dependence of microhardness on the cutting regimes and processing method correlates with the results of non-destructive testing

4. At present, investigations are being carried out using the methods of optical and transmission electron microscopy to establish the dependencies between the modes and processing methods and the parameters of the structural-phase state.

References

[1] Matalin A. A. Tekhnologiya mekhanicheskoy obrabotki. L., «Mashinostroyeniye» (Leningr. otd-niye),1977. 464 p.

[2] Suslov A.G., Gorlenko A.O., Simkin A.Z. Povysheniye dolgovechnosti kulachkovykh par treniya//Treniye i iznos. T. 18. № 3. 1997. p. 395-398.

[3] Tekhnologicheskaya nasledstvennost' v mashinostroitel'nom proizvodstve/A.M. Dal'skiy, B.M. Bazrov, A.S. Vasil'yev i dr.; Pod red. A.M. Dal'skogo. M.: Izd-vo MAI, 2000. 364 p.

[4] Blyumenshteyn V.YU., Smelyanskiy V. M. Mekhanika tekhnologicheskogo nasledovaniya na stadiyakh obrabotki i ekspluatatsii detaley mashin. M.: Mashinostroyeniye-1, 2007. 400 p.

[5] Pimonov M.V. Strukturno-analiticheskaya model' formirovaniya parametrov strukturnogo sostoyaniya poverkhnostnogo sloya v protsessakh mekhanicheskoy obrabotki // Innovatsii v mashinostroyenii: trudy 4-oy Mezhdunarodnoy nauchno prakticheskoy kon-ferentsii, pod. red. KH.M. Rakhimyanova. Novosibirsk: Izd-vo NGTU, 2013. p. 181–186.

[6] Diagnostika, povrezhdayemost' i remont barabanov kotlov vysokogo davleniya / Ababkov N. V., Kashubskiy N. I., Knyaz'kov V. L., Knyaz'kov A. F., Kozlov E. V., Koneva N. A., Makarov N. M., Murav'yev V. V., Popova N. A., Smirnov A. N., Fol'mer S. V. M.: Mashinostroyeniye. 2011. 256 p.

[7] GOST R ISO 6507-1-2007. Metally i splavy. Izmereniye tverdosti po Vikkersu. Chast' 1. Metod izmereniya. – Moskva: Izd-vo Ctandartinform, 2008. 16 p.

[8] Gradiyentnyye struktury pri obrabotke metallov rezaniyem / A.N. Smirnov, N.V. Ababkov, E.V. Kozlov, N.A. Koneva, N.A. Popova, A.S. Glinka, M.S. Makhalov, S.A. Ryabov, M.V. Pimonov. Kemerovo, 2013. 179 p.

[9] Blyumenshteyn V.YU., Miroshin I.V., Petrenko K.P. Otsenka vzaimosvyazey deformatsii i zapasa plastichnosti s parametrami signalov akusticheskoy strukturoskopii // Innovatsii v mashinostroyenii sbornik trudov VIII Mezhdunarodnoy nauchno-prakticheskoy konferentsii. Novosibirsk, 2017. p. 510-522.

[10] GOST 9450-76 (ST SEV 1195-78) Izmereniye mikrotverdosti vdavlivaniyem almaznykh nakonechnikov. – Moskva: Izd-vo IPK Izdatel'stvo Standartov, 1993. 34 p.

IOP Conf. Series: Earth and Environmental Science **115** (2018) 012043 doi:10.1088/1755-1315/115/1/012043

Research in Area of Longevity of Sylphon Scraies

Natalia Y Golovina, Svetlana Y Krivosheeva

Industrial University of Tyumen, the Branch of IUT in Surgut,
Entuziastov Street, 38, Surgut, 628400 Russia

E-mail: ntgolovina@rambler.ru

Abstract. The proposed method of determining the sensitivity threshold for bellows expansion compensator by fitting a power polynomial of the empirical curves of distribution of durability for a large number of samples, followed by statistical analysis. For a reliable justification of resource products with regard to dispersion of their durability is of great importance the evaluation of the sensitivity threshold in cycles. Currently, the appointment of a guaranteed resource is made with high strength requirements. The proposed method allows to define sensitivity threshold value for the cycles between the curves of the distribution of longevity as having a distinct threshold, and not having it. Method is in good agreement with other known methods. Thus, by statistical processing of results of fatigue tests of joints obtained values of the sensitivity threshold in cycles and confidence interval for mathematical expectation of the magnitude of the threshold.

Introduction

Setting the reliable indicators is determined largely by a distribution function of operating time to failure. The value of operating time to failure in the increasing order reflects the empirical distribution function of endurance. For reliable study of items resource with a view of the dispersion of its endurance it is important to estimate the threshold of sensitivity to N_0 cycles. Currently, the appointment of the safe life is made without taking into account the values of N_0. Consequently a large endurance margin n_N is necessary in order to account the dispersion of endurance. Safe life based on the average rated life is determined from the expression:

$$N_{\text{safe life}} = \frac{N_{\text{aver rated}}}{n_N} \tag{1}$$

As a rule such setting of safe life leads to increased strength properties of items and to increasing of laboriousness, metal consumption and excessive mass. Determination N_0 values allows us to justify the safe life of items. Researchers have found that the endurance has lower and upper boundaries under constant maximum voltage [8]. The lower boundary of endurance N_0 denotes the threshold of sensitivity to cycles, i.e. the largest possible number of cycles, σ_{max} scarcely causes damage at a given voltage. The value N_0 separates insensitive zone of operating time to changing load.

It is difficult to estimate the N_0 empirically, as a considerable number of tests is required. Even more difficult it is to determine the threshold of sensitivity at relatively low voltages due to the large test duration.

Materials and methods for solving the problem

There are currently several methods for determining the threshold of sensitivity: graphic, symmetrical form bringing, least squares, maximum likelihood.

However, the effectiveness of these methods is not resolved until now. Application of these methods to the results of fatigue tests of syphons has not given results:

1. The method of reducing to the symmetrical form is applicable only to a comparatively large volumes of samples (more than 60), so it was not considered.

2. The method of least squares, as well as its graphic expression – a graphical method gives a solution only for the curve distribution with clearly defined lower boundary of endurance and not clearly upper limit. If endurance curves do not comply at least with one condition it means that the values N_O, calculated on these methods become unrealistic and even negative. Figure 1 shows the distribution curves of endurance [9]. The method of least squares does not give solutions for the curves 1, 2, 3, the solution for the curve 4 matches the value $N_O = 6.1...10^5$ cycles.

3. The maximum likelihood method does not give convergence solutions for samples less than 36, including curves (Figure 1). The need for solving this problem is recognized by the authors [1].

4. An approximate method [7] does not take into account the conditions of existence of the threshold of sensitivity; therefore it gives lower results (straight lines in Figure 1).

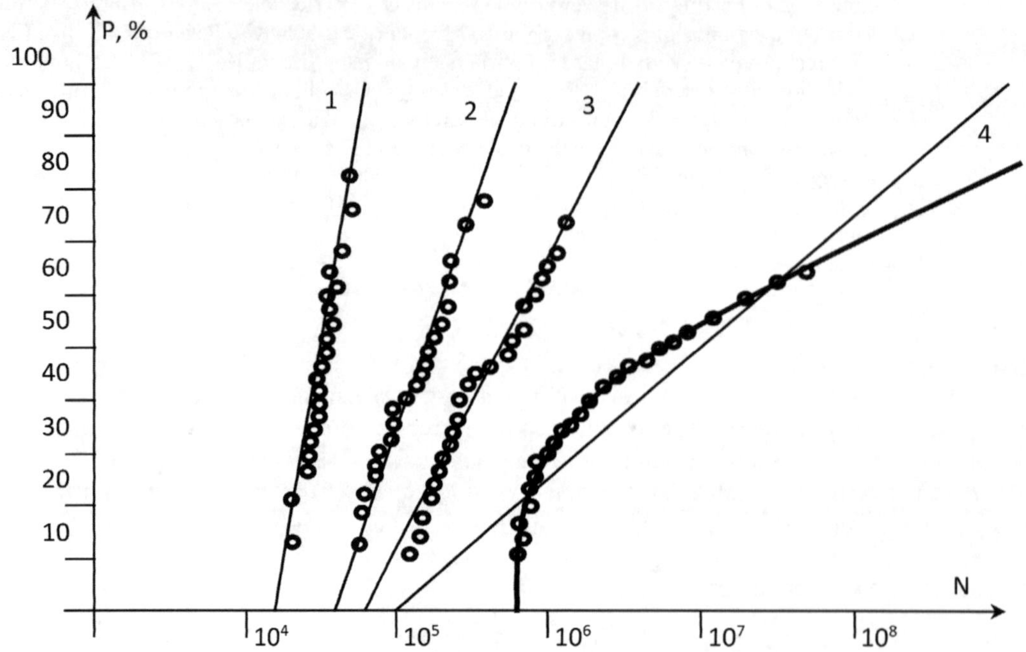

Figure 1. Distribution curves of endurance

In this research, the sensitivity threshold determination for compensator made of steel 12X18H10T, is performed by approximation of exponential polynomial of empirical curves distribution of endurance for a large number of samples, followed by statistical processing.

When the approximation is used under condition of sensitivity threshold display on the lower and upper boundaries of the empirical distribution curves.

Endurance test results of the products at a level of voltage σ_{max} were located in the variation series in ascending order of endurance and were grouped by voltages with a deviation from one another not more than 10–15 MPa. The value of the probability of destruction *P*, estimated at the accumulated frequency, was calculated by the formula [10]

$$P_i = \frac{i - 0.5}{n} \tag{2}$$

i – item number in the variational series, *n* – the total number of tested items.

IOP Conf. Series: Earth and Environmental Science **115** (2018) 012043 doi:10.1088/1755-1315/115/1/012043

As noted in [2-4], sign of the sensitivity threshold N_0 is the curvature of the curve P – lgN, built on probability paper, bump up. Analysis of the data presented in [6], shows that the distribution curves of endurance in the limit at probabilities of failure $P \to 0$ and $P \to 1$ form a right angle with *lgN* axis. Thus, the problem reduces to a dot approximating of the empirical distribution curves with zero derivative at the boundary points that are sought. To increase the accuracy, the approximation was made with the transition to a uniform probability scale. And the perpendicularity of distribution curve to the axis *lgN* in the limit is persists.

The solution of the problem can be represented in the form of approximating m – dimensional polynomial.

$$y(x) = \sum_{k=0}^{m} a_k x^k \tag{3}$$

x represents the values of P. We consider at $P = 0$, $x = 0$, $N_a = a_0$.

Naturally, on the segment $N_1 < N_2 < \ldots < N_0$ empirical distribution function has a range of values, that means that the values of the desired distribution curve were set with an error. To solve the problem it is necessary to construct a function that gradually took place nearby the predetermined values. Polynomial is used in this function, at which the minimum of the functional is reached [5],

$$S[f(x_k) = \alpha \int_{x_0}^{x_{n+1}} [f''(x_k)]^2 dx + \sum_{k=0}^{n+1} P_k [f(x_k) - y_k]^2 \tag{4}$$

α – smoothing parameter; P_k – mass coefficient ($P_k > 0$); y_k corresponds to the values N_k.

Boundary condition for sought function will be:

$$f'(x_0) = \sum_{j=0}^{m} j\, a_j x_0^{j-1} = 0\,, \qquad f'(x_{n+1}) = \sum_{j=0}^{m} j\, a_j x_{n+1}^{j-1} = 0 \tag{5}$$

After differentiating the general expression for a system of equations has the following form:

$$\frac{\partial}{\partial a_i} S[f(x_k)] == \alpha \sum_{i=0}^{m}\sum_{j=0}^{m} \frac{2i(i-1)j(j-1)}{i+j-3} a_j + 2\sum_{k=0}^{n+1} P_k \left(\sum_{j=0}^{m} a_j x_k^j - y_k \right) x_k^i = 0_k \tag{6}$$

Thus, for determining $m + 3$ the unknowns $a_0, a_1, \ldots a_m, y_0, y_{n+1}$, we have $m + 3$, equations. Taking into consideration that $x_0 = 0$, $x_{n+1} = 1$, the final system of equations for determining the coefficients in a general form can be written as:

$$\alpha \sum_{i=0}^{m}\sum_{j=0}^{m} \frac{ij(i-1)(j-1)}{i+j-3} a_j + \sum_{k=0}^{n+1} P_k \sum_{j=0}^{m} a_j x_k^{i+1} - y_0 x_0^i - y_{n+1} x_{n+1}^i = \sum_{k=1}^{n} y_k x_k^i \tag{7}$$

$$a_1 = 0$$

$$\sum_{j=0}^{m} j a_j = 0 \tag{8}$$

Accuracy of approximation of a function to set values depends on the parameter α and mass coefficient P_k. The higher the mass coefficient P_k, the greater contribution the interpolation conditions make into the functional and the closer the smoothing function passes to the set values. Since the tests were conducted on identical equipment, with the same error, it is natural to take the values of the coefficients P_k equal at all points of the empirical distribution curve.

Selecting α smoothing parameter is a problem. At a low α value the smoothing will be insignificant, at inflated α values the function will be extremely smooth. In this case, the choice of α parameter determines by the behavior of the first derivative of the approximating polynomial, which should not be less than zero.

As a result of the approximation of the empirical distribution curves of the fifth degree polynomial the corresponding values of sensitivity thresholds N_0 were obtained. Figure 2 shows the results of calculations N_0 values, depending on the voltage σ_{max}. N_0 values can be viewed as a random variable having a range of possible values.

In order to get the averaged curve $lgN_0 = f(\sigma_{max})$, having a probability of $P = 0.5$, these presented results are approximated by a polynomial in the eighth power (dash-dot line).

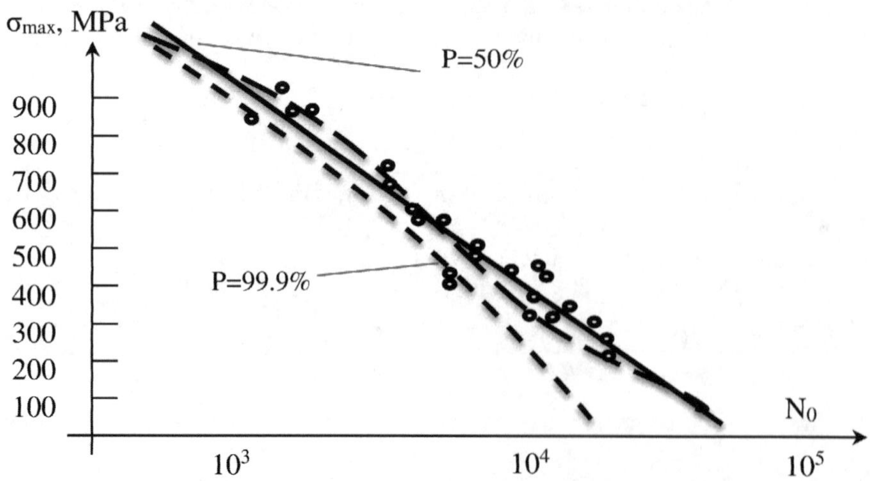

Figure 2. The empirical regression line for the sensitivity threshold N₀ and its bounder of the confidence region

It is obvious that the resulting dependency without a large error can be replaced by a linear dependence, that is, the connection between the normally distributed random variable $y = lgN_0$ and the value of $x = \sigma_{max}$ can be established by means of linear regression analysis. The empirical regression line has the form

$$Y = a + b\,(x-\bar{x}) \qquad (9)$$

where Y - the evaluation of the conditional expectation value of $y = lgN_0$ for a set value x; $\bar{x}$- sample average value of x.

Parameter estimation of the regression line was calculated from the formula:

$$\bar{x} = \frac{1}{m}\sum_{i=1}^{m} x_i \qquad (10)$$

$$a = \bar{y} = \frac{1}{m}\sum_{i=1}^{m} y_i$$

$$b = \frac{\sum_{i=1}^{m}(x_i - \bar{x})y_i}{\sum_{i=1}^{m}(x_i - \bar{x})^2}$$

m – the number of voltage levels.

Finally, for $m = 29$, we have:

$$Y = 4.2754 - 0.0238\,(x - 47.8896)$$

or:

$$lgN_0 = 5.4152 - 0.0238\,\sigma_{max} \qquad (11)$$

The regression line corresponding to the resulting equation is shown in Figure 2 (solid line).

At probabilistic calculations of item strength to enhance their reliability it is not appropriate to use the average sampled values of the random variable, but the values of the boundaries of the confidence intervals, in particular, we should take the lower limit of the confidence interval instead of the sample average values of N_0. The boundaries of the confidence interval Δ for the expectation N_0 are determined by the expression:

$$Y - t_{\alpha,k} S_y < \Delta < Y + t_{\alpha,k} S_y \qquad (12)$$

$t_{\alpha,k}$ - student criterion for the significance level α and the number of degrees of freedom K; S_y – the variance of expected value estimation.

On Figure 2 the lower limit 99.9% of confidence region with a significance level $\alpha = 0.001$ is marked by the dashed line. The upper limit of endurance was not viewed as it had no practical interest.

Results

The proposed statistical processing method allows to determine the N_0 value for the distribution curves of endurance with a distinct sensitivity threshold, and without it. The method is consistent with other known methods, there is a pronounced tendency of the distribution curve to N_e. For example, for distribution curve 4 (Figure 1) $N_0 = 6.3\ldots10^5$ cycles (least squares $N_0 = 6.1\ldots10^5$ cycles).

Conclusion

Thus, by means of statistical processing of the compensators fatigue test results we obtained the values of sensitivity threshold on cycles and the confidence interval for the expectation values of N_0.

Disclosure statement

No potential conflict of interest was reported by the authors.

References

[1] Bezenov SA (2011) Methodical aspects of assessment of the fatigue resistance of metallic materials characteristics. *New Materials and Technologies in Metallurgy and Machinebuilding* 1: 27-31.

[2] Golovina NY (2014), Sustainability issues forced transverse vibrations of flexible metal pipes. *Nauchnoe obozrenie* 10: 63-66.

[3] Golovina NY (2005), New approach in the study of forced transverse vibrations of flexible metal pipes for stability. *Proceedings of the higher educational institutions. Oil and gas.* 1: 70-74.

[4] Golovina NY, Krivosheeva SY (2015), The impact of energy dissipation in the durability of flexible metal pipes. *Nauchnoe obozrenie* 12: 106-108.

[5] Marchuk GI (1980) Methods of Computational Mathematics, 456.

[6] Roessle, M.L., Fatemi, A.: Strain-controlled fatigue properties of steels and some simple approximations. Int J Fatigue 2000;22:495–511.

[7] Serensen SV (1975) Bearing capacity calculations and machine parts for durability, 488.
[8] Stepanov MN (2005), Statistical methods for processing the results of mechanical tests (directory), 399.
[9] Stepanov M.N., Giatsintov EV (1973) Fatigue light structural alloys, 320.
[10] Weibull W (1964), Fatigue testing and analysis of results, 275.

A Vibrating Jaw Crusher with Auteresonant Electric Motor Drive of Swinging Movement

E A Zagrivniy and D A Poddubniy [a]

Saint-Petersburg Mining University, 21-st line, Saint-Petersburg, 199106, Russian Federation

E-mail: [a] poddubniy.da@yandex.ru

Abstract. The article relates to a vibrating jaw crusher with pendulum vibrating exciter auteresonant electric motor drive and with elastic element rational force distribution, with limited peak-to-peak swing. Its design and its math model are presented. Also disclosed is the operating principle of a vibrating jaw crusher and the control algorithm for controlling the crushing jaw for maintaining the operating mode at resonant frequency.

1. Introduction

Traditional vibrating jaw crushers are provided with electromechanical vibrating exciters based on self-synchronized unbalanced rotors with the alternating current electric drive. Such vibrating jaw crushers are inefficient due to power losses in the behind-resonance zone, for operating those at the resonant frequency of a mechanical system is not possible because of Sommerfeld effect. Therefore, electric motor power 2-5 times exceeds the calculated one, thus making the drive efficiency low [1,2,3].

2. Materials and methods

The Mining University in Saint-Petersburg developed vibrating jaw crushers with autoresonant reciprocating motion electric motor drive of pendulum exciters that do not have the abovementioned disadvantage. The crusher comprises a body hosting two moving jaws and spring suspensions, two pendulum exciters, two petal couplings, two electric motors and elastic elements that may be torsional, or an elastic system based on several torsion springs.

A torsion used as an elastic element may result in increased load torque on petal coupling, for it links a pendulum vibrating exciter and a motor shaft, which, in turn, has its opposite end connected to the torsion attached at one side to the crusher body. A crusher of this type has the torque acting on the petal coupling in all operating modes regardless technological load, reaching maximum value, when rotor speed curve passes zero, to be defined by multiplying the acceleration and the sum pendulum mass moment and the electric motor drive torque. This is because the elastic element (torsion) and the pendulum vibrating exciter are located opposite relatively the petal coupling.

To eliminate the abovementioned disadvantages, the vibrating jaw crusher is designed, as presented at the figure 1.

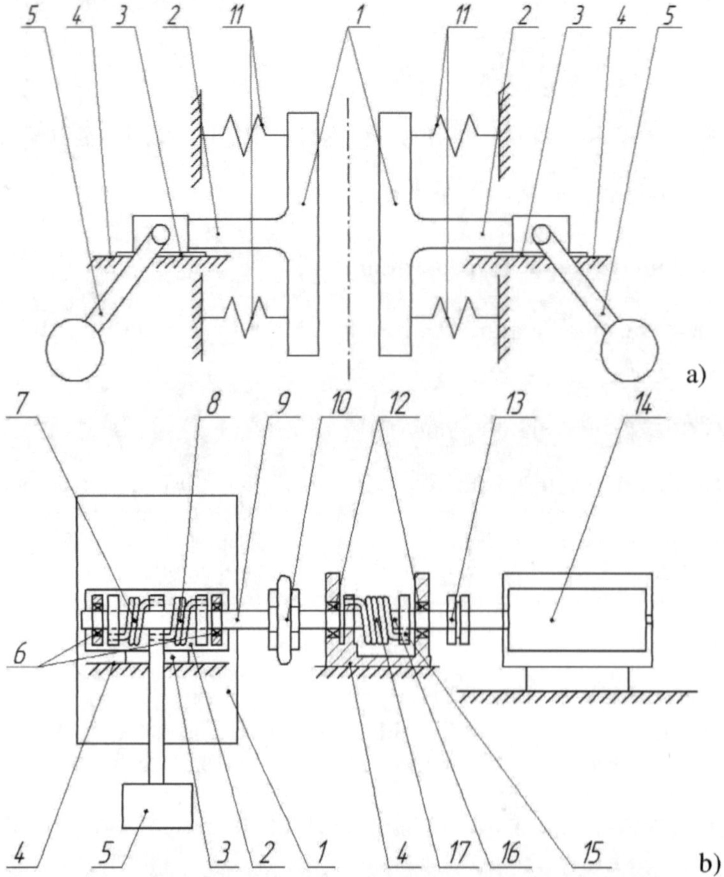

Figure 1. Idealized structural scheme of a vibrating jaw crusher with pendulum vibrating exciter auteresonant electric motor drive: a) front view; b) lateral view.
1 – moving jaws; 2 – jaw ram; 3 – ram guides; 4 – body; 5 – pendulum vibrating exciter; 6, 12 – bearing nodes; 7, 8, 17 – torsional springs (7, 8 – elastic element of pendulum vibrating exciter; 17 – elastic element of the electric motor rotor); 9 – pendulum shaft; 10 – connecting element (petal coupling); 11 – spring suspensions; 13 – connecting element (coupling); 14 – electric motor; 15 – fixture ring; 16 – intermediate shaft

The proposed vibrating jaw crusher with pendulum vibrating exciter auteresonant electric motor drive is capable of operating at the resonant frequency. Pendulum vibrating exciters of each crushing jaw 5 are connected to electric motor 14 rotor shaft via petal couplings 10 by the intermediate shaft 16 and the coupling 13. Each pendulum vibrating exciter 5 is firmly attached to two torsional springs (7 and 8). Those springs have one end firmly attached to the jaw, and the second end attached to the vibrating exciter. Also, every intermediate shaft 16 is firmly attached to the fixture ring 15, which, in turn, is fixed to one end of the third torsion spring 17 (an elastic element of the electric motor rotor), and the second end is fixed to the body 4. Using an elastic system like this one allows cutting notably the mass moment at each coupling, resulting from torque by the pendulum vibrating exciter [4,5]. Because of symmetry, calculating the crusher necessitates considering only one half of the selected structure scheme.

The electromechanical system of the vibrating jaw crusher disclosed herein comprises three vibration systems work table equal own frequencies being excited by the same source (figure 1). The first mechanical vibration system is created by the moving jaw 1 of the crusher, with the ram 2, the pendulum vibrating exciter 5 with the shaft 9, torsional springs 7-8 and the body 4 of the crusher. The weight of the jaw with the ram and the pendulum vibrating exciter m1 and the elastic pendulum element with stiffness factor C1 provide the defined own vibration frequency:

$$\omega_{r1} = \sqrt{\frac{C_1}{m_1}} . \tag{1}$$

The second, exciting, mechanical vibrating system is created by the rotor of the electric motor 14 with the intermediate shaft 16, the petal half-coupling 10, the connecting coupling 13, the elastic element of the electric motor 17 rotor. The own vibration frequency of the system:

$$\omega_{r2} = \sqrt{\frac{C_2}{J_2}} , \tag{2}$$

where C_2 is the stiffness factor of the torsion spring, N·m/rad; J_2 is the sum mass moment of the system's rotating elements, kg·m^2.

The third vibrating system is created by the pendulum vibrating exciter 5, the pendulum shaft 9, the petal half-coupling 10 and the elastic element of the pendulum vibrating exciter (7, 8). The own vibration frequency of the system is in accordance with:

$$\omega_{r3} = \sqrt{\frac{C_3}{J_3}} , \tag{3}$$

where C_3 is the stiffness factor of the torsion ring, N•m/rad; J_3 is the sum mass moment of the system's rotating elements, kg•m^2.

Therewith, all own frequencies of the three systems approximately equal:

$$\omega_{r1} \approx \omega_{r2} \approx \omega_{r3} . \tag{4}$$

Such a device and such settings provide string inertial association between the vibrating systems with minimum power spending to be taken from the power mains, when operating at the resonance frequency, with high efficiency factor [6]. It is worth mentioning separately, that such layout provides minimal loading torque onto the petal coupling, which is to be defined by the technological load only. In addition, having the three vibrating systems expands the operating resonant frequency range.

3. Mathematical model

Building the system math model needs the second type Lagrange equations to be used [7,8]:

$$\frac{d}{dt}\left(\frac{\partial T}{\partial \dot{q}_i}\right) - \frac{\partial T}{\partial q_i} + \frac{\partial F_i}{\partial \dot{q}_i} + \frac{\partial \Pi}{\partial q_i} = Q_i , \; i = 1,...,n , \tag{5}$$

where T and Π represent kinetic and potential power of the system, correspondingly; F is the dissipative function; q is the generalized system coordinates; Q_i is the generalized external force; n is the number of freedom degrees.

With q_1=x and q_1=φ, as the generalized coordinates, the system (5) would be as follows:

$$\begin{cases} \frac{d}{dt}\left(\frac{\partial T}{\partial \dot{x}}\right) - \frac{\partial T}{\partial x} + \frac{\partial F_X}{\partial \dot{x}} + \frac{\partial \Pi}{\partial x} = 0 \\ \frac{d}{dt}\left(\frac{\partial T}{\partial \dot{\varphi}}\right) - \frac{\partial T}{\partial \varphi} + \frac{\partial F_\varphi}{\partial \dot{\varphi}} + \frac{\partial \Pi}{\partial \varphi} = Q_\varphi \end{cases} . \tag{6}$$

A number of allowances had been accepted in connection therewith (figure 2): elastic elements are linear; electromagnetic transitional processes are omitted because they take too little time compared to mechanical vibration period; power losses in the electric drive are defined by the electric motor efficiency factor; friction force in bearings and during ram motion is zero.

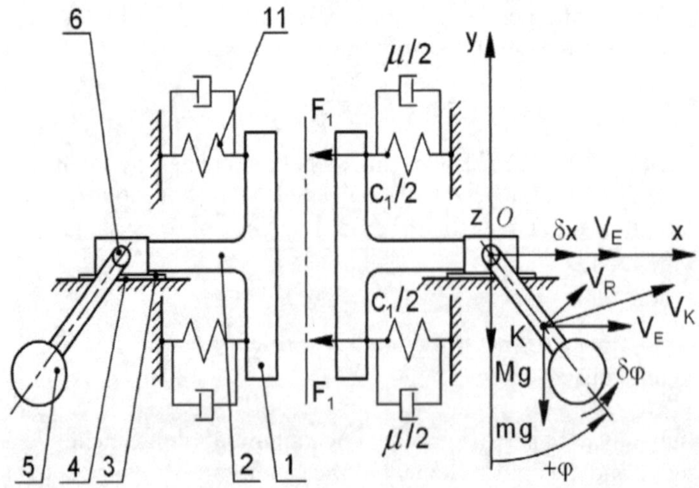

Figure 2. The dynamical calculation diagram of vibrating jaw crusher with pendulum vibrating exciter electric motor drive:
(1-11th elements' numeration matches the one at the figure 1).μ – viscous friction coefficient; F_1 – force of elasticity of suspension elements; c_1 and c_2 – stiffness factors of the elastic elements of the suspension of the crushing jaw and the rotor with a pendulum exciter; M – weight of crushing jaw with a ram and a shaft of the pendulum vibrating exciter; m – weight of a pendulum vibrating exciter; g – acceleration of gravity; J_1 and J_2 – moment of inertia of the pendulum vibrating exciter and rotor; OK is the distance equal to a, where K – center of gravity of the pendulum exciter.

The kinetic power of the system is defined by the expression:

$$T = T_1 + T_2 + T_3, \quad (7)$$

where T_1 is the kinetic power of the crushing jaw 1 with the ram 2 and the shaft 9 of the pendulum vibrating exciter 5; T_2 is the kinetic power of the pendulum vibrating exciter 5; T_3 open source the kinetic power of the rotor of the driving motor 14.

The crushing jaw with the ram moves progressively, therefore

$$T_1 = \frac{1}{2} \cdot M \cdot \dot{x}^2. \quad (8)$$

The kinetic power of the pendulum vibrating exciter is defined in accordance with the expression

$$T_2 = \frac{1}{2} \cdot m \cdot \dot{x}^2 + m \cdot a \cdot \dot{x} \cdot \dot{\varphi} \cdot \cos\varphi + \frac{1}{2} \cdot J_1 \cdot \dot{\varphi}^2. \quad (9)$$

The kinetic power of the rotor is

$$T_3 = \frac{1}{2} \cdot J_2 \cdot \dot{\varphi}^2. \quad (10)$$

The kinetic power of the system is equal to

$$T = \frac{1}{2} \cdot \dot{x}^2 (M + m) + m \cdot a \cdot \dot{x} \cdot \dot{\varphi} \cdot \cos\varphi + \frac{1}{2} \cdot \dot{\varphi}^2 (J_1 + J_2). \quad (11)$$

The potential power of the system is

$$P = P_1 + P_2 + P_3 + P_4, \quad (12)$$

where P_1 is the potential power of the pendulum vibrating exciter 5; P_2 is the potential power of the elastic element of the pendulum (7 and 8); P_3 is the potential power of the elastic element of the electric motor 17 rotor; P_4 is the potential power of the elastic elements 11 of the jaw suspension.

For the pendulum vibrating exciter

$$P_1 = m \cdot g \cdot a \cdot (1 - \cos \varphi). \tag{13}$$

For the elastic element of the pendulum vibrating exciter

$$P_2 = \frac{1}{2} \cdot C_3 \cdot \varphi^2 . \tag{14}$$

For the elastic element of the electric motor rotor

$$P_3 = \frac{1}{2} \cdot C_2 \cdot \varphi^2 . \tag{15}$$

For the jaw suspension elastic elements

$$P_4 = \frac{1}{2} \cdot C_1 \cdot x^2 . \tag{16}$$

Therefore, the potential power of the system is

$$P = m \cdot g \cdot a \cdot (1 - \cos\varphi) + \frac{1}{2} \cdot C_3 \cdot \varphi^2 + \frac{1}{2} \cdot C_2 \cdot \varphi^2 + \frac{1}{2} \cdot C_1 \cdot x^2 . \tag{17}$$

By putting the results into the Lagrange equation (6) and making appropriate conversions, we obtain the differential equations describing the system motion:

$$\begin{cases} (M + m) \cdot \ddot{x} + \mu \cdot \dot{x} + C_1 \cdot x = m \cdot a \cdot (\dot{\varphi}^2 \cdot \sin\varphi - \ddot{\varphi} \cdot \cos\varphi) \\ (J_1 + J_2) \cdot \ddot{\varphi} + C_2 \cdot \varphi + C_3 \cdot \varphi = M_d - m \cdot a \cdot (g \cdot \sin\varphi + \ddot{x} \cdot \cos\varphi) \end{cases} . \tag{18}$$

The distinctive features of the system include using nontraditional pendulum vibrating exciter auteresonant electric motor drive that allows obtaining autoresonant vibrations of the rotor with the peak-to-peak values of 60°, 120° and 180°, when using three-phase stator with one pair of poles, and 30°, 60° and 90° with two pairs of poles [9,10].

Peak-to-peak limitation feature allows creating vibrating machines capable of working in all modes without exceeding the limits of mechanical tensions in heavy-loaded systems.

Worth mentioning is the useful feature of the system disclosed – the capability to automatically decrease the amplitude of the vibrations by the pendulum vibrating exciter maintaining the amplitude of vibrations by the jaw when the load open source reduced – something that is the property of systems with two degrees of freedom (vibration dynamic attenuation effect with $\mu \to 0$, $\varphi \to 0$).

4. Conclusion

Providing resonant vibrations of the moving jaws and pendulum vibrating exciters uses such a method for exciting and regulation of the autoresonant vibrations that each vibration half-period has the speed modified, and when the rotor speed curve passes zero relatively the stator, the voltage is applied to the stator windings to form electromagnetic torque to be modified in a co-phase manner with the rotor vibration speed, and the defined amplitude value controlled by changing the voltage via negative feedback by the vibration speed amplitude value during every half-period of the vibrations. Implementing those conditions relies on the reciprocating motion electric drive. It may, for example, be the one with three phase stator from the regular asynchronous electric drive equipped with the magnetic rotor with a single pair of poles. The electric drive allows obtaining fixed angle of the rotor vibrations, wherein the value may depend on particular winding connection scheme.

References

[1] Shestakov V.M., Epishkin A.E. Dynamics of automated electromechanical systems of vibrating units/ Under the general editorship by prof. V.M. Shestakov. SPb.: Publishing house SPSPU, 2005. 94p.

[2] Koptev V.Y., Kopteva A.V. Mining business transportation system structure optimization. 2016/ International Journal of Applied Engineering Research. International Journal of Applied Engineering Research ISSN 0973-4562 Volume 11, Number 11 (2016) pp 7402-7405

[3] Andrievsky B.R., Blekhman I.I., Bortsov Yu.A. Control of mechatronic vibration units. SPb.: Publishing House “Nauka”, 2001. 278 p.

[4] Patent of the Russian Federation No. 2492931. Vibrating jaw crusher // Zagrivniy E.A., Poddubniy D.A. // Bul. No. 26, 09.20.2013.
[5] Zagrivniy E.A., Poddubniy D.A., Basin G.G. Vibrating jaw crusher with Auteresonant electric drive of pendulum vibrational excavators and the rational distribution forces of elastic elements / Materials of the 10th International Scientific and Practical Conference, "Development of Mineral Resources of the North: Problems and Solutions", Vorkuta 2012
[6] Mandelshtam L.I. Lectures on the theory of oscillations. Moscow: Nauka, 1972. 471 p.
[7] Timoshenko S.P. Vibration problems in engineering. 2nd edition, M.: Science, 1967. 444 p.
[8] Gavrilov Yu.A. Mathematical and simulation models of the inertial resonance jaw crusher with an electric drive of the reciprocating-rotary motion / Journal of Mining Institute. T.182, 2009, pp. 69-73.
[9] Gavrilov Yu.A. Increasing of the productivity of the vibrational jaw crusher based on resonant mode of its work // Materialy XLVIII Sesji Pionu Gorniczego, Krakow, 2007, P. 212.
[10] Patent of the Russian Federation No. 2410826. Method to excite and control autoresonance vibrations in electric drive of swinging motion // E.A. Zagrivniy, Ju.A. Gavrilov // Bul. No. 3, 27.01.2011

Processing of Building Binder Materials to Increase their Activation

R S Fediuk [a], I S Garmashov, D E Kuzmin, N Yu Stoyushko, N A Gladkova

Far Eastern Federal University, Vladivostok, Russia

E-mail: [a] roman44@yandex.ru

Abstract. The paper deals modern physical methods of activation of building powder materials. During mechanical activation a composite binder active molecules cement minerals occur in the destruction of the molecular defects in the areas of packaging and breaking metastable phase decompensation intermolecular forces. The process is accompanied by a change in the kinetics of hardening of Portland cement. Activated concrete has a number of features that are used as design characteristics of structures and are due to the structure of the activated binder and its contacts with concrete aggregates. These features also have a significant impact on the nature of the destruction of concrete under load, changing the boundaries of its microcracks and durability.

1. Introduction

It is known that to obtain high-quality new-generation concrete it is necessary to provide it more dense monolithic structure. This requirement is achieved by observing certain techniques and rules:

- extremely low water-cement ratio [1];
- increased requirements for the quality of concrete aggregates [2-3];
- application of modifying additives, providing obtaining a dense structure of concrete [4];
- especially careful mixing and compaction of the concrete mixture [5].

In this paper, studies have been carried out on the activation of binder. Mineral binders - is finely crushed mineral powders, which form when mixed with water, plastic mass, which over time under the influence of physical and chemical processes passes in stone condition [6-7]. This property is used for binders for producing the artificial stone materials (concrete, etc.). In this case, the mechanical process of natural raw materials is increasingly replaced by chemical - a simple, efficient and cost-effective.

The purpose of study is to determine the influence of methods of mechanical and chemical activation to the properties of composite binders.

2. Results and Discussion

Activation of cement-water suspensions, mortar and concrete mixtures is a variety of physical, chemical and complex physical and chemical methods of exposure, both to individual components and to their compositions. Activation leads to intensification of the processes of structure formation, modification of the structure and properties of composites. The activated state of a substance is a certain critical intermediate state, through which a process that passes through time passes. One of the promising areas of practical application of activation in the building materials industry is the joint use of mechanical and chemical methods for processing binders in order to increase their useful properties.

Activation leads to an increase in the specific surface of astringents, a change in the surface structure of particles, an appearance of physical defects in sublattices and lattices of minerals that accelerate the elementary interactions of the surface layer with water. There is reducing of the time to achieve cement strength and to ensure a more complete use of the chemical energy of the binder. Activated concrete has a number of features that are used as design characteristics of structures and are due to the structure of the activated binder and its contacts with aggregates of concrete. These features also have a significant impact on the nature of the destruction of concrete under load, changing the boundaries of its microcracks and durability. The established fundamental difference in the nature of the destruction of activated concrete is the formation of an extensive pre-destruction zone and the explosive release of compressive stress energy. The concrete based on the activated binder has a homogeneous structure, which reduces the concentration of stresses on the boundary between the aggregate and the cement stone, so the deformation of such concrete under load takes quite a long time without micro-fractures. The purpose of mechanical chemistry is to use or prevent those chemical reactions that are caused or accelerated by mechanical activation. Part of the mechanical energy supplied to the solid during activation is assimilated to it in the form of a new surface, linear and point defects.

It is known that the chemical properties of crystals are determined by the presence of defects in them, their nature and concentration. The mechanical activation of the mixture is numerically equal to the total change in the free energy of the system under the action of mechanical forces. One of the main provisions of mechanical activation is that there can be mechanical activation without a grinding, but can not be a grinding without an activation. From this it follows that, first, it is impossible to separate grinding and activation. Any grinding is activation, because by influence of external forces the energy reserve of the ground substance increases, at least due to an increase in the surface energy. Secondly, any grinding machine is a mechanical activator. The most complete achievement of the joint mechanical and chemical activation of the binder is possible in a vario-planetary mill. In a vario-planetary mill, the rotational speeds of the grinding jars and the support disc can be set completely independently of one another. By varying the gear ratio, the movement and trajectory of grinding balls can be influenced so that the balls strike horizontally on the inner wall of the grinding jar (high impact energy), approach tangentially (high friction) or simply roll over the inner wall of the grinding jar (centrifugal mills). All intermediate stages and combinations between pressure friction and impact can be freely installed (figure 1).

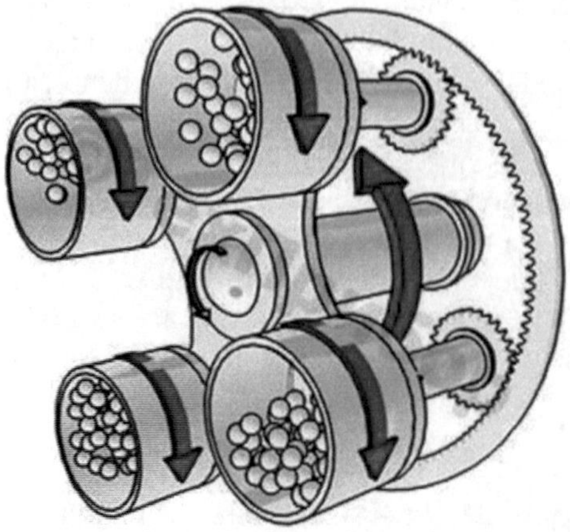

Figure 1. Operating principle of a vario-planetary mill

Grinding by any apparatus gives the activation of the processed material more or less. The effect of mechanical activation of components of a concrete mixture consists in the transition of the passive (inactive) surface of both astringent and inert materials to a chemical active state, which is expressed in an increased ability to react during subsequent technological operations. The increase in the specific surface area of the cement, its reactivity (activity) has a significant effect on the formation of the structure of the concrete, the rate of hardening and its strength characteristics. Using of activated cement makes it possible to form a denser and more uniform structure of concrete, which allows one to obtain a sharp increase in the one-day strength and increase it after 28 days of hardening (figure 2).

Figure 2. Microphotography of cement stone neoplasms based on an activated composite binder

Particularly relevant is the mechanoactivation of components for the production of foam concrete, polystyrene concrete, when the quality and stability of the characteristics of the constituents of the mixture is of particular importance. Increase in the activity of materials is achieved as a result of grinding, dispersing (grinding) in special energy-intensive grinding units (mills). One of the leading factors affecting the degree of structural change in minerals is the destruction method, determined by the type of shredder. Conventionally, several basic methods for grinding materials using energy-stressed fine grinding aggregates can be distinguished: grinding by crushing, abrasion and splitting (the free-blow method), and a combination of these methods. Activation of astringent and inert components of a concrete (mortar) mixture by the free-impact method and subsequent vibration activation in a turbo mixer-vibration activator makes it possible to save an expensive binder without reducing the strength characteristics of finished products and increasing their cost price, improve frost resistance, and improve resistance to all types of wear.

Scientists for several decades studied and developed the theory and practice of joint mechanical and chemical activation of solid materials [8-13].This area of science and technology concerns the conduct of solid-phase reactions in grinding machines and contains a huge innovative potential. Solid-phase synthesis is attractive in that it provides a comparative simplicity of the process, the ability to conduct reactions in the absence of solvents, which is important from an environmental view point. Despite the destruction in the crystal structure and defectiveness, the chemical composition during activation does not change. Cement is a fine-grained powder with high specific surface area (100-600 m^2 / kg). This causes intensive condensation on the cement grains of vaporous moisture and gases from the surrounding space. It is known that, in spite of the high specific surface area of the cement, their granulometric composition is far from uniform, and a significant part of the grains (40-50%) are larger than 50-60 μm. In the process of increasing the strength of cement stone, the main role is played by a fraction of 3-30 μm in size. Grains of cement 40-60 μm or more remain unhydrated, and only after six months the thickness of the cement stone layer reaches 15 μm. The incompleteness of the use of cement is aggravated by the difficulties in achieving an even distribution of water between the individual particles of the binder, which, due to adsorption and molecular bonding forces, are aggregated into floccules, which prevent uniform wetting.

Increasing in the specific surface is considered uneconomical, so it is rational to activate the binder during preparation of the concrete mixture. On the other hand, high rates of hydration of fine cement fractions are associated not only with their high specific surface area, but also with the highest dislocation density and concentration of defects on the surface of fine cement particles. During increase defectiveness of the particles, a transition to a nonequilibrium state occurs, which leads to a decrease in chemical stability and the intensification of a whole series of physical and chemical processes, including the hydration activity of clinker minerals. This fact is explained by the fact that as a result of grinding the clinker, the grain surface contains a number of defects in the form of submicro- and micro- cracks. Destruction of cement grains at the beginning of hydration occurs and develops on defects, and is accompanied by dislocation movement. Speed of dislocation movement is determined by the physical and chemical nature of the surface of clinker minerals, the boundaries of their phases, and the content of impurity elements in crystals. In turn, crack growth is equivalent to a continuous distribution of dislocations in the volume of the solid phase. Clinker particles with defects are in a state of higher energy of interaction during hydration than structurally perfect minerals.

The activation of binders is also caused by an increase in temperature and pressure, the addition of special additives, chemical and mechanical dispersion of individual components and their mixtures, etc. The rate of most homogeneous reactions increases by a factor of 2-4 with a temperature increase of 10°C. Addition of surfactants and other chemical additives promotes the formation of additional crystallization centers and stimulates their growth. Using of chemical additives in construction shows that the largest share belongs to plasticizers and superplasticizers. They can significantly reduce the water requirement of the concrete mix (by 20% or more), and also reduce the consumption of binders and significantly improve the strength characteristics of concrete, to use when building concrete and reinforced concrete structures, cast self-compacting and non-dispersible concrete mixtures. The negative moment of application of such additives in concretes is their synergy with cement and a significant increase in the cost of the final product.

With the development of technologies, new opportunities appear to affect the structure and properties of water, there is the possibility of purposeful management of the process of structure formation and the properties of cement composites, which represent a complex hierarchical system including the nanoscale. Scientists are conducting studies on the modification of the mixing water by carbon fulleroidal nanoparticles [14-15] (figure 3).

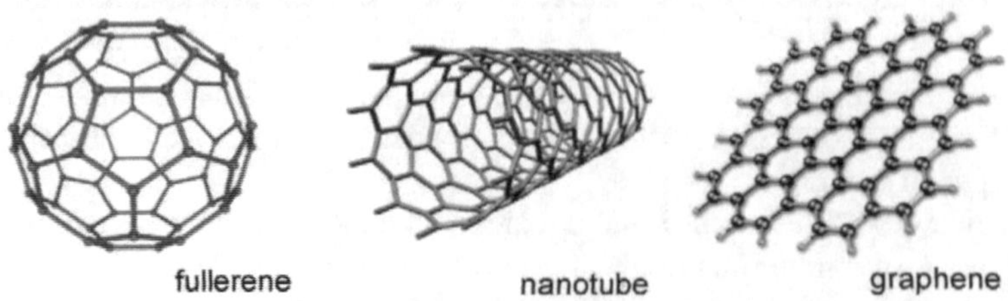

Figure 3. Carbon nanoparticles

The method of modifying (activating) mixing water allows, by reducing the cost of expensive components (cement and additives) to lower the cost of concrete, while the physical and mechanical properties of the final product do not deteriorate. Research is developing on electromagnetic activation methods aimed at intensifying the hydration of individual clinker minerals, regulating the basicity of hydrosilicates by applying an alternating or discrete constant electric field of a certain frequency [16-17].Electromagnetic activation of binders compositions is less energy-intensive than joint mechanical and chemical activation, which is associated with an increase in the activity of components of astringent compositions. Interesting results were obtained by activating the cement test with high energy sources, aero hydro-dynamic radiators, and also with magneto-mechanical, electro-hydraulic and thermoelectric effects on solutions and concretes.

The technology of magnetic activation of mixing water has been developed for a long time, but it still does not find wide application in construction practice. One of the reasons for this is the problem of obtaining a stable level of water activation. This leads to variability in the manifested water properties and poor repeatability of the results. Magnetic treatment consists in passing the flow of water through a magnetic field. Strength of products made using magnetic activated water, statistically significantly increases. Magnetic treatment of mixing water affects the hardening process: the setting speed and the plastic strength of the cement paste change, the sizes of the cement granules decrease, the hydration process becomes active. The mixing of concrete mixtures with magneto-activated water intensifies the processes of cement dissolution and hydration in the early periods of hardening and accelerates the release of smaller crystals, which leads to a decrease in porosity, an increase in density and frost resistance of concretes (figure 4).

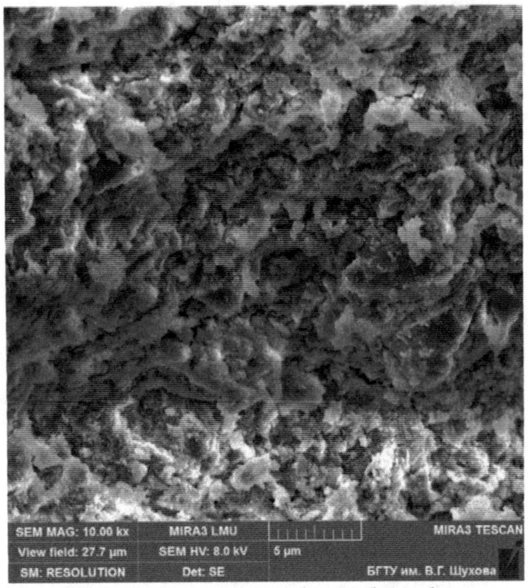

Figure 4. Microphotography of cement stone neoplasms based on magnetic activated composite binder

Among non-reactive methods of activation of building mixtures and their components, one of the promising is the treatment with high-voltage electric discharges. It is necessary to recognize the priority of electroactivation against magnetic activation, since the mechanism of influence of the first, at least at the phenomenological level, lends itself to logical interpretation. The study of the mechanism of the effect of the basic factors of a high-voltage electric discharge on cement-water systems and the influence of the discharge parameters on the physical characteristics of activated concrete with a view to reducing energy consumption showed that for the presence of the activation effect, it is necessary to observe the initial conditions:

$E_{sp} \geq E_i$,

$E_{sp} \geq E_c$,

where E_{sp} - specific electric energy introduced into the volume; E_i - energy that ensures ionization of the mixture; E_c - energy, which ensures the destruction of the crystal lattice of the cement conglomerate.

The ionization energy is the resultant value, the term of several components, the total value of which is sufficient for dissociative-associative phenomena to occur in the medium being processed. This facilitates the involvement of water layers located outside the electrical discharge channel to various oxidation reactions. Consequently, the initial impulse leads to a significant change in the ionic composition of the suspension and the appearance of polarized groups in the water. Mechanical dispersion of cement leads to an increase in its specific surface, morphological uniformity, increase in density and strength of concrete. Thus, the alternate execution of the first or second condition is forming two stages of the activation process. At the first stage of the processing, water-cement systems are activated by ionization, and in the second stage, the cement slurry is dispersed. As a result of electroimpulse activation, the water of mixing and the cement slurry become chemically active, acquiring the state of the ionic solution, and can be used as an accelerator for hardening concretes. The selected processing regime, close to the critical one, helps increase the concentration of aqueous solution. As a result of electro-physical processing, the rheological properties of building mixtures benefit greatly and the strength characteristics of cement stone are increased.

System analysis of the structure gives a real picture of the change in density and porosity of activated concrete. As a result of active hydration processes, the amount and volume of pores is reduced. Cement stone has a more uniform morphological composition, in which particles with a smaller size predominate.

Recently, along with other areas, there has been an interest in the possibility of carrying out technological processes that require thermal activation under microwave (dielectric) heating conditions. Dielectric heating effect mainly consists in the absorption material energy microwave electromagnetic field (HF or UHF) band and converting this energy into heat. The thermal power emitted by the material depends on its dielectric characteristics and field parameters and provides a number of advantages over other thermal activation methods: high process speed, lack of heat carriers, dynamic temperature control, selective activation of individual components in composite systems, etc.

The activation of the cement dough also is by ultrasonic processing. It causes the effect of cavitation, the dispersion of solid particles, microcracks in crystals, which contributes to the dissolution of cement particles and their more complete hydration. Unlike high-frequency vibration, the relative increase in strength occurs with an increase in the water-cement ratio by ultrasonic action. Dense and strong crystal hydrate structure of cement stone is formed by influence of wave pressure arising in an acoustic field. Intensification of the hardening process of concrete is achieved by joint effect of an acoustic field with a frequency of 10-16 kHz and an elevated temperature. Cement paste can be preprocessed in an aero-hydrodynamic activator, followed by stirring with heating aggregates and pouring at 60-65°C. Thermo-acoustic activation of the concrete mixture is also possible with agitation in the heated state in mixers with acoustic radiators. The combination of activation with preheating allows about a 1.5-fold increase in the 28-day strength of concrete.

3. Conclusion

Activation leads to an increase in the specific surface of astringents, a change in the surface structure of particles, the appearance of physical defects in sublattices and lattices of minerals that accelerate the elementary interactions of the surface layer with water. There is a reduction in the time to achieve cement strength and ensure a more complete use of the chemical energy of the binder. Activated concrete has a number of features that are used as design characteristics of structures and are due to the structure of the activated binder and its contacts with concrete aggregates.

Despite the foregoing, the common drawbacks of all physical methods of activation are: the laboriousness of finding quantitative parameters characterizing the degree of activation of the aquatic environment in production conditions; the need to retrofit technological lines with special equipment for activation; the need for refinement, and in some cases, for rewrighting of technical normative documents and technological regulations, etc.

References

[1] Lesovik V S, Zagorodnyuk L K, Mestnikov A E, Kudinova A I, Sumskoi D A 2015 *International Journal of Applied Engineering Research* **10(5)** 12383-12390

[2] Murtazaiev S-AY, Lesovik V S, Bataiev D K-S., Chernysheva N V, Saidumov M S 2015 *Modern Applied Science* **9 (4)** pp 233-245.

[3] Suleymanova L A, Lesovik V S, Lukuttsova N P, Kondrash E V, Suleymanov K A 2015 *International Journal of Applied Engineering Research* **10 (5)** pp 12399-12406

[4] Lesovik R V, Klyuyev S V, Klyuyev A V, Netrebenko A V, Kalashnikov N V 2014 *World Applied Sciences Journal* **30 (8)** pp 964-969.

[5] Ranjbar N et al. 2016 *Journal of Cleaner Production* **112** pp 723-730

[6] Fediuk R S 2015 *IOP Conf. Series: Materials Science and Engineering* **125** 012019.
[7] Fediuk R S 2016 *IOP Conf. Series: Materials Science and Engineering* **116** 012020.
[8] Fediuk R S, Yushin A M 2016 *IOP Conf. Series: Materials Science and Engineering* **116** 012021.
[9] Ibragimov R 2016 *ZKG International* **69 (6)** pp 34-39.
[10] Zagorodnjuk L H, Lesovik V S, Volodchenko A A, Yerofeyev V T 2016 *International Journal of Pharmacy and Technology* **8(3)** pp 15146-15155.
[11] Ibragimov R A, Pimenov S I 2016 *Magazine of Civil Engineering* **62 (2)** pp 3-12.
[12] Glagolev E, Suleimanova L, Lesovik V 2016 *International Journal of Environmental and Science Education* **11(18)** pp 12383-12389.
[13] Buyantuev S L, Urkhanova L A, Lkhasaranov S A, Stebenkova Y Y, Khmelev A B, Kondratenko A S 2016 *IOP Conf. Series: Materials Science and Engineering* **168 (1)** 012046.
[14] Buyantuev S L, Urkhanova L A, Kondratenko A S, Shishulkin S Y, Lkhasaranov S A, Khmelev A B 2016 *IOP Conf. Series: Materials Science and Engineering* **168 (1)** 012048.
[15] Semenov A P, Smirnyagina N N, Urkhanova L A, Kanakin S V, Lkhasaranov S A, Semenova I A, Tsyrenov B O, Dasheev D E, Khaltarov Z M 2016 *IOP Conf. Series: Materials Science and Engineering* **168 (1)** 012059.
[16] Fediuk R S, Yushin A M 2015 *IOP Conf. Series: Materials Science and Engineering* **93** 012070.
[17] Fediuk R S, Smoliakov A K, Timokhin R A, Stoyushko N Yu, Gladkova N A 2017 *IOP Conf. Series: Materials Science and Engineering* **66** 012026.

Structural Degradation of the Welded Joint of the Gas Main after a Long-Term Operation in Sub-Acid Soil

P O Maruschak[1], R T Bishchak[2], O V Maruschak[2, a], S V Panin[3,4]

[1] Ternopil National Ivan Pul'uj Technical University, Ruska 56 str., Ternopil 46001, Ukraine
[2] Ivano-Frankivsk National Technical University of Oil and Gas, Karpatska 15 str., Ivano-Frankivsk, 76000, Ukraine
[3] Institute of Strength Physics and Materials Sciences SB RAS, 634021, Tomsk, Russia
[4]National Research Tomsk Polytechnic University, 30, Lenin Avenue, Tomsk, 634050, Russia

E-mail: [a] Maruschak.tu.edu@gmail.com

Abstract. The authors investigated the main regularities in the structural and mechanical degradation of steel and a welded joint of the "Soyuz" gas main in operation. It is established that the defect accumulation kinetics in welds are determined both by technological disruptions in their manufacture and by hydrogenation of the weld during a long operating time. A weakening of the mechanical properties of the weld after a prolonged operation is shown, which in some cases led to the appearance of cracks and fracture of the gas main.

1. Introduction

Gas mains of Ukraine have been in operation for a long time, which causes degradation of their material, and especially welded joints. In addition, since the time Ukraine's gas mains were commissioned, the precision of diagnostic devices has increased significantly, due to which it is now possible to detect both in-service defects of gas mains, and technological imperfections of the process of their manufacturing. The structural heterogeneity of welds makes them a "weak point" in terms of resistance to the initiation of defects and long-term fracture of the gas main [1-2].

At present, there are many approaches to assessing the degradation of the gas main material, in particular, pipes of strength class K46-K65, whose impact toughness does not exceed 125 J/cm^2. It is known that welded joints, especially after a prolonged exploitation, become the most vulnerable parts of welded structures [3]. An important component of the above considerations is the assessment of the actual condition of the gas main metal after a prolonged exploitation, including welded joints, which is the basis for predicting their capacity for work.

Significant differences in the degradation mechanisms of the gas main steel lead to a variety of operational defects and damage to the structure of welds [4-6]. Consequently, the existing approaches to assessing their strength and ability to work, as well as the methods for the prevention of unpredictable fracture need to be improved and concretized. In addition, investigations into the defects of the long exploited pipe steels are the basis for the creation of high strength steels with different types of microstructures. It should be noted that the systematization of operational damage is an important scientific and engineering task, and its description requires further research [7-9].

The purpose of this paper is to analyze the degradation mechanisms of the weld of "Soyuz" gas main after a long-term operation.

2. Ecological features of operation of gas mains

To ensure a reliable and safe operation of underground gas mains, it is necessary to consider the corrosion activity of soils, in which they are laid, in addition to ensuring a high-quality protective insulating coating. Therefore, we should consider carefully the environmental indicators that characterize the corrosive activity of the soil. It is caused by the presence of hydrogen ions, whose concentration is expressed in pH. The pH of soil varies depending on the total mineralization of groundwater and the presence of carbonate and mineral acids, acid and basic salts in it. According to the pH, soils are classified into: strongly acid (3.0-4.5), acid (4.5-5.5), sub-acid (5.5-6.5), neutral (6.5-7.0), mildly alkaline (7.0-7.5), alkaline (7.5-8.5), strongly alkaline (more than 8.5).

In all regions of the Ternopil oblast, except Zalishchitsky, Zbarazhsky and Chortkivsky, the area of acid soils has increased in recent years as compared to previous studies. The average weighted pH in the region varies from 5.5 (slightly acid reaction) to 6.7 (neutral). In the last five years, the average pH dropped by 0.1 and is 6.0 (the reaction is close to neutral). However, according to some predictions, the soil solution in the region will tend to acidification in the coming years.

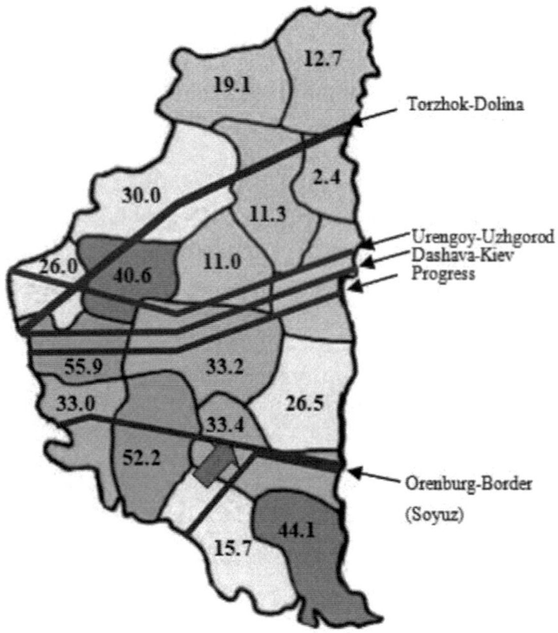

Figure 1. The map of gas mains running through the Ternopil oblast and percentage (%) of acid soils (the place of cutting the template is indicated with a red arrow)

3. Research technique

In order to analyze the reasons for the formation of a weld defect and find out a mechanism of crack propagation from it, a complex research of a fragment of the gas main was performed. In particular, we used a fragment of the "Soyuz" gas main with a diameter of 1420 mm and a wall thickness of 16 mm, which was taken near the village of Kolindiany, Chortkivsky region, Ternopil oblast, Fig. 2a.

The research included testing of the mechanical properties of the weld zone, visual inspection, and a metallographic analysis on a microscope "AXIOVERT 40- MAT". Specimens were cut from a pipe fragment in a zone of a weld, next they were ground to attain a gradual decrease in grain size, and polished by a diamond paste on the USHPO-0 setup. After that, the prepared surface of specimens was etched in 4.0% nitric acid solution in ethyl alcohol.

Mechanical tests of the specimen cut along the weld joint of the pipeline with operational defect were performed, Fig. 2 c, d under uniaxial static stretching on a hydraulic test machine R-100. A fractographic analysis of the specimen fracture was performed on the REM-106 microscope.
Measurements of microhardness were made at a load of 200 g on a microhardness tester PMT-3, based on the average result of three measurements in the selected area of the weld.

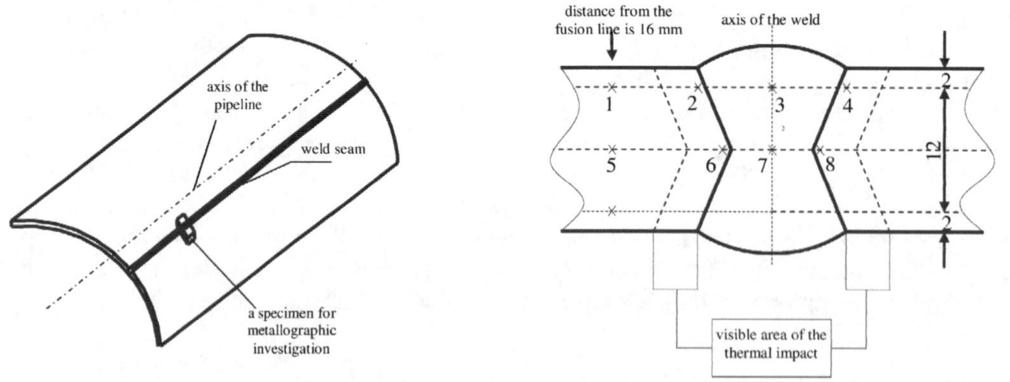

Figure 2. A general view of a pipe fragment (a) and a weld with a crack (b)

4. Analysis of the weld cross-section

The analyzed welded joint was destroyed due to the propagation of a cold crack, which was formed in the area adjacent to the outer surface of the welded joint after welding. That is, it was a technological defect that propagated during exploitation. To study the reasons, a metallographic specimen was made from the part of the welded joint that remained intact, and its hardness was measured on a hardness tester TK-2M according to the scheme of Fig. 2 b, which is related to a number of mechanical and physical properties of materials, in particular, the density of dislocations in small-angle boundaries. According to the measurement results, an increase in hardness up to 78 HRB was found around the metal of the welded joint, and its reduction to 73 HRB in the zone of thermal influence.

Table 1 Results of hardness measurement of the welded joint of "Soyuz" gas main

Hardness	Measurement points, see Fig. 2b							
	1	2	3	4	5	6	7	8
HRB	72	73	78	73	74	73	78	74
HB*	132	134	146	134	136	134	146	136

*data obtained by transferring HRB to HB [10].

According to the hardness measurements, a conventional strength limit was calculated [10]:

$$\sigma_{us} = 3.45HB \quad (1)$$

$$\sigma_{us} = 3.29HV - 47 \quad (2)$$

$$\sigma_{us} = \frac{(HV - 1.837)}{0.304} \quad (3)$$

Values of the relative strength of the weld of the gas main "Soyuz" from steel X65 (Japan, TU-1975 TECHNICAL CONDITIONS [11]) are calculated by equations (1-3).

It should be noted that the welded joint had reduced strength properties, since the main metal of this pipe was characterized by values σ_{us} = 620 MPa, σ_{ys} = 590 MPa [11]. In the analysis of the weld, an initial concentrator was discovered, from which fracture began. Degradation of the weld zone compared to the initial material was observed both in the zone of the weld and in the zone of thermal impact. This lowered the load-bearing capacity of the weld and partly its crack strength. In addition, the heterogeneity of the grain substructure is typical of the welded joint. The presence of harmful impurities, including those prone to segregation along the grain boundaries, caused the destruction of the weld. Thus, the crack originated in the sub-surface layer of the weld, and propagated from the outer surface to the inner one.

5. Metallographic analysis of the welded joint

The microstructure of the welded joint of the gas main "Soyuz" cut out in different directions was investigated. The areas of ferrite and fine perlite were found, which, in general, gives grounds to believe that the structure belongs to the ferrite-perlite class. The grain size was 9-11 points according to GOST 5639-82. Micropores were found on the surface of metallographic specimens, which are formed at the boundary "matrix-ferrite", Fig. 3

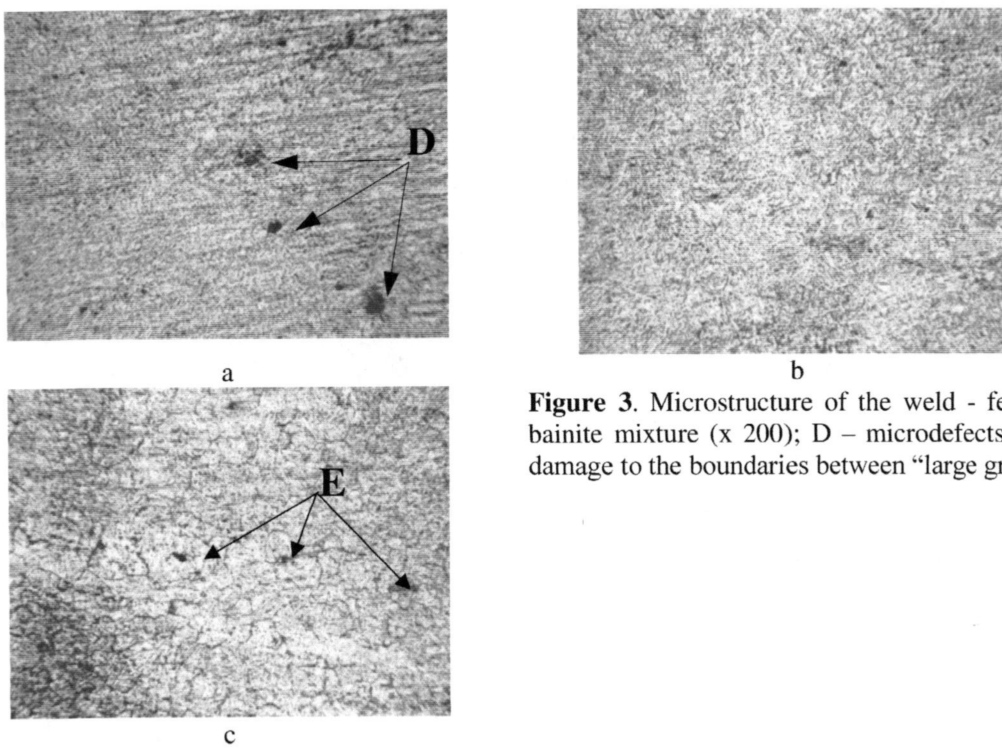

Figure 3. Microstructure of the weld - ferrite-bainite mixture (x 200); D – microdefects, E – damage to the boundaries between "large grains"

The obtained metallographic images of the metal of different zones of the weld show that small crystallites with an "ordered" structure are caused by welding technology. The structure of the metal is a ferrite-bainite mixture, Fig. 3b. Large grains are found mainly in the near-surface part of the weld. In our opinion, they result from crystallization of the metal. This structure indicates a low impact strength and generally low crack resistance [12, 13]. It is this site that served as a zone for the crack start in a weld. Also, changes in the initial structure were detected by the selection and coagulation of carbides along the grain boundaries, with a simultaneous formation of microscopic damage.

The appearance of pores and the destruction of the interstices between them, with the formation of intergranular cracks in the axial section of the weld, occurred over a long period of time, and the hydrogenation of the metal facilitated their merging with each other. Similar defects were discovered in the vicinity of a transverse cross-sectional crack in the Urengoy-Pomari-Uzhhorod gas main with a diameter of 1420 mm [14]. In our opinion, the technological reason for the defect formation is a very rapid and uneven cooling of the weld.

6. Discussion

Since the fragment of the gas main was cut in the territory of the Chortkivsky region of the Ternopil oblast, which is characterized by acid soils, one should take into account the possible complex nature of the damage, which is due to the features of the material microstructure, the conditions of mechanical loading, and the action of the aggressive environment. It is known that in the process of a prolonged exploitation of structural steels, the resistance to brittle fracture [1-8] deteriorates most, which is dangerous due to the uncontrolled growth of cracks, especially in aggressive corrosive-hydrogen environments. Their role is twofold, because they: a) accelerate fracture particularly of the most degraded material; b) contribute to the operational degradation of the material "in volume", which is associated with its hydrogenation. The latter factor is often neglected, as they usually consider only stresses caused by the mechanical loading of structures.

In our opinion, a crack was formed in the area of the weld due to defects accumulated on grain boundaries [12-14]. The said defects were formed in the first moments after welding, however, they were scattered and did not coalesce, because the level of welding stresses in this area was insufficient for their growth.

After mounting the pipe into the gas main, a local growth and merging of microdefects occurred under the influence of operational factors, including hydrogenation. In our opinion, such technological influence could be the "superimposition and summation" of welding stresses with technological ones, which caused the localization of microdeformation processes. Thus, according to the measurements of hardness, as well as fractographic and metallographic analysis of this defect, we can conclude that the defect was formed due to the influence of negative structural and mechanical factors: the size of austenitic grains, a significant gradient of grain structure in the weld zone, and the "coarsening of grain".

7. Conclusions

The investigated crack of the weld has a technological and operational origin, its occurrence is a consequence of the effect of welded microdefects formed during the manufacture of the weld. An additional factor influencing the propagation of the defect is the hydrogenation of the pipe wall. Fractographic and metallographic studies have allowed revealing the zone of crack origin and micromechanisms of crack propagation. The analysis of microstructural features of the weld zone and its comparison with the morphology of the crack growth allowed determining the causes of its occurrence and stages of origin and growth of the macrodefect.

Acknowledgments

The reported study was funded by the Ministry of Education and Science of Ukraine, according to research project No. 0116U006908.

References

[1] Semenov S E, Mandel'berg S L, Goritskii V N, Bolotov A S, Anenkov N I 1987 *Welding International* **1** 610-614.

[2] Min Sung Joo, Kyung-Min Noh, Wan-Keun Kim, Jin-Ho Bae, Chang-Sun Lee 2015 *Metallurgical and Materials Transactions E* **2** 119-130.

[3] Hashemi S H, Mohammadyani D 2012 *Int. J. of Pressure Vessels and Piping* **98** 8-15.

[4] Curiel F F, García R, López V H, González-Sánchez J 2011 *Corrosion Science* **53** 2393-2399.

[5] Zachrisson J, Börjesson J, Karlsson L 2013 *Science and Technology of Welding and Joining* **18** 603-609.

[6] Rahman S, Ghadiali N, Wilkowski G M, Paul D 1997 *Int. J. of Pressure Vessels and Piping* **70** 209-221.

[7] Ostsemin A A 1991 *Strength of Materials* **23** 406-410.

[8] Gorkunov E S, Zadvorkin S M, Putilova E A, Savrai R A 2014 *The Physics of Metals and Metallography* **115** 949-956.

[9] Khotinov V A, Arabei A B, Pyshmintsev I Yu, Farber V M 2013 *Russian Metallurgy (Metally)* **2013** 336-343.

[10] Lee K-S, Song J-H 2006 *Int. J. of Fatigue* **28** 386-400.

[11] X65 (Japan, TU-1975 TECHNICAL CONDITIONS).

[12] Brahim Belkessa, Djamel Miroud, Naima Ouali, Billel Cheniti 2016 *Acta Metallurgica Sinica (English Letters)* **29** 674-682.
[13] Erofeev V V, Aimetov F G, Shakhmatov M V, Smetanin F E, Baldin V M 1993 *Strength of Mater.* **25** 928-935.
[14] Maruschak P O, Bishchak R T, Shlapak L S, Panin S V 2017 *IOP Conf. Series: Materials Science and Engineering* **177** 012114 doi:10.1088/1757-899X/177/1/012114

IOP Conf. Series: Earth and Environmental Science **115** (2018) 012047 doi:10.1088/1755-1315/115/1/012047

Analysis of Fracture Characteristic of a Gas Main Pipe on the Basis of the Additive Mathematical Model of the Cyclic Random Process and Polynomial Function

I V Lytvynenko[1], P O Maruschak[2, a], S V Panin[2,3], A P Sorochak[1]

[1] Ternopil National Ivan Pul'uj Technical University, Ruska 56, Ternopil 46001, Ukraine
[2] Institute of Strength Physics and Materials Sciences SB RAS, 634021, Tomsk, Russia
[3] National Research Tomsk Polytechnic University, 30, Lenin Avenue, Tomsk, 634050, Russia

E-mail: [a] Maruschak.tu.edu@gmail.com

Abstract. The use of a mathematical model to describe the phenomenon of dynamic fracture of the gas main in a broad range of crack propagation rates is substantiated, taking into account the stochastic and cyclic nature of this process. Particular attention is paid to the reliability of the proposed approach, which is evaluated based on comparison with known natural experiments.

1. Introduction

Experimental studies of the dynamic fracture of oil and gas mains allow establishing the basic regularities in deformation processes in steels of different strength classes at high crack propagation rates [1]. However, at the moment, a single standard has not yet been developed, which would consider the main effects that are crucial for describing the elastic-plastic properties and the dynamic strength of materials under impact loads.

That is why a mathematical model to describe the dynamic deformation and fracture, evaluate the kinetics of crack propagation in gas mains is being actively developed, in order to create simple and effective models that would allow for a physically correct and simple description of these complex phenomena [2, 3]. Particular attention is paid to the choice or development of effective numerical methods and algorithms for their optimization in order to ensure the reliability of calculations.

However, as well as in case of other complex phenomena, the issue of identifying parameters and substantiating the reliability of the proposed models causes some difficulties in the mathematical description of these processes. This work is aimed at extending and supplementing the existing research [5, 6], in order to provide a more accurate description of the kinetics of crack propagation in case of dynamic fracture of the gas main based on a deeper analysis of the rate fluctuations caused by branching of the crack tip as it propagates along the body of the pipe.

The proposed approach allows establishing the physical relationship between the elastic-plastic fracture of a pipe and rate fluctuations at different distances from the place of origin of a crack of dynamic fracture. The development of such methods for modeling the kinetics of crack propagation in the pipe is the basis for the development of new technological approaches for the manufacture of high strength pipes with the preservation of high ductility of the material [6, 7].

The purpose of this work is to use a mathematical model of the cyclic random process to describe the dynamic fracture of a gas main pipe, taking into account the stochastic and cyclic nature of this phenomenon.

IOP Conf. Series: Earth and Environmental Science **115** (2018) 012047 doi:10.1088/1755-1315/115/1/012047

2. Description of the crack propagation kinetics by a polynomial

A peculiarity of the proposed approach is the interpretation and model description of the regularities in dynamic fracture of the pipe taking into account the relaxation mechanisms at the crack tip due to the nonlinear behavior of the main defect, as well as structural-mechanical relaxation, inelastic deformation, and plastic flow.

So, several stages of fracture of the gas main pipe are known:

- initiation of the primary crack and its start - a rapid propagation of the crack under the influence of internal pressure in the pipe;
- crack growth, at which its length increases and the propagation rate decreases, due to the change in the stress-strain state;
- "stabilization" of the crack propagation rate due to the "self-organization" of the fracture process of the pipe material.
- loss of stability and ultimate fracture of the pipe.

However, since the description of the dynamic fracture of the pipe must be physically grounded, on the one hand, and, on the other hand, provide an acceptable accuracy and correctness, it was assumed that all of these sections of the pipe fracture could be described by a single power dependence. This allows, in addition to the above conditions, to provide sufficient simplicity of computing and avoid determining a significant number of parameters. It should be emphasized that such simplification is a compulsory step, necessitated by the complexity of the macro- and micromechanisms of the dynamic fracture of the pipe.

Figure 1 shows the approximation of experimental data by a polynomial. The proposed approach consists in using an additive mathematical model in the following form:

$$V(\omega,l) = f(l) + \xi(\omega,l)\,,\ \omega \in \mathbf{\Omega},\, l \in \mathbf{R}\,, \qquad (1)$$

where $\xi(\omega,l)$ is the cyclic random process as a model of the cyclic component of the signal; $f(l)$ is the deterministic function, which reflects the trend of the pipe fracture process,

$$f(l) = \sum_{n=0}^{3} c_n \cdot l^n\,,\, l \in \mathbf{R}\,, \qquad (2)$$

where c_n is the coefficient of polynomial function (2).

Function (2) describes well the experimental data along the entire length of the analyzed signal.

3. Approach to the analysis and modeling of the dynamic propagation of the crack

We considered the kinetics of the dynamic crack propagation in a gas main pipe in the form of an additive model - the sum of the polynomial and the cyclic random process. This allowed us to apply the methods of statistical analysis of the crack propagation kinetics in the gas main pipe for the purposes of its analysis, modeling and possible further forecasting.

Experimental data on the gas main fracture (F. Oikonomidis et al.), published in [2, 3], were taken as input data at the first stage of processing the experimental dependence "the rate of dynamic propagation of the crack - the length of fracture". Of these, a trend component $\{f(l), l \in \mathbf{R}\}$ and a cyclic component $\{\xi(\omega,l), \omega \in \mathbf{\Omega}, l \in \mathbf{R}\}$ were identified. The procedure for extracting the trend consisted of two steps: describing the data by a polynomial and finding the corresponding coefficients.

The cyclic component was determined by subtracting the trend function $\{f(l), l \in \mathbf{R}\}$ from the signal of the input process $V(\omega,l)$, in particular:

$$\xi(\omega,l) = V(\omega,l) - f(l)\,,\ \omega \in \mathbf{\Omega},\, l \in \mathbf{R}\,, \tag{3}$$

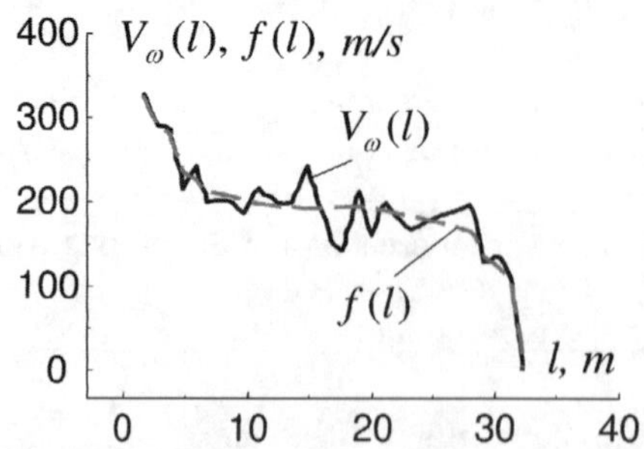

Figure 1. Experimental dependence of a decrease in the growth rate of the crack along the trajectory of its growth [2, 3] and its approximation by a polynomial.

4. Some results of statistical processing of the data on the gas main fracture.

$V_\omega(l)$ is the input realization of the investigated process of the crack growth rate. $\xi_\omega(l)$ is the realization of the cyclic component of the investigated process of the crack growth rate.

To analyze this component, it is necessary to use the methods of segmentation [8], define the discrete function of the rhythm, interpolate it [8, 9], and use the obtained results when applying the methods of statistical evaluation [10].

Methods of statistical analysis of cyclic random processes are adapted to the problems of analyzing the fracture rate of the gas pipeline to evaluate such probabilistic characteristics as the initial moment function of the first order (mathematical expectation), and the central moment function of the second order (dispersion).

The realization of the estimation of mathematical expectation is given in the form:

$$\hat{m}_\xi(l) = \frac{1}{M}\sum_{n=1}^{M} \xi_\omega\left(l + L(l,n)\right),\, l \in \mathbf{W}_1 = \left[L_1, L_2\right), \tag{4}$$

where $L_1 \neq 0$ - in the general case, M is the number of cycles in the cyclic process (cyclic component), $\xi_\omega(l)$ is the realization of the cyclic component of the crack propagation rate (the realization of a cyclic random process), $L(l,n)$ is the function of the rhythm of a cyclic random process, $\mathbf{W}_1$ is the definition area of the first cycle of the process.

The realization of the dispersion was defined as:

$$\hat{d}_\xi(l) = \frac{1}{M-1}\cdot\sum_{n=1}^{M-1}\left[\xi_\omega\left(l + L(l,n)\right) - \hat{m}_\xi\left(l + L(l,n)\right)\right]^2,\, l \in \mathbf{W}_1 = \left[L_1, L_2\right). \tag{5}$$

Thus, according to the simulation results, a description of the fracture process of the gas main pipe is proposed, which is based on in-depth study of the results of measuring the kinetics of the main crack propagation. For this, continuum-kinetic assumptions were used, in which fracture was considered using certain averaged parameters, such as the continuous process of accumulation and coalescence of defects that cause crack propagation. At the same time, a decrease in the intensity of deformation processes was noted in the local sections of the material in the vicinity of the crack tip, which propagates in conditions of dynamic fracture of the gas pipeline. This reduces fluctuations of the crack propagation rate, causing changes in the parameters presented in Fig. 2a.

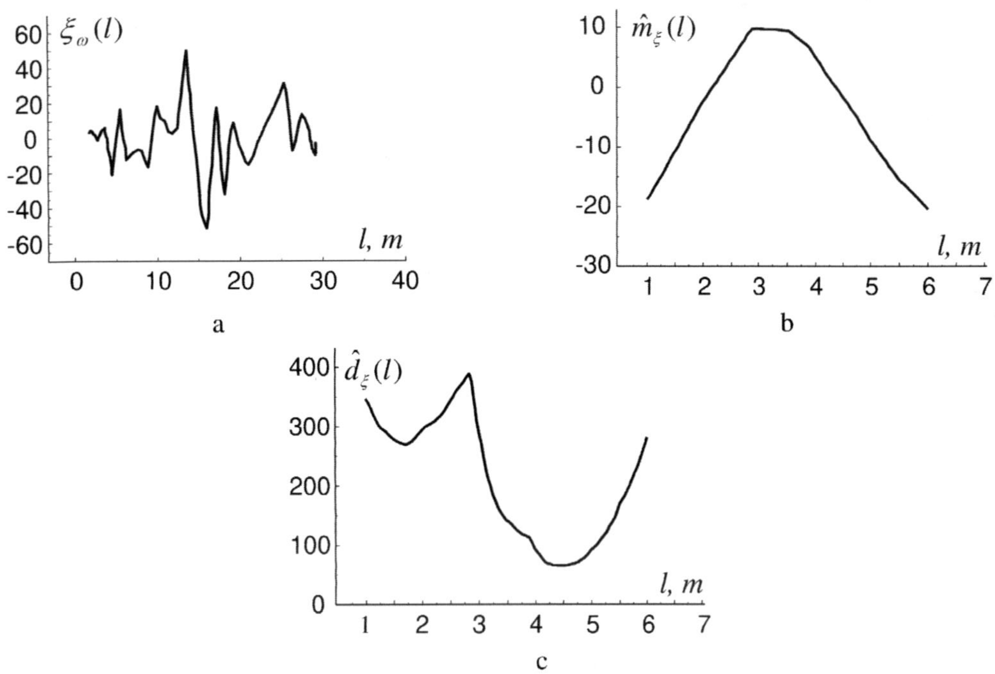

Figure 2. Cyclic component of the crack growth (a), statistical estimates of its mathematical expectation (b) and dispersion (c) in the analysis of the crack propagation rate in the main pipe on the basis of the proposed model

It should be emphasized that the obtained estimates of the mathematical expectation and dispersion, Fig. 2b,c are sensitive to fluctuations in the fracture rate of the main gas pipeline, which indicates the possibility of using these numerical features in automated systems for technical diagnostics.

In order to approbate the mathematical model and the method of analysis used in this work, a series of simulation experiments was carried out to compare actual and simulated realizations of fracture rate processes that occur in the main gas pipeline, taking into account the obtained statistical estimates and the trend component.

5. Imitation modeling of the dynamic propagation of the crack

In order to predict fracture of the gas pipeline and develop the methods to arrest the dynamic propagation of the crack, it is necessary to perform an unambiguous mathematical description of the results obtained on the natural structures. The results of estimating changes in the fracture rate of the pipe allowed to quantitatively analyze the fracture process and indirectly estimate the coordinates of defects along the trajectory of the crack propagation, which is important for the analysis of macro- and micromechanisms of fracture. The data of computer simulation of the fracture rate of the main gas pipeline by the proposed method were compared with the experimental ones. As is seen from Fig. 3a, b, the proposed methods of statistical processing of the fracture rate of the gas pipeline describe very well the nature of growth of the main crack [2, 3].

It is established, Fig. 3, that the relative mean square error of simulation does not exceed (in experiments) 15%, which is sufficient for engineering evaluation. It should be noted that the comparative analysis of the accuracy of reproducing the fracture rate of the pipe is individual for each realization of the dynamic crack propagation process. Figure 3 shows the results of realization of the crack propagation rate in the main pipe: the experimental and simulated data.

IOP Conf. Series: Earth and Environmental Science **115** (2018) 012047 doi:10.1088/1755-1315/115/1/012047

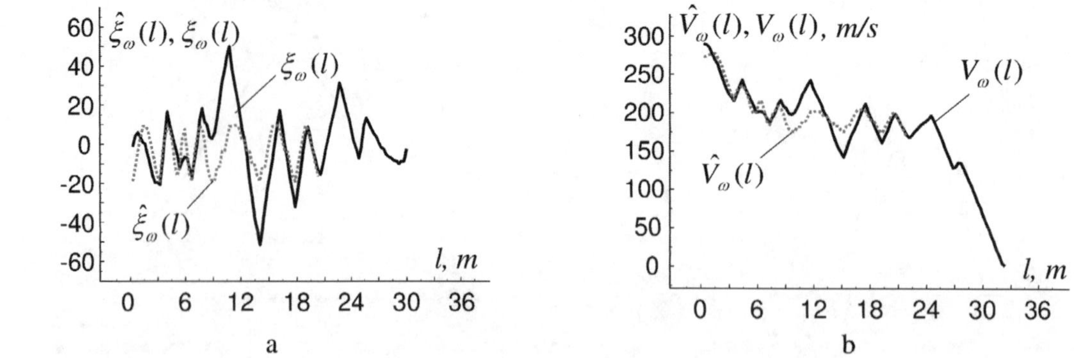

Figure 3. Comparison of simulation results (a, b) of the crack propagation rate in the main pipe with experimental data: $\hat{\xi}_\omega(l)$ is the simulated cyclic component, which takes into account statistical estimates; $\xi_\omega(l)$ is the cyclic component defined from the input signal; $\hat{V}_\omega(l)$ is the simulated realization of the crack propagation process in the main pipe, taking into account statistical estimates of the cyclic component and the trend component; $V_\omega(l)$ is the input realization of the crack growth rate process.

The results of simulation of the dynamic fracture of the main gas pipeline are satisfactorily consistent with the experimental data. They allow to numerically describe the rate variations along the trajectory of the crack propagation, which result from fracture localization in the conditions of dynamic accumulation of defects.

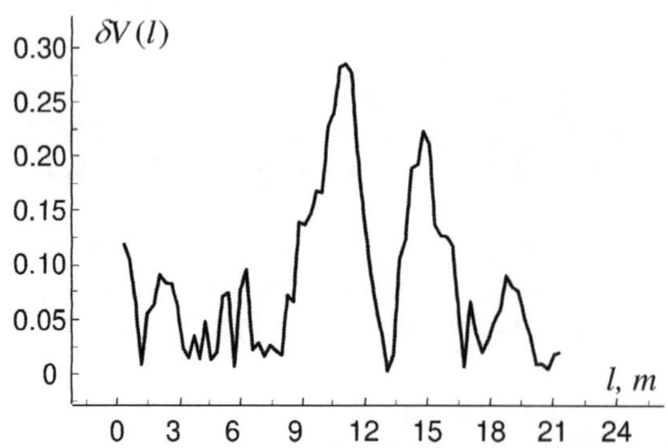

Figure 4. Relative mean square error of the simulation results of the crack propagation rate in the main pipe compared to the experimental data.

The results obtained in the work are the basis for developing the method of computer simulation of the gas main fracture. They are also used to develop the criteria that prevent the dynamic fracture. These criteria are then used in specific techniques of fracture prediction based on the results of the preliminary evaluation of the crack growth kinetics in natural structures [8-10]. The fact of repeatability (approximate to cyclic repeatability) of statistical estimates obtained for probability characteristics of processes and the variability of their rhythm functions is confirmed experimentally. This indicates the adequacy of the developed mathematical model and the method, as well as the appropriateness of their use in the problems of modeling, analysis and forecasting of processes of dynamic fracture of the gas main pipe.

6. Conclusions

A mathematical model is proposed, which allows taking into account the structural and mechanical features of fracture of gas mains. This model reflects the relationship between the nonlinearity of the crack growth rate in a pipe metal and relaxation mechanisms, and stages of its dynamic fracture. The fracture model of the main gas pipeline is considered as the sum of the deterministic polynomial function and the cyclic random process. Computer software is created based on the proposed mathematical model for analyzing the two-dimensional crack growth structure along its trajectory.

The proposed methods and approaches are substantiated on the basis of comparison with literary experimental data. The parameters are identified and the model is verified based on comparison with known natural experiments. Relative mean modeling error does not exceed 15 %.

In further research, it is planned to study the application of the exponential function as a model of the trend component, and compare the simulation results of the crack propagation rate in the main pipe on the basis of two approaches using the polynomial and exponential trend components.

Acknowledgments

The reported study was funded by the Ministry of Education and Science of Ukraine, according to research project No. 0116U006908.

References

[1] Nordhagen H O, Kragset S, Berstad T, Morin A, Dorum C, Munkejord S T *Computers and Structures*, 2012, **94-95** 13-21.
[2] Oikonomidis F., Shterenlikht A., Truman C. E. *Int. J. of Pressure Vessels and Piping* 2013, **105-106** 60-68.
[3] Oikonomidis F., Shterenlikht A., Truman C. E. *Int. J. of Pressure Vessels and Piping* 2014 **122** 15-21.
[4] Demofonti G., Mannucci G., Spinelli C.M., Barsanti L., Hillenbrand H.G. Large-diameter X 100 gas line pipes: Fracture propagation evaluation by full-scale burst test, Pipeline Technology, Vol. 1, R. Denys (Editor), Elsevier Science, 2000, 509-520.
[5] Zhuang Zhuo, O'Donoghue P E *Int. J. of Fracture* 2000 **101** 269-290.
[6] Ren Z J, Ru C Q *Engineering Fracture Mechanics* 2013 **99** 214-222.
[7] Mirzaei M, Najafi M, Niasari H *Int. J. of Impact Engineering* 2015 **85** 27-36.
[8] Lytvynenko I, Maruschak P, Lupenko S, Panin S *Applied Mechanics and Materials* 2015, **770** 288-293.
[9] Lytvynenko I V *J. of the Ternopil National Technical University* 2016 **84** 131-138.
[10] Lytvynenko I V, Maruschak P O, Lupenko S A, Popovych P V 2016 *Mat. Sci.* **52** 113-122.

IOP Conf. Series: Earth and Environmental Science **115** (2018) 012048 doi:10.1088/1755-1315/115/1/012048

Modification of Structure and Tribological Properties of the Surface Layer of Metal-Ceramic Composite under Electron Irradiation in the Plasmas of Inert Gases

V E Ovcharenko[1, 2, a], K V Ivanov[1], A A Mohovikov[2], B Yu[3], Yu Xu[4], L Zhong[4]

[1] Institute of Strength Physics and Materials Science SB RAS, pr. Akademichesky 2/4, Tomsk, 634055, Russia
[2] National Research Tomsk Polytechnic University, pr. Lenina 30, Tomsk, 634050, Russia
[3] Institute of Metal Research CAS, 72 Wenhua Road, Shenyang, Liaoning, 110016, China
[4] Xi'an Technological University, No. 5 South Jinhua Road, Xi'an, Shaanxi, 710048, China

E-mail: [a] ove45@mail.ru

Abstract. Metal-ceramic composites are the main materials for high-load parts in tribomechanical systems. Modern approaches to extend the operation life of tribomechanical systems are based on increasing the strength and tribological properties of the surface layer having 100 to 200 microns in depth. The essential improvement of the properties occurs when high dispersed structure is formed in the surface layer using high-energy processing. As a result of the dispersed structure formation the more uniform distribution of elastic stresses takes place under mechanical or thermal action, the energy of stress concentrators emergence significantly increases and the probability of internal defects formation reduces. The promising method to form the dispersed structure in the surface layer is pulse electron irradiation in the plasmas of inert gases combining electron irradiation and ion bombardment in one process. The present work reports upon the effect of pulse electron irradiation in plasmas of different inert gases with different atomic mass and ionization energy on the structure and tribological properties of the surface layer of TiC/(Ni-Cr) metal-ceramic composite with the volume ratio of the component being 50:50. It is experimentally shown that high-dispersed heterophase structure with a fraction of nanosized particles is formed during the irradiation. Electron microscopy study reveals that refining of the initial coarse TiC particles occurs via their dissolution in the molten metal binder followed by the precipitation of secondary fine particles in the interparticle layers of the binder. The depth of modified layer and the fraction of nanosized particles increase when the atomic number of the plasma gas increases and ionization energy decreases. The wear resistance of metal-ceramic composite improves in accordance to the formation of nanocrystalline structure in the surface layer.

1. Introduction

The metal-ceramic alloys on the base of refractory and hard chemical compounds (carbides, nitrides, carbonitrides, oxides) and metallic binder are promising composite materials exhibiting high mechanical and tribotechnical properties such as strength, hardness, crack- and wear resistance, fracture energy, thermal stability [1,2]. So they are applied as functional materials in the critical components of high-load tribomechanical systems including working parts of cutting tools.

Effectiveness and operational life of metal-ceramic parts under high-speed friction and abrasive action on their surface at high temperatures in aggressive environment are determined by a complex of tribological and strength characteristics of surface layers with the thickness of up to 100 μm. One of efficient methods of improving the performance characteristics of the surface layers of metal-ceramic composites is modification of their structure and phase composition by the methods of high-energy treatment including high frequency current heating or plasma, laser, ion and electron irradiation. The given methods of surface treatment are aimed at creating structure with a significant volume fraction of grains or phases less than 100 nm in size. The effect of the nanostructured state of the surface layer on the properties of the material is due to the more uniform distribution of elastic stresses in the surface layer as compared to coarse-grained counterpart when it is subjected to external mechanical or thermal action [3-6]. As a result the energy of stress concentrators emergence in the surface layer significantly increases and the probability of internal defects formation reduces.

The unique opportunities for structure modification of the surface layers of metallic materials are provided by the method of pulse electron irradiation [7, 8] because electron beam has a number of advantages when it is used as a treatment tool. It includes (i) the high degree of electric energy transformation into the energy of electron beam (over 90 %); (ii) low coefficient of the electron beam reflection from the irradiated surface (less than 10 %); (iii) significant cross-section area of the electron beam on the surface of the irradiating material (over 10 cm^2); (iv) high electron-beam power density (over 10^6 W/cm^2); (v) comparatively large depth of electron penetration into the irradiated material (several dozens of micron); (vi) high beam current pulse repetition rate (~10 s^{-1}). Within the millisecond range of pulse duration the mentioned above properties of the electron beam allow achieving ultrahigh heating rates (up to 10^9 K/s) to heat the surface layer to the prescribed temperature with further cooling due to the heat removal to the bulk material at a rate of up to 10^6 K/s creating the thermokinetic conditions for formation of non-equilibrium structure including refining grain or phase size down to nanosize [9]. To obtain nanostructured state in the surface layers of composite materials containing particles of hard chemical compounds it is promising to combine pulse electron irradiation with surface layer bombardment by inert gas ions [10].

The aim of the present work was studying the influence of pulse electron irradiation in the inert gas plasmas with varying values of ionization energy and atomic mass upon the microstructure and the tribological properties of the surface layer of metal-ceramic alloy.

2. Material and methods

Metal-ceramic alloy TiC/(Ni-Cr) with the volume ratio of 50:50 between the ceramic and the metallic components was used as a studied material. The samples of the alloy were irradiated with electron beam in Ar, Kr or Xe plasmas (Fig. 1). The energy of electrons was 25 KeV, pulse duration was 100, 150 and 200 μs, pulse energy density was 40 and 60 J/cm^2 and pulse rate was 10 s^{-1}.

The microstructure of the surface and cross-section of the surface layers was studied by SEM using Carl Zeiss Supra 55 Sapphire at accelerating voltage of 20 kV. TEM studies were carried out using JEM-2100 microscope at accelerating voltage of 200 kV. Thin foils in the cross-section of the surface layer were obtained by ion thinning using EM-09100IS ion slicer.

The wear resistance of the irradiated samples was studied using CSEM pin-on-disc tribometer. The load on the diamond cone was 5 N, the total number of sample rotations was 2500. The cross-section profile of the groove from diamond counterbody cutting the surface of the samples with numerical determination of the cutting depth were obtained u 3D profilometer "MICRO MEASURE 3D station".

3. Results and their discussion

TiC particles in the metal-ceramic composite in the ac-received state are approximately equiaxed. Their size distribution is presented in Fig. 2. The average and maximum particles size is 4.0±2.5 and 12.6 μm, respectively.

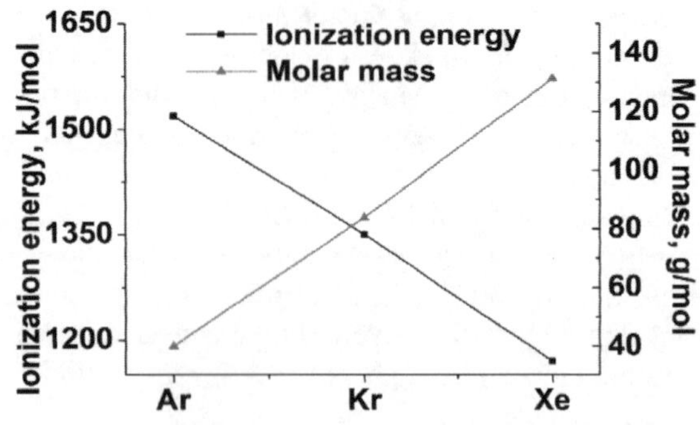

Figure 1. Values of ionization energy and atomic masses of the plasma gases

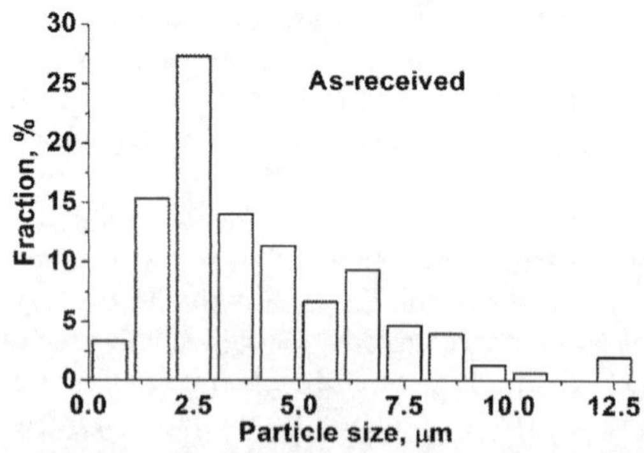

Figure 2. Size distribution of titanium carbide particle in the as-received composite

Electron irradiation of the metal-ceramic alloy in the inert gas plasmas radically changes the microstructure of the surface layer. The character of the obtained microstructures depends on the type of the used inert gas. Irradiation in argon plasma forms a two-phase structure on the surface of the alloy consisting of titanium carbide particles partly dissolved in the molten metal binder (Fig. 3a, shown with the arrows), areas where carbide particles dissolved in the molten binder (the areas in Fig.3a delineated by the black line). and areas with nanosized particles of secondary carbide in the binder (the area in Fig.3a delineated by the white line). When krypton plasma gas is used the refining of initial TiC particles down to 100 nm and less occurs forming a polycrystalline structure with dendrite-like framework in every crystal. One can see the remains of initial carbide particles in the center of each grain (Fig. 3b, shown by the arrows). Irradiation in xenon plasma also refines the initial carbide particles down to nanosize forming a surface polycrystalline structure consisting of nanosized metal-ceramic grains (in Fig. 3c the arrows show the remaining parts of the initial TiC particles in the metal-ceramic grains).

Investigation of the structure of the surface layers of the composite subjected to pulse electron irradiation in inert gas plasmas showed similarity in structure heterogeneity but revealed the influence of the plasma gas on the quantitative parameters of their structure. SEM images of the cross-sections of the surface layers show that they consist of a set of sublayers with characteristic microstructures at various depths. The depth of electron irradiation influence upon the structure of the composite increases as electron irradiation in argon plasma changes to that in krypton plasma and then to xenon one (Fig. 4).

Let us consider the surface layer irradiated in xenon plasma as an example (Fig. 4c). The mentioned set of sublayers looks as follows: the top sublayer which borders closely to the surface is characterized by columnar titanium carbide crystals oriented perpendicular to the irradiated surface of the sample (Fig. 4c, sublayer 1). The size of the crystals is 1.2±0.73 and 0.39±0.205 μm for the longitudinal and transverse directions, respectively. Under this layer there is a thicker sublayer with oriented columnar structure (sublayer 2 in Fig. 4c). The next sublayer has a structure of dendrite type. It is a transition sublayer to the initial structure of the metal-ceramic composite (sublayer 3 in Fig. 4c).

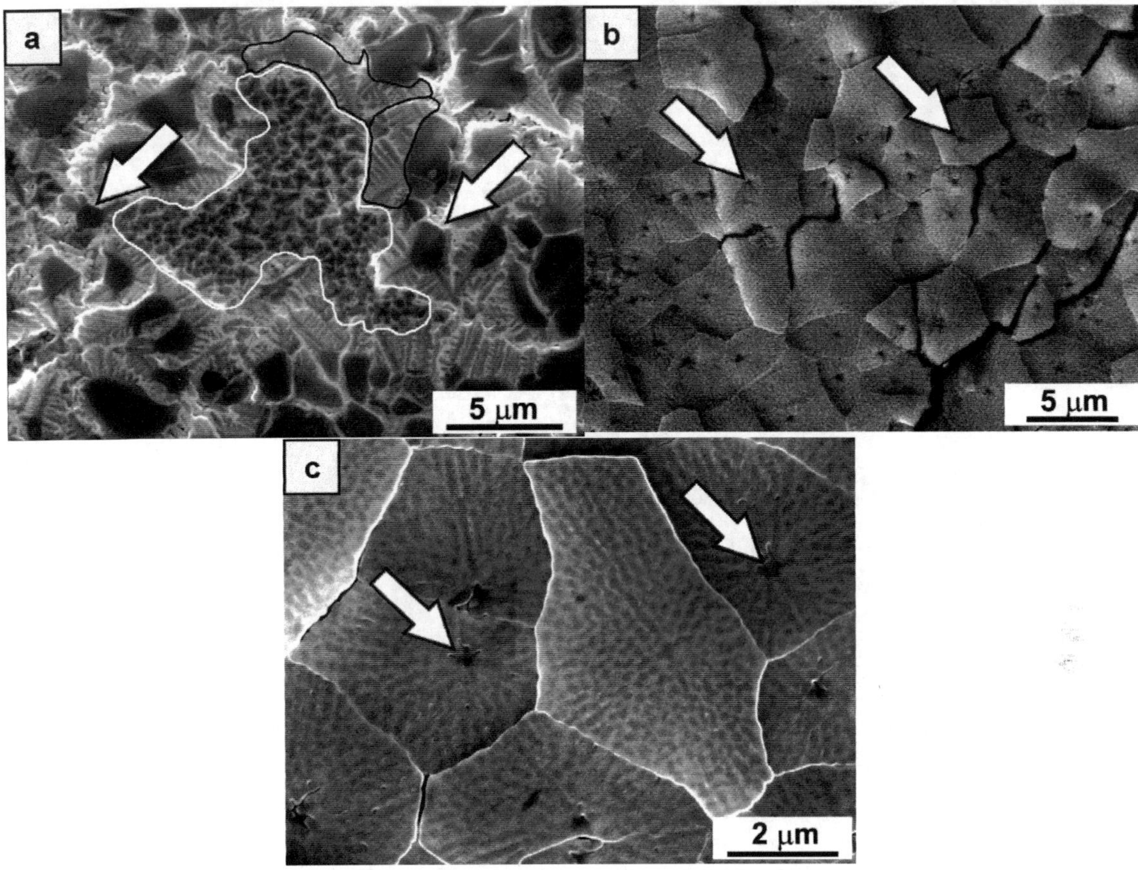

Figure 3. SEM images of the structure of the composite surface after pulse electron irradiation in the plasmas of argon (a), krypton (b) and xenon (c) (40 J/cm^2, 150 μs, 15 pulses)

Size reduction of titanium carbide particles is a distinctive feature of the structure of the surface layers of the composite after electron irradiation in the plasmas of inert gases. The degree of reduction depends on the type of plasma gas. Titanium carbide particles of micron size remain in the top sublayer 1 with columnar submicron sized structure after the irradiation in argon plasma (Fig. 4a). Micron sized carbide particles are found only in the next sublayer 2 with oriented columnar structure (Fig. 4b) after irradiation in krypton plasma. There are practically no micron sized carbide particles the modified surface layer after irradiation in xenon plasma (Fig. 4c).

IOP Conf. Series: Earth and Environmental Science **115** (2018) 012048 doi:10.1088/1755-1315/115/1/012048

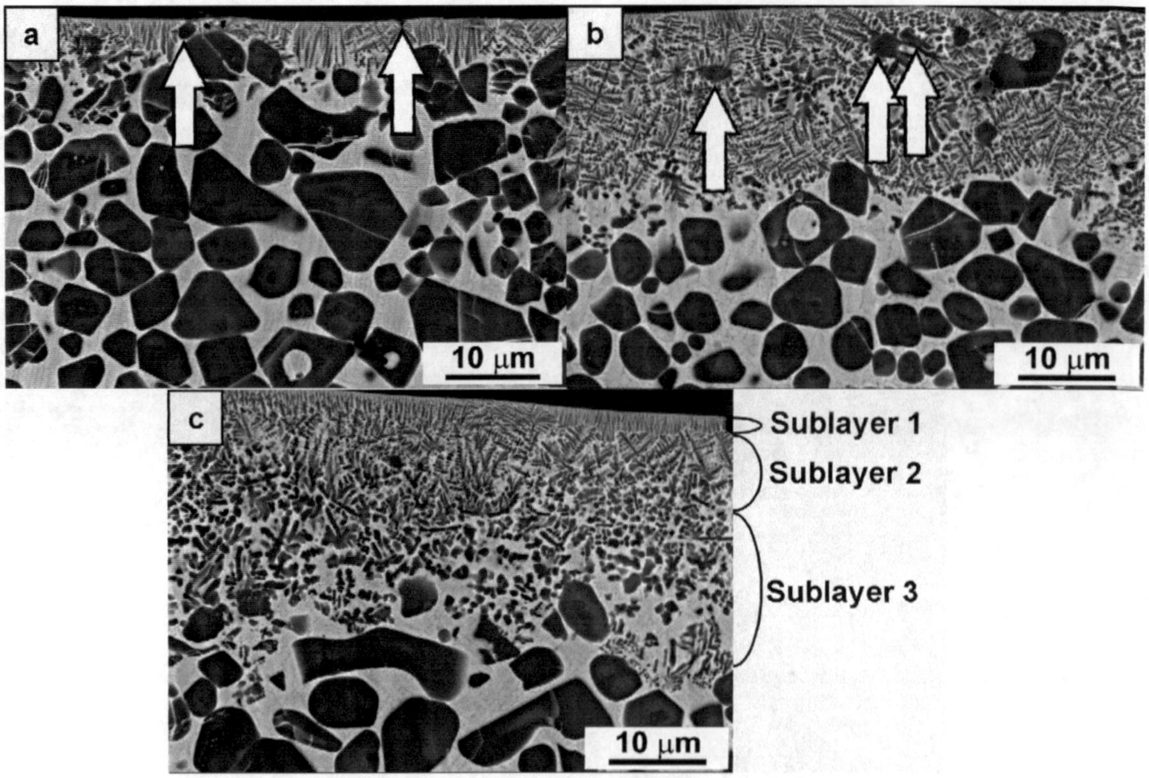

Figure 4. SEM images of the structure of the modified surface layer of the composite (cross-section) after pulse electron irradiation in the plasmas of argon (a), krypton (b) and xenon (c) (40 J/cm^2, 150 μs, 15 pulses). The coarse micron-sized titanium carbide particles are indicated by the arrows

The refining mechanism of carbide particles in the surface layer under the electron irradiation of the composite is illustrated in Fig. 5. The nanosized particles of titanium carbide are formed as precipitates from the supersaturated solid solution of carbon and titanium in the molten metallic binder right on the surface of initial coarse particle. The secondary carbide nanoparticles are oriented in the surface layer due to the temperature gradient occurrence during and just after irradiation. As a result a columnar and dendrite structures appeared in the top and lower sublayers, respectively.

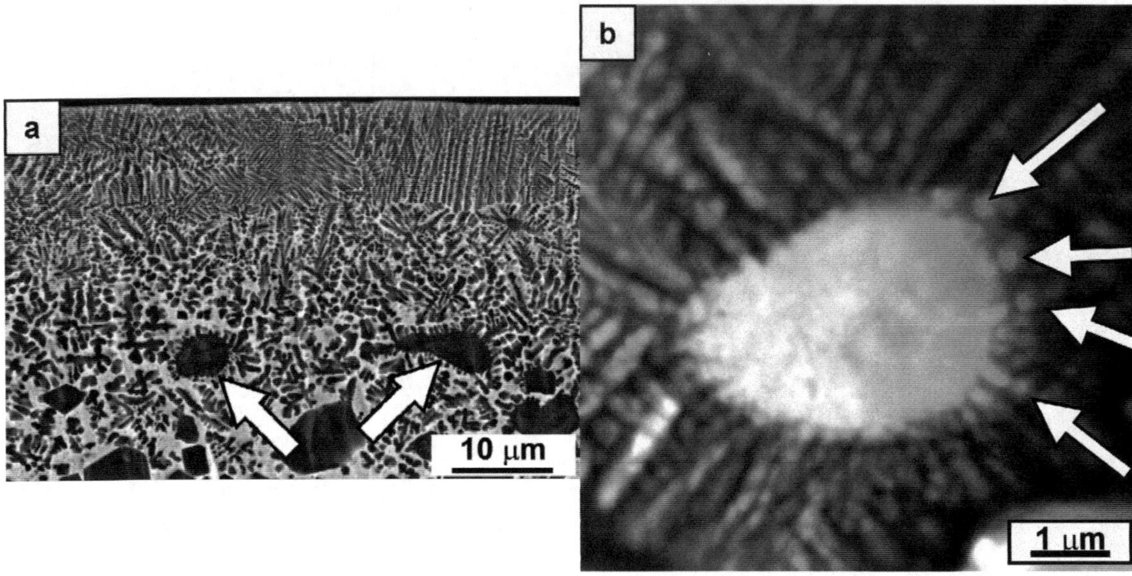

Figure 5. SEM (a) and TEM (b) images of the structure of the modified surface layer of the composite (cross-section) after pulse electron irradiation demonstrating formation of nanoparticles of secondary carbide nearby the surface of the initial coarse titanium carbide particles (indicated by the arrows). Krypton plasma, 60 J/cm^2, 150 μs, 15 pulses (a) and xenon plasma, 60 J/cm^2, 100 μs, 15 pulses (b)

Structural investigation suggests that reduction of the ionization energy of the used inert gas (and simultaneous increase of its atomic mass) increases amount of the entered energy into the surface layer of the composite increasing the depth and the temperature gradient in it. As a result, the depth of the modified surface layer increases and the size distribution of titanium carbide particles in the top, middle and bottom sublayers changes when passing from irradiation in argon plasma to irradiation in krypton plasma and then to that in xenon plasma (Figs. 6, 7). It is evident from Fig. 7 that titanium carbide particles become coarser as the distance from the surface grows. The coarsening is least demonstrated when the composite is irradiated in xenon plasma.

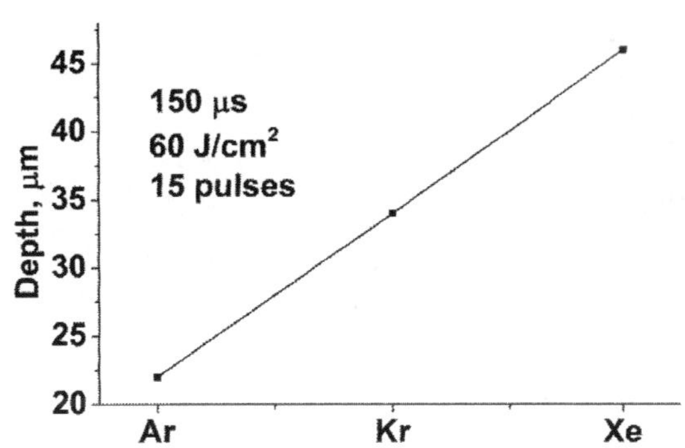

Figure 6. Dependence of the depth of modified surface layer on the type of plasma gas

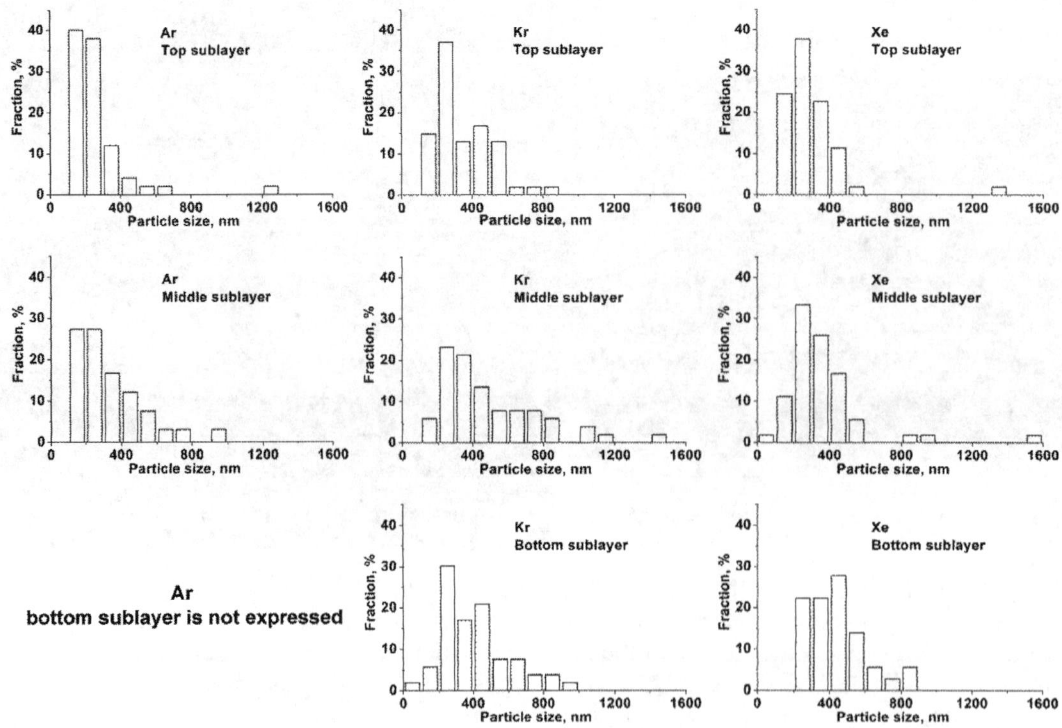

Figure 7. Size distribution of carbide particles in the top, middle and bottom sublayers of the modified surface layer after irradiation in the plasmas of argon, krypton and xenon (40 J/cm^2, 150 μs, 15 pulses)

Fig. 8 shows the dependence of the groove depth cut with a diamond cone on the surface of irradiated composite samples. One can conclude that modification of the structure of the surface layer of the composite as a result of electron irradiation in the plasmas of inert gases increases its wear resistance for the all used gases. Wear resistance increases when passing from irradiation in argon plasma to irradiation in krypton plasma and then to irradiation in xenon plasma.

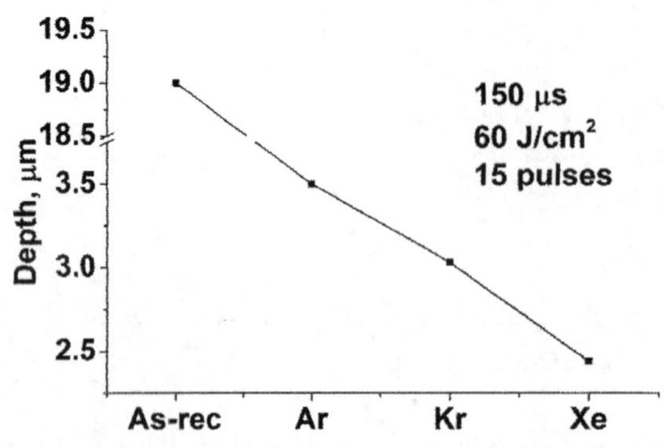

Figure 8. Dependence of the groove depth cut with a diamond cone on the surface of the irradiated composite on the type of plasma gas

4. Conclusion

Nanostructured heterophase structure is formed in the surface layer of metal-ceramic TiC/(Ni-Cr) alloy as a result of pulse electron irradiation in the plasmas of inert gases with different atomic mass and ionization energy. As the inert gas atomic mass increases together with the decrease of ionization energy the depth of the modified surface layer grows and fraction of submicron sized particles in the surface layer increases. The wear resistance of metal-ceramic composite improves in accordance to the formation of nanocrystalline structure in the surface layer.

Acknowledgements

This work was supported by the Program for Basic Scientific Research at the State Academies of Sciences for 2013–2020 (Project No. 23.2.2). Access to the experimental equipment in the "Nanotech" Common Use Center (ISPMS SBRAS) is acknowledged.

References

[1] Mortensen A and Llorca J 2010 *Ann. Rev. Mater. Res.* **40** 243
[2] Casati R and Vedani M 2014 *Metals* **4** 65
[3] Sanaty-Zadeh A 2012 *Mater. Sci. Eng. A* **531** 112
[4] Chawla N and Chawla KK 2006 *Metal matrix composites* (New York: Springer) p 137
[5] Chawla N and Chawla KK 2006 *J. Mater. Sci.* **41** 913
[6] Romanova VA, Balokhonov RR and Schmauder S 2009 *Acta Mater.* **57** 97
[7] Proskurovsky DI, Rotshtein VP, Ozur GE, Markov AB, Nazarov DS, Shulov MA, Ivanov YuF and Buchheit RG 1998 *J. Vac. Sci. Technol. A* **16** 2480
[8] Potshtein V, Ivanov Yu and Markov A 2006 *Materials surface processing by direct energy techniques. Chapter 6: Surface treatment of materials with low-energy, high-current electron beams* (Amsterdam: Elsevier) p 205
[9] Ovcharenko VE, Psaknye SG, Ivanov YuF, Mokhovikov AA, Ignat'ev AS, Yu B, Zhao Y, Xiyou JC and Yu G. 2015 *Rare Metal Mater. Eng.* **44** 1
[10] Ovcharenko VE, Ivanov KV, Ivanov YuF, Mokhovikov AA and Yu B 2017 *Russ. Phys. J.* **59** 2114

Impact of High-Temperature, High-Pressure Synthesis Conditions on the Formation of the Grain Structure and Strength Properties of Intermetallic Ni_3Al

V E Ovcharenko [1, 2, a], K V Ivanov[1], E N Boyangin[1], T A Krylova[1], A P Pshenichnikov[1]

[1] Institute of Strength Physics and Materials Science of Siberian Branch of Russian Academy of Sciences, Pr. Akademichesky 2/4, 634055, Tomsk, Russia
[2] National Research Tomsk Polytechnic University, pr. Lenina 30, Tomsk, 634050,Russia

E-mail: [a] ove45@mail.ru

Abstract. The impact of the preliminary load on 3Ni+Al powder mixture and the impact of the duration of the delay in application of compacting pressure to synthesis product under the conditions of continuous heating of the mixture up to its self-ignition on the grain size and strength properties of the synthesized Ni_3Al intermetallide material have been studied. The grain structure of the intermetallide synthesized under pressure was studied by means of metallography, transmission electron microscopy and EBSD analysis, with the dependence of ultimate tensile strength on the grain size in the synthesized intermetallide having been investigated at room temperature and at temperatures up to 1000°C. It is shown that an increase in the pressure preliminarily applied to the initial mixture compact results in reduced grain size of the final intermetallide, whereas an increase in pre-compaction time makes the grain size increased. A decrease in the grain size increases the ultimate tensile strength of the intermetallide. The maximum value of the ultimate tensile strength in the observed anomalous temperature dependence of this strength exhibits a shift by 200°C toward higher temperatures, and the ultimate strength of the synthesized intermetallide at 1000°C increases roughly two-fold.

1. Introduction

Intermetallic compound Ni_3Al provides the basis for a broad spectrum of alloys with improved properties intended for use under thermal and force loads as forging dies, turbocompressor and internal-combustion-engine components, turbine blades in rocket engines, combustion-chamber components, heat exchangers, microreactors, etc. [1]. Enhancement of the application efficiency of intermetallic compound Ni_3Al as the main component of heat-resisting alloys is restricted because of the high susceptibility of the material to brittle fracture; indeed, polycrystalline Ni_3Al suffers brittle rupture along grain boundaries both at low and elevated temperatures [2-5]. Increasing the plasticity and the strength of intermetallic compound Ni_3Al is possible either via doping of the material with various impurities such as boron [6] or via grain-structure refinement [7]. Practical implementation of boron-doping process for intermetallides by traditional methods encounters certain difficulties [8], and enhancement of the mechanical strength of polycrystalline intermetallide samples via grain-structure refinement achieved using severe plastic deformations realized at temperatures as high as the melting point also seem to be impossible because of the high brittleness of the intermetallide at equally high temperatures.

An alternative method for obtaining intermetallic Ni_3Al and alloys based on this intermetallide is the self-propagating high-temperature synthesis (SHS) of the intermetallic compound in a powder mixture of initial elements taken in desired proportion [8-12]. Thermophysical conditions of bulk-uniform synthesis reaction of the intermetallide in the initial powder mixture of pure elements guarantee synchronous proceeding of phase transformations throughout the entire volume of the powder sample. A unique feature of SHS process is the possibility to consolidate individual structural fragments within the synthesis product at the moment the thermo-reacting mixture reaches a homogeneous structural-phase composition at the stage of material crystallization and grain-structure formation [13].

The high rates of phase transformations during the bulk-uniform synthesis reaction in a powder mixture of initial elements yielding the intermetallide entail considerable difficulties in controlling the formation process of the structural-phase state of the synthesis product. The main drawback of known solutions [14,15] hampering the mentioned control is the initiation of high-temperature synthesis of the intermetallide in the initial powder mixture in frontal burning mode, followed by subsequent application of pressure to the synthesis product after the burning front has already traversed the entire volume of the mixture. As a result, during the compaction the synthesis product at its different sections occurs in different structural-phase states, and this circumstance does not allow one to obtain a grain structure with desired grain sizes in synthesized intermetallide. A solution to this problem can only be gained via implementation of synchronization of the bulk-uniform exothermal-reaction processes of intermetallide formation in the powder mixture of initial elements with the dynamic compaction of the synthesis product.

In the present publication, we report on the results of a study of the impact of synthesis conditions on the structural-phase state and strength properties of the intermetallic compound Ni_3Al synthesized under pressure from an initial 3Ni+Al powder mixture whose volume was continuously heated until a self-ignition followed by subsequent dynamic compaction of the synthesis product was reached.

2. Material and methods

The initial materials used for the high-temperature synthesis of intermetallic Ni_3Al samples were mixtures of nickel and aluminum powders in which particle sizes ranged in the intervals 1...3 and 5...10 μm, respectively. The high-temperature synthesis of intermetallic Ni_3Al samples in stoichiometric powder mixture was conducted under a pressure implemented using an automated hydraulic press. The hydraulic press was additionally equipped with a facility for rf heating of the steel mould, with an indicator showing the current heating temperature of the powder compact, with a digital manometer that was used for monitoring the pressure in the main-press-cylinder hydrosystem, and with a timer serving the function of working-stroke initiation (Fig. 1).

The phase composition of synthesized intermetallide samples was studied by means of the X-ray diffraction implemented on a DRON-7 X-ray diffractometer using Co K_α radiation at an accelerating voltage of 35 kV and an electron-beam current of 20 mA. The grain structure of the material was investigated on a Neophot 32 microscope using intermetallic samples sized 40×15×5 mm (these samples were cut out from central parts of the pressure-synthesized pellets). The metallographic sections of the samples were prepared by mechanical grinding performed with gradual reduction of diamond abrasive size down to 1 μm. The grain structure in the samples was revealed by etching the samples with argon ions at an accelerating voltage of 0.6 kV. For evaluation of grain sizes, the random secant technique with the averaging performed over 150 measurements was used.

The TEM study of the material structure was carried out on a JEM-2100 electron microscope at an accelerating voltage of 200 kV. Thin foils were prepared using the ion thinning process implemented on an EM-09100IS (JEOL) facility at an accelerating voltage of 8 kV.

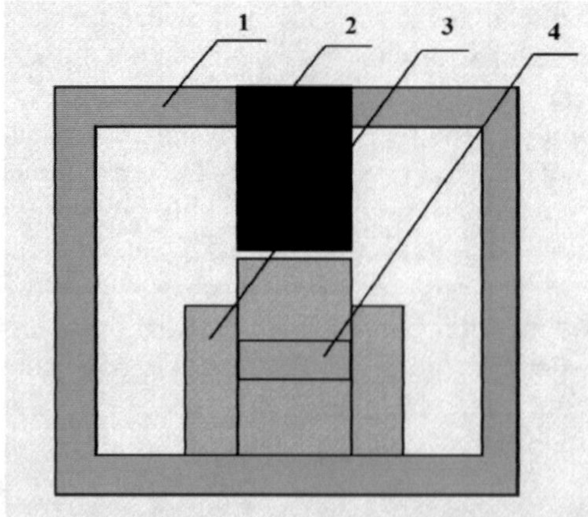

Fig. 1. Schematic of the experimental facility for high-temperature, high-pressure synthesis of chemical compounds and alloys based on such compounds to be conducted in thermal explosion mode during heating of the initial powder compact in a steel mould: 1- press volume, 2 – main press cylinder, 3 – steel mould, 4 – powder compact.

The EBSD analysis of the material structure of the samples was performed on a Quanta 200 3D tungsten-hot-cathode microscope operated at an accelerating voltage of 30 kV. The step size was 0.2 μm. Poor indexed points with the confidence index (CI) values less than 0.1 were discarded from consideration. The cross-sectional cuts of the samples under examination were polished electrolytically at 10°C temperature and 80-V voltage in an electrolyte of the following composition: CH_3COOH – 78%, $HClO_4$ – 22%.

The tensile tests of the intermetallic samples at temperatures 20…1000 °C were performed on LFM-125 walter + baiag testing machines, Switzerland. The deformation rate during the tests was 3×10^{-4} s^{-1}, the dimensions of the gage parts of the test samples being 10×3×1.2 mm.

3. Results and Discussion

The key parameters of the synthesis process of intermetallic compound Ni_3Al held under pressure and proceeding through a bulk-uniform exothermal synthesis reaction are the magnitude of the initial load applied to the powder mixture of nickel and aluminum and the duration of the delay in application of compacting pressure to the synthesis product after ignition of the mixture. Those parameters define the formation kinetics of the grain structure in synthesized products. Programming of the values of both parameters is based on the control of the load being preliminarily applied to the powder compact in the mould exerted using preliminary fixation of a set pressure value in the main-press-cylinder hydrosystem and a set duration of the delay in the initiation of the press working stroke used to compact the synthesis product. Figure 2a shows a diagram illustrating the variation of the pressure applied to the powder compact in the steel mould. This diagram was obtained by recalculating the pressure values in the main-press-cylinder hydrosystem in coordinates "pressure-time" over the entire cycle of the high-temperature synthesis process of the intermetallide in the powder mixture continuously heated in the steel mould. At an initial pressure P_o implemented during continuous heating of the mixture, the plastic deformation of the powder compact, the melting of the aluminum component, and the initiation of the exothermal synthesis reaction all resulted in a reduction of the pressure applied to the powder compact down to a pressure P_1. Further scenario implies either an immediate compaction of the synthesis product under the pressure P_2 implemented due to initiated press working stroke or preliminary storage of the mixture during some time under pressure P_1 with subsequent compaction of the synthesis product. Figure 2b shows the experimental time dependences of the pressure applied to the powder compact being continuously heated in the steel mould until reaching self-ignition with various pre-compaction times. In accordance with the indicated regimes of the high-temperature synthesis of the intermetallide under pressure, intermetallic Ni_3Al samples were obtained at different magnitudes of the preliminary load exerted on the initial powder mixture and at different times during which the synthesized product was held prior to the moment of its compaction.

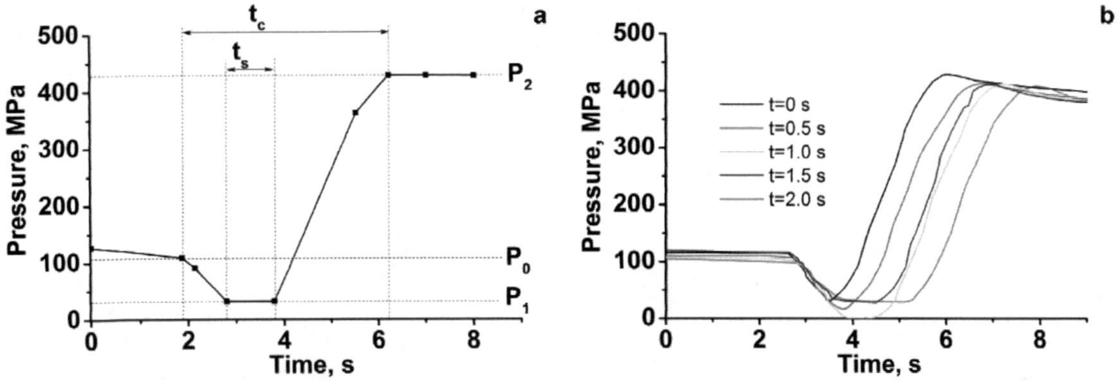

Fig. 2. A diagram of pressures applied to 3Ni+Al powder compact plotted in coordinates "pressure-time" over the entire process cycle of the high-temperature synthesis of Ni_3Al intermetallic compound from an initial powder compact being continuously heated under a preliminary load in the steel mould with a time delay in application of compacting pressure (a) and experimentally obtained diagrams illustrating the variation of pressure during the entire programmed cycle of the synthesis process of Ni_3Al intermetallide under continuous heating of the powder mixture implemented with various pre-compaction times (b).

3.1. Impact of the magnitude of the preliminary load on the phase composition and grain size in the pressure-synthesized Ni_3Al intermetallide

X-ray data obtained for intermetallic Ni_3Al samples synthesized at extreme pressure values in the examined range of initial pressures applied to the powder compact, 33 and 115 MPa, without any delay in time in application of the minimum compacting pressure (diffractograms 1 and 2, respectively) are shown in Fig. 3.

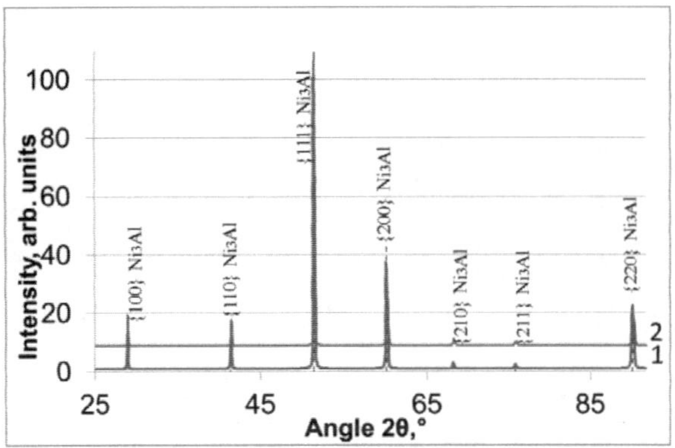

Fig. 3. Diffractograms of intermetallic Ni_3Al samples synthesized under pressure from an initial powder mixture continuously heated in the steel mould at different pressures applied to the 3Ni+Al powder compact, 33 MPa (diffractogram 1) and 115 MPa (diffractogram 2) with zero pre-compaction time.

A comparison between the diffractograms of Fig. 3 shows that the change in the magnitude of the initial pressure applied to the initial powder mixture does not affect the phase composition of the synthesized intermetallic samples, - in both cases, the phase composition corresponds to the intermetallic compound Ni_3Al.

The change in the magnitude of the initial pressure has an impact on the grain size of the pressure-synthesized Ni_3Al intermetallide. Figure 4 shows the general appearance of the grain structure in intermetallic Ni_3Al samples synthesized under pressure at different values of the pressure that was preliminarily applied to the initial powder compact kept in the mould. Numerical estimations of the dependence of the grain sizes in the pressure-synthesized intermetallide samples show that, with increasing the initial pressure, the grain size in synthesized intermetallides diminishes from 8.5 μm down to 1.5 μm (see Fig. 5).

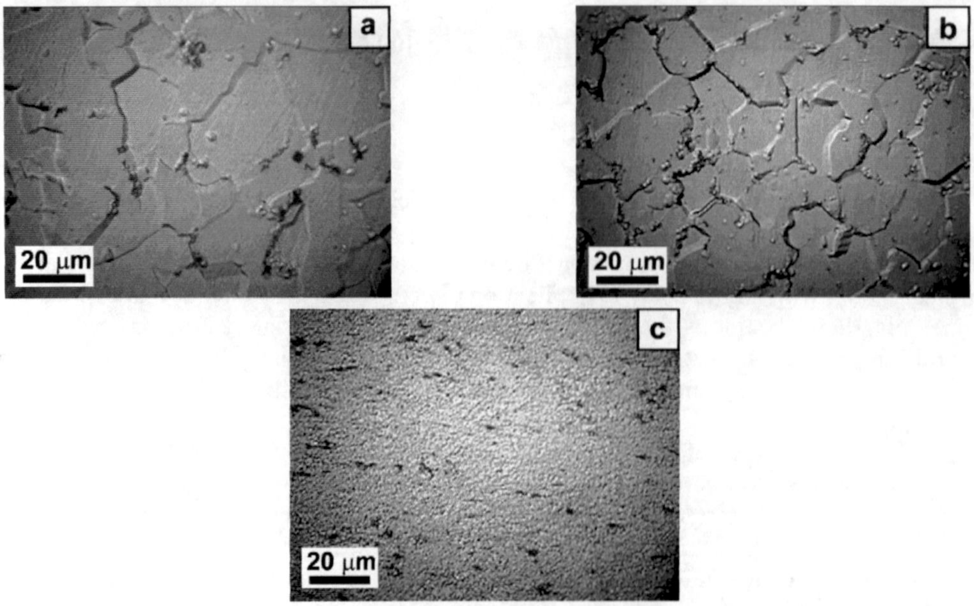

Fig. 4. The grain structure of intermetallic Ni_3Al samples synthesized under pressure at different values of the pressure that was preliminarily applied to the initial powder mixture: a – P_o= 33 MPa, b – P_o= 76 MPa, c – P_o= 115 MPa.

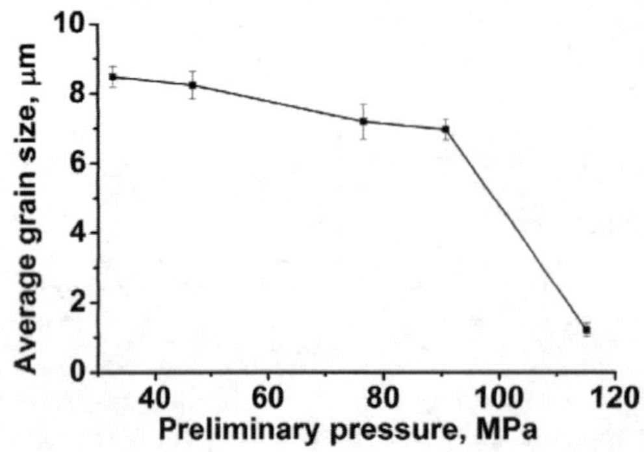

Fig. 5. The average grain size in the pressure-synthesized intermetallic Ni_3Al samples versus the magnitude of the pressure preliminarily applied to the initial powder mixture.

The TEM study of the microstructure of pressure-synthesized Ni_3Al intermetallides and their EBSD analysis have confirmed the data gained while performing the metallographic study of the grain doi:10.1088/1755-1315/115/1/012049

structure, - it was found that the grain sizes in the grain structure formed in the intermetallide under preliminary 115-MPa loading of the initial powder mixture fell into the micrometer range (see Fig. 6).

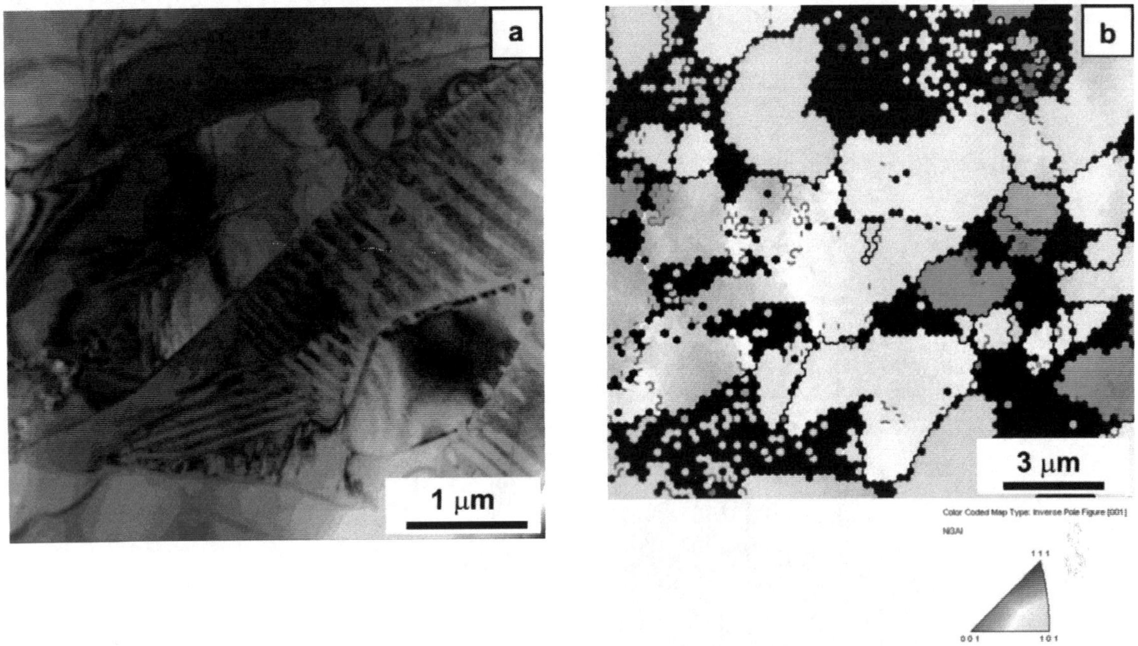

Fig. 6. The grain structure of the intermetallic Ni_3Al synthesized under pressure at 115-MPa pressure preliminarily applied to the initial powder mixture (as revealed by TEM) (a) and the inverse pole figure demonstrating the grain structure of the intermetallide. The black and red lines show the high-angle and low-angle boundaries, respectively (b).

3.2. Impact of pre-compaction time on the phase composition and grain size in pressure-synthesized Ni_3Al intermetallide

Diffractograms of intermetallic Ni_3Al samples synthesized under pressure at different pre-compaction times are shown in Fig. 7.

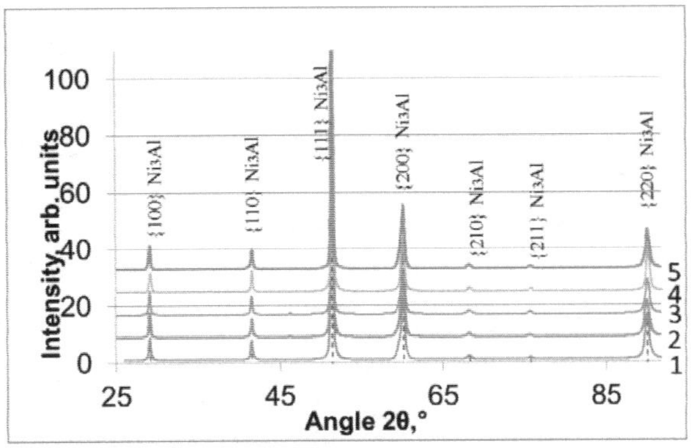

Fig. 7. Diffractograms of intermetallic Ni_3Al samples synthesized under pressure at different pre-compaction times: 1 – 0 s, 2 – 0.5 s, 3 – 1.0 s, 4 – 1.5 s, 5 – 2.0 s.

The increase in pre-compaction time from 0 to 2.0 s has no influence on the phase composition of the synthesized intermetallic material, - indeed, the intermetallides synthesized at all pre-compaction times presented a single-phase Ni_3Al product. However, with increasing the pre-compaction time the average grain size of the intermetallide clearly increases (see Figs. 8 and 9).

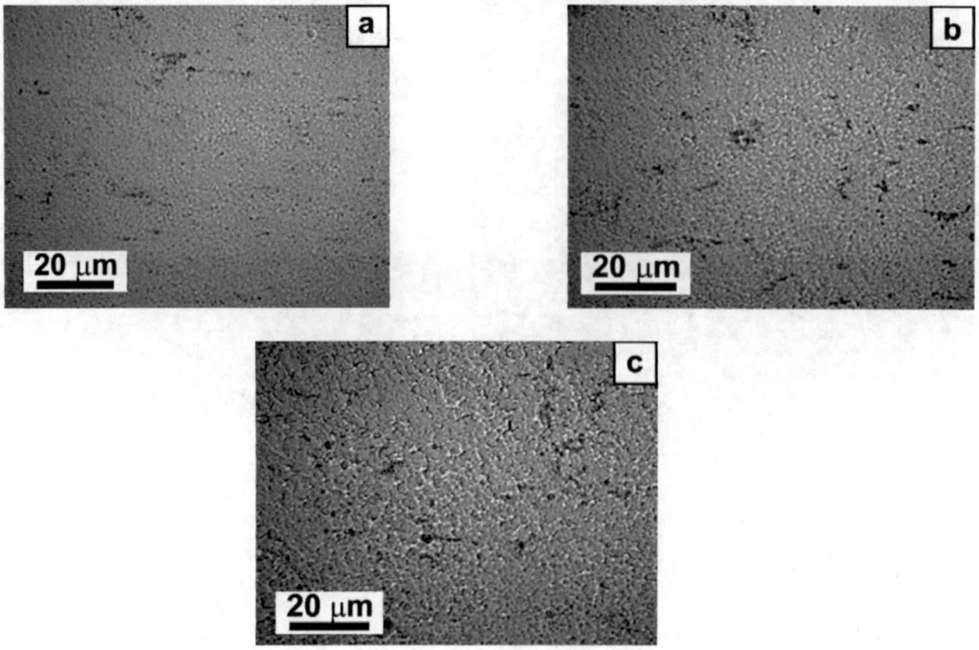

Fig. 8. The grain structure of intermatallic Ni_3Al samples synthesized under pressure at different pre-compaction times: a – 0 s, b – 1 s, c – 2 s.

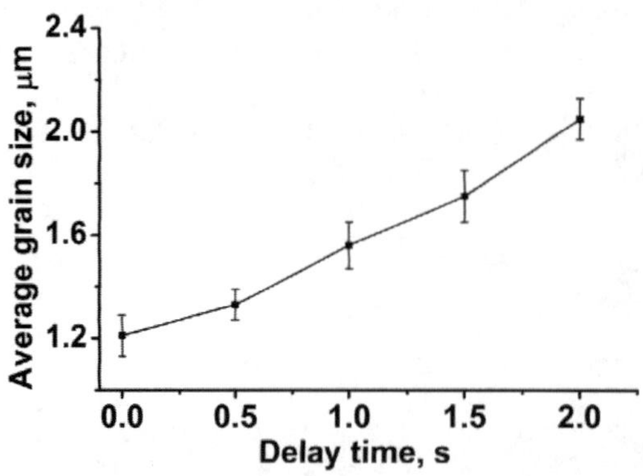

Fig. 9. The grain size in intermetallic Ni_3Al samples synthesized under pressure versus the pre-compaction time.

3.3. Impact of grain sizes on the strength properties of pressure-synthesized Ni_3Al intermetallides

The variation of grain sizes largely affects the strength properties of pressure-synthesized Ni_3Al intermetallides. The ultimate strength of intermetallides under tension versus the grain size in the material and the temperature dependences of the ultimate tensile strengths in the materials with grain sizes 1.2 μm and 8.5 μm are shown in Fig. 10 (a and b, respectively).

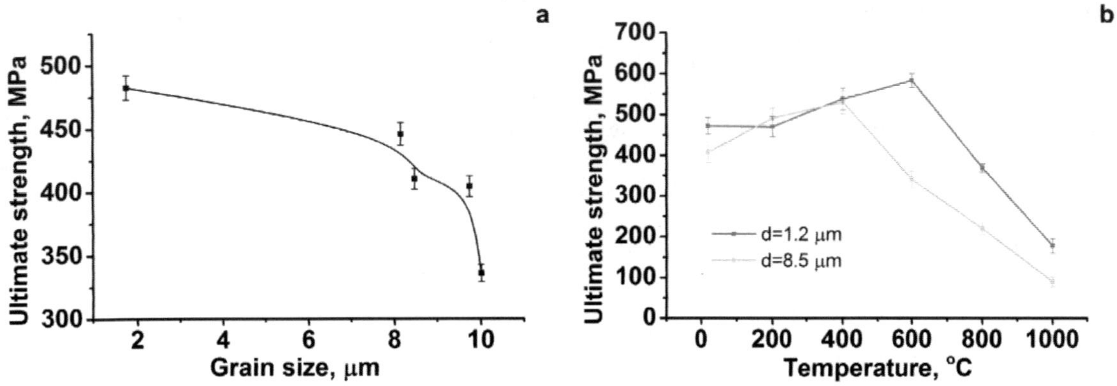

Fig. 10. The ultimate tensile strength of pressure-synthesized Ni_3Al intermetallides versus the grain size (a) and versus the temperature at which the tests were performed (in the materials with grain sizes 8.5 μm (green) and 1.2 μm (red) (b).

The decrease of the grain size in pressure-synthesized Ni_3Al intermetallides from 8.5 down to 1.2 μm increases the ultimate strength of the intermetallide under tension from 336 to 482 MPa (i.e., by a factor of 1.4). In the temperature dependence, the maximum ultimate strength increases from 538 to 583 MPa, and it exhibits a shift by 200 °C toward higher temperatures. At a temperature of 1000 °C, the ultimate strength of the intermetallide under tension increases from 90 to 178 MPa (i.e. roughly two-fold).

4. Conclusion

Programming of main process parameters of the high-temperature, high pressure synthesis of Ni_3Al intermetallide under continuous heating of the initial 3Ni+Al powder mixture prior to reaching its self-ignition provides an efficient means for controlling the formation process of the grain structure in the synthesized intermetallides. Indeed, an increase in pre-compaction pressure decreases the grain size in the synthesized intermetallide. On the contrary, an increase in pre-compaction time increases the grain size in the synthesized product.

A decrease in the grain size increases the room-temperature ultimate strength of the intermetallide under tension. The maximum ultimate tensile strength of the intermetallide in the anomalous temperature dependence of this strength exhibits a shift by 200 °C toward higher temperatures, and the ultimate tensile strength of the intermetallide at a temperature of 1000 °C shows roughly a two-fold increase.

Acknowledgements

This work was performed within the framework of the Program for Basic Scientific Research at the State Academies of Sciences for 2013–2020 (Project No. 23.2.2). Access to the experimental equipment in the “Nanotech” Common Use Center (ISPMS SBRAS) and Tomsk Material Science Common Use Center (Tomsk State University) is acknowledged.

References

[1] Jozwik P, Polkowski W and Bojar Z 2015 *Mater.* **8** 2537
[2] Stoloff NS, Liu CT and Deevi SC 2008 *Intermetallics* **8** 1313
[3] Sikka VK, Deevi SC, Viswanathan S, Swindeman RW and Santella ML 2000 *Intermetallics* **8** 1329
[4] Pope DP and Ezz SS 1984 *Int. Met. Rev.* **29** 136
[5] Deevi SC and Sikka VK 1996 *Intermetallics* **4** 357

[6] Kim MS, Hanada S, Watanabe S and Izumi O. 1988 *Trans. Jap. Inst. Met.* **29** 274
[7] Valiev RZ, Estrin Y, Horita Z, Langdon TG and Zehetbauer MJ 2006 *JOM* **58** 33
[8] Merzhanov AG 1995 *Ceram. Int.* **21** 371
[9] Lebrat JP and Varma A 1992 *Combust. Sci. Technol.* **88** 211
[10] Moore JJ and Feng HJ 1995 *Prog. Mater. Sci.* **39** 243
[11] Hibino A, Matsuoka S and Kiuchi M 2001 *J. Mater. Process. Technol.* **112** 127
[12] Yeh CL and Sung WY 2005 *J. Alloys Compd.* **390** 74
[13] Ovcharenko VE, Lapshin OV and Ramazanov IS 2006 *CESW* **42** 302
[14] Stolin AM 1992 *Int. J. Self-Propag. High-Temp. Synth.* **1** 135
[15] Podlesov VV, Radugin AV, Stolin AM, Merzhanov AG 1992 *J. Eng. Phys. Thermophys.* **63** 1065

IOP Conf. Series: Earth and Environmental Science **115** (2018) 012050 doi:10.1088/1755-1315/115/1/012050

Calculation and research of electrical characteristics of induction crucible furnaces with unmagnetized conductive crucible

M A Fedin[a], A B Kuvaldin, A O Kuleshov, I Y Zhmurko, S V Akhmetyanov

National Research University "Moscow Power Engineering Institute", Krasnokazarmennaya 14, 111250 Moscow, Russia

E-mail: [a] FedinMA@mpei.ru

Abstract. Calculation methods for induction crucible furnaces with a conductive crucible have been reviewed and compared. The calculation method of electrical and energy characteristics of furnaces with a conductive crucible has been developed and the example of the calculation is shown below. The calculation results are compared with experimental data. Dependences of electrical and power characteristics of the furnace on frequency, inductor current, geometric dimensions and temperature have been obtained.

1. Introduction

Induction crucible furnaces (ICF) are widely used in the industry for melting ferrous and non-ferrous metals and alloys, which require special purity, homogeneity and accuracy of chemical composition. Depending on the electrical properties of the crucible material, one can distinguish ICF with a non-conductive and conductive crucible.

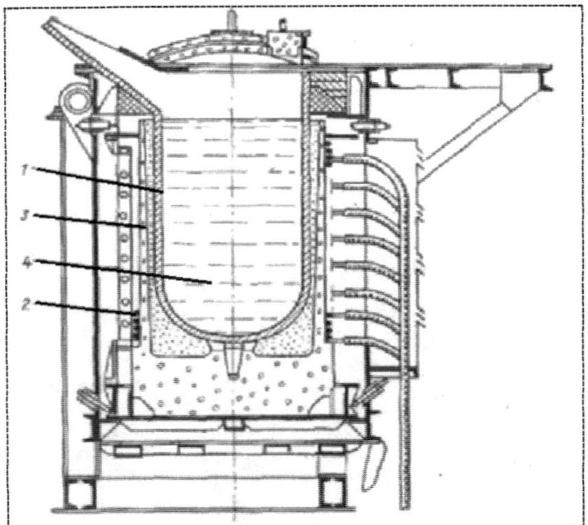

Figure 1. ICF with a conductive crucible: 1- conductive crucible; 2 - inductor; 3 - charge; 4 - thermal insulation

In ICF with a non-conductive ceramic crucible the metal (charge) is heated by the electromagnetic field produced by the inductor current. The crucible is almost equivalent to the air gap because of the large value of the electrical resistivity. In ICF with a conductive crucible (further ICF with CC) made from steel, graphite or graphite-chamotte in the electromagnetic field both the crucible and the charge are heated (Fig.1). ICF with CC are used for melting copper and magnesium alloys, uranium, gold, as well as non-conductive materials [1 - 4].

If the thickness of the crucible wall exceeds 2-3 times the penetration depth of the current into the crucible material, one can consider that the induced current is concentrated in the crucible wall. The load is heated only by heat transfer and may not have electrical conductivity. With a smaller wall thickness of the crucible the electromagnetic field gets into the charge, and energy is also released in it. In this case it is possible to speak about induction heating of a double-layer body which is formed by the crucible and the charge. Each layer has its own values of electrical resistivity and relative magnetic permeability, which affects the distribution of the released energy. The fact that under certain conditions the charge of ICF with CC represents a two-layer body is a significant difference between calculation of the ICFs with CC characteristics and the calculation of conventional

IOP Conf. Series: Earth and Environmental Science **115** (2018) 012050 doi:10.1088/1755-1315/115/1/012050

ICFs with a non-conductive crucible. At the same time, the calculation of the electrical and energy characteristics of ICFs with CCs is substantially complicated because it must be taken into account that during the heating the electrophysical parameters of the crucible material and the charge which are depend on the temperature and the strength of the magnetic field also change.

ICFs with conductive crucibles are widely used in Russia and abroad. They are made of ferromagnetic steel for melting of magnesium with the capacity of 0.3-8 ton and the power of 120-1200 kW and are also made of graphite for melting copper with the capacity up to 6 tons and the power up to 1500 kW [5-7].

2. Calculation of electrical characteristics of the ICF with CC in the ELCUT package.

The ELCUT package allows us to calculate ICFs with CCs which have magnetic properties and also to take into account the edge effects, the presence of the bottom and collar, as well as its versatility and the possibility of further solving of related problems (thermal problem, calculation of forces). It can be concluded that FEM realized in ELCUT is the most appropriate solution for the tasks set out in the article.

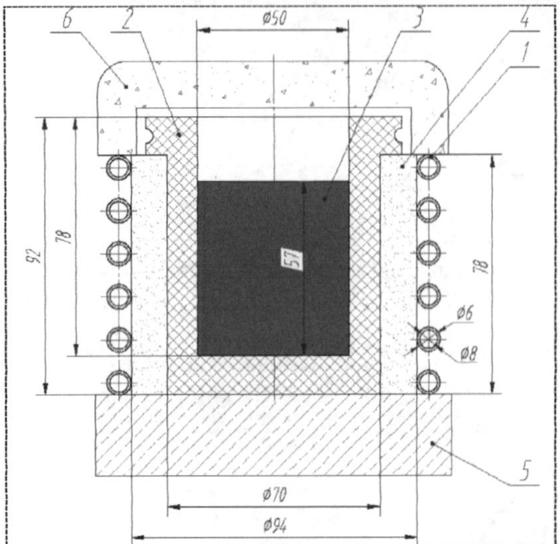

Figure 2. ICF with a conductive crucible: 1 - inductor; 2 - conductive crucible; 3 - charge; 4 - thermal insulation

As an example, ICF with a graphite crucible for melting copper with a mass of 1 kg was calculated. The furnace is powered by a high-frequency transistor converter with the power of 2.5 kW and an operating frequency of 22 kHz. Fig. 2 shows a drawing of the "inductor-charge" system of a furnace with geometric dimensions. The crucible is made of the FGG graphite.

Electric calculation of the furnace has been carried out in the universal finite-element program package ELCUT 6.0 (professional version) by solving the problem of magnetic field of alternating currents.

The analysis of magnetic field of alternating currents consists in electric and magnetic fields calculation excited by the applied variable currents (sinusoidally varying with time) or an external variable field. The change of the field with time is assumed to be sinusoidal.

The total current in the conductor is considered as the sum of the external current caused by the externally applied voltage and the eddy current induced by the alternating magnetic field

$$j = j_{ext.} + j_{e.c.} \cdot \quad (1)$$

Where the electrical conductivity γ and the components of the magnetic permeability tensor μz and μr are constant within each block of the model.

The problem is formulated as a partial differential equation relatively to the complex amplitude of the vector magnetic potential A (B = rot A, B is the vector of magnetic induction).

The magnetic induction vector is assumed to lie in the plane of the zr model (Figure 3), while the vector of the electric current density j and the magnetic vector potential A are orthogonal to it.

The equation for the axisymmetric problem can be written as:

$$\frac{\partial}{\partial r}\left(\frac{1}{r\mu_r}\cdot\frac{\partial}{\partial r}\mathrm{r}A\right)+\frac{\partial}{\partial z}\left(\frac{1}{\mu_z}\cdot\frac{\partial}{\partial z}A\right)-\mathrm{i}\,\omega\gamma\,A=-j_{ext} \quad (2)$$

Where the electrical conductivity γ and the components of the magnetic permeability tensor μz and μr are constant within each block of the model.

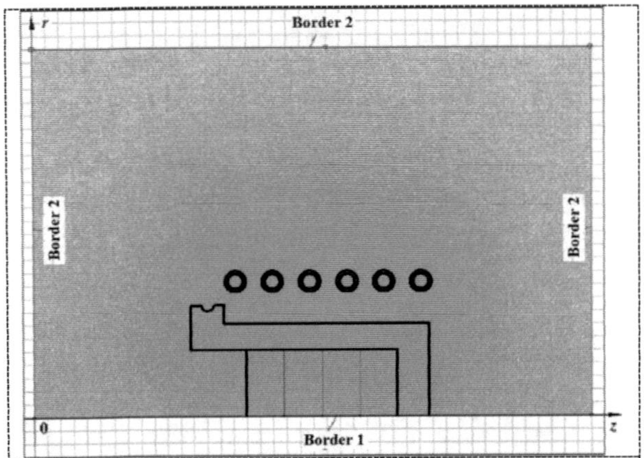

Figure 3. Calculation area in the ELCUT package

In Fig. 3 shows the calculated area divided by the finite elements. The number of grid nodes is about 350,000. Grid irregularities can be achieved by setting different values of the sampling step for different subareas of the calculating area and directly at nodes marked with dots in Fig. 3. The model allows performing calculations for the cases of a graphite crucible without metal and a crucible with different levels of metal by changing the properties of blocks. The Neumann boundary condition dH/dr = 0 (the absence of a tangential component of the magnetic field strength) is given on the axis of symmetry (boundary 1 in fig. 3); on the outer boundaries (boundary 2) is the zero Dirichlet boundary condition A = 0 (the value of the magnetic potential is assumed to be zero, i.e. the field is localized within the computational domain).

Table. 1. The main results of calculation of ICF with CC without metal and with copper

Inductor-charge" system parameter	Crucible without metal		Copper crucible	
	Var. 1	Var. 2	Var. 1	Var. 2
Inductor voltage U1, B	62.7	63.8	64.1	65.5
Power in the inductor P1, W	147	146	147	146
Power in crucible P2.1, W	1278	1286	1040	997
Power in copper P2.2, W	-		50	55
Total power in the load P2Σ, W	1278	1286	1090	1052
Total active power PΣ, W	1425	1432	1237	1198
Impedance of the system is z, Ω	0.348	0.354	0.356	0.364
Active resistance of the system R, Ω	0.044	0.044	0.038	0.037
Inductive resistance of the system xL, Ω	0.345	0.351	0.354	0.362
Electrical efficiency η	0.897	0.898	0.881	0.878
cos φ	0.126	0.125	0.107	0.102

Practice shows that the radius of the calculation area (the distance along the r axis in Figure 3) should be set 2 to 2.5 times greater than the internal radius of the inductor of the furnace, and the distances along the z axis from the bottom of the crucible to the right to boundary 2 and from the collar to the left to the boundary 2 should be set approximately equal to the internal radius of the inductor.
Practice shows that the radius of the calculation area (the distance along the r axis in Figure 4) should be set 2 to 2.5 times greater than the internal radius of the inductor of the furnace, and the distances along the z axis from the bottom of the crucible to the right to boundary 2 and from the collar to the left to the boundary 2 should be set approximately equal to the internal radius of the inductor.
As a result of solving the problem of the magnetic field of alternating currents the integral results have been obtained (shown in the Table 1) for the case of a graphite crucible without metal and a crucible with copper (copper mass about 1 kg) with the inductor current I1 = 180 A and the frequency f = 22 kHz. In Table 1, for each case, the values in the left column correspond to the real crucible; the values in the right column correspond to the crucible without the bottom and collar.

IOP Conf. Series: Earth and Environmental Science **115** (2018) 012050 doi:10.1088/1755-1315/115/1/012050

So, the results of the calculation show that if we do not take into account the influence of the bottom and the collar of the CC on the electrical and energy characteristics of the furnace with the CC, it makes a small error in the calculation - no more than 10%. Fig. 4 shows the calculated distributions of the volume power density of heat release for both these cases.

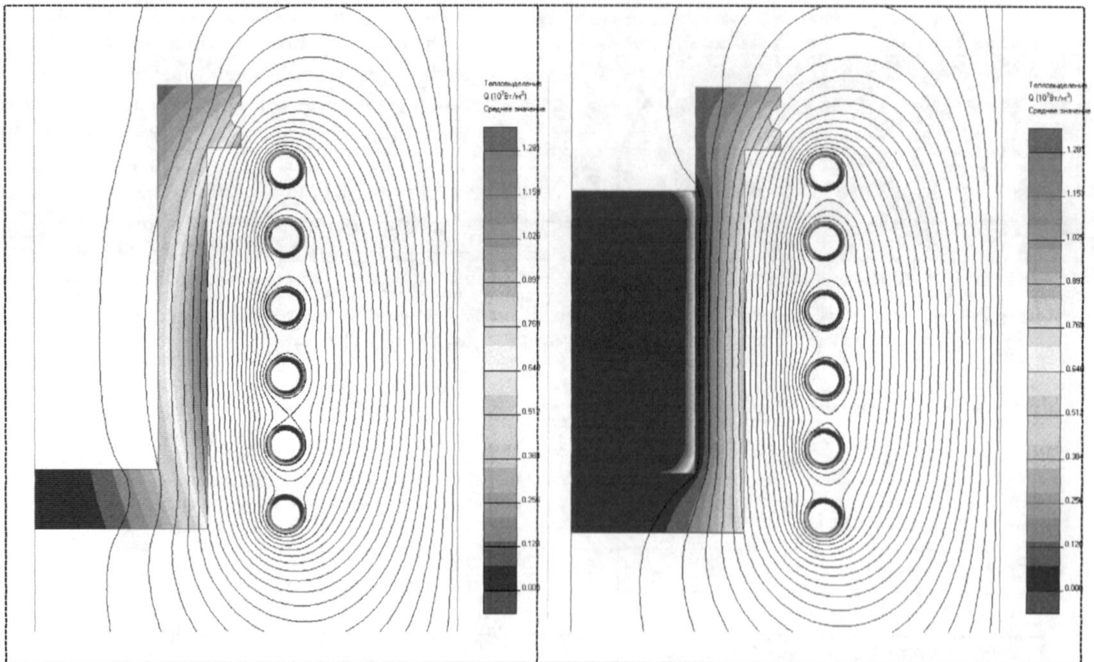

a b

Figure 4. Distribution of volume power density of heat release in the inductor and furnace charge: a - graphite crucible without metal; b - graphite crucible with copper.

Table 1 and Fig. 4 show that the presence of liquid copper in the graphite crucible (in the calculations, the resistivity ρ of copper was assumed to be $2.15 \cdot 10\text{-}7\ \Omega \cdot m$) has a significant effect on the parameters of the "inductor-charge" system. So, the heat release in a graphite crucible with copper in it decreases by almost 20% with the same inductor current, while the heat release in copper itself is insignificant - about 50 W.

This can be explained by the fact that the depth of electromagnetic wave penetration into the wall of a graphite crucible 10 mm thick, the inductor current of 22 kHz, is about 13 mm. Therefore, the integral value of the induced current in copper is comparable with the total current in the crucible (about 400 A and 780 A, respectively). This leads to the appearance of a backward electromagnetic wave reflected from the boundary between media with different values of ρ. So the distribution of the magnetic field strength H in the thickness of the crucible wall is changed (Fig. 5a) [1, 3, 11]. Thereby, the values of the volume power density of heat release p_{0V} in each point along the wall thickness of the crucible decrease (Fig. 5b), which leads to a significant decrease of power in the crucible. In this case, the internal heat sources in the inner layer of the two-layer charge contribute insignificantly to the total active power in the charge as a consequence of relatively small values of ρ for the inner layer. Fig. 6 shows the area which is limited by two vertical lines corresponds to the wall of the crucible.

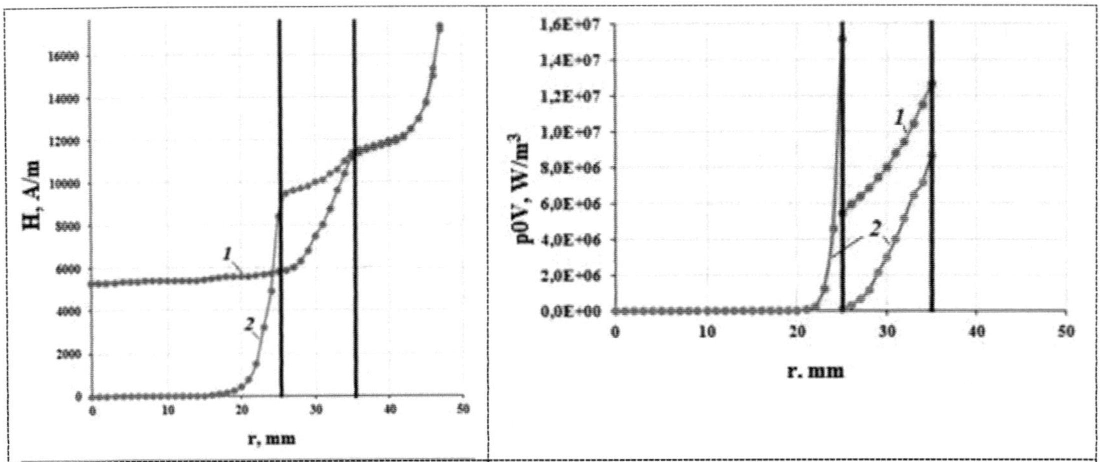

a	b

Figure 5. The distribution of the magnetic field strength H (a) and the volume power density p0V (b) along the radius r for the empty crucible (1) and the crucible with liquid copper (2) at the inductor current I1 = 210 A

As a result of the calculations, it was also found that the value of ρ copper present in the graphite crucible had little effect on the parameters of the "inductor-charge" system (the value of ρ varied from the value corresponding to liquid copper at the temperature of 1100 ° C, 72 · 10-8 Ω · m at 20 ° C), because it still remains substantially smaller in comparison with the value ρ of graphite, which varies slightly with heating. All this makes special specific requirements for the operation of the power supply - transistor voltage inverter [12 - 13].

Since at the initial stage of the technological process of metal smelting in ICF with CC the crucible charge represents a lump charge, some pieces of which are electrically unrelated, the induced currents are closed mainly within each piece, and the equivalent resistance of such lumpy charge in several times exceeds the analogous value for liquid copper. At this stage, the presence of a metallic (in particular copper) charge in the crucible can be neglected and one can use the results obtained for the crucible without metal. It was confirmed experimentally [12].

Calculation parameters of ρ graphite: 7.14 · 10-6 Ω · m at 20 ° C, 7.6 · 10-6 Ω · m - at 1100 ° C.

3. Conclusion

1. The review and comparison of the methods for calculating ICF with CC has been carried out. It is shown that the most appropriate option for calculation of the electrical characteristics of ICF with CC is MCE, implemented in ELCUT.
2. A methodology for determining the electrical characteristics of ICF with CC has been developed and corresponding studies have been carried out.
3. It is established that in the case when the CC is partially transparent to the electromagnetic wave, the total power of heat release in the crucible with the molten metal becomes less than the power of heat release in the CC without metal with the same inductor current and frequency.

The work was supported by the Ministry of Education and Science of the Russian Federation (an initiational scientific project 8.9608.2017 / BCh).

IOP Conf. Series: Earth and Environmental Science **115** (2018) 012050 doi:10.1088/1755-1315/115/1/012050

References

[1] Kuvaldin A, Fedin M, Generalov I Determination electrical parameters lumpy ferromagnetic charge when heated to the Curie point. Advanced Methods of the Theory of Electrical Engineering AMTEE'15– Třebíč, Czech Republic, vol 1 p I-2

[2] Kuvaldin A B, Fedin M A and Generalov I M 2015 *J. Induction heating* **1(28)** Physical model for definition of characteristics of the induction crucible furnace when heating ferromagnetic lumpy charge 3–8

[3] Fedin M A 2014 *J. Induction heating* **1(27)** Choice principle of regulation and control system development induction crucible furnace with a conductive crucible 24-28

[4] Kuvaldin A B, Fedin M A and Generalov I M 2016 *J. Industrial Power Engineering* **5** Increase in power efficiency of an electrotechnological complex with the induction crucible furnace when melting ferromagnetic lumpy charge 19 – 25

[5] Arkhipov V A and Berezikov A P 2008 *Fundamentals of the theory of engineering and physical experiment* (Tomsk: Tomsk Polytechnic University) p 103

[6] Gitgarz D A 1984 Automation of melting electric furnaces with the use of micro-computers (Moscow: Energoatomizdat) p 136

[7] Kruchinin A M, Mahmudov K M and et al 1990 Automatic control of electrothermal installations (Moscow: Energoatomizdat) p 416

[8] Kazantsev Y M 2000 *J. Electrical Engineering* **4** Direct synthesis of control in converter technics 31-36.

[9] Kazantsev Y M, Lekarev A F and Tikhonov E G 2004 **6** *J. Devices and systems. Management, control, diagnostics* Synthesis of control of servo inverters 20 – 25

[10] Kyo B 1986 Theory and Design of Digital Control Systems (Moscow: Science) p 448c.

[11] Tsypkin Y Z 1974 Relay automatic systems (Moscow: Science) p 575

[12] Shapiro S V, Zinin Y M, and et al 1989 Control systems with thyristor frequency converters for electrotechnology (Moscow: Energoatomizdat) p 166

[13] Galperin M V 1987 Practical circuitry in industrial automatic equipment (Moscow: Energoatomizdat) p 320

IOP Conf. Series: Earth and Environmental Science **115** (2018) 012051 doi:10.1088/1755-1315/115/1/012051

Enhancing Three-dimensional Movement Control System for Assemblies of Machine-Building Facilities

O N Kuzyakov and M A Andreeva

Industrial University of Tyumen, Russia

E-mail: andreevama@tyuiu.ru

Abstract. Aspects of enhancing three-dimensional movement control system are given in the paper. Such system is to be used while controlling assemblies of machine-building facilities, which is a relevant issue. The base of the system known is three-dimensional movement control device with optical principle of action. The device consists of multi point light emitter and light receiver matrix. The processing of signals is enhanced to increase accuracy of measurements by switching from discrete to analog signals. Light receiver matrix is divided into four areas, and the output value of each light emitter in each matrix area is proportional to its luminance level. Thus, determining output electric signal value of each light emitter in corresponding area leads to determing position of multipoint light emitter and position of object tracked. This is done by using Case-based reasoning method, the precedent in which is described as integral signal value of each matrix area, coordinates of light receivers, which luminance level is high, and decision to be made in this situation.

1. Introduction

Applying automation systems in machine-building industry results in increasing accuracy, efficiency and reliability of engineering facilities, decreasing labor cost to implementing operations and enhancing quality of a final product.

Machine-building facilities, constructive elements of which perform three-dimensional displacements, are highly spread in the industry. There is a need to track these displacements, measure bank and tilt angles of assemblies etc. Parallel kinematic machines are examples of such facilities.

Automation systems characterized by low response time, high accuracy and reliability are used for tracking three-dimensional movement of machinery. Systems based on contactless measurements are especially relevant in this case.

One of such systems known is three-dimensional movement control system based on device with optical principle of action shown at Fig. 1. The device consists of multi point light emitter 4, which includes a set number of light emitters 5, firmly attached to object tracked 1 at connection point 3 by a hinge 2 and light receiver matrix 6 located at a set radius around light emitter.

When all light emitters are on and object tracked is in a certain point in space, beams of light emitters cover a group of light receivers, which make up a matrix. When object tracked changes its location in space, tilts at some angle or rotates around its axis, the direction of light beam of each emitter changes, and each light beam covers another group of light receivers. Thus, actuation of certain group of light receivers is influenced by three-dimensional location of object tracked.

System previously elaborated uses Case-based reasoning (CBR) method to determine three-dimensional location of object. This method is based on searching a precedent in Case base (CB). Output discrete signal of each light receiver in matrix flows to microcontroller. The value of the signal defines if light receiver is covered by light beam. A group of light receivers covered by light beams is identified, based on which three-dimensional location is determined by applying CBR method.

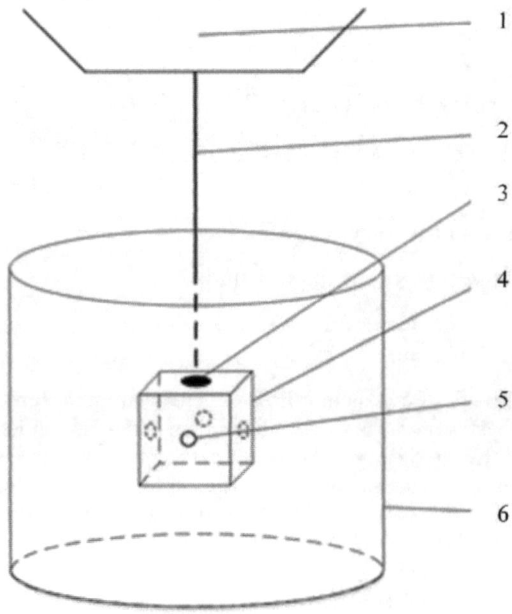

Fig. 1. Optical device for tracking three-dimensional movement

However, the algorithm for processing signals of light receivers described does not provide with high accuracy. This is because the output signal of each light receiver is in discrete form, and the output value of light receiver, one quarter of which is covered by light beam (for instance), is equal to the output value of fully illuminated light receiver. Nevertheless, two these values of light receiver illumination respond to different directions of light beams and, consequently, different location of object tracked.

For this reason, the objective is to increase accuracy of three-dimensional movement control system by changing algorithm of processing output signals of light receivers.

2. Method

Whereas a disadvantage of signal processing principle described is a discrete form of signals, it was decided to switch from discrete signals to analog signals. In this case, three-dimensional location of object tracked is defined not only by a group of light receivers actuated, but by a level of luminance of each light receiver in this group. It is considered that light receiver luminance is proportional to the value of its output signal.

In system elaborated each light receiver has an identifier and light receiver matrix is divided into four space areas (areas 1-4) which responds to number of laser light emitters. Each of four light emitters is oriented towards a certain area of light receivers.

A situation in which light beam of one light emitter covers the sensitive surfaces of a few light receivers is presented hereafter. In such situation, the output of each light receiver is an analog signal which value is proportional to the square of light receiver area, which is covered by light beam. Thus, the output values of light receiver areas 1-4 equal to the algebraic sum of electrical output signals of each light emitter covered by light beam and located in respective area:

$$U_{1-4} = \sum_{i=1}^{n} U_i \quad (1)$$

where U_1, U_2, U_3, U_4 [V] are values of output signals of light emitter areas 1-4 respectively, 1..n are light emitters located at areas 1-4 of light receiver matrix (it is considered that number of light emitters located at areas 1-4 is equal), U_i [V] is electrical output signal of light emitters 1..n located at light receiver areas 1-4 respectively.

Therefore, an electrical output signal which determines luminance level of whole light emitter matrix is calculated upon the formula:

$$U = U_1 + U_2 + U_3 + U_4 \quad (2)$$

and values U_1, U_2, U_3, U_4 indexed to certain light emitters, which are covered by a light beam from light emitters 1-4, determine tree-dimensional location of studied object. Three-dimensional coordinates x, y, z of studied object at this moment in time depend on electrical output signals of light emitter areas 1-4:

$$(x, y, z) = f(U_1, U_2, U_3, U_4) \quad (3)$$

3. Results and discussion

Functional structure of movement control system for studied object is presented at Fig. 2. The system includes microcontroller unit (MCU) as a base element.

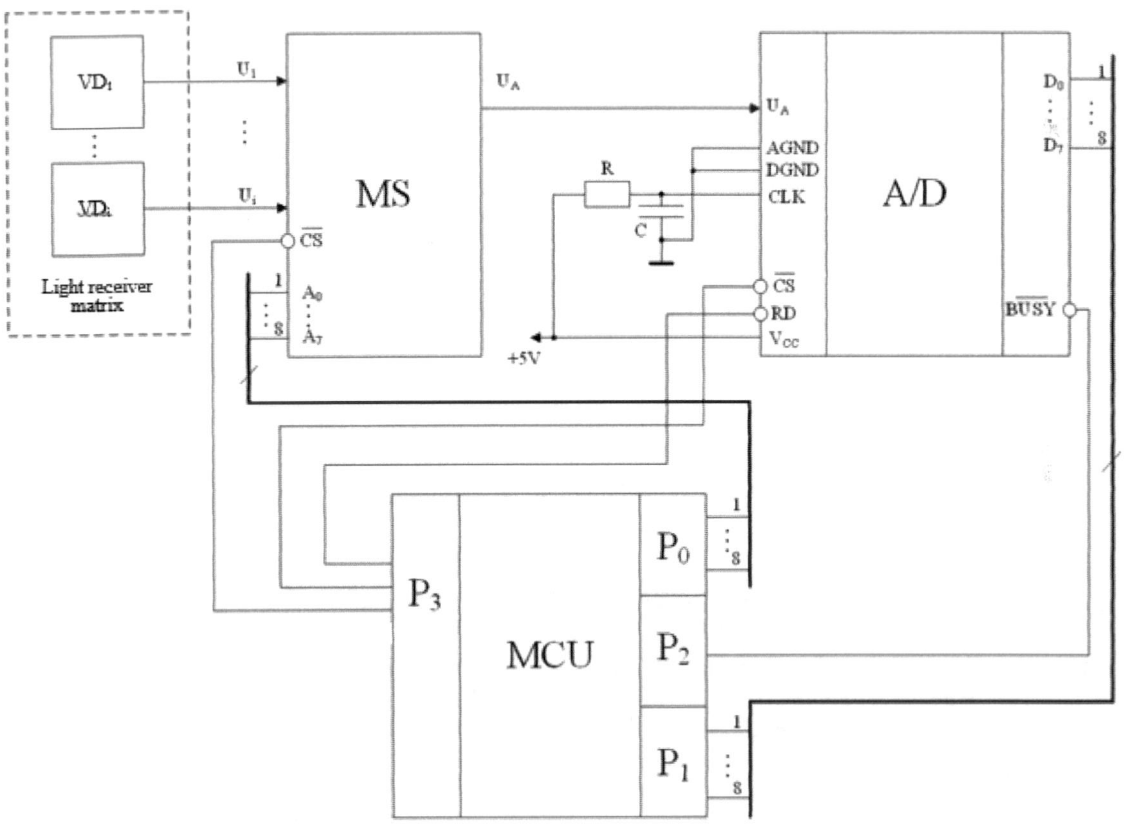

Fig. 2. System functional diagram

MCU controls light emitters scanning by using multiplexor if signals (MS) Analog-to-digital converter (A/D) consequently transforms input signals to digital code.

The value of output signal BUSY of A/D corresponds to the end of transformation. This signal flows to MCU port P2. After transformation is finished, the signal RD, which corresponds to reading, is sent by line P3 of MCU. This signal value start the procedure of reading data from outputs D0-D7 of A/D and writing digital code to MCU internal memory.

After signals, which correspond to condition (luminance level) of all light receivers, are written to MCU internal memory in digital codes, a digital data processing is performed: obtaining the resulting signal for active (covered by light beam) light emitters located at areas 1-4, application of precedent method for determining a precise location of light emitter block in order to determine location of studied object.

The elaborated algorithm presented at Fig. 3 describes system performance.

Initialization includes setting microcontroller (setting reference values and constant values), preparing light emitter block (setting it to reference position and switching on) and light receivers (resetting their values).

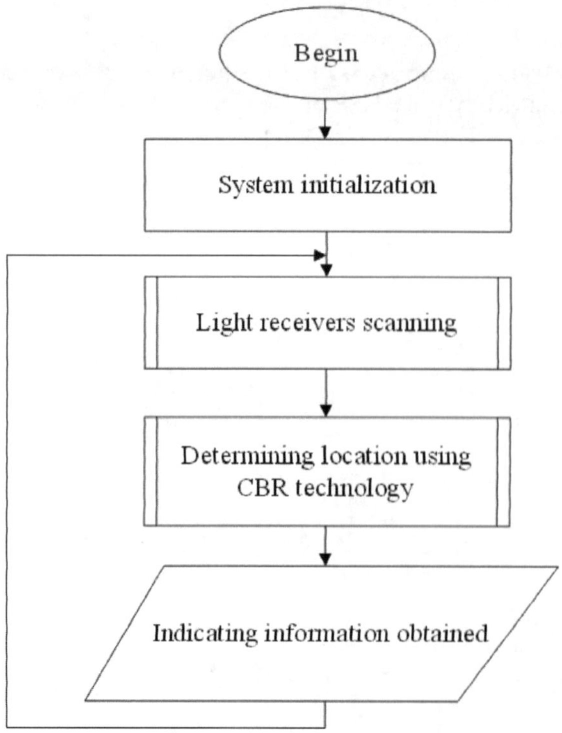

Fig. 3. System implementation algorithm

A consequent scanning of light receivers is performed thereafter. Scanning includes signal commutation (performed by MS), signal transformation from analog to digital form (performed by A/D) and uploading results to internal memory of MCU. After that data is processed in order to obtain information about three-dimensional location of studied object; information obtained is indicated in a previously set way. Data processing is performed with the use of precedent method.

System performance is cyclical, and all the procedures are implemented with preset periodicity.

Algorithm of forming decision based on precedent method also known as Case-based reasoning (CBR) technology [1-5] is given at Fig. 4.

A precedent in common case [6, 7] consists of information about current location of object tracked, associated decisions and other meaningful factors. A precedent is defined as follows:

$$\text{CASE} = \langle(\Sigma U_i,\ \Sigma(X,Y)_i), M\rangle \quad (4)$$

where ΣU_i are values of output signals of light emitter areas, $\Sigma(X,Y)_i$ are coordinates of light receivers which register signals from respective light emitters, *M* is a decision to be made according to information obtained.

In this instance, for four light receiver areas current situation is defined the following way:

$$CURRENT\_CASE = (U_1, U_2, U_3, U_4, (X, Y)_1, (X, Y)_2, (X, Y)_3, (X, Y)_4,) M \quad (5)$$

where U_1, U_2, U_3, U_4 are values of output signals of light emitter areas 1-4 respectively, $(X, Y)_{1\text{-}4}$ are coordinates of light receivers which register signals from respective light emitters, M is a decision to de made according to information on signal values and coordinates of light receivers obtained.

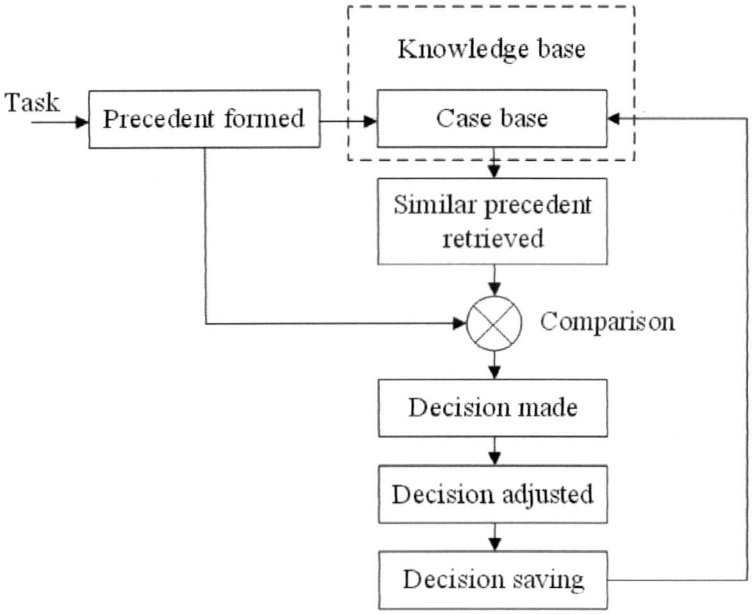

Fig. 4. CBR technology application algorithm

After that, a precedent mostly common to current case is searched in Case base (CB) [8-13], which is formed during system testing before commissioning. CB is also an element of a knowledge base. Meanwhile, a similarity degree parameter is taking into consideration. This parameter characterizes whether precedent from CB complies with current case and defines whether a decision, kept inside precedent from CB, is suitable for current case.

After similarity degree is determined for each precedent from CB, the precedent with higher similarity degree value is chosen for further processing.

A decision kept in precedent [14, 15] with higher similarity degree presents three-dimensional coordinates of studied object. After the decision is retrieved, it is adjusted and saved into CB. Thus, a result of CBR technology application to the system is a decision kept in this precedent.

4. Conclusion

Thus, a system enhanced performs three-dimensional movement control of assemblies of machine-building facilities. Accuracy of system previously developed was increased by switching from processing digital to analog signals, which is based on proportionality of luminance level and output signal value of each light receiver.

Signal processing is followed by application of CBR technology, coordinates of three-dimensional location of object tracked are the result of which.

References

[1] Kolodner J.L. Maintaining Organization in a Dynamic Long-Term Memory. Cognitive Science, 7(IV), 1983. PP. 243-80.
[2] Kolodner J.L. Reconstructive Memory: A Computer Model. Cognitive Science, 7(IV), 1983. PP. 281-28.
[3] Watson I.D., Marir F. Case-based reasoning: A review. The Knowledge Engineering Review, Vol. 9, No. 4, 1994. PP. 355-381.
[4] I.D. Watson and F. Marir 1994 Case-based reasoning: A review The Knowledge Engineering Review vol. 9, num. 4, pp. 355-381.
[5] J.L. Kolodner 1992 An Introduction to Case-Based Reasoning Artificial Intelligence Review vol. 6, 1992, pp. 3-34.
[6] Rodrigo G. C. Rocha, Ryan R. Azevedo, Ygor Cesar Sousa, Eduardo de A. Tavares, Silvio Meira 2014 A case-based reasoning system to support the global software development 18th International Conference on Knowledge-Based and Intelligent Information & Engineering Systems - KES2014. Procedia Computer Science 2014. №35. Pp. 194–202.
[7] T. Olsson, P. Funk 2012 Case-based reasoning combined with statistics for diagnostics and prognosis 25th International Congress on Condition Monitoring and Diagnostic Engineering IOP Publishing. Journal of Physics: Conference Series 364 012061 DOI:10.1088/1742-6596/364/1/012061
[8] L.E. Mujica, J. Vehı, J. Rodellar, P. Kolakowski 2005 A hybrid approach of knowledge-based reasoning for structural assessment Institute of physics publishing smart materials and structures, 14, pp.1554–1562, DOI:10.1088/0964-1726/14/6/048
[9] Gómez-Vallejo H.J., Uriel-Latorre B., Sande-Meijide M., Villamarín-Bello B., Pavón R., Fdez-Riverola F., Glez-Peña D. 2016 Case-based reasoning system for aiding detection and classification of nosocomial infections Decision Support Systems Vol. 84 Pp. 104-116.
[10] Douali, N.a , De Roo, J.b, Jaulent, M.-C.a 2012 Clinical Diagnosis Support System based on Case Based Fuzzy Cognitive Maps and Semantic Web 24th Medical Informatics in Europe Conference, MIE 2012; Pisa; Italy; 26 August 2012 through 29 August 2012 Volume 180, 2012, Pp. 295-299.
[11] López, B., Pous, C., Gay, P., Pla, A., Sanz, J., Brunet, J. 2011 EXiT CBR: A framework for case-based medical diagnosis development and experimentation Artificial Intelligence in Medicine Volume 51, Issue 2, Pp. 81-91.
[12] Sreeparna Banerjee, Amrita Roy Chowdhury 2015 Case Based Reasoning in the Detection of Retinal Abnormalities using Decision Trees Procedia Computer Science 46, pp. 402 – 408.
[13] Mohamed M. Marzouk, Rasha M. Ahmed. 2011 A case-based reasoning approach for estimating the costs of pump station projects Journal of Advanced Research 2, pp. 289–295.

IOP Conf. Series: Earth and Environmental Science **115** (2018) 012052 doi:10.1088/1755-1315/115/1/012052

Mechanical and Acoustic Characteristics of the Weld and the Base Metal Machine Part of Career Transport

Alexander N Smirnov, Victor L Knjaz'kov, Elena E Levashova, Nikolay V Ababkov [a], Maksim V Pimonov

T.F. Gorbachev Kuzbass State Technical University, 28 str. Vesennyaya, Kemerovo, Russia

E-mail: [a] n.ababkov@rambler.ru

Abstract: Currently, many industries use foreign-made machinery. There is no opportunity to purchase quality original spare parts for which machinery. Therefore, enterprises operating this equipment are looking for producers of analogues of various parts and assemblies. Quite often, the metal of such analog components turns out to be substandard, which leads to their breakdown at a much earlier date and the enterprises incur material losses.

Due to the fact that the complex of performance characteristics and the resource of products are laid at the stage of their production, it is extremely important to control the quality of the raw materials. The structure, mechanical, acoustic and magnetic characteristics of metal samples of such destroyed details of quarry transport as hydraulic cylinders and detail "axis" of an excavator are investigated. A significant spread of data on the chemical composition of metal, hardness and characteristics of non-destructive testing is established, which gives grounds to recommend to manufacturers and suppliers of parts is more responsible to approach the incoming quality control. The results of the investigation of metal samples by destructive and non-destructive methods of control are compared, which showed that the spectral-acoustic method of nondestructive testing can be used to control the quality of the responsible machine parts under conditions of import substitution.

Introduction

Ensuring the durability of machine parts is an important task of modern machine building, which is determined to a significant extent by the quality of the surface layer [1]. At the same time, it is necessary to solve the problems of improving the reliability of instruments, installations, improving their quality and efficiency, and, consequently, the issues of saving metals, fighting corrosion and wear of machine parts [1]. This is especially important now in conditions of import substitution. In many industries, foreign-made machinery is used, original spare parts for which there is no way to purchase. Therefore, enterprises that operate this equipment are looking for manufacturers of analog parts and components. Quite often, the metal of such analog components turns out to be substandard, which leads to their failure at a much earlier time [2].

In general, in the mining industry in the Kemerovo region, the share of equipping the mines with the main mining equipment of the imported production is: 76.6% of the face faces, 46% of the bottom holes, 63% of the equipment for the transportation of goods and personnel (diesel transport, electric locomotives, lifting installations), equipment ventilation - 30%, systems of air-gas control, communication, warning and automation of about 50% [3, 4].

High performance of mining equipment operating under extreme conditions is provided by a complex of mechanical properties such as resistance to shock loads at low temperatures, wear resistance of the friction surface, etc. [5, 6].

Thus, the purpose of this work is to investigate the microstructure, mechanical, acoustic and magnetic characteristics of the metal of the "axis" part, as well as to derive the dependencies between the measured

IOP Conf. Series: Earth and Environmental Science **115** (2018) 012052 doi:10.1088/1755-1315/115/1/012052

mechanical characteristics and the characteristics of non-destructive testing for the timely and accurate determination of the part boundary at the pre-destruction stage.

Methods and materials of experimental research

The "axis" part (Fig. 1) after 1 year of operation and having wear, and samples of the analogue metal of the piston rod of the hydraulic cylinder of a mining excavator were selected as objects of study.

The investigated "axle" part of the P & H 2800 XPC excavator is the most loaded element in the hub-axis system, which receives unevenly distributed loads during operation, which leads to their accelerated wear (Fig. 1). After the indicated service life (1 year), the wear value was 17 %, and the diameter decreased to 163 mm. The wear of the shaft axis is due to the fact that on the surface there are defects - scuffing. In the future, there are gaps, which leads to an accelerated destruction of the shaft-bushing pair. The technology of manufacturing the "axis" part is much more complicated and more expensive than the details of the bush, so it is more economical to restore the "axis" part.

a) *b)* *c)*

Fig. 1. Bucket of excavator P & H 4100 XPC (*a*) and the nature of the damage details "axis" (*b*) and details "bushing" (*c*)

The fragment in the form of a disc in the cross section of the part was cut out for research. Samples of metal of analogs of the details of the hydraulic excavator unit of the mining excavator have also been investigated in the present work (Fig. 2). The destruction of one of the hydraulic cylinders occurred along the eyelet, the destruction of the other hydraulic cylinder – according to the rod.

Fig. 2. External view of the hydraulic cylinder of the excavator

The chemical composition of the metal of the samples was determined by the spectral analysis method using the Q4 TASMAN optic emission spectrometer. The hardness of the metal was measured using a universal hardness tester DuraVision-30. Toughness tests were carried out at room temperature on a Izod Pendel pendulum coil, the type of samples with a U-shaped concentrator.

Spectral acoustic analysis, with the automated acoustic system ASTRON, was used to identify the zones with a changed structure, as well as the degree of its change, measurements were made on a transverse sample at 11 points 9 mm apart between the points. Using the surface acoustic wave sensor (Rayleigh waves), the following characteristics were determined:

- Surface acoustic wave delay time, ns;
- attenuation factor, 1 / μs;
- amplitude of the received signal, mm · 10^{-1}; [7–11]

For the "axis" part, the acoustic characteristics were measured in two stages: first, measurements were made on the surface of rotation (Fig. 3, *a*), then, after selecting the sections with the greatest deviations, along the cross-section of the part in the direction from the surface of rotation to the center of the part (Fig. 3, *b*).

a)

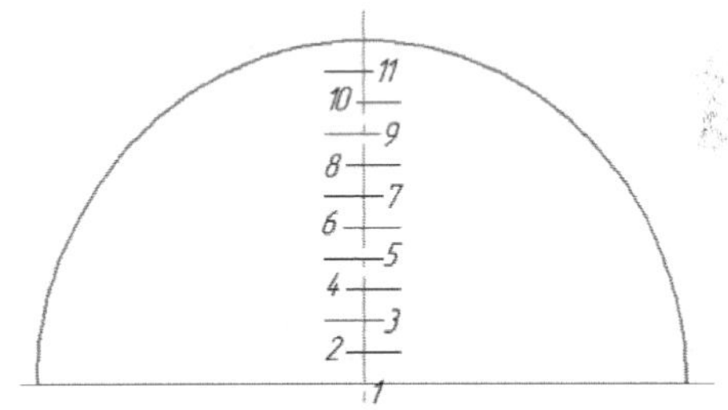

b)

Fig. 3. Scheme of measuring acoustic characteristics by the spectral-acoustic method of the "axis" component:
a – the surface of revolution; *b* – cross-section

The microstructure analysis [12] was performed with an optical microscope Axio Observer. To study the structure, one sample was cut from each fragment of the hydraulic cylinder, which were pressed into the resin. The samples pressed into the resin were polished, polished and etched in a 4 % solution of nitric acid and 2 % solution of nitric acid in ethyl alcohol. Microstructure studies were performed with an increase of ×130, ×500, ×800 for the sample metal of the "axis" part and with an increase of ×200 for samples of the hydraulic cylinder fragments.

The microhardness was measured on a PMT-3M device [13] with an increase of 100 times by indenting the tip (quadrilateral pyramid with a square base), under a load of 1.96 N applied for 20 s, the analysis of the data was carried out according to [14].

Results and its discussion

The results of the determination of the chemical composition of the metal of the "axis" part are presented in Table 1.

Table 1 – Chemical composition of weld metal and base metal "axis" details

Chemical composition of surfacing made by the manufacturer, %							
C	Si	Mn	Cr	Mo	Ni	V	W
0.34	0.37	1.20	5.00	0.90	1.00	0.15	0.90
Chemical composition of base metal, %							
C	Si	Mn	Cr	Mo	Ni	V	W
0.43	0.22	0.50	1.20	0.30	4.10	<0.01	<0.01

Fig. 4 and 5 show the results of investigations of the surface layer of a component by the spectral-acoustic control method.

At the first stage of measurements, the maximum deviations of the acoustic characteristics along the surface of rotation were established (Fig. 4). The areas with the maximum deviations from the average values were identified: 12, 15, 19, 28 and 30. However, for sections 12, 15 and 19, where the attenuation coefficient and the amplitude values were decreasing, further cross-sectional measurements were not performed, since defects or structural changes are not implied.

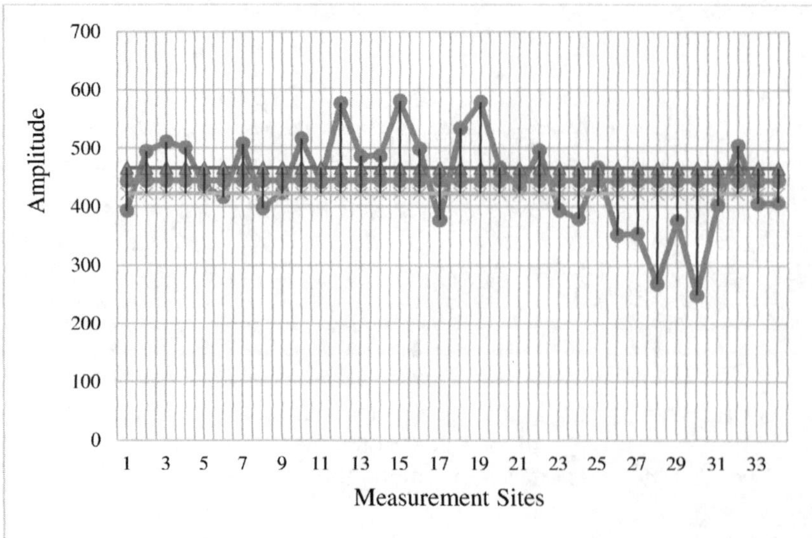

Fig. 4. Results of measuring acoustic characteristics by the spectral-acoustic method. Surface of rotation

For sections 28 and 30 (Fig. 4), acoustic characteristics along the cross section of the part were measured (Fig. 3, b). Analysis of the measurement results of the values of the damping factor and the amplitude of the received signal showed the presence of a transition between surfacing and the base metal, as well as significant differences in the values of the characteristics for the overlaying layer and the base metal of the part.

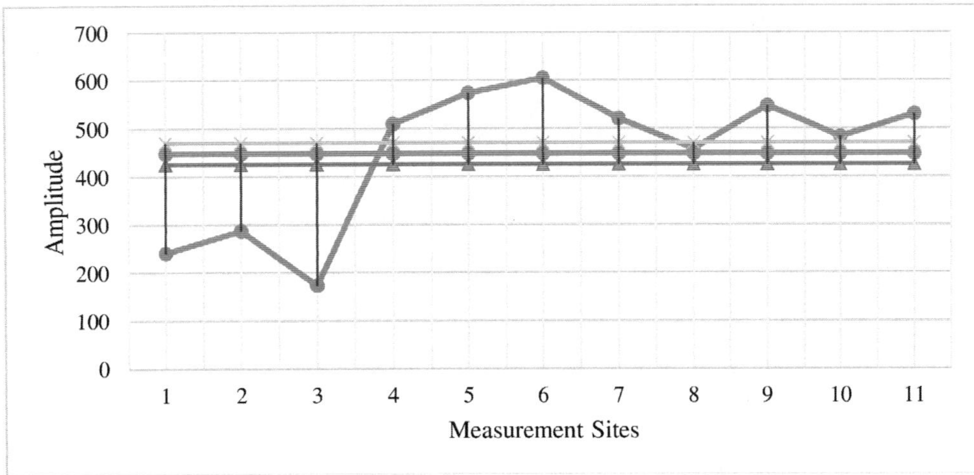

Fig. 5. Results of measuring acoustic characteristics by the spectral-acoustic method. Cross section

According to the results of measuring acoustic characteristics, zones were detected with a change in the amplitude of the received signal (Fig. 7) on the sample of the "axis" part. Samples were cut from these zones to study the microstructure and measure the microhardness.

In the metallographic study, the interfaces between the layers were identified, and the microstructure of the deposited and base metals was investigated. In the cross section, the initial sample of the "axis" part consists of several layers: a layer subjected to impact loads; a layer of welded metal and a section of HAZ – a transition to the base metal, which differ in their degree of etching, microstructure and grain sizes. The structure at the boundary of the HAZ and the base metal is a ferritic cement mixture with dendrites that pass into the deposited metal, the microhardness in this layer is from 305 to 501 HV. The boundary of the HAZ and the welded layer is characterized by a high heterogeneity of the microstructure, which is martensite, and the range of microhardness values, which is from 438 to 590 HV. High values of microhardness, relative to other areas of surfacing, are probably due to an increased cooling rate of the lower part of the liquid phase of the bath.

Microcracks are found in the HAZ – transition layer to the base metal, the layer differs by a sharp jump in the microhardness values from 305 to 501 HV (HAZ). Microcracks have the same orientation in the direction of tensile stress, but at different angles. In the deposited metal layer, microcracks were also detected, but in a smaller amount compared to the HAZ layer, a transition to the base metal. The microhardness values were from 438 to 540 HV.

In the layer subjected to impact loads, a deformed metal structure was detected, the microhardness values were from 416 to 499 HV.

Based on the results of experimental studies, a plot was plotted of the amplitude of the received signal of the surface acoustic wave from the microhardness for the metal of the "axis" part (Fig. 6).

It follows from the graph (Fig. 6) that as the microhardness values increase, the values of the amplitude of the received signal of the surface acoustic wave decrease.

Thus, according to the results of metallographic studies, microcracks were detected in areas with a change in the amplitudes of the received signal of surface acoustic waves, and also the boundaries of the section to be selected were determined.

For the destroyed parts of the hydraulic cylinders (rod and eye), differences in the chemical composition of the metal between the original and analogues are established. According to the passport, the original is made of steel SM45C (steel 45). Of the analogues, the closest was the version made from the material most appropriate to steel 50. Thus, the deviation in content of C between the original counterparts was more than 5 %.

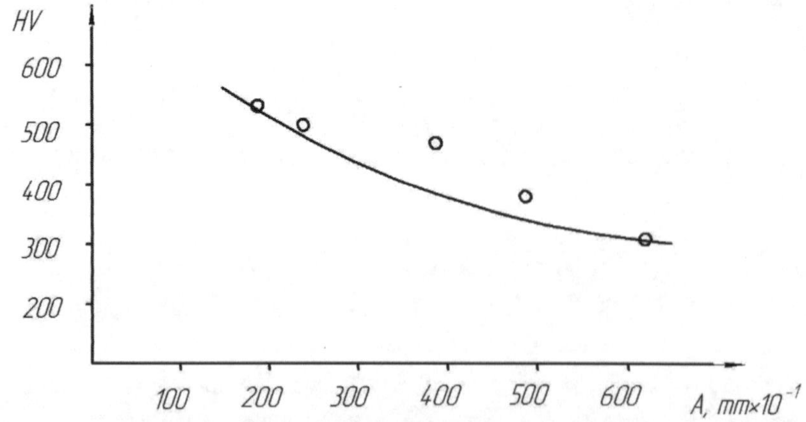

Fig. 6. Dependence of the amplitude of the received signal of surface acoustic wave from the microhardness for metal parts "axis"

The hardness of significant differences was not revealed. The results of the toughness tests showed the following: the metal of the originals corresponds to KCU valuesof 70–80 J/cm^2, then the value of KCU 18 J/cm^2 was obtained for the sample metal of the fragment of the cylinder destroyed by the rod, and for the sample metal of the fragment of the cylinder destroyed by the eye – 16 J/cm^2.

When analyzing the microstructure, it was revealed that the metal structure of the samples of the fragments of hydraulic cylinders consists of a ferrite-pearlite mixture, the structure is fine-grained, the grains have an elongated character. In all the samples studied, defects in the form of nonmetallic inclusions and microcracks in the metal of a specimen of a fragment of a cylinder destroyed by a rod were observed (Fig. 7). The size of nonmetallic inclusions is not the same.

The evaluation of metal by the spectral-acoustic method was performed with respect to the delay time of the surface acoustic waves. The values of this characteristic varied from 4691 ns in the original metal to 4752 ns in the analog metal. Based on the results of a comparison of the study of metal samples with destructive and non-destructive testing methods, a plot of the shock viscosity versus the time delay of surface acoustic waves was constructed (Fig. 8).

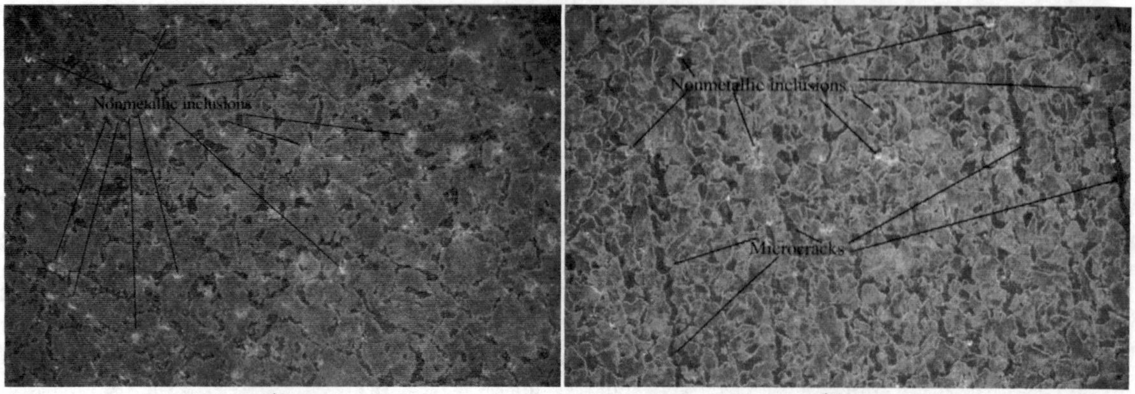

a) *b)*

Fig. 7. Structure of the metal samples of the fragments of the cylinder: *a* – destroyed by eye, ×200; *b* – destroyed by the rod, ×200

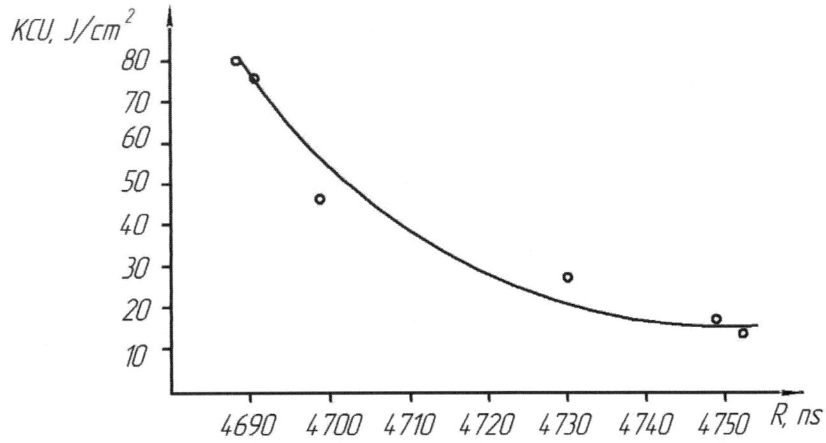

Fig. 8. Dependence of the toughness KCU (J/cm^2) on the delay time of the surface acoustic wave, R (ns)

The graph shows (Fig. 8) that with a decrease in toughness occurs growth of the surface acoustic wave delay time values of values. This relationship can be used to assess the state of equipment on the characteristics of non-destructive testing.

Conclusions

1. Analysis of the equipment of coal enterprises with equipment of domestic and imported production showed that at present the share of domestic equipment is about 20 %, while that of imported equipment is 80 %.

2. Samples of originals and analogues of such details as the piston of the hydraulic cylinder and the "axis" of conjugation "axis-bush" were destroyed by destructive and non-destructive methods of control. It is established that for these details a considerable scatter of data on the toughness, microhardness and characteristics of non-destructive testing is observed. This gives grounds for recommending manufacturers and suppliers of parts to take a more responsible approach to incoming quality control.

3. Comparison of the results of the study of metal samples with destructive and non-destructive methods of control was conducted, which showed that the spectral-acoustic method of nondestructive testing can be used to control the quality of the responsible machine parts under import substitution conditions.

Acknowledgement

This work was supported by a grant from the RSF, agreement No. 14-09-00724-P and a grant from the President of the Russian Federation to support young candidates of sciences, MK-1341.2017.8.

References

[1] Gorkunov, B. M. Analiz metodov i ustroystv dlya kontrolya uprochnennogo sloya metallicheskikh izdeliy / B. M. Gorkunov, A. A. Tishchenko // Sbornik nauchnykh trudov «Vestnik NTU "KHPI»: Yelektroyenergetika ta peretvoryuval'na tekhníka, №12 – Vestnik NTU «KHPI», 2010. p. 128–135.

[2] Murav'yev, V. V. Kontrol' ostatochnykh napryazheniy v bandazhakh lokomotivnykh koles metodom akustouprugosti / V. V. Murav'yev, L. V. Volkova, Ye. N. Balobanov // Defektoskopiya, 2013. 7. p. 22–28.

[3] Trubitsina N. V. Obosnovaniye kontseptual'nykh podkhodov k razrabotke programmy importozameshcheniya oborudovaniya ugol'nykh shakht Kemerovskoy oblasti. // N. V. Trubitsina, A. S. Yarosh // Nauchno-tekhnicheskiy zhurnal «Vestnik Nauchnogo tsentra po bezopasnosti rabot v ugol'noy promyshlennosti» №1-2014. Izd-vo: OOO «VostEKO» (Kemerovo). p. 6–14.

[4] Dubov, G.M. Sravnitel'nyy analiz osnashchennosti ugol'nykh predpriyatiy gornoshakhtnym oborudovaniyem otechestvennogo i importnogo proizvodstva // sbornik trudov Mezhdunarodnoy nauchno-prakticheskoy konferentsii "Innovatsii v toplivno-energeticheskom komplekse i mashinostroyenii (TEK-2017)". – 2017. p. 45–50.

[5] Sleptsov O.I. Perspektivy razvitiya metallurgii v Yakutii v svyazi s sozdaniyem novykh pokoleniy staley severnogo ispolneniya/ B.S. Yermakov, O.I. Sleptsov, P.P. Petrov// NAUKA I OBRAZOVANIYe. – 2015. №3. p. 59-63.

[6] Kobernik, N. V. Sovremennyye predstavleniya o modifikatsii naplavlennogo me-talla i metalla shva nanorazmernymi chastitsami / N. V. Kobernik, R. S. Mikheyev, A. S. Pankratov i dr. // Svarka i diagnostika. –2015. № 5 p. 13–18.

[7] Criteria for the evaluation of the technical state of the long-lived metal of HPP equipment based on acoustic structuroscopy / Smirnov A. N., Ababkov N. V., Muravev V. V., Folmer S. V. // Russian Journal of Nondestructive Testing. 2015. T. 51. № 2. p. 94-100.

[8] Conceptual model of resource assessment of fuel and energy complex long working equipment on the basis of structural criteria / Smirnov A. N., Ababkov N. V. // IOP Conference Series: Materials Science and Engineering 6. Cep. "6th International Scientific Practical Conference on Innovative Technologies and Economics in Engineering" 2015. p. 012005.

[9] Analysis of acoustic and physical-mechanical parameters Of Cr-Mo-Va steel in the vicinity of the crack / Smirnov A., Logov A., Ababkov N. // AIP Conference Proceedings 2. "Advanced Materials in Technology and Construction, AMTC 2015: Proceedings of the II All-Russian Scientific Conference of Young Scientists "Advanced Materials in Technology and Construction"" 2016. Ch. 020009.

[10] Substructure, grain boundaries, and microcracks in long-operating metal / Smirnov A. N., Kozlov E. V., Koneva N. A., Popova N. A. // Metal Science and Heat Treatment. 2005. T. 47. 3-4. p. 155-161.

[11] Mitenkov, F.M. O novom metode kontrolja povrezhdaemosti materiala oborudovanija JAJEU i apparatno-programmnyh sredstvah dlja ee realizacii / Mitenkov F.M., Uglov A.L., Pichkov S.N., Popcov V.M. // Problemy mashinovedenija i nadezhnosti mashin, 1998. – №3. p3–9.

[12] GOST R ISO 6507-1-2007. Metally i splavy. Izmereniye tverdosti po Vikkersu. Chast' 1. Metod izmereniya. Moskva: Izd-vo Ctandartinform, 2008. 16 p.

[13] GOST R ISO 6507-4-2009. Metally i splavy. Izmereniye tverdosti po Vikkersu. Chast' 4. Tablitsy opredeleniya tverdosti. Moskva: Izd-vo Ctandartinform, 2010. 25 p.

[14] GOST 5639-82 Stali i splavy. Metody vyyavleniya i opredeleniya velichiny zerna. Moskva: Izd-vo IPK Izdatel'stvo Standartov, 2003. 21 p.

Modeling the Technological Process for Harvesting of Agricultural Produce

S D Shepelev[1], V D Shepelev[2, a], Z V Almetova[2], N P Shepeleva[2], M V Cheskidov[1]

[1] South Ural state agrarian university 75, Lenin Avenue, Chelyabinsk, 454080, Russia
[2] South Ural state university, 76, Lenin Avenue, Chelyabinsk, 454080, Russia

E-mail: [a] shepelevvd@susu.ru

Abstract. The efficiency and the parameters of harvesting as a technological process being substantiated make it possible to reduce the cost of production and increase the profit of enterprises. To increase the efficiency of combine harvesters when the level of technical equipment declines is possible due to their efficient operating modes within daily and every season. Therefore, the correlation between the operational daily time and the seasonal load of combine harvesters is found, with the increase in the seasonal load causing the prolonged duration of operational daily time for harvesters being determined. To increase the efficient time of the seasonal load is possible due to a reasonable ratio of crop varieties according to their ripening periods, the necessary quantity of machines thereby to be reduced up to 40%. By timing and field testing the operational factor of the useful shift time of combine harvesters and the efficient modes of operating machines are defined, with the alternatives for improving the technical readiness of combine harvesters being identified.

1. Introduction

The competitiveness of agricultural enterprises is possible to improve due to the substantiation of rational parameters of technological processes of grain production on the basis of mathematical modeling. The need for technical means depends on the harvesting period duration both during the shift and the season. The operating hours of a technological machine during the day are well known to be limited by the moisture level of the grain mass during the morning and evening hours, as a rule, because of dew. In [2, 3] it is determined that the time of the daily work of combine harvesters should not exceed twelve hours under the conditions of the Southern Urals. It is necessary to bear in mind that the work duration will be affected by the level of technical equipment involved in harvesting, which, in its turn, requires reviewing the matter. In the authors' papers [4, 5] it is recommended to increase the efficient harvesting time due to the sowing of different varieties of wheat according to the length of its growing seasons in equal proportions, but it is known that early varieties have yields lower than late ones, thus, it is necessary to find a reasonable proportion between the sowing area (taking into account the efficiency of machines) and cost indicators. One of the most effective ways to reduce production losses when the level of technical equipment is rather low is to increase the sowing terms of wheat [6].

2. Results and discussion

As for the Ural region of Russia in the State Register there are three groups of wheat varieties according to its earliness: middle-early, mid-season and middle-late [7, 8]. To substantiate the proportion between the areas of different wheat varieties there has been developed an economic and mathematical model with the criterion of the maximum profit taken as the basic one [9, 10]:

$$P(Q_{me}, Q_{ms}) = C_y - L_{me}(Q_{me}) + L_{ms}(Q_{ms}) + L_{ml}(Q_{ml}) \rightarrow \max \quad (1)$$

where:

C_y is the yield cost, RUB/ha; L_{me}, L_{ms}, L_{ml} are the losses on the area of middle-early (Q_{me}), mid-season (Q_{ms}) and middle-late (Q_{ml}) wheat varieties.

The expanded form of the model is as follows:

$$P(Q_{ms}) = (P_{ml}(Q_s - Q_{me} - Q_{ms}) + P_{ms} \cdot Q_{ms}) \cdot C_p - \\ - \frac{K_{r.i} \cdot K_l \cdot P_{ms}}{Q_d}\left[P_{ml}(Q_s - Q_{ms})(Q_s - Q_{ms} - Q_{me}) + P_{ms} \cdot P_{ms} Q_{ms}^2\right] \to \max \qquad (2)$$

where:

P_{ml}, P_{ms}, P_{me} are the basic productivity of middle-late, mid-season and middle-early varieties, t/ha; C_p is the price of products RUB/t; Q_s is the seasonal load on a harvester, ha; $K_{r.i.}$ is the coefficient of ripening irregularity ($K_{r.i.} = 0.5$); K_l is the production loss factor, share/day; Q_d is daily harvester capacity, ha/day.

On the basis of the obtained expression it is established that in the present conditions it is reasonable to cultivate about 20 % of middle-early varieties, 35 % of mid-season and about 45 % middle-late varieties (Fig. 1).

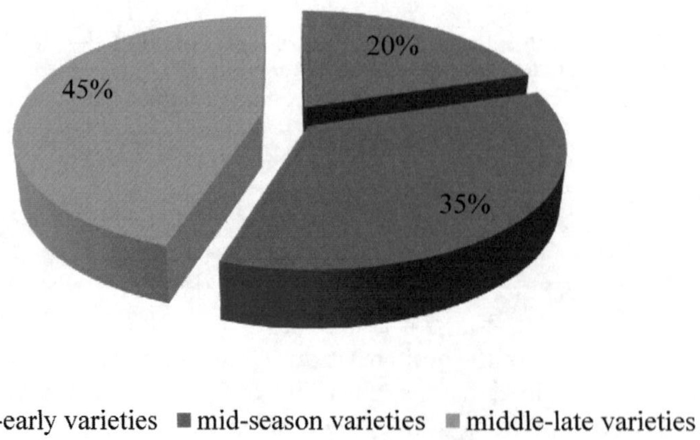

Figure 1. Rational ratio of wheat varieties in precocity

In general, the objective function to substantiate the load for a combine harvester, with the production losses during the day [11] and the season [12] being considered, will be:

$$U_k(Q_s) = \frac{B_h \alpha \gamma_i}{Q_s \cdot P} + \frac{\sum_{i=1}^{n} T \cdot \gamma_j}{Q_s P} + K_{r.i} K_{crop} K_{var} K_{w.cond} C_p \frac{Q_s}{Q_d} + \\ + 0.01 P C_p \left(L_m + k \frac{T_o^{\beta}}{\beta + 1} \right) \to \min \qquad (3)$$

where:

B_h is the balance price of a combine harvester, RUB; α is the coefficient to include the costs caused by involved machines; i, j is the utilization share of the i-th machine of the j-th combine harvester for the given type of operations; P is the crop productivity, t/ha; K_l is the production loss factor, share/day; Q_s

IOP Conf. Series: Earth and Environmental Science **115** (2018) 012053 doi:10.1088/1755-1315/115/1/012053

is the seasonal load on a harvester, ha; τ is the operation factor of the shift time; T_o is the time of combine operation, h; H_w is the width of the harvesting machine, m; K_{crop} is the loss factor reduction due to the combination of crops; $K_{var.}$ is the loss factor reduction due to the combination of varieties; $K_{w.cond.}$ is the factor of weather conditions; T_o is the efficiency rate of the machine operator's work, RUB; L_m is the minimum loss caused by a combine harvester, %; k, β are the factors that reflect pattern changes of current losses during the day.

The daily output of grain harvesters is determined by the expression [13-15]:

$$Q_d = 0.1 H_w V \tau T_o K_{w.cond} \quad (4)$$

where:
H_w is the width of the harvesting machine, m, V is speed of motion, km/h; τ is the operation factor of the shift time, T_o is the time of combine operation, h, $K_{w.cond.}$ is the factor of weather conditions during the harvesting period.

For modeling the technological process the statistical data are collected. The daily output amounts to 17 ha/shift, the degree of unevenness of fluctuations and the standard deviation being 1.7 and 6.7, respectively. The coefficient of variation is 38.115.
The average shift is found to be 14 hours, with the maximum being 16 hours 15 minutes and the minimum – 7 hours 45 minutes. The mean-square deviation is 84 minutes. As a result of timing the operation factor of the shift time of combine harvesters is found to be on average 0.62. The economic-mathematical modeling makes it possible to find the efficient time of daily and seasonal operating modes of combine harvesters (Fig. 2, 3).
The modeling shows that the efficiency when using combine harvesters will be significantly higher if the duration of their operation will be within the range of 14–16 hours a day (Fig. 2). With the load of a combine harvester of the 3rd Class being 200 hectares, its operating time is found to be 12...14 hours. When its seasonal load increases up to 300 hectares, the operating time is to be 15...16 hours, when the seasonal load of a combine harvester is 400 hectares, its operating time should be 17...19 hours.
The modeling has found that the combination of three groups of varieties that differ in ripening with their reasonable ratio can increase their seasonal productivity for a combine harvester of the 5th Class from 400 to 600 hectares, when using early maturing crops up to 700 hectares, and when using combine harvesters of the 3rd Class up to 400 hectares.

IOP Conf. Series: Earth and Environmental Science **115** (2018) 012053 doi:10.1088/1755-1315/115/1/012053

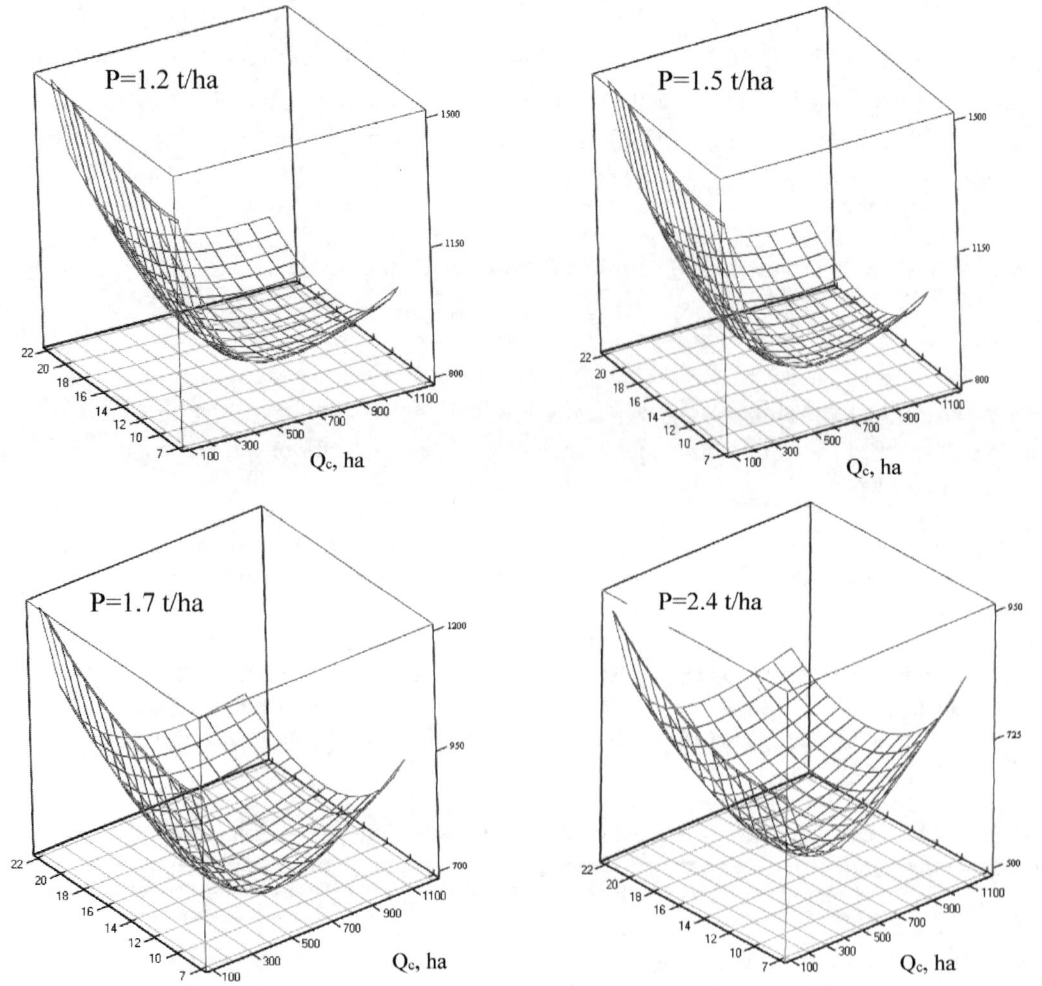

Figure 2. Dependence of the operating time of the combine harvester of the 3rd Class on the productivity of fields (C_p =6800 RUB/t; τ=0.45; H_w = 6 m)

Thus, when planning the technological processes of sowing and harvesting of crops, there is a need for combining different varieties of spring wheat and early maturing crops. This will reduce the loss of production and a rational amount of technical and human resources up to 40%.

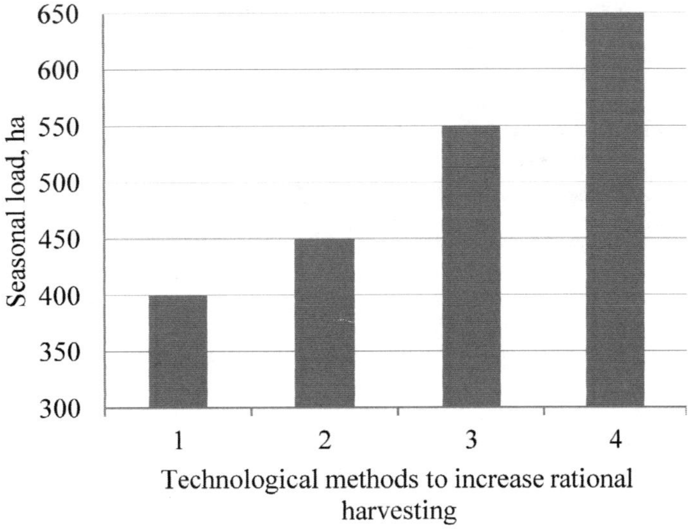

Figure 3. Dependence of the seasonal load of combine harvesters with the capacity of 8-9 kg/s along with a combination of wheat varieties according to their ripening periods and cultures (1 – one-crop; 2 – the ratio of two varieties; 3 – the ratio of three varieties, 4 – the ratio of varieties and crops)

3. Conclusion

The field tests have shown that by changing the structure of sown areas, the use of various crops and varieties that differ in ripening the rational seasonal load on a combine harvester increases up to 725 hectares, which makes it possible to reduce the required amount of technical and human resources up to 40%.

It is found that the largest share of downtime is due to eliminating the consequences of technical failures caused by the reaper elements, thresher and undercarriage of a combine harvester. To reduce the time to eliminate the consequences of technical failures, it is necessary to increase the efficiency of machines due to the aggregate method of repairing in the field and to reduce the delivery time of spare parts due to exchange funds [15-20].

Thus, the economic-mathematical modeling of the technological process of grain crops harvesting made it possible to determine the rational modes for operation of technical means, labour resources and their amounts, taking into account the biological characteristics of agricultural crops.

References

[1] Shepelev S, Shepelev V and Cherkasov Y 2015 *Procedia Engineering.* **129** 161.
[2] Pexa M 2011 *Research in Agricultural Engineering.* **57(1)** 1.
[3] Redreev G V, Myalo O V, Prokopov S P, Solomkin A P and Okunev G A 2017 *IOP Conference Series: Materials Science and Engineering.* **221(1)** 012016.
[4] Kroulík M., Mašek J., Kvíz Z., Jošt B., Prošek, V 2011 *Research in Agricultural Engineering.* **57(2)** 51.
[5] Läänemets O 2011 *Agronomy Research.* **9(SPPL. ISS. 2)** 421.
[6] Oko C.O., Diemuodeke E.O., Akilande I.S 2010 *Research in Agricultural Engineering.* **56(2)** 53.
[7] Redreev G. V 2016 *Materials Science and Engineering.* **142(1)** 012085
[8] Almetova, Z., Shepelev, V., & Shepelev, S 2016 *Procedia Engineering.* **150** 1396.
[9] Akbarnia A 2014 *Research in Agricultural Engineering.* **60(4)**142.
[10] Bracacescu, C., Pirna, I., Sorica, C., Popescu, S., & Stan, O 2012 *Engineering for Rural Development.* **11** 16.

[11] Qiao, J., Han, Z., Hong, K., Chen, H., Hao, Y., & Zhang, X 2016 *Nongye Gongcheng Xuebao/Transactions of the Chinese Society of Agricultural Engineering.* **32(7)** 43.
[12] Risius, H., Prochnow, A., Ammon, C., Mellmann, J., & Hoffmann 2017 *Biosystems Engineering.* **156** 120.
[13] Shepelev S., Shepelev V., Almetova Z 2016 *Procedia Engineering*. **150** 1258.
[14] Tavakoli H., Mohtasebi S.S., Jafari A 2009 *Research in Agricultural Engineering*. **55** 45.
[15] Wang, G., Yi, Z., Chen, C., & Cao, G 2016 *Nongye Gongcheng Xuebao/Transactions of the Chinese Society of Agricultural Engineering.* **32(2)** 36.

Institute of Physics
Dirac House, Temple Back
Bristol BS1 6BE UK

CURRAN ASSOCIATES INC.
proceedings.com

ISSN: 1755-1307
ISBN 978-1-5108-5834-3
9 781510 858343